Technology Meets Research

60 Years of CERN Technology: Selected Highlights

ADVANCED SERIES ON DIRECTIONS IN HIGH ENERGY PHYSICS

ISSN: 1793-1339

Published

Vol. 1 – High Energy Electron–Positron Physics (*eds. A. Ali and P. Söding*)
Vol. 2 – Hadronic Multiparticle Production (*ed. P. Carruthers*)
Vol. 3 – CP Violation (*ed. C. Jarlskog*)
Vol. 4 – Proton–Antiproton Collider Physics (*eds. G. Altarelli and L. Di Lella*)
Vol. 5 – Perturbative QCD (*ed. A. Mueller*)
Vol. 6 – Quark–Gluon Plasma (*ed. R. C. Hwa*)
Vol. 7 – Quantum Electrodynamics (*ed. T. Kinoshita*)
Vol. 9 – Instrumentation in High Energy Physics (*ed. F. Sauli*)
Vol. 10 – Heavy Flavours (*eds. A. J. Buras and M. Lindner*)
Vol. 11 – Quantum Fields on the Computer (*ed. M. Creutz*)
Vol. 12 – Advances of Accelerator Physics and Technologies (*ed. H. Schopper*)
Vol. 13 – Perspectives on Higgs Physics (*ed. G. L. Kane*)
Vol. 14 – Precision Tests of the Standard Electroweak Model (*ed. P. Langacker*)
Vol. 15 – Heavy Flavours II (*eds. A. J. Buras and M. Lindner*)
Vol. 16 – Electroweak Symmetry Breaking and New Physics at the TeV Scale (*eds. T. L. Barklow, S. Dawson, H. E. Haber and J. L. Siegrist*)
Vol. 17 – Perspectives on Higgs Physics II (*ed. G. L. Kane*)
Vol. 18 – Perspectives on Supersymmetry (*ed. G. L. Kane*)
Vol. 19 – Linear Collider Physics in the New Millennium (*eds. K. Fujii, D. J. Miller and A. Soni*)
Vol. 20 – Lepton Dipole Moments (*eds. B. Lee Roberts and William J. Marciano*)
Vol. 21 – Perspectives on Supersymmetry II (*ed. G. L. Kane*)
Vol. 22 – Perspectives on String Phenomenology (*eds. B. Acharya, G. L. Kane and P. Kumar*)
Vol. 23 – 60 Years of CERN Experiments and Discoveries (*eds. H. Schopper and L. Di Lella*)
Vol. 24 – The High Luminosity Large Hadron Collider: The New Machine for Illuminating the Mysteries of Universe (*eds. O. Brüning and L. Rossi*)
Vol. 25 – Massive Neutrinos: Flavor Mixing of Leptons and Neutrino Oscillations (*ed. H. Fritzsch*)
Vol. 26 – The Standard Theory of Particle Physics: Essays to Celebrate CERN's 60th Anniversary (*eds. L. Maiani and L. Rolandi*)
Vol. 27 – Technology Meets Research: 60 Years of CERN Technology: Selected Highlights (*eds. C. Fabjan, T. Taylor, D. Treille and H. Wenninger*)

Forthcoming

Vol. 8 – Standard Model, Hadron Phenomenology and Weak Decays on the Lattice (*ed. G. Martinelli*)

Advanced Series on
Directions in High Energy Physics — Vol. 27

Technology Meets Research

60 Years of CERN Technology: Selected Highlights

Editors

Christian Fabjan, Thomas Taylor,
Daniel Treille and Horst Wenninger

with members of the Editorial Group

Cristoforo Benvenuti, Giorgio Brianti,
Kurt Hübner, Pier Giorgio Innocenti,
Philippe Lebrun, Romeo Perin
and Christine Sutton

and with important contributions from
many other CERN experts

World Scientific

NEW JERSEY • LONDON • SINGAPORE • BEIJING • SHANGHAI • HONG KONG • TAIPEI • CHENNAI • TOKYO

Published by

World Scientific Publishing Co. Pte. Ltd.
5 Toh Tuck Link, Singapore 596224
USA office: 27 Warren Street, Suite 401-402, Hackensack, NJ 07601
UK office: 57 Shelton Street, Covent Garden, London WC2H 9HE

Library of Congress Cataloging-in-Publication Data
Names: Fabjan, C. W., editor. | Taylor, Thomas (Physicist), editor. | Treille, Daniel (Physicist), editor. | Wenninger, Horst, editor. | European Organization for Nuclear Research, honouree.
Title: Technology meets research : 60 years of CERN technology : selected highlights / editors, Christian Fabjan (Vienna University of Technology, Austria & Austrian Academy of Sciences, Austria), Thomas Taylor (CERN, Switzerland), Daniel Treille (CERN, Switzerland), Horst Wenninger (CERN, Switzerland).
Other titles: Advanced series on directions in high energy physics ; v. 27.
Description: Singapore ; Hackensack, NJ : World Scientific, [2017] | Series: Advanced series on directions in high energy physics, ISSN 1793-1339 ; vol. 27 | Includes bibliographical references and index.
Identifiers: LCCN 2017014301| ISBN 9789814749138 (hardcover) | ISBN 9814749133 (hardcover)
Subjects: LCSH: Particles (Nuclear physics)--Research. | Particles (Nuclear physics)--Technological innovations. | European Organization for Nuclear Research--History.
Classification: LCC QC776 .T43 2017 | DDC 539.7/3--dc23
LC record available at https://lccn.loc.gov/2017014301

British Library Cataloguing-in-Publication Data
A catalogue record for this book is available from the British Library.

Contents

Foreword

Throughout 2014, CERN and its Member States celebrated "60 Years of Science for Peace" to mark this major anniversary of the Organization's foundation in 1954. As the founding convention states, the main purpose in establishing a European Organization for Nuclear Research was "to provide for collaboration among European states in nuclear research of a pure scientific and fundamental character".

CERN has since amply fulfilled the hopes and dreams of the founders, while its field of research has transformed into modern high energy particle physics. Over more than 60 years, the Laboratory has transported European physicists to the frontier of the field, where, together with colleagues from countries far beyond the Member States, they continue to make vital contributions to our understanding of fundamental physics.

This success is rooted in the succession of accelerators at CERN, from those envisaged in the original convention to the world-famous Large Hadron Collider, as well as in experiments that have made remarkable discoveries and exquisite measurements. Underlying it all, the crucial interplay between physics and technology has led to the innovative developments and the many technological achievements described in this book. These accomplishments have been essential in enabling CERN to fulfil its scientific mission. They have also had a significant impact beyond the laboratory, touching many different areas of science, industry and society.

While much in the world has changed since 1954, the collaborative and innovative spirit of CERN — so essential to its pursuit of science — is more important than ever.

Fabiola Gianotti
Director-General, CERN

Preface

New directions in science are launched by new tools more often than by new concepts. The effect of a tool-driven revolution is to discover new things that have to be explained!
Freeman Dyson, *Imagined Worlds* (Harvard University Press, 1997)

This book illustrates the interplay between research and technology, which has led to numerous achievements during the different periods of CERN. It describes the context that has given rise to important technological innovations, often due to individuals or small teams, illustrated with selected highlights. Abundant references are provided for readers who are inspired to explore particular topics in more detail.

The introduction to each chapter provides the setting for a specific period, motivates the selected technological highlights, and places them in the overall context. The highlights themselves describe important examples having both a novel character and a substantial impact on physics and/or technology. They are either "CERN firsts" or CERN-based breakthroughs in the development of specific technology. Most of the authors are contemporary witnesses, who were directly involved in the activities described. Some highlights relate to developments by larger, frequently international teams, where CERN assumed a leading or major role.

This book is addressed to scientists, engineers, technicians and administrators, and indeed to all those who may be interested in the evolution of technology at a leading European scientific organization. The style and the level of the technical detail aim to tell in accessible terms the remarkable story of why and how these achievements were accomplished.

The book traces CERN's development from the "learning" stage to its present position amongst the leading laboratories in high energy particle physics research. It describes the Organization's ever-broader programme, its style of management and its relationship to society.

We extend our thanks to the members of the Editorial Team and to all authors for their contributions to this volume.

The Editors

List of Authors

Brian Allardyce	CERN†
Paul Baillon	CERN†
Alan Ball	CERN†; N.C.S.R.D., Athens (GR)
Amalia Ballarino	CERN
Cristoforo Benvenuti	CERN†; SRB Energy Research Sàrl, Geneva (CH)
Werner Beusch	CERN†
Giorgio Brianti	CERN†
René Brun	CERN†
Helmut Burkhardt	CERN
Jean-Paul Burnet	CERN
Michael Campbell	CERN
Fritz Caspers	CERN†; E.S.I., Archamps (F)
Michel Chanel	CERN†
Enrico Chiaveri	CERN†; Manchester University (U.K.)
Vinod (Nick) Chohan	CERN†
Donald (Don) Cundy	CERN†; University and INFN, Milano (I)
Carmelo D'Ambrosio	CERN
Heiko Damerau	CERN
Pierre Darriulat	CERN†; VATLY, Hanoi, Vietnam
Hans Dijkstra	CERN
Niels Doble	CERN†; University and INFN, Pisa (I)
Michael Doser	CERN
Christian Fabjan	CERN†; OeAW and TU Vienna (A)
Philippe Farthouat	CERN
Giuseppe Fidecaro	CERN†
François Fluckiger	CERN†
Lau Gatignon	CERN
Simone Gilardoni	CERN
Massimo Giovannozzi	CERN
Claus Gößling	Formerly CERN; Tech. Univ., Dortmund (D)
Jean-Pierre Gourber	CERN†
Steven Hancock	CERN
Erik Heijne	CERN†; Czech Technical University (CZ)
Frédéric Hemmer	CERN
Alain Hervé	CERN†; Univ. Wisconsin-Madison (USA)
Reinhold Hohbach	CERN†
Albert Hofmann	CERN†
Kurt Hübner	CERN†
Pier Giorgio Innocenti	CERN†
Pierre Jarron	CERN†; University and INFN, Torino (I)
Peter Jenni	CERN†; Albert-Ludwigs-Univ., Freiburg (D)
Jasper Kirkby	CERN†; J.-W.-Goethe-Univ., Frankfurt (D)
Philippe Lebrun	CERN

†Retired

Paul Lecoq	CERN
Giorgio Matthiae	CERN[†]
Michel Mayoud	CERN[†]
Tapio Niinikoski	CERN[†]
Markus Nordberg	CERN
Romeo Perin	CERN[†]
Achille Petrilli	CERN
Guido Petrucci	CERN[†]
Werner Pirkl	CERN[†]; Tokyo University, Japan
Emanuele Quercigh	CERN[†]
Helge Ravn	CERN[†]
Jean-Pierre Revol	CERN[†]; Centro Enrico Fermi, Roma (I)
Werner Riegler	CERN
Les Robertson	CERN[†]
Luigi Rolandi	CERN
Leonardo Rossi	Formerly CERN; University and INFN, Genova (I)
Lucio Rossi	CERN
Bernard Sadoulet	Formerly CERN; Univ. of California, Berkeley (USA)
Fabio Sauli	CERN[†]; TERA Foundation (I)
Karlheinz Schindl	CERN[†]
Daniel Schulte	CERN
Ben Segal	CERN[†]
Charles Steinbach	CERN[†]
Bent Stumpe	CERN[†]
Christine Sutton	CERN[†]
Igor Syratchev	CERN
Thomas Taylor	CERN[†]
Lars Thorndahl	CERN[†]
Daniel Treille	CERN[†]
Olav Ullaland	CERN[†]
Tejinder (Jim) Virdee	Formerly CERN; Imperial College, London (UK)
Horst Wenninger	CERN[†]; GSI, Darmstadt (D)
Crispin Williams	Formerly CERN; University and INFN, Bologna (I)
Edmund (Ted) Wilson	CERN[†]

Acronyms

AA	*A*ntiproton *A*ccumulator (at CERN)
AC	*A*ntiproton *C*ollector (at CERN)
AD	*A*ntiproton *D*ecelerator (at CERN)
ADLO	*A*LEPH *D*ELPHI *L*3 *O*PAL — the LEP experiments as a group
ADS	*A*ccelerator-*D*riven *S*ystems
AEGIS	*A*ntihydrogen *E*xperiment: *G*ravity, *I*nterferometry, *S*pectroscopy
AFS	*A*xial *F*ield *S*pectrometer — an experiment at the ISR
AGS	*A*lternating *G*radient *S*ynchrotron — an accelerator at BNL, USA
ALEPH	*A*pparatus for *LE*P *PH*ysics at CERN — an experiment at LEP
ALICE	*A*n *L*HC *I*on *C*ollider *E*xperiment — an experiment at the LHC
ALPHA	*A*ntihydrogen *L*aser *PHysics* *A*pparatus
AMS	*A*lpha *M*agnetic *S*pectrometer — an experiment on the ISS
APD	*A*valanche *P*hotodiode *D*etector
ASACUSA	*A*tomic *S*pectroscopy *A*nd *C*ollisions *U*sing *S*low *A*ntiprotons
ASIC	*A*pplication-*S*pecific *I*ntegrated *C*ircuit
ATHENA	*A*n*TiH*ydrog*EN* *A*pparatus
ATRAP	*A*ntihydrogen *TRAP*
ATLAS	*A* *T*oroidal *L*HC *A*pparatu*S* — an experiment at the LHC
AWAKE	*A* proton-driven plasma *WAKE*field acceleration experiment
BASE	*B*aryon *A*ntibaryon *S*ymmetry *E*xperiment
BC	*B*ubble *C*hamber [Box 3.3]
BCS theory	*B*ardeen–*C*ooper–*S*chrieffer theory of superconductivity [Box 8.1]
BEBC	*B*ig *E*uropean *B*ubble *C*hamber (co-financed by France and Germany)
BEH	*B*rout–*E*nglert–*H*iggs mechanism [Box 8.2]
BES	*BE*ijing *S*pectrometer at the Beijing Electron Positron Collider, China
BETEL	*B*roadband *E*xchange over *T*rans-*E*uropean *L*inks (informatics)
BEVALAC	*BEVA*TRON + superHI*LAC* — Bevatron with superHILAC as injector
BEVATRON	*B*illions of *eV* synchro*TRON* — weak focusing synchrotron at LBL, US
BGO	*B*ismuth *G*ermanium *O*xide — a scintillating crystal
BIC	*B*usiness *I*ncubation *C*entre
BITNET	*B*ecause *I*t's *T*ime *NET*work (informatics)
BPM	*B*eam *P*osition *M*onitor
BSCCO	*B*ismuth *S*trontium *C*alcium *C*opper *O*xide — a superconductor (HTS)
BSM	*B*eyond the *S*tandard *M*odel
CALICE	*CA*lorimeter for the *L*inear *C*ollider *E*xperiment
CAS	*C*ERN *A*ccelerator *S*chool
CC	*C*ERN *C*ouncil
CC	*C*loud *C*hamber [Box 3.3]
CCC	*C*ERN *C*ontrol *C*entre, from where the accelerators are operated
CCD	*C*harge-*C*oupled *D*evice
CEDAR	*C*h*E*renkov *D*etector with *A*chromatic *R*ing focus
CESAR	*C*ERN *E*lectron *S*torage and *A*ccumulation *R*ing
CESR	*C*ornell *E*lectron–positron *S*torage *R*ing, Cornell University, NY, USA
CLIC	*C*ompact *L*inear *C*ollider

CLOUD	*C*osmics *L*eaving *OU*tdoor *D*roplets — an experiment at the CPS
CM	*C*entre of *M*ass
CMOS	*C*omplementary *M*etal-*O*xide *S*emiconductor — integrated circuit
CMS	*C*ompact *M*uon *S*olenoid — an experiment at the LHC
CNGS	*C*ERN *N*eutrinos to *G*ran *S*asso, a facility used for sending neutrino beams through the Earth from CERN to experiments at LNGS, Italy
CNAO	*C*entro *N*azionale di *A*droterapia *O*ncologica, Pavia, Italy
COMPASS	*Com*mon *M*uon and *P*roton *A*pparatus for *S*tructure and *S*pectroscopy — an experiment at the SPS
CORE	*CO*st *RE*view
COTS	*C*ommercial *O*ff-*T*he-*S*helf — as of components
CP	*C*harge-*P*arity symmetry [Box 3.4]
CPT	*C*harge-*P*arity-*T*ime reversal symmetry [Box 2.2]
CPS	*C*ERN *P*roton *S*ynchrotron — a particle accelerator [Box 2.1]
CT	*C*omputerized *T*omography
DAQ	*D*ata *A*c*Q*uisition system
DELPHI	*De*tector with *L*epton, *Ph*oton and *H*adron *I*dentification — an experiment at LEP
DRDC	*D*etector *R&D* *C*ommittee
DNP	*D*ynamic *N*uclear *P*olarization
EARN	*E*uropean *A*cademic and *R*esearch *N*etwork (informatics)
ECFA	*E*uropean *C*ommittee for *F*uture *A*ccelerators
EDG	*E*uropean *D*ata *G*rid (informatics)
EGS	*E*lectron *G*amma *S*hower — Monte Carlo package
ELENA	*E*xtra *L*ow-*EN*ergy *A*ntiproton ring (at CERN)
Elettra	Storage ring at the Elettra Sincrotrone Trieste Research Centre, Italy
EMC	*E*uropean *M*uon *C*ollaboration — an experiment at the SPS
EMI	*E*xternal *M*uon *I*dentifier
ENLIGHT	*E*uropean *N*etwork for *LIG*ht ion *H*adron *T*herapy
EPA	*E*lectron *P*ositron *A*ccumulator
EPI	*E*xternal *P*article *I*dentifier
ESD	*E*vent *S*ummary *D*ata (informatics)
ESO	*E*uropean *S*outhern *O*bservatory
ESR	*E*lectron *S*pin *R*esonance
EUDet	*EU Det*ector R&D towards the International Linear Collider
EW	*E*lectro*W*eak theory [Box 2.3 & Box 6.4]
FADC	*F*lash *A*nalog-to-*D*igital *C*onverters
FAIR	*F*acility for *A*ntiproton and *I*on *R*esearch, Darmstadt, Germany
FC	*F*inance *C*ommittee — a body of the CERN Council
FCC	*F*uture *C*ircular *C*ollider
FEAT	*F*irst *E*nergy *A*mplifier *T*est
FEE	*F*ront-*E*nd *E*lectronics
FET	*F*ield-*E*ffect *T*ransistor
FLUKA	*FLU*ktuierende *KA*skade — Monte Carlo simulation program
FORTRAN	*FOR*mula *TRAN*slation — programming language (informatics)
FPGA	*F*ield *P*rogrammable *G*ate *A*rray

Gargamelle	HLBC — supplied to CERN by CEA, France
GANIL	*G*rand *A*ccélérateur *N*ational d'*I*ons *L*ourds, Caen, France
GBAR	*G*ravitational *B*ehaviour of *A*ntihydrogen at *R*est — experiment at AD
GEM	*G*as *E*lectron *M*ultiplier [Box 4.6]
GÉANT	Sequence of EU-sponsored networking projects (GN1 to GN4) interconnecting National Research and Education Networks (NRENs)
Geant1 to 4	*G*eneration of *E*vents *AN*d *T*racks — toolkit for the simulation of the passage of particles through matter using Monte Carlo Methods
GeantV	Geant*V*ectorized (read Geant5) — follow-up toolkit to Geant4
GNU	*G*NU's *N*ot *U*nix — free computer operating system (informatics)
GRID	Collection of computer resources at multiple locations (informatics)
GriPhyN	*Gri*d *Phy*sics *N*etwork (informatics)
HEP	*H*igh *E*nergy *P*hysics
HERA	*H*adron *E*lektron *R*ing *A*nlage — electron–proton collider at DESY, Germany
HLBC	*H*eavy-*L*iquid *B*ubble *C*hamber
HL-LHC	*H*igh *L*uminosity-*LHC* — project to upgrade the LHC
HLT	*H*igh-*L*evel *T*rigger
HOM	*H*igher *O*rder *M*ode (as in an RF cavity)
HPD	*H*ybrid *P*hoton *D*etector
HSM	*H*ierarchical *S*torage *M*anagement — data retrieval system (informatics)
HTc	*H*igh *Tc* (high critical temperature, as of a superconductor)
HTML	*H*yper*T*ext *M*arkup *L*anguage (informatics)
HTTP	*H*yper*T*ext *T*ransfer *P*rotocol (informatics)
HTS	*H*igh *T*emperature *S*uperconductor [Box 8.4]
IANA	*I*nternet *A*ssigned *N*umbers *A*uthority (USA) (informatics)
IB	*I*ntermediate *B*oson — particle mediating the weak interaction
ICE	*I*nitial *C*ooling *E*xperiment
ICFA	*I*nternational *C*ommittee for *F*uture *A*ccelerators
ILC	*I*nternational *L*inear *C*ollider
IP	*I*nternet *P*rotocol (informatics)
IP	*I*nteraction *P*oint (of a collider)
IPF	*I*nternal *P*icket *F*ence
ISO	*I*nternational *O*rganization for *S*tandardization (a semi-acronym)
ISOLDE	*I*sotope *S*eparator *O*n *L*ine *DE*vice — a facility at CERN
ISR	*I*ntersecting *S*torage *R*ings — world's first hadron collider, at CERN
ISRC	*ISR* *C*ommittee
ISS	*I*nternational *S*pace *S*tation
ITER	*I*nternational *T*hermonuclear *E*xperimental *R*eactor, Cadarache, France
KT	*K*nowledge *T*ransfer
L3	*LEP* *3*rd experiment
LAA	Detector R&D programme at CERN, initiated and financed by Italy
LAr	*L*iquid *Ar* (argon)
LKr	*L*iquid *Kr* (krypton)
LCG	*L*HC *C*omputing *G*rid (informatics)

LEAR	*L*ow *E*nergy *A*ntiproton *R*ing
LEP	*L*arge *E*lectron *P*ositron collider at CERN
LEP2	*LEP* upgrade from about 50 to 104.5 GeV/c per beam
LEP3	*LEP* with sufficient centre-of-mass energy for a Higgs factory (study)
LEPC	*LEP C*ommittee
LHC	*L*arge *H*adron *C*ollider
LHCC	*LHC C*ommittee
LHCb	*LHC b*-physics experiment
LHCf	*LHC f*orward physics experiment
LIL	*L*EP *I*njector *L*inac
LINAC	*LIN*ear *Ac*celerator [Box 2.1]
LNGS	*L*aboratorio *N*azionale del *G*ran *S*asso, Italy — underground laboratory
LRPC	*L*ong *R*ange *P*lanning *C*ommittee
LSO	Lu_2SiO_5 — Lutetium Oxyorthosilicate — a scintillating crystal
LSP	*L*ightest *S*USY *P*article (candidate for dark matter)
LTS	*L*ow *T*emperature *S*uperconductor [Box 4.3]
MDM	*M*agnetic *D*ipole *M*oment [Box 2.2]
MedAustron	Centre for hadron therapy and research, Wiener Neustadt, Austria
MoEDAL	*Mo*nopole & *E*xotics *D*etector *a*t the *L*HC
MPGD	*M*icro-*P*attern *G*as *D*etectors [Box 4.6]
MRI	*M*agnetic *R*esonance *I*maging
MTE	*M*ulti-*T*urn *E*xtraction
MVA	*M*ulti-*V*ariate *A*nalysis
MWPC	*M*ulti-*W*ire *P*roportional *C*hamber [Box 4.4 & Box 4.5]
n_TOF	*n*eutron *T*ime-*O*f-*F*light experimental facility at the PS
NA	*N*orth experimental *A*rea at CERN
NC	*N*eutral *C*urrents
NE	*N*uclear *E*mulsion [Box 3.3]
NEG	*N*on-*E*vaporable *G*etter — getter that is regenerated by heating [Box 7.3]
NMR	*N*uclear *M*agnetic *R*esonance
NODAL	Programming language for first true multi-computer operation (SPS)
NOMAD	*N*eutrino *O*scillation *MA*gnetic *D*etector — an experiment at the SPS
NREN	*N*ational *R*esearch and *E*ducation *N*etwork (informatics)
OMEGA	Spectrometer facility for experiments using beams from the PS / SPS
OOP	*O*bject-*O*riented *P*rogramming (informatics)
OPAL	*O*mni-*P*urpose *A*pparatus for *L*EP — an experiment at LEP
OPERA	*O*scillation *P*roject with *E*mulsion t*R*acking *A*pparatus – experiment at LNGS, Italy
PAM	*PA*tchy *M*aster file (informatics)
PATCHY	Source code management system (informatics)
PAW	*P*hysics *A*nalysis *W*orkstation (informatics)
PEP	*P*ositron *E*lectron *P*roject — circular e^+e^- collider at SLAC, USA
PETRA	*P*ositron-*E*lektron-*T*andem-*R*ing-*A*nlage, e^+e^- collider at DESY, Germany
PET	*P*ositron *E*mission *T*omography

PETS	*P*ower *E*xtraction and *T*ransfer *S*tructure (CLIC)
PFA	*P*article *F*low *A*lgorithm
PID	*P*article *I*dentification *D*etector [Box 5.2]
PS	*P*roton *S*ynchrotron [Box 2.1]
PSB	*PS B*ooster — pre-accelerator of the PS
PT	*P*olarized *T*arget
PU	*P*ick *U*p electrode
QCD	*Q*uantum *C*hromo*D*ynamics [Box 4.2 & Box 6.4]
QED	*Q*uantum *E*lectro*D*ynamics [Box 2.3]
QFT	*Q*uantum *F*ield *T*heory [Box 2.3 & Box 6.4]
QM	*Q*uantum *M*echanics
RAL	*R*utherford *A*ppleton *L*aboratory, Didcot, UK
RBE	*R*elative *B*iological *E*ffectiveness
REBCO	*R*are *E*arth *B*arium *C*opper *O*xide — a superconductor (HTS)
RF	*R*adio *F*requency [Box 2.1 & Box 3.1]
RFQ	*R*adio *F*requency *Q*uadrupole
RFQD	*R*adio *F*requency *Q*uadrupole for *D*eceleration
RHIC	*R*elativistic *H*eavy *I*on *C*ollider – a collider at BNL, USA
RICH	*R*ing *I*maging *CH*erenkov counter
RIOS	*R*emote *I*nput *O*utput *S*tation (informatics)
RIPE	*R*éseaux *IP E*uropéens — European equivalent of IANA (informatics)
RISC	*R*educed *I*nstruction *S*et *C*omputer (informatics)
ROOT	Program and libraries for treating HEP data
RPC	*R*emote *P*rocedure *C*all (informatics)
RPC	*R*esistive *P*late *C*hamber
ROTCO	*ROT*ary *CO*ndenser — rotating capacitor used at the SC
RRB	*R*esources *R*eview *B*oard
SC	*S*ynchro*C*yclotron — a particle accelerator at CERN [Box 2.1]
SC	*S*park *C*hamber [Box 3.3]
SCRF	*S*uper*C*onducting *RF* cavity
SCT	*S*emi*C*onductor *T*racker
SFM	*S*plit-*F*ield *M*agnet — an experimental facility at the ISR
SGML	*S*tandard *G*eneralized *M*arkup *L*anguage (informatics)
SHIFT	*S*calable *H*eterogeneous *I*ntegrated *F*acili*T*y (informatics)
SiPM	*Si*licon *P*hoto-*M*ultiplier
SIS	*S*cientific *I*nformation *S*ervice at CERN
SLC	*S*tanford *L*inear *C*ollider — an e^+e^- collider at SLAC, USA
SLD	*S*LAC *L*arge *D*etector at the SLC
SM	*S*tandard *M*odel of particle physics [Box 4.2 & Box 6.4]
SPC	*S*cientific *P*olicy *C*ommittee of CERN
SPC	*S*tatistical *P*rocess *C*ontrol
SPEAR	*S*tanford *P*ositron *E*lectron *A*ccelerating *R*ing, CA, USA
SPS	*S*uper *P*roton *S*ynchrotron — an accelerator (and $p\bar{p}$ collider) at CERN
SPSC	*SPS C*ommittee
SSC	*S*uperconducting *S*uper *C*ollider — a proton collider with a planned beam energy of 20 TeV (cancelled), Texas, USA

STAR	*S*olenoidal *T*racker *A*t *R*HIC – an experiment at BNL, USA
SUSY	*SU*per*SY*mmetry — an extension of the Standard Model
TARC	*T*ransmutation by *A*diabatic *R*esonance *C*rossing
TCC	*T*rack *C*hamber *C*ommittee
TDC	*T*ime-to-*D*igital *C*onverter
TCP/IP	*T*ransmission *C*ontrol *P*rotocol/*I*nternet *P*rotocol (informatics)
TEVATRON	*TeV*-level high energy accelerator/collider at FNAL, USA
TMAE	*T*etrakis di*M*ethyl*A*mino *E*thylene — a gas used in RICH
TOF	*T*ime-*O*f-*F*light
TOTEM	*TOT*al cross-section, *E*lastic scattering, diffraction and dissociation *M*easurement — an experiment at the LHC
TPC	*T*ime *P*rojection *C*hamber
TRD	*T*ransition *R*adiation *D*etector [Box 3.2]
TRISTAN	*T*ransposable *R*ing *I*ntersecting *S*torage *A*ccelerator in *N*ippon — an electron–positron collider with up to 30 GeV beams at KEK. Japan
TST	*T*rack *S*ensitive *T*arget
TT	*T*ransfer *T*unnel
UA1, UA2	*U*nderground *A*reas 1 & 2, housing experiments at the SPS $p\bar{p}$ collider
UHV	*U*ltra-*H*igh *V*acuum
URI	*U*niversal *R*esource *I*dentifier (informatics)
URL	*U*niversal *R*esource *L*ocator (informatics)
VELO	*VE*rtex *LO*cator — the vertex detector in the LHCb experiment
WA	*W*est experimental *A*rea at CERN
WLCG	*W*orldwide *L*HC *C*omputing *G*rid (informatics)
WWW	*W*orld *W*ide *W*eb — system to link hypertext documents

A general glossary of the vocabulary of high energy physics can be found at:
http://www.fnal.gov/pub/interactions/glossary/index.html

A listing of all the laboratories conducting particle physics research can be found at:
http://www-elsa.physik.uni-bonn.de/accelerator_list.html

Boxes with Supplementary Information

Chapter 1

Introduction

Christine Sutton and Horst Wenninger

The fruitful interaction between research in physics and technological development in the field of particle accelerators and detectors is the main theme of this book. For accelerators, research is driven by the need for ever-higher energy and intensity, and to supply an increasingly diverse range of beams, from electrons and protons and their antiparticles to heavy or radioactive nuclei. For experimentation, measurement capabilities need to be continuously improved, taking advantage of an ever stronger symbiosis between detectors, electronics and computing technology.

Research, technology and innovation at CERN have been part of global developments in nuclear and particle physics for the past six decades (1954–2014) [1–5]. The basic motivation has always been to advance the understanding of the elementary particles and forces that shape our Universe. This, however, is not the main theme of this book. Instead, it presents some of the major, even spectacular, technological advances, and the framework and conditions that enabled them to propel this European organization into the league of world-leading physics laboratories (Fig. 1.1).

The field of particle physics has its roots in the pioneering studies of the atom that began towards the end of the 19th century, and the discovery of the most familiar particle, the electron. The first decades of the 20th century saw the development of quantum mechanics and special and general relativity on the theoretical side, while experimental studies showed — astonishingly — that the atom is mainly empty space, with electrons surrounding a tiny nucleus, generally consisting of protons and neutrons. At the same time, studies of the cosmic rays that rain down through the atmosphere began to reveal new, short-lived particles — for example, muons, pions and kaons — as well as a whole new realm of antimatter.

It was to take these investigations further that the first particle accelerators were developed. The European Organization for Nuclear Research — known as CERN after the Conseil Européen pour la Recherche Nucléaire that preceded it — was founded in 1954 for just this purpose. Since the early days of its first accelerator, the Synchrocyclotron, the Organization has made many significant contributions to the present understanding of the world of elementary particles, as embodied in the current Standard Model of particle physics [Box 6.4]. Over the decades, research at CERN and similar laboratories around the world has probed deep into the structure of matter to the level of the basic constituents — the quarks — that make up protons, neutrons, pions and other particles known collectively as hadrons. Figure 1.2 shows the families of quarks and leptons (which include the electron) that are basic elements of the Standard Model. These constituent particles are called elementary because they appear point-like and without sub-structure on the scale of 10^{-17} cm. They span more than 12 orders of magnitude in mass — from the tiny neutrino at < 0.23 eV/c^2 to the top quark at 173 GeV/c^2.

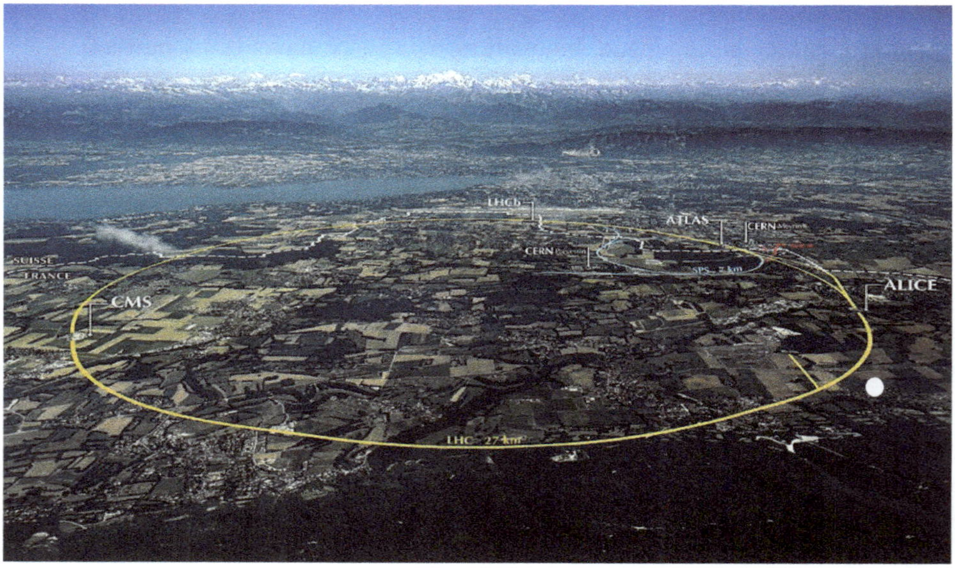

Fig. 1.1. Aerial view of the CERN laboratory today, indicating the location of most of the underground accelerator complex (PS (red), SPS (blue) and LHC (yellow)), and the four major experiments at the LHC (ALICE, ATLAS, CMS and LHCb). (Source: CERN).[a]

[a] Unless noted otherwise, CERN is the source of all photos and figures in this publication.

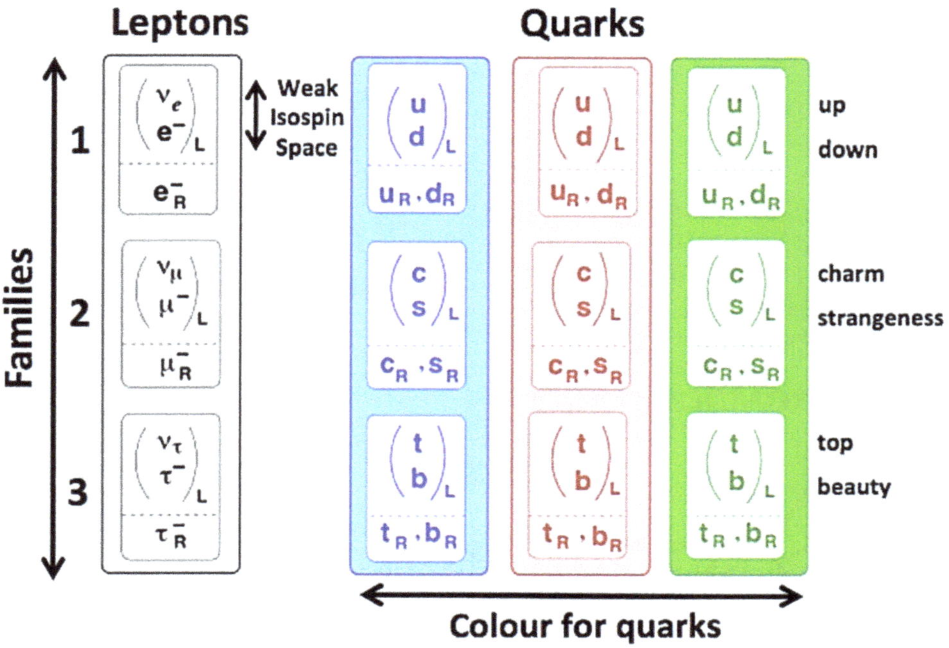

Fig. 1.2. The constituent leptons and quarks, all with spin ½ (fermions). They appear in three families, the reason for which remains mysterious. Quarks also come in three varieties or "colours" [Box 4.2]. L and R refer to the constituent "handedness" [Box 2.2], treated very differently in the Standard Model.

These investigations have also determined that the structure of matter as a whole depends critically on the way that these constituent particles interact through fundamental forces. Three forces — electromagnetic, weak and strong — are important at subatomic scales. A major success of the Standard Model has been to bring together the electromagnetic interaction (electrodynamics) and the weak interaction in a unified description in terms of an electroweak interaction, while the strong force is described in terms of interactions between quarks with a "strong charge" known as colour [Box 4.2] (chromodynamics). Within the Standard Model, the constituent particles with spin ½ interact through these forces via the exchange of a spin-1 particle, as illustrated in Fig. 1.3. A recent highlight at CERN was the discovery at the Large Hadron Collider (LHC) in 2012 of the last missing piece of the Standard Model, the spin-0 particle related to the Brout–Englert–Higgs mechanism through which particles acquire mass [Box 8.2]. This particle will be referred to as the Higgs particle throughout this book.

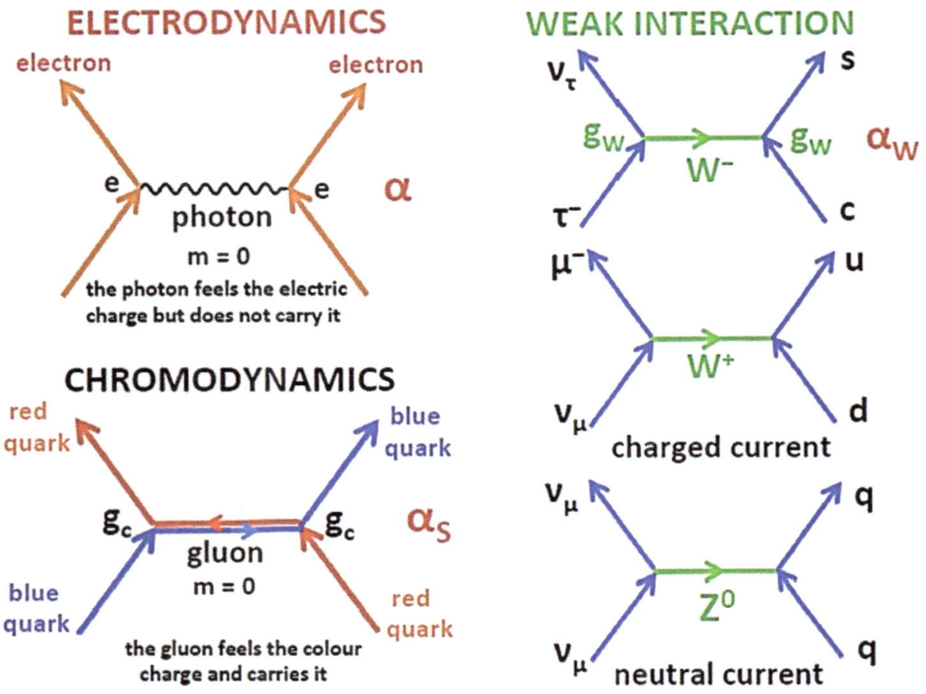

Fig. 1.3. These diagrams give an intuitive idea of the way the spin-½ fermions interact, by exchanging a spin-1 boson. This is a common feature of all three interactions (electrodynamic, chromodynamic and weak), the relevant force-carrying boson being indicated in the figure. These interactions are described in more detail in Boxes 2.3, 4.2 and 6.4.

A fourth force, gravity, influences matter on a truly cosmic scale. While for the time being, this force remains outside the Standard Model, there is nevertheless an intimate relationship between particle physics and cosmology, in the context of an expanding Universe. With the higher energies of particles at the LHC and, previously, at the Large Electron Positron (LEP) collider, experiments at CERN have reached back in time to cast light on the moments soon after the Big Bang, when our currently observable Universe was much smaller, hotter and denser than today. Figure 1.4 places the LHC and LEP accelerators on its timeline, in terms of the energy density of the collisions they produce. At about 10^{-11} s after the Big Bang, the Higgs field permeating the Universe underwent a "phase transition" giving mass to the elementary particles — a perfect illustration of the intimate relationship between particle physics and cosmology.

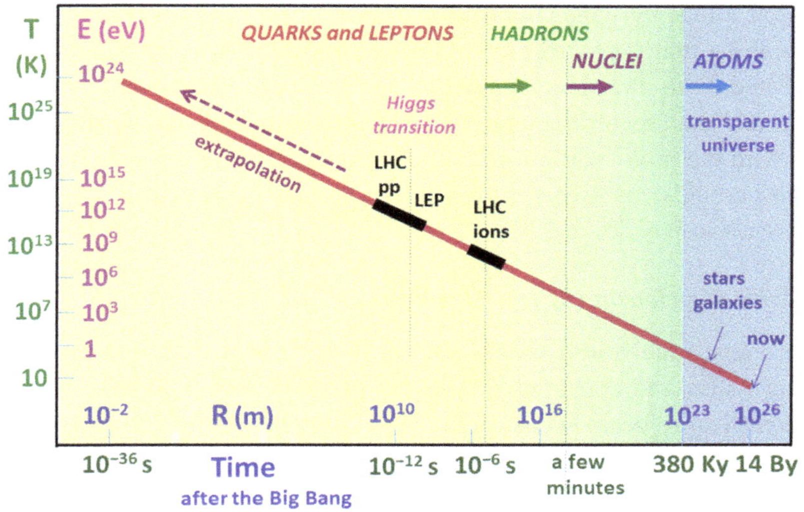

Fig. 1.4. Timeline of the Universe (logarithmic scale) with temperature T versus age. Observational astronomy is limited to the period after the Universe became transparent to photons (at an age of 380 000 years). Nuclear and particle physics experiments reproduce, on a microscopic scale, the energy per particle E (from Boltzmann's relation E = kT) that prevailed at much earlier times, as shown for LEP and LHC. The corresponding radius R of the observable Universe is also shown (courtesy F. Pauss).

1.1 CERN's First 30 Years: From Fixed Targets to the First Colliders

Until the late 1960s, the high energy frontier of nuclear and particle physics at CERN, as elsewhere, was explored with the great proton synchrotrons in which protons were directed at internal or external targets to produce secondary particles. However, it was recognized early on that head-on collisions between particle beams would be more effective than collisions between a beam and a fixed target as they exploit the full energy of the beams — e.g. for the production of new particles. The 1960s saw the construction of several successful electron–positron colliders in France, Italy, the Soviet Union and the US, but for proton machines the challenge was to produce colliding-beam intensities high enough to yield a useful interaction rate. CERN's technological contributions and inventions for solving this problem led in the early 1970s to the Intersecting Storage Rings (ISR) [6] and, a decade later, to the proton–antiproton collider [7].

On the experimental side, bubble chambers were prominent detectors in high-energy physics for several decades [8]. They were used to visualize individual interactions in the liquids they contained, but were slow as they recorded the interactions — "events" — on photographic film that had to be scanned and measured, often by hand. In parallel, experiments based on counter techniques were making major progress [9]. This transition to faster tracking detectors was accelerated with the development of wire "spark chambers" with electronic readout and data recording, and gathered important momentum at the end of the 1960s with the multi-wire proportional chambers (MWPCs) developed at CERN [10], in particular by Georges Charpak, who was honoured by the award of the Nobel Prize in Physics in 1992.

The Synchrocyclotron (SC) — Chapter 2

CERN's first accelerator, the Synchrocyclotron (SC), designed by CERN in its learning phase and constructed with the participation of industry in the 1950s, provided a record beam energy at the time. From the start it allowed important and innovative experiments with detectors based on developments in nuclear and cosmic-ray physics. Technological highlights included the SC Rotary Capacitor (ROTCO), which enabled beams of ions other than protons to be accelerated in the SC and opened up a long-lasting programme for the Isotope Separator On Line Device (ISOLDE). Pioneering experiments on pion-decay and the measurement of the anomalous magnetic dipole moment of the muon, are some of the early CERN success stories.

The CERN Proton Synchrotron (PS) — Chapter 3

The CERN Proton Synchrotron (PS), constructed after a daring design-switch to the alternating-gradient principle, underwent a remarkable evolution with increased proton intensity, allowing the production of beams of deuterons, alpha-particles, antiprotons, lead ions and CERN's first neutrino beam with the concept of the magnetic horn. The PS is still at the heart of CERN's accelerator network.

From the start, extracted beams from the PS were used with counter detectors and bubble chambers. These early experimental programmes allowed European physicists to enter increasingly into a fair competition with colleagues at well-established laboratories in the Soviet Union and the USA. Experiments with bubble chambers progressed from the 30 cm liquid-hydrogen chamber to the Big European Bubble Chamber (BEBC), and in the early 1970s, a first-generation neutrino experiment using Gargamelle, the large heavy-liquid bubble chamber built at Saclay, discovered weak neutral currents — a major breakthrough that helped to establish the Standard Model of particle physics.

Simultaneously, the introduction of Cherenkov counters, time-of-flight counters, and spark chambers, allowed for early electronic "non-bubble-chamber" experiments. Following the invention of MWPCs, large-area installations of these detectors were first used inside the OMEGA spectrometer, which was equipped with a large superconducting dipole magnet with forced cooling — itself a remarkable design by CERN engineers. Experiments at the PS continue today, e.g. at ISOLDE, n_TOF and CLOUD.

The Intersecting Storage Rings (ISR) — Chapter 4

The pioneering work on the Intersecting Storage Rings (ISR), the world's first proton–proton collider, is another success story of CERN's accelerator builders. This collider was the birthplace of many enabling technologies relevant for the future, such as ultrahigh vacuum, superconducting beam-focusing quadrupole magnets and, most notably, stochastic cooling, the ingenious technology for achieving high luminosity collisions in CERN's proton–antiproton collider.

A new era for experiments also began at this pioneering collider, ranging from the difficult detection of secondary particles emerging at small angles with respect to the beams (e.g. Roman pots) to detectors with nearly full 4π angular coverage (e.g. the Axial Field Spectrometer). With the large-scale deployment of MWPCs, the introduction of calorimeters, and the use of transition-radiation detectors, the ISR was the cradle of developments for the next step, the proton–antiproton collider experiments.

The Super Proton Synchrotron (SPS) — Chapter 5

In 1974, a new series of fixed-target experiments started at CERN with the first operation of particle beams at the Super Proton Synchrotron (SPS). Construction of the SPS was based on larger-scale application of the alternating-gradient principle, with refinement (separated function), together with installation in a deep underground tunnel as opposed to the shallow cut-and-fill galleries used previously. The first remote controls using computers were introduced to operate the large accelerator, the infrastructure and novel accelerator/operator interfaces were developed, and ingenious beam lines (e.g. muons and kaons) were set up.

At the same time, CERN adapted the West experimental area and constructed the large North area with a variety of primary, secondary and tertiary beams including a high energy neutrino beam. SPS beams served many experiments with a range of detectors, including the Big European Bubble Chamber (BEBC), streamer chambers, new variations of Cherenkov counters, and new kinds of calorimeter, e.g., with liquid krypton for experiments on CP violation or on a new scale for the giant electronic neutrino detectors. The first applications of micro-strip detectors and polarized targets also formed part of the early SPS physics programmes. The versatile PS–SPS complex also delivered high energy heavy-ion beams, opening the era of relativistic heavy-ion physics.

The proton–antiproton collider — Chapter 6

Only a few years after the start-up of the SPS, CERN launched the audacious proton–antiproton collider, leading to the experimental proof of the existence of the weak interaction mediator bosons — W and Z — in 1983 and the Nobel Prize in Physics in 1984 for Carlo Rubbia and Simon van der Meer.

The idea to use the SPS as a proton–antiproton collider triggered a chain of new technological challenges and inventions. In particular, it called for the development of an antiproton source, which in turn required a rapid test of proton-beam cooling in the Initial Cooling Experiment (ICE), made possible by CERN's development of the practical application of stochastic cooling. It also required the construction of the Antiproton Accumulator (AA) and the Antiproton Collector (AC), and led later to the Low Energy Antiproton Ring (LEAR) and the Antiproton Decelerator (AD). Experiments at the proton–antiproton collider called for the design and construction of large 4π detectors, together with online computing and data handling on an unprecedented scale.

1.2 CERN's Second 30 Years: The LEP and LHC Story

In 1981, following years of intensive discussions among physicists, and consensus formulated by the recommendation of the European Committee for Future Accelerators (ECFA), approval was given for the construction of the Large Electron Positron (LEP) collider. Unlike previous CERN projects such as BEBC, the ISR and the SPS, in the case of LEP the large-scale infrastructure, the 27 km tunnel and the collider itself, had all to be planned and constructed within the regular budget with no additional project funding. To free adequate resources, the CERN scientific community and management had to close major facilities, such as the ISR, as well as physics programmes. The increased collaboration and participation of universities in the design and construction of the large detectors needed for LEP went together with a major regrouping of technical and engineering teams at CERN. This restructuring allowed the organization of professional support on a new scale for the construction both of the collider and of four major detectors: ALEPH, DELPHI, L3 and OPAL.

By 1986, ideas, germinating since the LEP discussions, for a proton–proton collider to be built in the same tunnel had become much more concrete, making a long-term R&D programme mandatory. CERN's participation in this programme, at a time when most of its resources were committed to the construction phase of LEP, was facilitated through the LAA project [11]. This special programme, initiated and financed by Italy, approved by CERN Council, and implemented and executed at CERN, contributed to a broad detector R&D activity. In addition, LAA funds helped to recruit experts, particularly in the area of microelectronics, contributing to the success of the experiments at the Large Hadron Collider (LHC).

The technological challenges for the LHC machine and detectors to meet the expectations of research were different from those that faced the accelerator and detector builders when preparing LEP. One of the main challenges set by the energy envisaged for the LHC in the existing LEP tunnel concerned the dipole magnets needed to steer the proton beams around the ring. In order to stimulate interest for supplying the large number of high-field superconducting dipoles that would be required to provide the maximum possible bending power, CERN launched several collaborations with industry. In parallel, new detector concepts and experimental methods were required for the beam intensities and energies expected at the LHC. CERN responded with a focused detector R&D programme under the auspices of a CERN committee, the DRDC [12], which was set up in collaboration with physicists from participating institutions and sought funding support from the EU framework programmes.

With the emerging global collaboration on accelerator and detector development, communication became a key issue. CERN's answer was an Internet-based networking strategy. This brought a revolution in communication technologies, best exemplified by the invention of the World Wide Web at CERN in 1989. It came "just in time", enabling the computing boom, working tools and methods without which the LEP and LHC programmes could not have been accomplished.

Formally approved in 1994, the LHC would dominate CERN activity after the closure of LEP in November 2000. Following the first physics runs in 2009, its high energy proton collisions have taken research at CERN a step nearer to the origin of matter in the early Universe. In 2012, experimental observations by the LHC collaborations established the existence of the Brout–Englert–Higgs mechanism — a key ingredient of the Standard Model — leading to the award of the 2013 Nobel Prize in Physics to its originators.

The Large Electron Positron Collider (LEP) — Chapter 7

The choice of the size of the LEP tunnel was based on a careful balance between economic feasibility and physics discovery potential. A ring with a circumference of 27 km was adopted [13], largely driven by the need to minimize losses through synchrotron radiation — a consequence of the electron's small mass — and therefore power consumption. Compensating these losses required a powerful system of radiofrequency (RF) cavities and klystrons, while the large circumference for the design energy implied novel low-field bending magnets (dipoles), with superconducting quadrupole magnets needed for the final focus.

Although a smaller machine would have sufficed for the initial physics goal, i.e. to produce and study the neutral mediator of the weak interaction, the Z boson, the second goal, i.e. the production of W pairs and Z pairs and the search for the Higgs boson, called for higher energy. The large size of the ring enabled a gradual push of the LEP beam energy to 105 GeV, the maximum possible with the installed set of powerful niobium-clad superconducting cavities (which exceeded their design performance). However, the potential of LEP as determined by the limit of the bending magnets (125 GeV) was not fully exploited.

For the experiments at LEP, the particle-tracking chambers were either time projection chambers (TPCs) or of the advanced "jet chamber" type; particle identification was based on measurements of particle velocity with novel Cherenkov counters and the decay topology of short-lived particles; energy measurements were based on crystal or sampling calorimeters; and finally large solenoids were employed to determine particle momentum through measurements of the curvature of particle tracks in the magnetic field. The interactions of electrons — which, so far, seem to be structure-less — produce "clean" events. This motivated the LEP experiments to develop vertex detectors with high spatial resolution to reveal details close to the interaction point. Read-out electronics in this era progressed from electronic transistors to microchips on printed boards, and later to integrated microelectronic circuits incorporating event selection and data-recording techniques.

The Large Hadron Collider (LHC) — Chapter 8

The LHC complex, which includes the 27 km superconducting collider (thanks to the existence of the LEP tunnel) and several huge particle detectors, is the largest and most sophisticated research instrument ever built [14]. Chapter 8 outlines the main features and the highlights of the developments that paved the way from its conception in the early 1980s to the first years of exploitation from 2009 to 2013.

To reach its physics goals, the LHC had to achieve the highest collision energies and luminosity technically possible. Constrained by the size of the existing tunnel, this implied the use of the highest possible magnetic bending fields, given that for the massive protons of the LHC, synchrotron radiation is less of a problem than with lightweight electrons. Thus, CERN prepared for the LHC with ambitious R&D on superconducting cables and magnets. In addition, the envisaged high beam intensity, with the concomitant large beam power, required effective beam collimation, a new level of equipment protection, and reliable solutions for dumping the beams. The research agenda called for about one billion collisions per second. The interactions of protons, which are composite particles, produce hundreds of particles per collision. The experiments required the highest possible time and spatial resolution for vertex detection; radiation-tolerant materials and electronics; very large superconducting solenoidal/toroidal magnets; and novel technologies for event triggers, data processing and storage. Carrying out the experiments and recording the events seemed impossible to many experts during the early discussions in the 1980s. The international collaborations already mentioned, coordinated by the CERN DRDC, made it reality.

Data handling and communication — Chapter 9

To support its research activities, CERN's information technology and computing tools progressed over the decades from the Ferranti Mercury of the 1950s based on vacuum tubes, via the IBM 7090 in the 1960s with punched cards, magnetic tapes, and FORTRAN programming, to mainframes such as the CDC and CRAY computers (as well as online DEC PDP and NORD computers), before reaching the Internet-based computing of today. Chapter 9 traces this journey, and highlights include computing clusters and data storage, local data networks, high-speed worldwide networking, and simulation, as well as the World Wide Web and the CERN Computing Grid.

Knowledge and technology transfer — Chapter 10

The wealth of experience and knowledge acquired in fulfilling its research programme is a major asset of CERN, and transferring this expertise beyond the boundaries of the Laboratory is an important mission for the Organization. Since its founding, CERN has profited from the accumulation of its expertise in accelerator physics, RF engineering, superconductivity and cryogenics, instrumentation and controls, sensors, electronics and detector design, data acquisition and advanced informatics. In all of these areas, the research requirements have led to the innovations outlined in this book — many of which have gone to find applications not only at other, similar laboratories, but also in quite different fields. Chapter 10 describes the processes for knowledge and technology transfer that CERN has developed over the decades. The highlights focus on innovations that contribute to a wider benefit for society, in particular for medicine and the environment.

Managing the laboratory and large projects — Chapter 11

Chapter 11 gives an appraisal of CERN's past and present management approach to foster creativity on the scale discussed in this book, demonstrating an evolution that has led to what is sometimes referred to as the "CERN Model". The highlights illustrate the association with industry as applied to the accelerators and worldwide collaboration as applied to the LHC experiments. Compared to laboratories in the USA, where funding is approved on a yearly basis, CERN has benefitted greatly from regular funding and 5-year plans that have enabled it to follow through on long-term projects. However, the restrictions on the staff complement and the scale of the current collider experiments have made large worldwide collaborations the norm, albeit with strong backing from CERN. The responsibility for providing the beams remains squarely with the Laboratory, but sharing of tasks among associated institutes has also grown in the accelerator sector. The cost-effective use of accumulated infrastructure has been an important element and asset for CERN, but the true driving force resides in the endeavour to accomplish a common goal, which fosters team spirit and loyalty to the missions of the Organization.

R&D for the future — Chapter 12

CERN's scientific projects are generally characterized by timescales of the order of 30 years, from first discussions and planning through design and technological preparation, the proposal and approval, to construction and full exploitation, and finally completion. It is therefore not too early to start thinking about the next steps after the LHC, summarized in Chapter 12. Results from the LHC, the High Luminosity LHC and other experiments will ultimately determine the future of experimental particle physics. Preparing this future while keeping options open requires a multi-pronged approach in accelerator R&D. While the present focus at CERN is on a linear electron–positron collider in the Compact Linear Collider (CLIC) study, and on a proton–proton collider, the Future Circular Collider (FCC), other alternatives are also being actively pursued [15].

Acknowledgements

For more than 60 years, the interplay between physics and technology at CERN has led to the technological innovations that are the main subject of this book. The following chapters describe the context that gave rise to these achievements during the different periods of the Organization. They are illustrated by highlights that have been selected for their novel character and impact on physics and/or technology. Either "CERN firsts" or CERN-based breakthroughs, they were due to individuals or small teams as well as, particularly in later years, to work within larger, often international collaborations with other institutes, where CERN had a leading role. The selection does not purport to be exhaustive. Only Nobel laureates, CERN directors-general and leaders of the major accelerator projects are named where appropriate; in other cases the names of those involved can be found through the abundant references that allow interested readers to dig deeper into these many fascinating topics. The generous support of society at large is gratefully acknowledged, without which research is not possible. In addition to activity financed via the CERN budget, the national funding agencies are supporting scientists and engineers at the home institutes and universities in their ambitious research programmes at the CERN facilities. Today, more than 12 000 such scientists and engineers are using these facilities: they are the motivators, shakers and shapers of the Laboratory.

References

1. A. Hermann *et al.*, *History of CERN*, Vol I, (North-Holland, 1987).
   A. Hermann *et al.*, *History of CERN,* Vol II, (North-Holland, 1990).
   J. Kriege, *History of CERN,* Vol III, (North-Holland, 1996).
2. L. van Hove and M. Jacob, Highlights of 25 years of physics at CERN, *Phys. Rep.* **62** (1980).
3. J.R. Ellis *et al.* (ed.), *CERN: The second 25 years* (Elsevier, 1994).
4. K. Hübner, G. Veneziano, D. Treille, and D.O. Williams, Fifty years of research at CERN from past to future (2006), https://cds.cern.ch/record/1177861/files/CERN-2006-004.pdf.
5. H. Schopper and L. Di Lella (eds.), *60 Years of CERN Experiments and Discoveries* (World Scientific, Singapore, 2015).
6. K. Hübner and T. Taylor, The Birth and Development of the First Hadron Collider, The CERN Intersecting Storage Rings (ISR) (2013); http://cds.cern.ch/record/1626809/files/CERN-ACC-2013-0248.pdf.
7. L. Evans, The Proton-Antiproton Collider, CAS Lecture (1987), https://cas.web.cern.ch/cas/John%20Adams'%20Lectures/Evans.pdf.
8. G. Harigel *et al.*, BUBBLE 40, *Nucl. Phys. B Proc. Suppl.* **36** (1994); http://cerncourier.com/cws/article/cern/50796.
9. T. Massam, T. Muller, M. Schneegans *et al.*, A new electron detector with high rejection power against pions, *Nuovo Cim.* **39**, 464 (1965).
10. G. Charpak and F. Sauli, High-resolution electronic particle detectors, *Ann. Rev. Nucl. Part. Sci.* **34,** 285 (1984).
11. LAA Programme of Activities, CERN Council document CERN/1641 (1986), https://cds.cern.ch/record/30538/files/CM-P00076385-e.pdf.
12. The CERN DRDC - detector research and development committee (1991–1999); http://library.web.cern.ch/archives/history_CERN/Scientific_committees#DRDC.
13. H. Schopper, *LEP - The Lord of the Collider Rings at CERN 1980-2000* (Springer, 2009).
14. L. Evans (ed.), *The Large Hadron Collider: a Marvel of Technology* (EPFL Press, 2010).
15. O. Brüning and S. Myers (eds.), *Challenges and Goals for Accelerators in the XXI Century* (World Scientific, 2016).

Chapter 2

The 600 MeV Synchrocyclotron: Laying the Foundations

Brian Allardyce and Giuseppe Fidecaro

2.1 Introduction

On 15 February, 1952, the agreement was signed constituting a "Council of Representatives of European States for Planning an International Laboratory and Organizing Other Forms of Co-operation in Nuclear Research." I. Rabi, Nobel Prize 1944, considered this "The official birth of the project fathered in Florence" with a resolution submitted to the Fifth General Conference of UNESCO in June 1950. While the UNESCO resolution was deliberately vague and abstract, it lent authority to the ensuing debates among the leading scientists, spearheaded notably by E. Amaldi, P. Auger and N. Bohr (Nobel Prize 1922), on the possible mission of such an international laboratory, considering accelerator-based fundamental ("nuclear") physics as the most attractive choice. The signing of the February 1952 Agreement set in motion a sequence of events unfolding with astounding swiftness and purposefulness. Barely three months later, in May 1952, the first meeting of the (provisional) Council agreed on a detailed and prescient "business plan" for the future laboratory, laid down the major lines for two accelerators and established the corresponding study groups in several European countries. Council also initiated and sponsored a conference, to be held in June 1952 chaired by Bohr to evaluate scientific topics related to the planned laboratory.

The second Council meeting was held in June 1952, following the two-week Copenhagen Conference. At that meeting, W. Heisenberg (Nobel Prize 1932) summarized the key conclusions of the conference in a remarkable *tour d'horizon* of particle physics and accelerators. This led to the recommendation that one group should design a 600 MeV synchrocyclotron (SC) and a second group should undertake a feasibility study of a powerful proton synchrotron (PS) [Box 2.1]. The larger one would be a frontier proton machine with energies in the 10 to 20 GeV range, while the smaller one should be based on well-established principles and provide beams as soon as possible. This would allow the laboratory and the community of users to acquire expertise in the techniques required to handle a variety of particle beams (protons, pions, neutrons, muons) [1, 2]. In short, the small machine should be ready early, it had to be successful, and it had to be meaningful for physics.

Council adopted the report on the Copenhagen Conference. At the same meeting, C. Bakker was appointed to lead the group responsible for the design and construction of the SC. By October 1952 the group had divided their work into five main sectors and distributed the responsibilities among laboratories in several European countries. Six months later cost estimates for the major items were at hand; by the end of 1953 the group was ready to start the tendering process with industry. Formal construction started in June 1955, as shown in Fig. 2.1.

Fig. 2.1. On 1 June 1955 F. Bloch, CERN's first Director-General, watched by M. Petitpierre, President of the Swiss Confederation, laid the foundation stone of the SC.

In parallel with the technical work, steps were taken to establish the permanent organization. In October 1952 Geneva was chosen to be the site for the laboratory. In April 1954 F. Bloch, Nobel Prize 1952, was designated as first director-general of the yet-to-be established permanent Organization. In May 1954 earth was broken in the fields just outside Geneva. By September 1954 all documents for the ratification of the Convention establishing the permanent laboratory were deposited at UNESCO and the provisional CERN Council disbanded. In the following month the first permanent Council appointed C. Bakker to lead the SC and J. Adams the PS construction. However, already in September 1955 C. Bakker was named Director-General, succeeding F. Bloch, who, on leave from Stanford University, could only serve the Organization during its first year. W. Gentner was appointed head of the SC. He recruited engineers and physicists, building the team responsible for the SC construction and working in an atmosphere of academic freedom. CERN's by now traditional dual role of building new accelerators at the frontier of technology and participating in cutting-edge research owes much to the effort and vision of Gentner.

Although the SC was the biggest European accelerator at the time of construction, it was not intended to be a 'pioneering machine' — in contrast to the PS, which was designed in parallel and opened new accelerator territory. It was inspired by and conceived as a scaled-up version of the Chicago 450 MeV machine where Enrico Fermi and his group had done their fundamental work on pion-nucleon interactions. The choice of 600 MeV, possibly influenced by a suggestion attributed to Fermi, was a compromise between the wish for the highest possible energies and the need to keep costs safely under control. The magnet, which is huge and a major cost driver, determined the energy. With magnetic fields of 1.8 T, achievable with good-quality, low-carbon steel, protons with a kinetic energy of 600 MeV have circular orbits with a radius of 2.25 m, setting the scale for the diameter D of the SC magnet, D = 5 m. The magnet was a classic window-frame design with a 45 cm gap (Fig. 2.2). The iron magnet yoke and the coils were the first major items to be ordered, because they required long manufacturing times, would present transport problems and would condition the assembly of all other elements. The pole-faces were manufactured in France, made from more than 50 pieces, each weighing more than of 50 tonnes. The two coils for the SC magnet, each weighing 60 t and measuring 7.2 m in diameter, were fabricated in Belgium. For reasons of cost, weight and relative ease of manufacturing, an aluminium alloy was chosen.

Accelerators **Box 2.1**

These devices accelerate charged particles such as protons, antiprotons, ions, electrons and positrons. They come in two main types: linear and circular.

Linear accelerators (linacs) work by passing the beam through a linear array of accelerating radio frequency (RF) fields interspersed with focusing elements and diagnostics. At CERN they are used to accelerate heavy particles (protons, ions) after the source, serving as injectors to the next stage (and also accelerating radioactive ions right up to the experiment). Protons reach 50 MeV (160 MeV) kinetic energy in present (future) injectors. Top energy is limited by the difficulty to adapt to particle velocity and diminishing returns on investment. For higher energies synchrotrons are more suitable, but linacs must be used for accelerating light particles (e^+ and e^-) to energies beyond the reach of conceivable large synchrotrons (e.g. E_{CM} = 0.35 TeV of FCC-ee, see 12.1) where radiation loss makes synchrotrons inefficient [Box 4.1].

Circular accelerators come in two types: *cyclotrons* and *synchrotrons*. *Cyclotrons* have a single large cylindrical magnet producing a constant vertical field, with particle injection from a source at the centre. The particles are accelerated by a RF field of fixed frequency, which twice per turn provides a kick to the particles. These spiral outwards due to the combined action of magnetic field and acceleration until they emerge at the edge of the magnet. Higher energies are reached with an improved version, the *synchrocyclotron*. If E_k and $E_0 = m_0 \cdot c^2$ are respectively the kinetic and rest energy of the particle, the RF frequency $\propto E_0 / (E_k + E_0)$. For cyclotrons $E_k \ll E_0$ and the RF frequency can be constant. This is not the case at higher energies, where the RF must vary to ensure synchronism between the electric field and the particles. This technique extended the range of cyclotron-type machines, but to attain still higher energies the magnet becomes unrealistically large. The way out is the *synchrotron*, where an array of relatively small dipole magnets deflects the particles so that they follow a quasi-circular orbit. This structure has been adopted for most of CERN's accelerators, storage rings and colliders. The particles are accelerated by RF cavities located between the magnets. The magnetic field is ramped so that particles receiving the accelerating kick in the cavities stay on the same orbit. The RF is synchronized with the magnetic field such that the particles, grouped in bunches, always experience an accelerating electric field, with the decelerating phase occurring in gaps between bunches. The beam is focused with quadrupoles, interspersed among the dipoles. They provide a magnetic field that increases linearly across the aperture, focusing in one plane but defocusing in the orthogonal plane. Alternating focusing and defocusing quadrupoles results in overall focusing. This *alternating gradient principle* has been adopted for all synchrotron-type accelerators at CERN. The quadrupole fields can be superimposed on the dipole field of the bending magnets, so-called *combined-function magnets*. If the quadrupoles are separate, the array of magnets is a *"separated-function lattice"*. This provides greater flexibility for tuning the focusing and is used in most recent synchrotrons. Magnets producing non-linear magnetic fields are required at certain points in the lattice for special purposes, e.g. control of beam instabilities, and pulsed dipole magnets with short rise and fall times are required for injection and ejection.

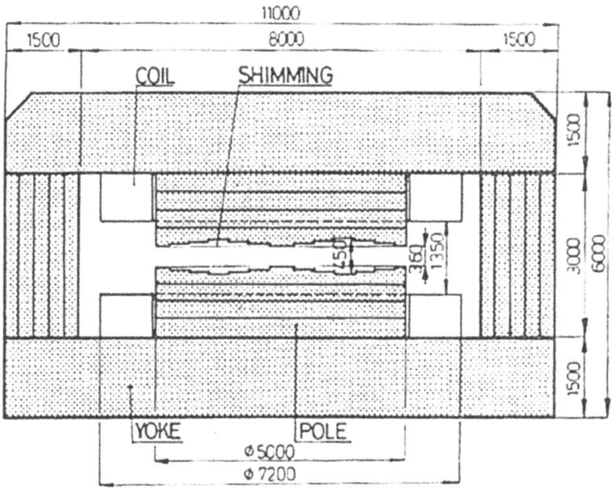

Fig. 2.2. Cross section of the SC [2]. Transport of the 7.2 m diameter coils to Geneva after shipment from Belgium to Bale was an adventure. Passing through the villages required tilting the coil on the lorry, leaving 3 cm clearance between the coil and the walls of the houses [2, 3].

The radiofrequency system (RF) providing the accelerating electric field for the protons was arguably the most challenging item for the CERN engineers and their industrial partners. The frequency of the electric field accelerating the particles has to change to stay in tune with the revolution frequency, which decreases due to the relativistic effects at energies higher than a few tens of MeV. This is done by changing the value of a capacitor in the resonant circuit. Most synchrocyclotrons had adopted mechanical, rotating capacitors, reminiscent of the tuning capacitors of old-fashioned radio receivers. To avoid recurrent difficulties encountered with rotating capacitors, arising from operation in high vacuum, overheating, broken bearings and sparking, the CERN team adopted a solution, as bold as it was elegant: a vibrating capacitor in the form of a tuning fork, manufactured from a carefully chosen aluminium alloy. The self-oscillating operation at 55 Hz was driven from the base of the fork, via an electromagnet (Fig. 2.3). Feedback circuits assured the control of the amplitude. During the intense development period there were many problems, e.g. parasitic vibrations and metal fatigue, and unconventional solutions adopted. Not surprisingly, the tuning fork was on the critical path for timely completion of the SC. With one week to go to the official inauguration, to which ministers, government officials and dignitaries had already been invited, the RF system was still not ready. But the problems were solved, and operation of the accelerator started on 1 August 1957, after a remarkably short construction time of three years [4] (Fig. 2.4).

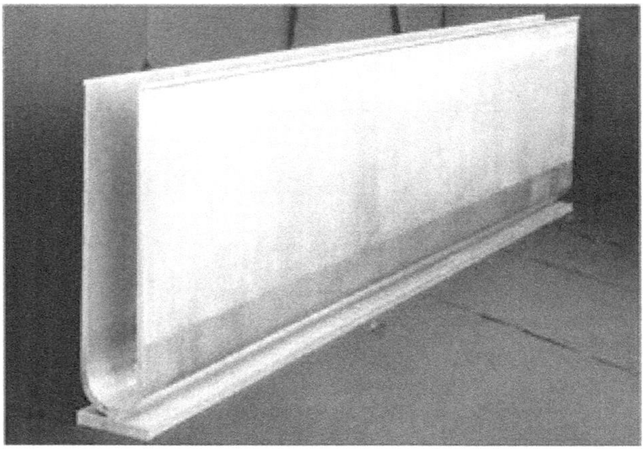

Fig. 2.3. The 2 m-wide aluminium alloy tuning fork, part of the variable capacitor used to modulate the frequency of the RF system for the SC.

Fig. 2.4. The SC in December 1957, ready for physics.

The successful completion of the SC was a milestone for CERN: it gave confidence to its staff and proof to the governments that large international collaborations between scientists and industry can be made to work. It provided the European physics community with an instrument for research on a par with those in the USA.

Next on the agenda was optimizing the SC for physics use and learning to develop the research programme. Beams of high energy particles were developed and their intensity gradually increased. Initially, such beams could only be produced from internal targets that could be remotely positioned at different radii and azimuthal angles. One of the earliest such devices was a mobile chariot of a type first used at the Chicago synchrocyclotron, invented by Fermi, hence aptly called the 'Fermi trolley'. In this way, beams of neutrons, pions and muons were directed into the Neutron Hall on one side of the accelerator, while a magnetically shimmed channel allowed an extracted proton beam on the other side, from which secondary pion beams could be delivered into the Proton Hall. An internal proton beam of about 0.5 μA was available, of which about 5% could be extracted. One SC speciality was an innovative muon beam line, consisting of a series of closely spaced quadrupoles inside the 6 m thick concrete shielding wall.

One SC development aimed at improving the conditions for experimentation. Short spills of a secondary beam lasting approximately 150 microseconds were produced after the acceleration cycle every 55 milliseconds on a target of the 'Fermi trolley'. These rather short pulses were not optimal for most experiments using counter techniques. Ingenious ways were developed to "stretch" the secondary beam and to achieve a more favourable ratio of beam pulse length to acceleration cycle or "duty cycle". In one method a vibrating target intercepted the internal proton beam. Clever, but being a mechanical device, its reliability proved to be problematic. It was superseded by an elegant manipulation of the internal proton beam at the end of the cycle, where an auxiliary accelerating field gently moved the beam into the target, achieving a duty cycle of 30%.

The development of the various beams and SC operating conditions had to reflect the changing emphasis in physics research. Initially intended "to do good meson physics" the research surpassed by far these modest aspirations [5]. One reason was that during the start-up of SC operation in 1957, the physics world was shaken by a monumental revolution: in weak interactions (i.e. the force responsible for the radioactive decay of certain nuclei), the laws of physics are not symmetric under reflection in a mirror: "parity" is not conserved [Box 2.2].

Spin and mirrors **Box 2.2**

Spin is an intrinsic form of angular momentum carried by an elementary particle, a concept of quantum mechanics (QM). Any classical image to visualize spin, such as a spinning top, is quite misleading. The property 'Spin' is characterized by the spin quantum number, s, with $s = n/2$ and n any non-negative integer. Particles with half-integer spin, e.g. $s = 1/2$, are called *fermions* while those with integer spin, e.g. $s = 0, 1$, are called *bosons* [Box 6.4]. They have very different roles in the Standard Model (SM): fermions are the constituents of matter; bosons are the force carriers between them. SUSY [Box 7.2] would bring some symmetry between the two species.

The spin of a charged particle is associated with a magnetic dipole moment (MDM). MDM is proportional to spin, related by the particle's g-factor. For charged fermions, e.g. the muon, the Dirac equation gives $g = 2$. But loop corrections [Box 5.1] involving all three types of interaction, induce a slight departure from 2. Calculating and measuring the tiny $g - 2$ value is one of the triumphs of particle physics [Highlight 2.4].

QM requires the component of spin measured along any direction to take discrete ("quantized") values. Choosing as direction the particle momentum, fermions having spin $s = 1/2$ can be along the direction of momentum or opposite to it. This defines a right-handed (RH) or a left-handed (LH) fermion. A normal clockwise corkscrew (RH) pointed at a mirror has a LH image, a perfectly conceivable instrument, preferred by LH drinkers. Not only corkscrews, but also all physical systems, were thought to have a mirror image representing a possible system. Our right hand is the mirror image of our left hand, hence the term of *handedness*. This expresses the symmetry of physical systems under mirror reflection, or "parity operation (P)". This "sacred belief" was shattered in 1957, when it was shown that the weak interaction (WI) violates P and does so in a radical way: e.g. the neutrino was found to be a LH particle, but its mirror image, a RH neutrino, does not exist. Each charged fermion exists in both its LH and RH form, but couples differently to the carriers of the weak force. The W boson couples only to LH fermions. The Z couples to both species, but in different ways. The SM gives an explicit expression for these couplings, where the weak mixing angle intervenes. This is the point where the subtleties of SM testing start (Chapter 7 and Box 7.1).

With evidence for parity violation, physicists postulated that CP, the combined operation of charge conjugation C (changing a particle to its antiparticle) and P was a "good mirror". But in 1964 it was found that neutral K-mesons also violate ("break") the CP mirror in the world of weak interactions. A vast program followed, to study kaons and later beauty particles, aimed at identifying and measuring the different sources of CP violation, in particular the "direct" CP violation in meson decay (Chapter 5 and Box 3.4). In the SM the possibility of CP violation is found to be linked to the existence of three families of fermions. Another crucial fact is the dominance of matter over antimatter, which requires CP violating processes during the evolution of the Universe. However, the degree of CP violation as described by the SM does not explain the observed dominance of matter: another reason to expect Physics beyond the SM.

At LEAR an experiment has given direct proof of the non-invariance of the WI under time reversal T, and a precise demonstration of the validity of the TCP symmetry.

Good reading: R. K. Adair, A flaw in the Universal Mirror, *Scientific American*, February 1988.

Quickly a new theory of the weak interaction (technically called V-A theory), was formulated by Feynman (Nobel Prize 1965) and Gell-Mann (Nobel Prize 1969), accounting for parity violation. Weak interaction studies were to take the centre stage of SC research, leaving a legacy of several very important results. One longstanding issue was the decay of the pion into electron and neutrino, $\pi \rightarrow e\ \nu$. Assuming universality of the coupling of the pion to electrons and muons, this decay was calculated to be rare (a spin effect), at the level of 10^{-4} relative to the dominant $\pi \rightarrow \mu\ \nu \rightarrow e\ \nu\ \bar{\nu}$ decay. Several experiments searched for this decay. They failed to observe it, posing an apparent serious obstacle to the understanding of the weak interaction and to the V-A theory. One SC Team took up the challenge and succeeded, because it developed imaginative ways to suppress the background of electrons produced in the dominant decay of the pion and quickly discovered the rare decay mode at the predicted level [Highlight 2.3] [6]: a world-class experiment in the first year of SC operation! These early years saw other related fundamental studies, such as the first observation ever of the extremely rare decay-mode $\pi^+ \rightarrow \pi^0 + e^+ + \nu$, a further crucial test of the theory of weak interactions.

The muon (μ) was another preoccupation at that time (and still is today). As said famously by I. Rabi: 'Who ordered that?' Is there more to it than being a 200 times heavier copy of the electron? In a visionary experiment the decay mode $\mu \rightarrow e + \gamma$ was searched for, but not observed. Today we know this mode is 'forbidden' in the Standard Model, but it is once again the centre of interest: a finite, albeit tiny probability of such a decay would be a "smoking gun" for new physics.

Another series of studies, pioneered at the SC, made it into the physics textbooks. Electrons and muons have besides their electric charge a magnetic dipole moment involving the g-factor [Highlight 2.4], which could be precisely calculated in the newly formulated quantum field theory of electromagnetic interactions, Quantum Electrodynamics, QED [Box 2.3]. Precision measurements of the magnetic moment of the electron confirmed the theory. Measurements of the magnetic moment of the muon would either confirm the QED calculations or, if at variance, point to differences between the electron and muon. Very early on, SC researchers used muons from the decay $\pi \rightarrow \mu\ \nu$ in a pion beam in a very clever experiment. This was the first of a series of famous CERN $(g - 2)$ experiments, where the value $(g - 2)$ expresses the impact of quantum field effects on g, as explained in highlight 2.4 [7].

The experimental techniques built on the lessons learned from cosmic ray studies, with heavy emphasis on electronic counters to take full advantage of the high particle rates offered by the SC. The newly developed "plastic scintillators", sheets of a plastic doped with a fluorescent chemical, which would produce a tiny light flash, when hit by a charged particle, became dominant. With it came a new timescale, nanoseconds (10^{-9} s), rather than the microseconds characteristic of the Geiger–Müller tubes of the cosmic ray age. Photomultipliers, which register these tiny scintillation light flashes and convert them into electrical pulses, became ubiquitous. Vacuum tube based electronic instruments, amplifying, registering electronic pulses, forming logic operations ("coincidences" and "anti-coincidences") on signals from several detectors, found their way into the experiments. Information about time sequences of particles produced in a collision and their subsequent decays were displayed on traces on an oscilloscope and filmed. Films were then analysed with methods borrowed from the bubble chamber analysis. In those days the experimenters were truly "Renaissance Physicists", having to master all the skills from designing, constructing the experimental equipment and electronics to analysis and Monte Carlo calculations, mostly done after the experiment!

A pioneering spirit, rewarded by an unexpectedly successful research programme, characterized these initial SC years: CERN was off to a flying start thanks to the SC!

With the steady increase in intensity of the accelerator, novel programs could be imagined. One such project was ISOLDE (*I*sotope *S*eparator *O*n *L*ine *DE*tector), put into operation in 1967 [Highlight 3.8] [8]. An underground area had been constructed nearby to which an approximately 80 m long beamline delivered the extracted proton beam onto a variety of targets, e.g. uranium or other heavy elements. In the spallation or fission of these targets exotic nuclei far from stability were produced; these of course had short lifetimes, and so the design of the target had to allow the products to escape rapidly and be delivered via an analysing magnet to the detectors. ISOLDE thus started what would become a long and exciting series of experiments observing the properties of nuclei far from stability. Although a very successful venture, it was somewhat hampered by lack of proton intensity. By the early 1970s it was clear that an SC Improvement Programme (SCIP) was necessary [9], and for which the machine was shut down from June 1973 to January 1975. The aim was to increase both the internal intensity and the extraction efficiency by a factor 10. A new ion source was provided and measures taken to improve the intensity and duty cycle of the extracted beam. The tuning fork of the RF system did not meet the new requirements and had to be replaced by a rotating capacitor [Highlight 2.2]. The internal beam intensity rose to 8 μA and the extraction efficiency reached 75%: mission accomplished for the SCIP programme!

Quantum electrodynamics (QED) **Box 2.3**

The relativistic quantum field theory (QFT) of electrodynamics (ED), QED, describes the electromagnetic interaction. Its early very good — and later spectacular — agreement with experiment lent strong support to the QFT approach, ultimately culminating in the Standard Model (SM) [Box 6.4]. The theory goes back to the unified description of electricity and magnetism (Maxwell, 1861), special relativity (Einstein, 1905), and quantum mechanics (1920s). In 1928 Dirac developed the relativistic quantum equation describing the interaction of an electron (e) with a photon (γ). He treated the electromagnetic (e-m) field as an ensemble of harmonic oscillators and introduced the concept of operators creating and annihilating particles. The strength of the e-m interaction is characterized by the e-m 'coupling constant' $\alpha \sim 1/137$. As α is small, it should, in principle, be possible to compute any e-m process, expressing the result in a series expansion in α. However, it turned out that the computations were meaningful only to first order in α. Higher orders gave infinite and hence meaningless results, casting strong doubt on the validity of the theory. The breakthrough came by applying the concept of "*renormalization*", a well-defined mathematical procedure, which was used to connect these infinities to corrections of the mass and charge of particles in order to obtain their experimentally determined finite values. The infinities could *de facto* be absorbed into those constants, yielding finite results. It culminated in the relativistically invariant formulation of QED [1], finite to any order. Highlight 2.4 illustrates the marvellous agreement between theory and data. The measurement of the magnetic moment of the electron agrees with QED to better than 1 part in 10^{12}. Renormalizability is essential for any QFT to be acceptable [2], including the SM.

Three basic actions can be defined [1] as shown in Fig. 1: 1) a γ propagates from one space-time coordinate to another; 2) an e does likewise; 3) an e absorbs (or emits) a γ. Complex interactions of e and γ can be built up with the relevant actions (Fig. 2). These so-called "Feynman diagrams" visualize a process, but also represent "probability amplitudes", the square of which give the probability of a process to occur [1].

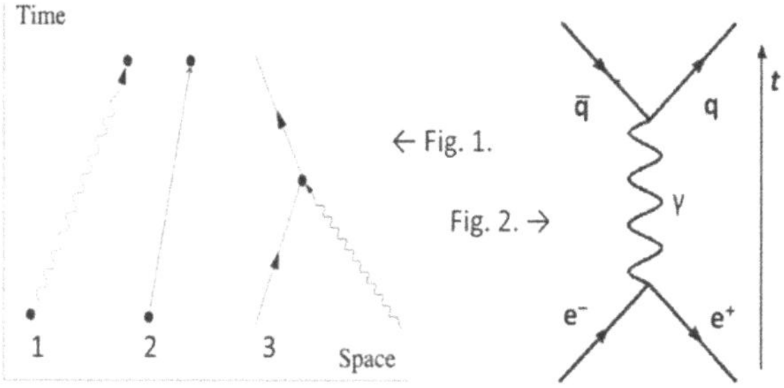

QED inspired the conceptors of the SM, and, unified with the weak interaction, it is the basis on which the model was built.

[1] R. Feynman, *QED: The Strange Theory of Light and Matter* (Princeton U. Press, 1985).
[2] G. t'Hooft, *Scientific American* **242**, 104 (1980).

The gradual development of the new characteristics of the machine continued throughout the following years, which also saw the acceleration of heavy ions and a gradual shift in the emphasis of the physics programme. New ion sources were installed to produce beams of carbon and other ions with energies of close to 100 MeV/n. Carbon and oxygen ions were accelerated in a partially stripped state and were then fully stripped on exiting the SC before delivery to the experiments in the Proton Hall. This new line of experimentation attracted immediate interest.

However, the heavy ion experimenters gradually moved to new machines such as GANIL in the early 1980s and the SC continued with ISOLDE as practically its only client, with two alternating target stations to give the more flexibility. Then, in the late 1980s a study was made of how ISOLDE might use the 800 MeV (later, 1 GeV) beam of the Booster (PSB) [Highlight 3.4], using free beam cycles that were not needed by the PS. This was found to be perfectly feasible.

With the advent of the much more powerful PS the emphasis changed from fundamental physics studies to nuclear physics and chemistry, solid state physics, using the muon as a probe of material properties, and an ever evolving program at ISOLDE. With the move of ISOLDE to the PS machine complex, the days of the SC were numbered. The machine delivered its last beam to ISOLDE on 17 December1990 after 33 years of service — far beyond its call of duty [10].

Conceived with modest aspirations, the SC was a watershed for the European physics community providing them with a state-of-the-art accelerator facility. It drove the transition from cosmic ray- to accelerator-based experimentation. This new source of high energy particles also required new tools for their study. With cosmic rays most of the fundamental discoveries were made with 'imaging' detectors. Cloud chambers and nuclear emulsion stacks provided the tracks, the image of new particles and their interaction. While these techniques still played a role in the early SC research, they were inadequate to take full advantage of the factor 10^{12} intensity increase compared with Cosmic Rays, as Heisenberg remarked to the 2nd CERN Council. Work at the SC required and spearheaded a revolutionary technology: the image was replaced with digital information and logical decisions. Many highlights in this book are witness to this transformation.

Throughout its distinguished 33 years of operation the SC worked remarkably well, constantly adapting to the changing research imperatives and delivering a rich physics harvest. Today the SC has been refurbished and turned into an exhibit for the public visiting CERN. As a fitting tribute to the pioneering spirit of its ancestors the European Physical Society, EPS, has recognized it as a "Historical site of the EPS".

2.2 The Rotary Capacitor: Tuning Acceleration

Reinhold Hohbach and Kurt Hübner

The orbital frequency of the protons in the synchrocyclotron decreases during the acceleration because of the relativistic increase of the mass of the particles and the radial decrease of the magnetic field required to focus the outwardly spiralling particles. The decrease of orbital frequency calls for a large frequency sweep of the accelerating electric field (this being the potential difference between a D-shaped metal box, called the Dee, and the grounded vacuum chamber). The Dee is located in the magnet gap where it covers about 150° of the orbital plane. The particles experience the accelerating field at the entrance and exit of the Dee. The required modulation of the radio frequency (RF) accelerating field from 30 to 17 MHz was initially obtained by means of a vibrating capacitor, the so-called tuning fork (Fig. 2.3) [11].

The synchrocyclotron (SC) was designed [11] to provide an average internal proton current of 1 μA at 600 MeV. In 1967 an improvement programme [12] was initiated to increase the current to 10 μA and improve the beam quality. Amongst other measures it was decided to raise the accelerating voltage from 20 to 30 kV and the repetition rate from 54 Hz to 500 Hz. The latter required a modification of the RF system: the tuning fork had to be replaced by a rotating capacitor, a device used in other synchrocyclotrons at that time. Figure 2.5 shows the layout of the new RF system [12, 13]. The acceleration time was 1.4 ms.

The power oscillator feeds the system through the constant value coupling capacitor C_{CL} and two rotary capacitors, for coupling (C_{cos}) and modulation with respect to earth potential (C_{mod}), both synchronized on the same shaft. Their capacitances vary as a function of the azimuthal position of the shaft. The function of C_{cos} is to present a constant load to the power oscillator, which provides 180 kW peak. The transmission line formed by the stub and the Dee has a carefully designed tapered impedance distribution to achieve the required variation of the resonant frequency for practical values of C_{mod} while keeping the voltage across the capacitors at acceptable levels.

The rotary capacitor C_{mod}, which the team affectionately dubbed Rotco (the old term for capacitor being condenser), consists of an aluminium alloy rotor of 1.5 m diameter rotating on a cantilevered shaft between earthed stator blades. It is driven by a variable speed motor and rotates in vacuum at speeds of up to 2600 revolutions per minute (rpm). Both rotor and stator blades are shaped and tapered to provide the required variation of capacitance. All blades are water-cooled and designed in such a way as to avoid mechanical resonances (Fig. 2.6).

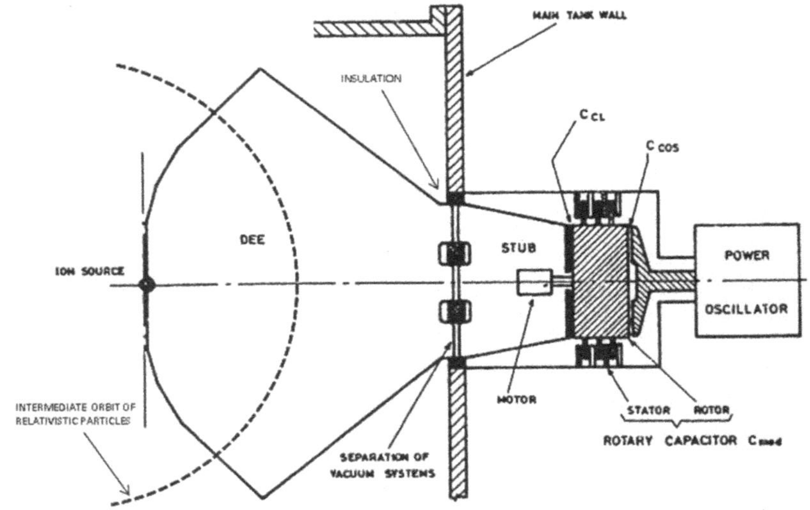

Fig. 2.5. Schematic plan view of the RF system [12].

The entire assembly was housed in an Al-alloy casing evacuated by a turbo-molecular pump that was later replaced by a cryo-pump. To reduce the risk of voltage breakdown due to oil vapour, a rotary-face seal retained the oil and a turbo-molecular pump mounted on the shaft removed any leaking oil. Despite these precautions, electrical breakdown caused by condensation of oil vapour remained a serious concern. A delicate element was the ceramic insulating bushing and the vacuum seal of the main bearing. Since the Rotco was also subject to a flux of activating high energy neutrons, two identical assemblies were constructed to provide a spare for reducing downtime after a breakdown.

The RF circuit design was initiated at CERN with modelling and simulation, and continued by industry where detailed design of the hardware was made to a functional specification and construction was started. But the project suffered significant delay and in 1973 it was agreed to transfer the partially built equipment to CERN, which undertook further development and finalised the construction. The CERN team had to carry out numerous repairs and improvements of the electrical and mechanical elements in order to get it working. The Rotco and its spare were put into operation in 1974. However, the design was intrinsically fragile and both assemblies were plagued by electrical and mechanical problems for several years, and the team had to make a continuous effort to gradually improve reliability. This has entailed the redesign of many components [14].

Fig. 2.6. View of the stator of the modulating capacitor seen from the generator side. The rotor is removed.

This project is an example of a technological development where the design and construction was entrusted to an industrial partner who in retrospect was over-ambitious and had taken on work that extended beyond its competence level. This was a hard lesson for CERN, signalling the need to better prepare in-house the design of complex technical apparatus, including the construction and thorough testing of models and prototypes.

2.3 Discovery of the $\pi \to e\nu$ Decay: Rare and Precious

Giuseppe Fidecaro

In 1948 at Berkeley, π-mesons were produced for the first time with an accelerator [15] and observed in its prominent decay $\pi \to \mu\nu$. Several searches for the $\pi \to e\nu$ decay remained however unsuccessful. These searches were motivated by the idea that all Dirac particles, such as electrons and muons, would have the same weak interaction constant ("electron-muon universality"). Assuming the Fermi description of weak decay and the symmetric coupling of electrons and muons to π-mesons [15] the ratios $R = (\pi \to e\nu) / (\pi \to \mu\nu)$ for the so-called Axial-Vector and Pseudo-Scalar interaction were calculated to be 1.28×10^{-4} and 5.49 respectively. For scalar, vector and tensor interaction the decay $\pi \to e\nu$ was expected to be forbidden.

This theoretical issue motivated intense searches for $\pi \rightarrow e\nu$. Neither three emulsion experiments, carried out in 1950 and 1951, nor the following electronic counter experiments at Nevis [16] and Chicago [17] succeeded in detecting this decay (although they did allow to rule out a Pseudo-Scalar interaction). The result of the Chicago experiment, using a powerful magnetic spectrometer, was particularly striking: the observed ratio $R = (-4.0 \pm 9.0) \times 10^{-6}$ led the authors [17] to conclude: "This appears to be statistically significant as it allows only a 1% probability that R is greater than 2×10^{-5}". Absence of the decay at the expected rate posed a real conundrum that theoretical physicists were trying hard to solve.

In 1954 at the Varenna Summer School the author learned first-hand about the Nevis experiment in a seminar by Jack Steinberger (Nobel Prize 1988). This was followed up by careful studies of the merits and drawbacks of the experiment and by gaining experience at Liverpool University with their brand new 400 MeV synchrocyclotron. In January 1958 he attended a meeting of the American Physical Society in New York, where R.P. Feynman (Nobel Prize 1965) presented an invited paper, "Theory of Beta Decay", dealing with the apparent absence of the $\pi \rightarrow e\nu$. Not convinced by the arguments advanced by Feynman and despite the prevailing view that the $\pi \rightarrow e\nu$ decay might be exceedingly rare (or non-existent), an initiative was launched for a new experiment at CERN [6]. In this experiment (Fig. 2.7) the 127 MeV positive pions from the 600 MeV SC were incident on the telescope $123\bar{4}$, consisting of four scintillation counters, and stopped in counter 3, which functioned as an active target and source of possible positrons. Veto counter $\bar{4}$ rejected contamination muons and positrons having the same momentum as the incoming pions. The positrons were detected by a range telescope, formed of scintillation counters 5–12 in coincidence, which also served as a variable thickness absorber to measure exploratory integral range curves (see below). With its 12 counters it was a rather complex and sophisticated apparatus for the time!

In principle, this was all that was required for the search. Tell-tale additional information was obtained by combining the output from counter 12 with that of counter 3 on the same oscilloscope trace. This direct visual coincidence of the signals generated by the same particle in the pion stop counter and in counter 12 clearly exposed the positron as the last particle in the signal sequence.

The novelty and the power of the experiment consisted in the complete recognition of the sequential signals of the particles of the two decay chains $\pi^+ \rightarrow e + \nu$ and $\pi^+ \rightarrow \mu + \nu \rightarrow e + \nu\bar{\nu}$, identified by two or three pulse sequences (Fig. 2.7). A single trigger tagged the two decay chains, which were distinguished exclusively by the absence or presence of the muon. The use of the active target to stop pions and to recognize a decay muon was a condition *sine qua non* for the success of the experiment.

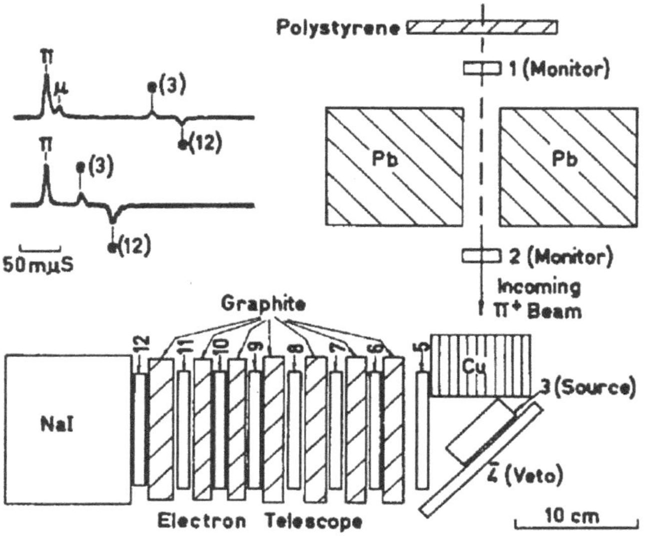

Fig. 2.7. Experimental layout and (insert): typical $\pi \rightarrow \mu \rightarrow e$ (top trace) and $\pi \rightarrow e$ pulse (bottom trace). This "electron telescope" detected positrons. Elements labelled 1 to 12 are scintillation counters [18].

The most important residual background consisted of $\pi \rightarrow \mu \rightarrow e$ events in which these $\pi \rightarrow \mu$ decays could not be resolved, resulting in so-called false $\pi \rightarrow e$ events. Their fraction was determined by measuring the e+ rate without absorbers in the positron telescope, which was then dominated by positrons from $\pi \rightarrow \mu \rightarrow e$, providing the correction factor of their misidentification. This correction, found to be in the range of 10%, was applied to the comparatively small number of $\pi \rightarrow \mu \rightarrow e$ events observed during the data taking (with absorber for the $\pi \rightarrow e$ decay search). Measurements with varying amounts of absorber confirmed the spectral distribution of e+ from the $\pi \rightarrow \mu \rightarrow e$ decay whereas the rate of the mono-energetic e^+ from $\pi \rightarrow e$ was unaffected within errors. To arrive at the absolute value for the branching ratio, counter efficiencies were evaluated with a Monte Carlo program (another First) written for the new CERN central computer.

The setting up of the experiment in the Neutron Hall started in February 1958 and data taking with first beam from the SC in August 1958. One month later, first results, based on 124 photographs displaying some 40 events, were presented in Geneva at "The Second United Nations International Conference on the Peaceful Uses of Atomic Energy" [18]. The local newspaper "Tribune de Genève" ran the headline "Découverte au CERN", the first such event for the Organization. The final result [19] for the ratio R $(\pi \rightarrow e\nu) / (\pi \rightarrow \mu\nu) = (1.22 \pm 0.30) \times 10^{-4}$ was perfectly consistent with theoretical calculations and definitively established the universality of the weak interaction. It opened the way to a more profound understanding, culminating in the discovery of the Weak Intermediate Bosons, the W and Z, at CERN 25 years later.

2.4 Measuring the Muon ($g - 2$): Precision with Precession

Guido Petrucci

A classical rotating top of mass m, angular momentum L and electric charge q behaves as a bar magnet of magnetic moment M. The ratio $(2m/q)\ M/L$ is the g-factor. If charge and mass have the same distribution, the value of g is precisely one, $g = 1$. A charged lepton has also a magnetic moment and the angular momentum is replaced by its quantum analogue called "spin" [Box 2.2]. Quantum Electrodynamics (QED) [Box 2.3] to first order and the Dirac theory predict $g = 2$, but due to quantum field effects [Box 5.1] g differs slightly, by about 2×10^{-3}, from this value. Hence, the accurate measurement of g forms a prime test of the validity of QED calculated with all corrections to highest order. A deviation from the theoretical value might also hint at fundamental differences between electrons and muons.

In 1958, at the CERN 600 MeV synchrocyclotron (SC), the first CERN "$g - 2$" experiment (1958–1964) started, aiming to measure the anomalous magnetic moment of the muon (μ), a charged lepton about 210 times heavier than the electron. This very successful experiment [7] was followed by a second and a third one, both at the PS, with improved experimental techniques. Each time a much more precise result confirmed the previous one. The most recent and a still more precise experiment done at BNL (US) in the late 1990s and involving some of the CERN pioneers, agrees with the CERN values, while enhancing the scope of this measurement: although consistent with QED at the 10^{-6} level, it indicates a tiny discrepancy between data and that expected from the full Standard Model (SM) [Box 6.4] calculation. If confirmed by the ongoing programme, it may be an indication of new physics beyond the SM.

The $g - 2$ measurement is performed using polarized muons (magnetic moment and spin aligned in the same direction) moving in circular orbits inside a magnetic field, in which the spin axis rotates, or precesses, around the field axis like a spinning-top around the vertical axis. The precession period is slightly higher than the muon period of revolution, by an amount linked to $g - 2$. The measurement consists of recording, as a function of the muons storage time, the electron produced in the muon decay, whose direction is correlated with the polarization of the parent muon. The electron decay rate versus time shows a typical beat curve

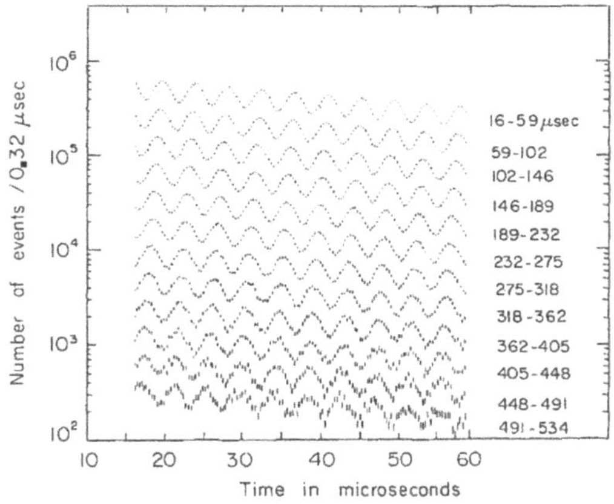

Fig. 2.8. The electron decay rate versus time in the muon storage ring 2 [20].

between the two rotation frequencies (Fig. 2.8) [20]. The precision of $g - 2$ increases with the degree of initial polarization, the observation time, hence the muon storage time, the number of electrons recorded and a better knowledge of the magnetic field, which is the main focus of this highlight.

In the first $g - 2$ experiment, the longitudinally polarized muons (100 MeV/c momentum) were produced on an internal target inside the SC and, after channelling through a focusing solenoid, were injected into a 6 m long straight magnet, with a field of 1.6 T, 52 cm wide poles and a 14 cm gap. A beryllium moderator at its entrance reduced the energy and hence the radius of the spiralling muons improving its observation. The poles faces were shimmed in order to superimpose a longitudinal movement along the magnet to the circular orbits of the μ (Fig. 2.9). The magnetic field was measured with NMR probes, Hall probes and a large moving coil. The muon trajectories were calculated with the first computer installed at CERN and the shims progressively adjusted to increase the storage time of the muons in the magnet (up to 2000 turns, about 2500 m of trajectory). The complete muon trajectory was in vacuum. Exiting the magnet, the muons were stopped in a non-depolarizing, non-conductive target.

A pulsed coil, flipping their spin by ± 90°, turned any transverse polarization into longitudinal. Backward or forward decay electrons were detected in counter telescopes, and tagged by the time spent in the magnet by the parent muons (2.0–6.5 μs). The measurement gave a value $a = (g - 2)/2 = (1162 \pm 3) \times 10^{-6}$, in agreement with QED up to the order calculated at that time, implying that the μ was behaving just like a heavy electron [7, 20, 21].

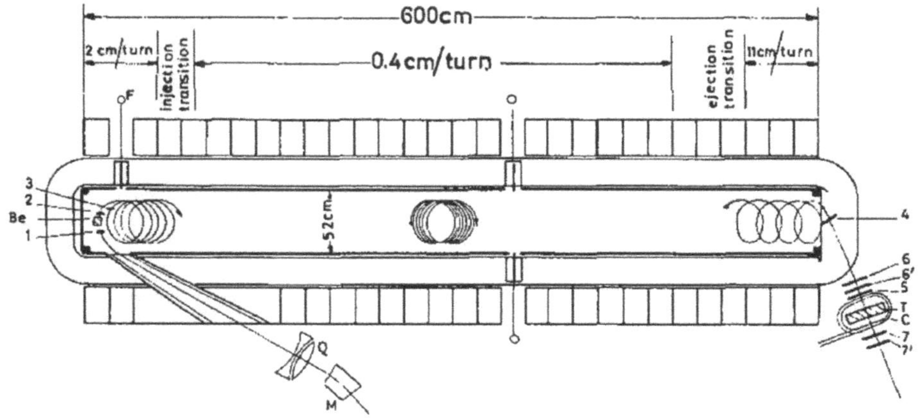

Fig. 2.9. The first $g-2$ experiment [7].

The second $g-2$ experiment (1962–1968) innovated by using for the first time a muon storage ring. Protons of 10.5 GeV/c momentum, fast-extracted from the PS (one to three radio frequency bunches, each 10^{-8} seconds (10 ns) long separated by 105 ns), hit a target located at the edge of the storage region of the magnetic ring. This ring, designed by S. van der Meer (Nobel Prize 1984), was of the weak focusing type [Box 2.1], with an orbit diameter of 5 m, a useful aperture of 4 cm (V) × 8 cm (H) and a magnetic field B = 1.711 T (Fig. 2.10). The muons were stored with a momentum of 1.28 GeV/c, resulting in a relativistically dilated muon lifetime $\gamma\tau$ of 27 μs, where $\gamma = \{1 - (v^2/c^2)\}^{-1/2}$, 12 times its lifetime τ at rest, allowing a much longer observation with correspondingly improved precision. The protons hitting the target produced a large background of particles, "blinding" the detection counters for the first 25 μs. Any pion crossing the storage volume and having an energy equal or higher than those accepted by the ring could produce stored muons. However, the muons born from parents of much higher energy were poorly polarized and the average μ polarization in this first ring was only about 30%. A fraction of the electrons produced by circulating muons were deflected inside the ring and detected by counters recording them versus time. The result of this second $g-2$ experiment was $a = (g-2)/2 = (116616 \pm 31) \times 10^{-8}$, almost two standard deviations above the theoretical value. As it turned out later, part of the discrepancy was due to the neglect of the $(\alpha/\pi)^3$ term in the QED calculation and motivated theorists to improve their estimates.

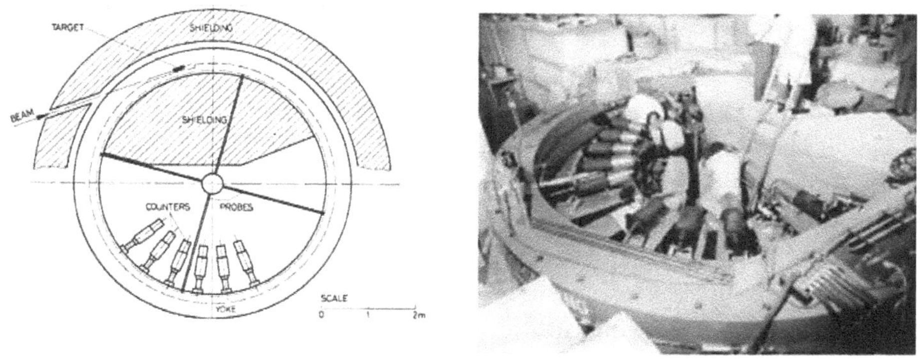

Fig. 2.10. Left: sketch of the second $g-2$ experiment. Right: view of the same experiment, using for the first time a storage ring [20].

The third CERN experiment (1969–1976) was a much improved version of the previous storage ring. The proton bunches extracted from the PS produced a secondary pion beam tuned to trap muons of 3.1 GeV energy. The ingenious new idea was the choice of this "magic" energy with its value $\beta^2\gamma^2 = 1/a$, for which the electric field has no influence on the spin motion. It was thus possible to focus the circulating beam with electrostatic fields (instead of magnetic fields) and design a more uniform magnetic configuration. Efforts were put in two directions: developing a suitable quadrupole electrostatic field and designing a B-field as uniform in space and constant in time as technically possible.

The electrostatic field had to be pulsed in order to avoid discharges caused by ions slowly accumulating in the vacuum chamber. This field provided vertical focusing, partially reducing the horizontal focusing effect of the circular motion (the ring was therefore weak focusing like the previous one). The electric field was used also to scrape the beam, both horizontally and vertically, after each injection to avoid subsequent beam losses that could introduce errors.

The magnet was mounted in a separate thermally stabilized hall. Its continuous, concrete support ring was itself kept at constant temperature by means of water circulating in pipes cast within. The magnet coils had independent supports; their very small deformations had no influence on the field. The iron yoke was made of 40 blocks, whose pole faces were very accurately ground by a specially designed, computer-assisted grinding machine. The field map finally showed variations in time and space of less than 1 ppm. During measurements, the field was monitored by 40 NMR probes with feedback loops acting on 40 compensating coils. The result of this third $g - 2$ experiment was $a = (g - 2)/2 = (1165924 \pm 8) \times 10^{-9}$ (a precision of 7 ppm). The validity of QED was confirmed up to the 6th order, i.e. terms $(\alpha/\pi)^3$. Further results from the Brookhaven $g - 2$ programme now hint at a slight discrepancy between the experimental result and the Standard Model prediction involving QED and the effects of weak and strong interactions.

As former Director-General J.B. Adams once put it: "$g - 2$ is not an experiment; it's a way of life". Indeed, one pioneer has participated and contributed greatly in all $g - 2$ experiments at CERN and BNL. A new experiment is planned at Fermilab starting in 2017. One awaits eagerly further results of this outstanding programme, which has seen a long lasting and stimulating competition between experiment and theory, a "tennis game with well-matched players on either side of the net," as one of the principal actors remarked. More information and a list of references can be found in [20, 22].

References

1. A. Hermann *et al.*, *History of CERN*, Vol I, (North-Holland, 1987).
2. A. Hermann *et al.*, *History of CERN,* Vol II, 95, (North-Holland, 1990).
3. http://cds.cern.ch/record/1662760.
4. F. Bonaudi, The CERN SC machine, *Phys. Reports* **225**, 31-43 (1993).
5. L. Di Lella, Elementary Particle Physics at the SC, *Phys. Reports* **225**, 45-66 (1993).
6. G. Fidecaro, CERN Discoveries: Rare pion decays at the synchrocyclotron, in *60 Years of CERN Experiments and Discoveries*, Advanced Series on Directions in High Energy Physics, Vol. 23, eds. H. Schopper and L. Lella, (World Scientific Publishing, 2016).
7. G. Charpak *et al.*, Measurement of the anomalous magnetic moment of the muon, *Phys. Rev. Lett.* **6**, 128 (1961).
   G. Charpak *et al.*, *Il Nuovo Cimento* **17**, 288 (1961).
8. A. Kjelberg and G. Rudstam (eds.), *The ISOLDE Isotope Separator Facility at CERN*, CERN-1970-003 (CERN, Geneva, 1970). http://dx.doi.org/10.5170/CERN-1970-003.1.
9. G. Brianti, Proposal for the SC improvement, SPC/249 (1967); https://cds.cern.ch/record/42523?ln=en.
10. G. Fidecaro (ed.), Thirty-three years of physics at the CERN synchro-cyclotron, *Phys. Rep.* **225**, 1-191 (1993).
11. W. Gentner *et al.*, The CERN 600 MeV synchrocyclotron at Geneva. *Philips Technical Review,* **22**(5), 1960/61, 141–180, (1961).
12. MSC Staff, presented by Vogt-Nilsen N. (1969). Status report on the CERN SC improvement programme, *Proc. 5th Int. Cyclotron Conf.* (1969) pp. 719–727.
13. H. Beger, Some essential features of the rotary capacitor in construction for the improved 600 MeV SC of CERN, *Proc. 4th IEEE Particle Accelerator Conference* (1971) pp. 307-310.
14. H. Beger *et al.*, Design, construction and operation of the new CERN SC RF system, *Proc. 7th Int. Conf. on cyclotrons and their applications* (1975) pp. 146–150.
15. M. Ruderman and R. Finkelstein, Note on the decay of the π-meson, *Phys. Rev.* **76**, 1458 (1949).

16. S. Lokanathan and J. Steinberger, Search for the β-decay of the pion, *Nuovo Cimento. (Suppl.)* **2**, 151 (1955).
17. H.L. Anderson and C.M.G. Lattes, Search for the electronic decay of the positive pion, *Nuovo Cimento* **6**, 1356 (1957).
18. T. Fazzini *et al.*, Electron decay mode of the pion, *Phys. Rev. Lett.* **1**, 247 (1958).
19. J. Ashkin, *et al.*, Electron decay mode of the pion, *Il Nuovo Cimento*, Serie X **13**, 1249 (1959).
20. F. Farley and E. Picasso, The muon $g - 2$ experiments, *Ann. Rev. Nucl. Part. Sci.* **29**, 243 (1979).
21. R.L. Garwin, The First CERN $g - 2$ Experiment, by G. Charpak, F.J.M. Farley, R.L. Garwin, T. Mueller, J. Sens, and A. Zichichi, https://fas.org/rlg/04_08_2014FirstCERNMuon.pdf.
22. F. Farley and Y. Semertzidis, The 47 years of muon $g - 2$, *Prog. Part. Nucl. Phys.* **52** (2004).

Chapter 3

The Proton Synchrotron (PS): At the Core of the CERN Accelerators

Donald Cundy and Simone Gilardoni

3.1 Introduction

The PS accelerator

After almost 60 years of honourable service, the history of the CERN Proton Synchrotron, PS, is marked by successes and discoveries, intimately linked to the CERN history. Today, the PS is the beating heart of the LHC (Large Hadron Collider) injector complex, the last of the countless successes of this incredible and versatile machine. At the origin of this longstanding success is the foresight of the CERN founding fathers who in the early 1950s took the risk to bet on a new technique in accelerator design called strong focusing [1]. Their "scientific and technological" audacity paved the road of CERN's future, but also encouraged others to embrace the same spirit of technological enterprise. Conceived as a prototype machine with a life time of only a few years the PS would not only exceed all expectations, but also inspire the design of all the modern circular machines. The PS was and still is an invaluable tool for beam physics studies, and was the cradle of high energy physics for many years, until more powerful machines like the ISR (Chapter 4) started in turn the era of the hadron colliders.

The PS is a synchrotron accelerating particles up to 28 GeV [2–5]. The energy was chosen based on physics considerations, i.e., study particle interactions at the GeV scale, but also economic ones. The initial energy was set well beyond the few GeV scale, up to 30 GeV, but the first machine design was considered too expensive and technologically risky. Fortunately for the future generations of physicists, CERN Council rejected the recommendation of W. Heisenberg to reduce the energy, thus the cost, to 20 GeV, and decided to keep the machine design that reached 25 GeV.

A synchrotron like the PS is a circular ring where particles during acceleration from the lowest to the highest energy are always kept on the same orbit with a constant radius of curvature [Box 2.1]. This is realized by increasing, synchronously with the energy change, the strength of the magnetic elements generating the force that counteracts the centrifugal force acting on the particles. Whereas the synchrotron concept was applied at other places before CERN, like the Cosmotron accelerator in the USA at BNL, the PS was unique with the first ever implementation of the concept of strong-focusing, i.e., the presence in the machine lattice of dedicated beam focusing elements, the quadrupoles. These are elements with properly shaped magnetic fields to control the physical size of the beam, thus limiting the maximum beam dimensions, much as in normal optics lenses would focus the light rays. This can be reached, as in classical light optics, by alternating focusing lenses with defocusing ones, from which the name of alternating gradient lattice was derived. The concept of strong focusing, also called alternating-gradient focusing, was invented in the early 1950s [1], and boldly adopted by CERN in 1952 for the future PS in an incredible gamble: while the cost of the construction could be significantly reduced, because the new focusing principle held the promise of smaller beam and therefore smaller beam elements, this novel idea had never been tested on a real accelerator. The PS was a resounding success, closely followed by its sister synchrotron in the USA, the AGS. All modern accelerators for high energy applications, including the LHC, are based on the strong focusing principle.

The implementation of this technique took a very particular form in the PS [2]: C-shape combined function magnets reaching 1.4 T at 28 GeV were chosen, which combine the functions of bending and focusing of the beam in the same magnetic element, a choice partially motivated by the heritage of the previous accelerators. However, the first-time use of a strong focusing lattice opened also a Pandora's box full of new issues in accelerator physics: having the beam circulating without being lost in the first few turns (the full accelerating cycle takes about 2.4 s, i.e. up to about 1 million turns) requires tight tolerances on magnet geometry and relative magnet alignment of the order of few 100 μm on a machine of 628 m circumference. The required magnetic field quality (i.e. precision) had never been achieved before. For example, the relative spread in deflecting strength of the magnets had to be less than 5×10^{-4}. Despite these challenges, measurements repeated in 2014 confirmed that the magnets, constructed in the first half of the 1950s, were actually better than specified, leaving a sense of awe and admiration for such work realized with the technology available at that time. Another example, the calculation of the beam stability to define the required tolerances, was realized using the first generation of computers available in the UK.

The machine took its final shape under the direction of J. Adams during a construction period between 1956 and 1959. The ring is composed of a sequence of 100 main magnets interleaved by 100 straight sections hosting (i) the auxiliary magnets, (ii) the elements for injecting and, later, extracting the beams, (iii) the accelerating radio-frequency (RF) cavities [Box 3.1]. The combined-function magnets shown in Fig. 3.1 are also equipped with a series of windings on their pole-faces to make fine adjustments to the field, a feature substantially enlarging the versatility of the PS.

The machine was assembled on top of a floating floor, mechanically separated from the CERN ground to avoid external disturbances to the machine alignment, e.g. arising from a deformation of the foundation by rain or even earthquake, which would adversely influence the beam orbit.

The first beam was accelerated to 26 GeV in a memorable night of the 24 November 1959, opening both a new era in accelerator physics and in accelerator-based high energy particle physics, an event celebrated in due form on the following day (Fig. 3.2).

Fig. 3.1. View of the PS ring showing the combined-function magnets.

RF Acceleration **Box 3.1**

The particles in synchrotrons and colliders [Box 2.1] are accelerated by one or more RF cavities. These devices provide an accelerating electric field to increase particle energy per turn or manipulate the bunches (e.g. splitting, merging). Their simplest version is a pill-box in which an oscillating electromagnetic (e-m) field is generated; this provides the longitudinal electric field (Fig. 1). The field does not propagate into the adjacent vacuum chamber; to do so would require a much higher frequency e-m field.

For all cavity types, the e-m field oscillates with a frequency f_{rf} (wavelength $\lambda = c/f_{rf}$) which is a multiple h of the revolution frequency f_0 of the particle $f_{rf} = h \cdot f_0 = h \cdot c \cdot \beta / C$, where $\beta = v/c$ and C the circumference of the ring. Since the RF field is oscillating, the beam cannot be continuous and the particles must come in groups, called bunches. The harmonic number h defines the maximum number of bunches in the beam. The bunch to bunch distance is nC/h or in time n/f_{rf} with $n = 1, 2, \ldots$ Gaps in the beam can be generated by not filling the positions of a group of bunches, such as when required for injection or extraction. To enable transfer from one circular accelerator k to the next k+1, C_{k+1}/C_k must be the ratio of two positive integer numbers. For example, LHC/SPS = 27/7, SPS/PS = 11/1. Bunch to bunch distance is preserved at transfer.

If the particle speed is varying during acceleration, which is the case until the particle energy E exceeds by far its rest energy, E_0, f_{rf} has to vary in synchronism with f_0 and the resonance eigenfrequency f_{ef} of the cavity needs to be tuned as f_{ef} has to be very close to f_{rf}. The difference between f_{ef} and f_{rf} is determined by stability criteria [Highlight 3.3]. When β approaches 1, or the cavity is used exclusively at a particular energy, only residual tuning is required to make f_{ef} close to f_{rf} to maintain stability. A remotely controlled mechanical device, the tuner, performs this task changing the shape of the cavity slightly. Depending on cavity use, the shape is optimized taking into account RF power requirements, RF power dissipation, peak surface electric fields, and space constraints. This can lead to shapes quite different from the pill-box (Fig. 2). In case powerful acceleration is required, many cavities (cells) are combined in a structure in which either a travelling wave (e.g. 200 MHz RF structure in the SPS) or a standing wave is set up. The RF power is injected via a coupling loop at a point along the structure and is transmitted by em coupling between the cells.

Fig. 1. (Left) Open PS 200 MHz 75 cm diameter ($\lambda/2$) pill-box cavity, with ceramic gap connected to the beam tube [5].

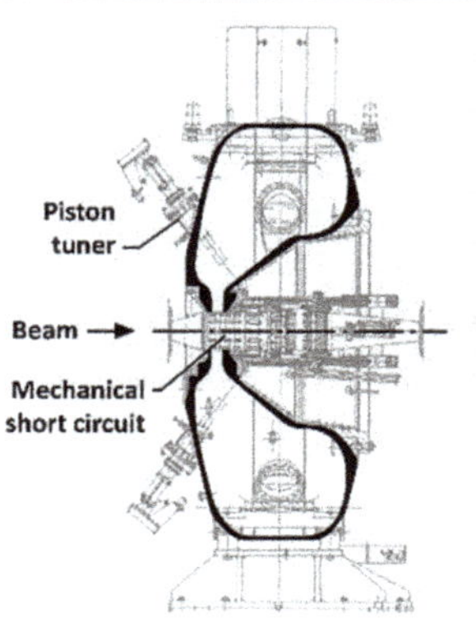

Fig.2. (Right) Cross-section of the PS 80 MHz, 1.8 m diameter cavity, shaped to house the short circuit [5].

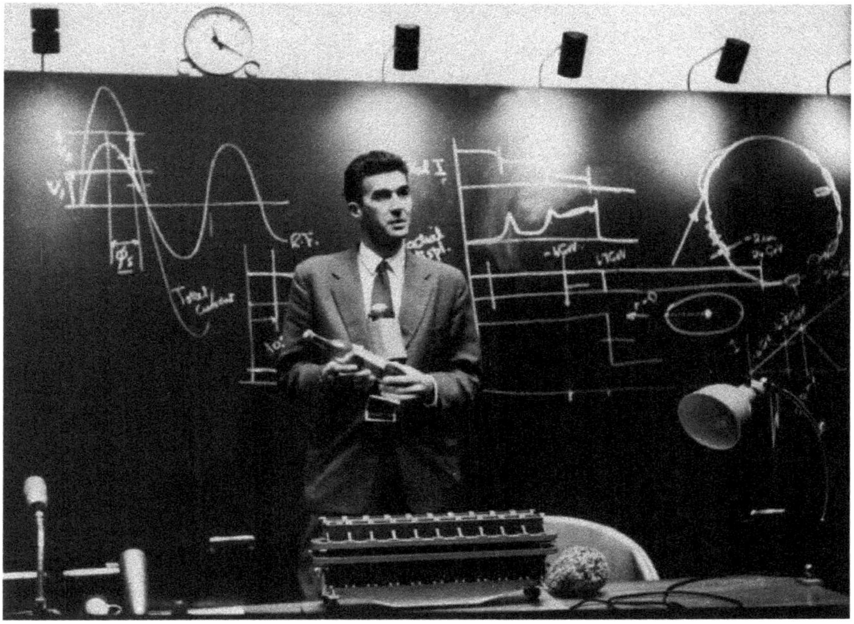

Fig. 3.2. J.B. Adams, the PS project leader, announcing acceleration to 24 GeV in the PS. The Vodka bottle in his hand was a gift from Dubna to be consumed on the occasion of the PS surpassing 10 GeV, the energy of the Dubna Synchro-Phasotron.

Once the basic features of the accelerator were understood, the PS and its injection systems underwent a long series of gradual improvements of performance, reliability and versatility leading to new applications. One indication is the gradual rise of the PS intensity over the years spanning from the design value of $\sim 10^{10}$ protons per pulse to more than 3×10^{13} achieved for the latest neutrino experiment (Fig. 3.3).

High energy physics research at the PS started first with internal targets and later in 1960 with dedicated beam lines receiving protons by fast or slow extraction. These novel beam extraction techniques were developed with the aim to suppress the originally installed internal targets, source of substantial beam loss, producing a high radiation dose to the machine elements, in particular to the main magnets, and a serious risk to the machine lifetime. Extraction of the beam reduced the internal losses. The external targets allowed higher impacting intensities and could be placed at positions optimal for the experiments [Highlight 3.2].

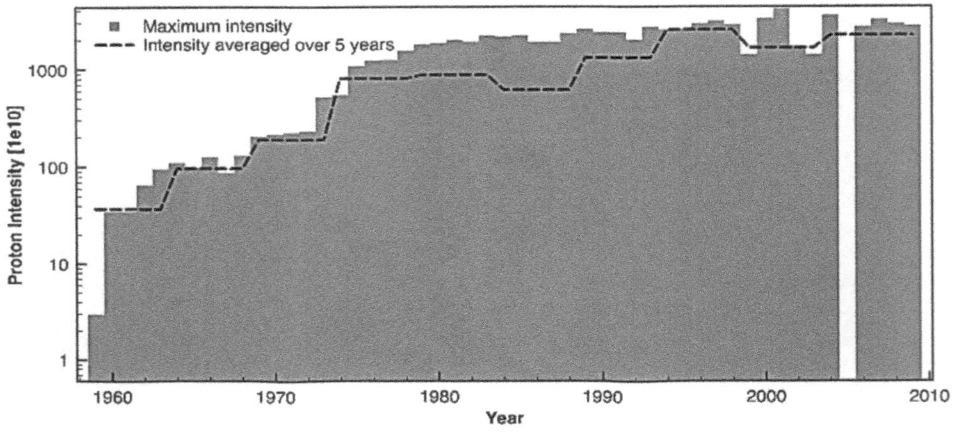

Fig. 3.3. Protons per pulse as a function of time. The PS was stopped for consolidation in 2005.

Special attention has always been paid to the PS injectors [6]. The first injector was a linear accelerator constructed by industry (Linac1) providing 50 MeV protons. In 1965 an improvement programme of the PS was launched together with the approval of the ISR (Chapter 4). The PS beam intensity was increased by inserting a small synchrotron between Linac1 and the PS, the PS Booster (PSB) [Highlight 3.4]. This novel type of accelerator of very compact design combines four synchrotrons in one ring with common bending and focusing magnets. Its potential turned out to be enormous, probably even a surprise to its designers, brought to life by a tenacious development since its first operation in 1972. Not only helped it to increase the intensity in the PS by combining the output of the four rings, but the 16 times higher beam energy compared to Linac1 made the beam stiffer moving the intensity limit at PS injection beyond the intensity provided by the PSB. This intensity limit was later further raised by successive steps in the PSB's output energy from the original kinetic energy of 0.8 GeV to 1.4 GeV. This was indispensable to catch up with the steadily increasing PSB output intensity, raised by a factor eight since 1972. However, the PS did not remain the only client of PSB: ISOLDE [Highlight 3.8] had been moved from the SC to the PSB in order to benefit from the higher proton intensities at the PSB, an elegant solution, because ISOLDE uses only those PSB pulses which are not injected into the PS. A new CERN-designed 50 MeV linac (Linac2) was added in 1978 to improve not only the performance but also the availability of the PSB. Linac1 had already earlier shown its versatility by accelerating deuterons and alpha particles for the ISR but when no longer needed for proton operation it was modified to provide oxygen and sulphur beams to the PS for fixed-target physics at the SPS. However, it was unsuitable for the acceleration of heavier ions so one resolved to replace Linac1 with Linac3 of modern design providing nowadays mainly lead ions to the LHC, but also indium and argon ions for the SPS.

The venerable PS still provides a variety of hadron beams including antiprotons (Chapter 6). It also supplied electrons and positrons of 3.5 GeV to the SPS for LEP (see Chapters 5 and 7). To this end, it was equipped with the appropriate acceleration systems, injection/extraction elements and a 600 MeV electron-positron linac as injector feeding a small accumulation ring preceding injection into the PS. In view of improving the LHC injector performances, Linac2 will be replaced by a new linac accelerating negative hydrogen ions to 160 MeV (Linac 4). In order the cope with the concurrent intensity increase, the PSB output energy will be raised in another step to 2 GeV, 2.5 times the design energy, another proof of the PSB's amazing inherent potential.

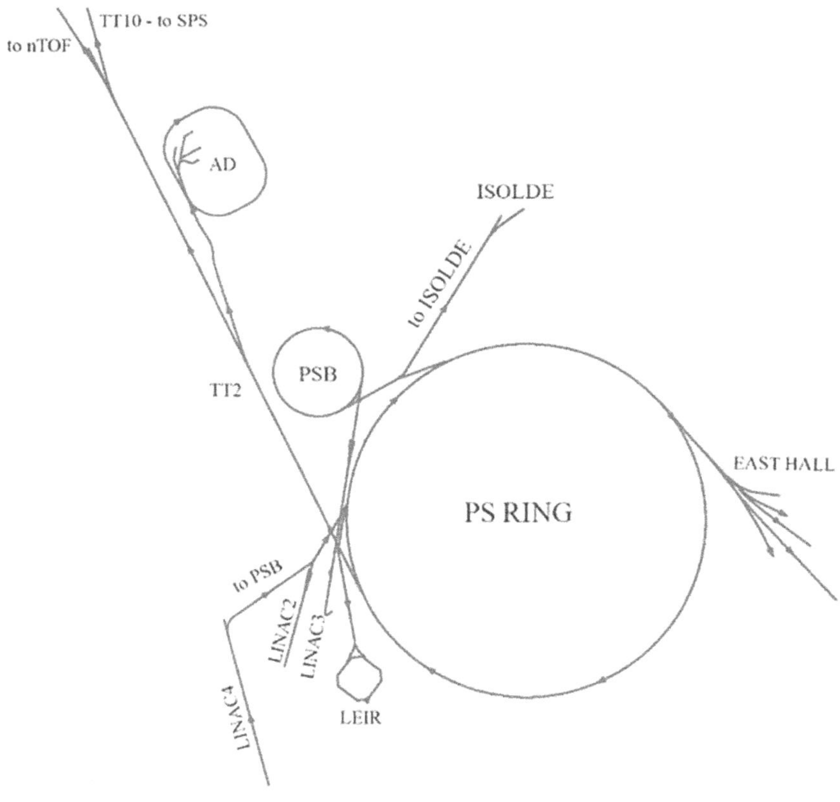

Fig. 3.4. Layout of the injectors and beam lines to the experiments.

The PS is still at the core of producing proton and ion beams as shown in Fig. 3.4 for fixed-target physics at the PS, the Antiproton Decelerator (AD), the SPS, and for the LHC. The LHC collider operates with two different bunch spacing, either 50 ns or 25 ns, the latter being the design configuration. Subtle manipulation of the beam by RF [Highlight 3.3] determines this bunch spacing in the PS and the downstream accelerators. Continuing a 60-year long tradition of attentive care, its infrastructure is maintained to keep up with the ever-increasing demand. A shining example is the novel solid-state main power converter with capacitive energy storage replacing the more than 30-years old rotating machine, which used a flywheel or energy storage to smooth the load to the mains [Highlight 3.5]. The new technology proved to be so successful that it inspired colleagues of the sister synchrotron AGS in the USA to consider also replacing their rotating machine.

A most flexible design, high-quality components and vigorous consolidation programmes for the accelerator and its infrastructure are the basis for the surprising durability of this synchrotron which proved, every time it was called upon a new task, its adaptability, versatility and reliability. This venerable machine stayed and will stay for many years at the heart of the CERN accelerator complex, a fine example of the continuous (re)use of CERN infrastructure.

The PS experimental programme

When the CERN PS was ready for physics in 1960, particle physics had entered a Golden Age of "particle discovery". Strange particles, collected painstakingly using cosmic rays, were now being produced copiously at the proton synchrotrons. The electron-neutrino ν_e had been detected. The antiproton and the antineutron had been discovered. Parity Violation (PV) in Weak Interactions (WI) was proven in 1956. However, theoretical guidance was still weak or absent: the quark hypothesis was made in 1964, followed by the electroweak theory in the late 1960s, the SM by the end of the sixties, and the theory of strong interactions (QCD) in 1973 [Boxes 4.2, 6.4]. The physics landscape and major PS results are reviewed in [7].

The bubble chamber (BC), invented in 1953, had become, with gradual improvements, the detector *par excellence* [Box 3.3 and Highlight 5.7]. The hydrogen BC, providing a proton target, lent itself to the detailed study of particle production, whilst the heavy liquid BCs proved very useful in the study of weak decays of the new particles. However, these BCs have a serious drawback: the repetition rate, i.e. their data taking rate is very low and they lack the capability of selecting specific events.

Particles interacting with matter **Box 3.2**

To be detected, a particle must transfer some of its energy to the detector material [Box 6.3]. The nature of the interaction — electromagnetic (e-m), strong, weak — depends on the particle charge and quantum numbers. A *charged particle* excites (*ionizes*) the medium through which it passes: this can produce a measurable signal, the basis for most detectors. Tracking detectors record *ionization* produced along the trajectory of charged particles [Highlight 4.8]. The ionization energy loss dE/dx depends on the particle velocity v and may provide a measure of v; if the particle momentum p is known, its mass can be determined [Box 5.2]. Some media exhibit the property of *scintillation*: a charged particle traversing matter leaves a trail of excited molecules which release part of this energy as photons. Although only a few percent of the total energy loss, it is sufficient to detect a particle. Besides colliding with atomic electrons, a particle, shaken by the e-m fields in the material, emits *radiation* at a rate roughly proportional to E/m^2. This process is characterized by the *radiation length* X_0 of the medium, over which an $e^\pm$ has lost all but $1/e$ of its energy, and a γ-photon has a 2/3 probability to convert into a $e^\pm$ pair. Typical values of X_0 are 1.8 cm (Fe) and 300 m (air 1 atm). Above a critical energy E_C, radiation effects (bremsstrahlung for $e^\pm$, pair conversion for γ's) dominate over ionization energy loss. For $e^\pm$, E_C is a few tens of MeV for most materials; heavier muons have a critical energy of several hundred GeV. Two metres of iron absorb most particles except muons, so muons are usually measured behind hadron calorimeters [Highlight 7.10]. *Neutral particles* are not sensitive to e-m effects. If unstable, they can be detected by tracking charged decay products [Highlight 8.6]; if stable or long-lived (e.g. neutrons), detection is ensured via strong interactions in devices such as calorimeters [Highlight 4.10]. *Hadrons* — baryons (e.g. the proton) and mesons (e.g. the pion) — interact strongly, producing a *hadron shower*, a process characterized by the *nuclear interaction length* λ (for iron, λ = 17 cm), the mean path length over which energetic hadrons have a ~2/3 probability to interact inelastically. About 10 λ absorbs a 100 GeV hadron. Among *leptons,* neutrinos (ν), neutral, stable and weakly interacting, are the most difficult to detect: a 10 GeV ν has only a 10^{-7} probability to interact when crossing the Earth! To observe ν interactions requires massive detectors and large ν flux. A ν escaping from an interaction can be inferred from its "*missing momentum*" vector by measuring all detectable particles. In e-m collisions a charged particle is deflected stochastically: the *mean deviation angle* is proportional to $\sqrt{t}/p$, where t is the thickness of the material in units of X_0. This impacts on the accuracy of tracking and explains the need for detectors with low total t. Other e-m energy loss mechanisms are (i) the *Cherenkov effect*: a charged particle traversing matter produces a trail of polarized molecules along its path, which depolarize and emit radiation. If the velocity $v > c/n$, n being the refractive index of the medium, there is a direction along which the radiation adds coherently, giving "Cherenkov light" (analogous to a sonic boom), emitted at the *Cherenkov angle* θ_C, where $\cos \theta_C = c/nv$ [Highlights 7.8, 8.10]; and (ii) *Transition Radiation (TR),* which occurs when a charged particle passes between media of different permittivity. It is used to identify ultra-relativistic electrons [Highlight 4.9].

This motivated the development of "electronic", triggerable detectors, such as spark chambers, scintillation counters and threshold Cherenkov detectors for particle identification [Boxes 3.2 and 6.3]. It culminated in the invention of the "Multi-wire proportional chamber (MWPC) in 1968 [Highlight 4.8]. Counter and emulsion experiments studied secondary particle production (π, K, antiprotons) from various targets located in the PS beam. These studies were to have strong impact on the design of separated beams that were soon to be built.

While the PS machine was built in a remarkably short time, CERN's start of its physics programme was rather more shaky. There were, it should be noted, not so many physicists participating in the research programme at CERN at that time, as the larger member states were still pursuing their national projects. Fortunately for the development of particle physics, there was still at that time strong political and financial support given to both nuclear and particle physics.

In 1960 one of the big puzzles in particle physics was the muon, which decays into an electron and two neutrinos. What prevents the muon from decaying into an electron and a gamma ray? An elegant hypothesis gives the electron and muon a hidden property ("quantum number"), which stipulates the existence of two neutrinos *species*, one related only to the muon, ν_μ, and another one associated to the electron, ν_e. Neutrino scattering it was argued, could test this hypothesis: $\nu_\mu + n \rightarrow p + \mu^-$ and $\nu_e + n \rightarrow p + e^-$.

A source of ν_μ is the pion, decaying essentially into a muon and a neutrino. Early in 1960 it had been argued that pions, produced at accelerators like the PS, would give enough ν_μ's to perform a conclusive experiment. CERN jumped on this idea with great enthusiasm, because of its conceptual simplicity and physics importance. In the South Hall pions were produced on an internal target in the PS, which subsequently decayed in a 20 m space, followed by some 20 m of iron and concrete shielding. The main detectors were the CERN and the Ecole Polytechnique BP3 heavy liquid bubble chambers containing a total of about a ton of CF_3Br Freon. The expected rate was around one event per day. However, a last minute beam survey showed that the pion flux would be an order of magnitude below design and the experiment was stopped. CERN lost its chance for a big discovery, the muon neutrino. Only one year later, in 1961, the two-neutrino hypothesis was confirmed in a Brookhaven experiment using the same method and a spark chamber detector!

Despite this sobering setback, neutrino physics would develop into a very important field of research at CERN and elsewhere. The radioactive decay of particles due to the weak interaction occurs at relatively low energies. In contrast, neutrino scattering would probe the weak interactions at much higher energies and

Imaging detectors **Box 3.3**

These provide an image of particle tracks, usually recorded photographically.

Nuclear emulsion (NE) is a photographic plate with a thick layer of emulsion of uniform grain size. Compact and dense, it has the best spatial resolution of all, at the μm level. A crossing charged particle sensitizes the grains. After development, tracks are visible and measured with a microscope. NEs have undergone a renaissance, coupled to fast precise electronic detectors, which point to the region of interest.

The *cloud chamber (CC)* is a sealed environment containing a supersaturated vapour of water or alcohol. Charged particles ionize the vapour and the ions act as nuclei for condensation. Mist forms around the ions to produce a visible track. In a *pulsed CC or Wilson chamber* (C.T.R. Wilson, Nobel 1927), a diaphragm expands the volume which cools and initiates condensation. Positrons and muons were discovered in cosmic rays using CCs. A *diffusion CC* is continuously sensitive, but needs cooling.

The *bubble chamber* (*BC*, A. Glaser, Nobel 1960), is similar in principle to a CC: a cylinder is filled with a liquid heated to just below its boiling point. A piston suddenly expands its volume, lowering the pressure and driving the liquid into a superheated metastable phase. Microscopic bubbles form along the ionization track. Their size grows as the BC expands, getting large enough to be photographed. Adding a magnetic field gives the momentum. More than 100 BCs were built [1], from very small to very large. The largest contained up to 40 m^3 of liquid hydrogen or heavier liquid (e.g. freon) and were used to record millions of images. Momentum resolution depends on the BC size, the magnetic field and the bubble size, ranging from typically 200 μm to 700 μm in diameter in BEBC [Highlight 5.7]. To study charm requires bubbles of 30 μm. Photographing such details limits the depth of field of the camera and the observable volume. This led to develop small rapid cycling BCs, e.g. LEBC [1].

The BCs were facilities, operated by large laboratories and used by many groups, leading to international collaborations, sharing the scanning and measuring of BC photos. Ever more elaborate analysis methods were developed, implying semi-automatic data recording and the use of powerful computers. These pictures, providing an intuitive view of physics events, helped to popularize the field. But restricted to fixed target physics, unable to be triggered and of low pulse rate, BCs could not give access to rare events and were progressively abandoned.

The *spark chamber* (*SC*) [2] uses the breakdown induced by a strong electric field locally around the particle trajectory. These "sparks" can be photographed or recorded electronically. A sequence of SCs provide a BC-like detailed view of the particle track. The two-neutrino US experiment (Chapter 3.1) used the first large SC set up. The largest one at CERN was the Omega spectrometer [Highlight 3.7]. *Streamer chambers* with shorter high-voltage pulses and image enhancement also used gaseous breakdown providing bubble-chamber type views of particle collisions.

From 0.01 Hz for CC, the pulse rate went to 0.3 Hz in BEBC and up to 20 Hz in SC. Discoveries made with these devices were essential in establishing the SM [Box 6.4].

[1] G.G. Harigel *et al.*, (ed.), Proc. Conf. on Bubble Chambers, *Nucl. Phys. B, Suppl.* **36** (1994).

[2] R.P. Schutt, (ed.), *Bubble and Spark Chambers* (Academic Press, New York 1967).

would also allow to investigate the interior of the nucleon. Neutrino scattering would shed light on the fundamental nature of this mysterious weak interaction. At high energies, it was understood, a new mechanism has to be invoked, a particle, technically called "Intermediate boson" (IB), had to exist to transmit or "mediate" this interaction. This idea was worked out in the 1960s, although its mass was a mystery, justifying early searches at quite low masses. The chase for the IB would become one of the grand successes of CERN (Chapter 6).

There were other questions: is the "lepton property", the so-called lepton number conserved? Could a neutrino interact with a nucleon without changing into a charged lepton, i.e. besides the known "Charged Currents" (CC) of muon and beta decay, did "Neutral Currents" (NC) exist?

Towards neutral currents

CERN had learned its lesson: To run a viable neutrino research programme neutrino beams of adequate intensity were required. Step 1 was to maximize the pion flux, achieved by extracting the PS proton beam towards an external target in the direction of the experiments. Step 2 was a real breakthrough: the ingenious invention by S. van der Meer (Nobel Prize 1984) of a pulsed magnetic horn, focusing the pions, produced with different momenta and angles at the target, towards the detector and hence maximizing the neutrino flux [Highlight 3.6].

The beam was ready in 1963; the detectors were the CERN Heavy Liquid BC (HLBC) and a 15 ton spark chamber (SC) setup. The BC team would study the neutrino reactions in detail and measure cross-sections of specific processes. The SC group was primarily searching for the possible production of the IB (W^+ or W^-) with a mass below 2 GeV in the strong electric field of a heavy nucleus:

ν_μ + Nucleus $\rightarrow \mu^-$ + Nucleus + W^+, with $W^+ \rightarrow \mu^+ + \nu_\mu$ or $W^+ \rightarrow e^+ + \nu_e$.

At the September 1963 Siena Conference, the two experimental teams presented their preliminary results, the BC with a couple of hundred events and the SC setup with many thousands. The former presented results on the cross-section measurements of the elastic process $\nu_\mu + n \rightarrow \mu^- + p$, the quasi-elastic process $\nu_\mu + p \rightarrow \mu^- + \Delta^{++} \rightarrow \mu^- + p + \pi^+$, on the energy dependence of the total cross-section and many other topics. The latter reported several di-muon candidates but could not say if they were W decays. The future was decided in a local trattoria: a very large heavy liquid BC with its splendid detail about the interactions was the way to move forward. It had to have a large target mass for a reasonable interaction rate and several interactions in length to distinguish pions from muons.

The proposal made to the French CEA was accepted and funded in 1964. In 1965 CERN agreed to host the heavy-liquid BC and to build a new neutrino beam. The BC was a cylinder 4.8 m long and 1.9 m in diameter with a volume of 12 m^3, bright field illumination, six cameras and a magnetic field of 2 T. It would be known as Gargamelle (Fig. 3.5, left). The intellectual father and project leader was A. Lagarrigue (Fig. 3.5, right).

While waiting for Gargamelle, the CERN HLBC, which had been doubled in size, was filled with propane and exposed to the new beam. About this time a US group studying neutrino electron scattering at a nuclear reactor had reported a single electron signal three orders of magnitude greater than theoretically expected. A careful search for single electrons in the HLBC propane photos did not find any event, placing new upper limits on the "neutral current" processes $\nu_\mu + p \rightarrow \nu_\mu + p$ and $\nu_\mu + p \rightarrow \nu_\mu + n + \pi^+$. The reported claim was eventually withdrawn.

When data taking started in 1971 with Gargamelle, priority was given to measure the CC total cross-section and to probe the nucleon structure. The reason was that electron scattering experiments at SLAC had shown the nucleon to contain point-like constituents. Combining neutrino data with the electron data would allow to measure the charge and net number of these point-like particles.

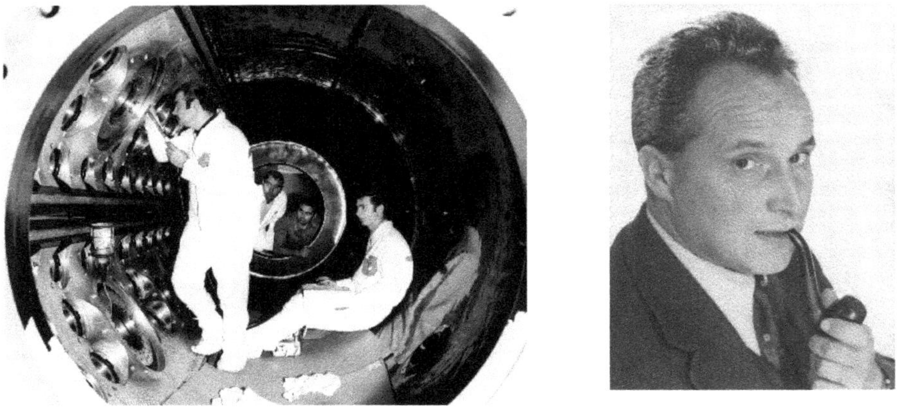

Fig. 3.5. Left: inside Gargamelle. Right: A. Lagarrigue, the driving spirit of Gargamelle.

However, early 1972, priorities were reversed. The Electroweak (EW) theory, unifying the weak and electromagnetic interactions, had been completed [8]. Crucially, it predicted the existence of a heavy *neutral* Intermediate Boson. Neutral Current (NC) events would therefore occur, revealed by a single recoil electron or by hadron production as in a CC event, but without a muon (Fig. 3.6). All efforts were devoted to scanning Gargamelle photos for single electrons and searching NC candidates with hadronic energy greater than 1 GeV. Studies on possible background events convinced the collaboration that neutrons coming from neutrino interactions in material around the detector were the only significant background, which however could be evaluated and controlled. In addition, one NC candidate event with a single electron had been found in the antineutrino exposure where this background was negligible.

A competing US experiment produced contradictory results and some confusion. However, sure about their results, the Gargamelle collaboration announced and published their evidence for the existence of neutral currents in July 1973 with the measured ratio NC/CC = 0.21 ± 0.03 for neutrinos and NC/CC = 0.45 ± 0.09 for antineutrinos [9]. This result was a turning point in the history of particle physics, leaving little doubt about the validity of the EW SM, allowing to predict the weak bosons masses with small errors and opening the modern era of precision tests of the SM. The premature death of A. Lagarrigue probably deprived the neutral current discovery of a Nobel Prize.

Other results followed. The total neutrino-nucleon cross-section was found to rise linearly with energy, consistent with point like objects in the nucleon, confirming with the weak probe the earlier SLAC observation with the electromagnetic probe. Combining both results showed that these constituents have indeed fractional charges.

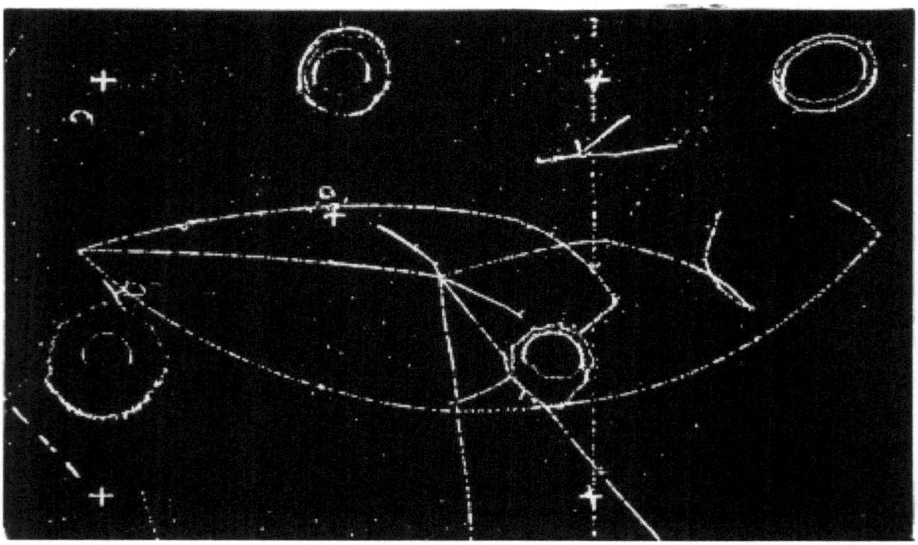

Fig. 3.6. A neutral current event in Gargamelle. The neutrino beam enters from the left. All three tracks from the collision vertex either undergo very large angle scattering or interaction, proving that no muon is produced.

Interestingly, a neutrino beam was again derived from the PS in the early 1980s, aiming at the West Area neutrino detectors in search of neutrino "oscillations", i.e. their "morphing" along their way, from a muon-neutrino to an electron- or tau-neutrino. As realized later, neutrino oscillations indeed exist, but with periodicities not experimentally accessible at CERN. Nevertheless, the experiments, e.g. PS191, were visionary and pioneering in their concept.

After its difficult start the PS neutrino programme had turned out to be quite remarkable!

Other programmes

In 1960 CERN reorganized their experimental activities in response to the changing nature of the experiments and their techniques. With the aim of strengthening the research programme the experiments using counter techniques and those with bubbles chambers were supported in separate dedicated Divisions.

Hydrogen bubble chambers at the PS

During the PS time BCs increased in size from 30 cm to 2 m size, and were exposed to an increasing variety of hadron beams.

Coping with the enormous number of BC photos led to R&D programmes for the machine inspection (scanning) of the BC pictures. Semi-automated or fully automated measurement equipment on digitizing tables were developed; computer programs were developed to analyse and interpret the events [10].

The list of new particles and resonances became ever longer. Cross-sections were measured with great precision and particle production mechanisms studied in detail. The CERN PS experiments made significant contributions. Various theoretical models came and went until it was eventually realized that this zoo of particles could be best described by the model of fractionally charged quarks by M. Gell-Mann and G. Zweig, then a visitor in the CERN Theory Division [11].

One of the most important spinoffs of the BC program was sociological: it attracted many European university teams in particle physics and contributed to fostering a culture of fruitful collaboration among different nationalities: The building of this aspect of the "CERN model" got on its way.

Experimentation with electronic detectors

Research, increasingly concentrating on phenomena occurring with very small cross sections and using a variety of particle beams, motivated a rapid growth and diversification of the electronic detectors. Scintillation counters, Cherenkov detectors and spark chamber became the "workhorses" of ever more sophisticated experiments based on these electronic techniques. The 1968 invention of the Multiwire Proportional Chamber (MWPC) by G. Charpak, Nobel Prize 1992, truly initiated this electronic revolution.

Initially, the spark chambers events were recorded photographically, soon to be replaced by acoustic, magnetic core and magnetostrictive readout. In 1964 a meeting was held at CERN on "Film-less Spark Chamber Techniques and Associated Computer Use" [12]. This led to the wire spark chamber, invented at CERN, in which the plates were replaced by wires. The "sparks" produced electrical pulses in the wires, which were directly logged into a computer. This was the start of using computers for monitoring and data recording of the experiments.

One experiment, using a combination of dipole and quadrupole magnets and an electrostatic separator, followed by a very sophisticated Time-of-Flight system discovered the production of anti-deuterons in proton–beryllium collisions, preceding a similar observation at BNL by a few weeks. The existence of the bound state of antinucleons was essential for the proof of the existence of antimatter, as Dirac liked to emphasize. It also lent credence to the concept of a fundamental symmetry, the so-called Charge–Parity–Time Reversal (CPT) symmetry [Box 2.2], fundamental to quantum field theories, such as the SM.

The rise of electronic detectors motivated a bold decision to develop an "electronic bubble chamber", a large 1.8 T magnetic field volume, instrumented with spark chambers, capable of selecting (triggering on) specific reactions. The device was known as OMEGA [Highlight 3.7]. At the PS it was used from 1971–76 to study meson spectrometry and rare modes of hadronic reactions, and subsequently in a productive program at the SPS, with spark chambers being replaced by MWPCs.

CP violation [Box 3.4] was discovered at Brookhaven in 1964 and the result quickly confirmed at CERN, eliminating some theoretical models and correcting some erroneous claims made elsewhere.

Experiments using the interference between the charged two-pion decay of K_S and K_L performed precise measurements of the K_S–K_L mass difference, the CP violation amplitude and phase [Highlights 5.5 and 5.6]. In 1969 one of these experiments began to use the first very large MWPC chambers. In 1972 a setup using wire chambers and a lead glass array proved that the ratio of the CP violating amplitudes of the two-pion charged and neutral decay of K_L was equal to unity to a 6% level. It would take another 20 years to know that they differ from unity at the per mille level (Chapter 5, SPS).

Kaon mixing and CP violation **Box 3.4**

The K^0 shows a puzzling behaviour: It decays via the weak interaction in ~10^{-10} s into two $\pi^{\pm}$ or two π^0, called K_1, and 600 times slower into three pions (π^+, π^-, π^0 or 3 π^0), called K_2! The explanation is that the K^0 can change into an anti-K^0 ($\overline{K}^0$), and vice versa, by virtual decay and recombination. The decay modes observed are the symmetric (K_2) and antisymmetric (K_1) quantum state combinations of K^0 and $\overline{K}^0$. This seems complicated, but is simple and has a classical analogue of two coupled pendula, Fig. 1.

For two identical pendula, coupled by a spring (Fig. 1), only two modes of oscillation can exist: parallel (P) and antiparallel (AP). One can ascribe a *symmetry* to the pendula under which the AP (P) mode is considered an even (odd) state. The spring is at work in the AP mode only, and hence it will have a slightly higher frequency than the P mode. If the spring is not perfect, it costs energy and the AP mode will decay away quicker than the P mode. Setting in motion e.g. the right pendulum (left of Fig. 2) the spring will transfer energy between the two pendula, such that the left one starts and the right one comes to rest, and vice versa. This is due to the interference (beating) between the two modes. The antiparallel mode is damped and will decay away first.

In our particle picture, assume the *motion* of the left (right) pendulum to represent a K^0 ($\overline{K}^0$), Fig. 2. The right pendulum describes an initial *pure* $\overline{K}^0$ state. The spring provides the interaction which changes the K^0 and $\overline{K}^0$ into each other. Fig. 2, right part, shows that the K^0-$\overline{K}^0$ system can be considered as the superposition of the P (odd) mode, representing a K_2, and the AP (even) mode, representing a K_1. The spring is not perfect and the antiparallel mode K_1 is damped first, leaving the long lived K_2. The two different K_1 and K_2 decay rates are reproduced. The decays were considered to be invariant under CP, i.e. CP to be a good or conserved symmetry in the weak interactions [Box 2.2]. The symmetry ascribed above to the pendula is the mechanical equivalent to CP symmetry seen in the kaon system.

If CP were a perfect symmetry, then the CP odd K_2 should not decay into a CP even 2π state: but in 1964 one found that it did at the 10^{-3} level, showing that CP is violated in K^0 mixing: K^0 and $\overline{K}^0$ do not change into each other at the same rate. This kind of CP violation is demonstrated with the coupled pendula by connecting the spring at different positions on the K^0 and $\overline{K}^0$ arms. Then the energy transfer between the two pendula is not equal: the long-lived mode contains a small part of K_1, is no longer a pure CP state, and is called K_L. The short-lived mode contains some K_2 and is called K_S.

Due to mixing the amount of CP violation seen in the decay mode into two $\pi^{\pm}$ should be the same as for the decay into two π^0. This is only true down to the 10^{-6} level, where so-called *Direct CP violation* is also observed [Highlight 5.5]. Related effects are observed in the decay of B-mesons, particles containing a beauty quark [Highlight 8.6].

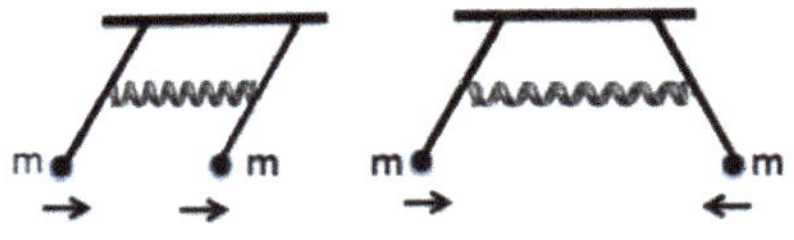

Fig. 1

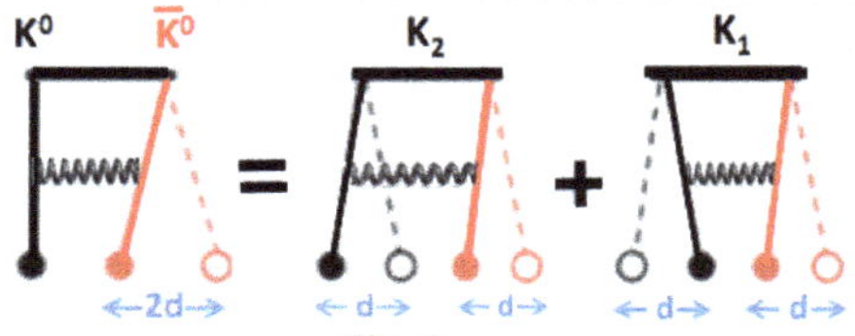

Fig. 2

The PS measurement of the anomalous magnetic moment of the muon, $g - 2$, [Highlight 2.4], successfully continued the tradition started at the SC. With a 5 m diameter storage ring experiment (1967–1970) an accuracy of 3 parts in 10^4 was reached, to be superseded with a new storage ring of 7 m (1972–76), which reached an accuracy of 1 part in 10^5. Today, higher order QED calculations, including weak and hadronic corrections, predict $g - 2$ with an accuracy of 1 part in 10^6 and ongoing experiments elsewhere are being pushed to achieve an even higher accuracy. An observed slight discrepancy between the theoretical and experimental value, if it persists, would be a "smoking gun" for new physics.

Other facilities

Besides supplying protons for the SPS and the LHC, the PS also feeds some other very important facilities, ISOLDE, n_TOF and CLOUD.

The PS Booster's high intensity 1.4 GeV proton beam is directed onto special thick targets of the ISOLDE facility [Highlight 3.8], from which beams of radioactive isotopes are obtained.

For the n_TOF facility 20 GeV/c proton pulses, 6 ns long, are directed onto a lead target producing spallation neutrons in a wide energy range [Highlight 3.9]. These neutrons travel ~ 20 or 200 m to the experimental areas depending on the flux required. The energy of the neutrons is measured by time-of-flight. Research ranges from stellar nuclear synthesis to studies of nuclear reactor fuel cycles.

An experiment involving atmospheric, cosmic ray and particle physicists, and chemists, the Cosmics Leaving Outdoor Droplets (CLOUD) project, is studying the very complex processes of cloud formation under the influence of cosmic rays. The PS supplies the particle beams that simulate cosmic ray flux [Highlight 10.7].

3.2 Extraction: Getting the Beam to Leave the Accelerator

Massimo Giovannozzi and Charles Steinbach

Originally, the PS had used only internal targets. However, high radiation damage, due to absorption of secondary particles in the accelerator components, and low efficiency motivated the development of beam extraction and external targets. Fast extraction provides a beam with a length of one turn or less in one shot for users requiring high intensity in a short pulse, e.g. for experiments with bubble chambers; slow extraction skims off the circulating beam over many turns and is the choice for counter experiments preferring a long spill length and low instantaneous intensity to match their limited time resolution [13]. In 1963, the PS was the first synchrotron with a fast extraction system in operation and equipped with a successfully tested slow extraction system, initiating a leading position of CERN in the development and improvement of these key techniques.

Fast extraction

Fast extraction has been used for experiments in the East, West and South-East experimental areas of the PS and for sending beam to downstream accelerators (ISR, SPS, AA, LEAR, AD).

The principle is that a device producing a short (in time) deflecting magnetic field, called fast kicker, imparts an oscillation to the circulating bunches to be ejected so that they enter a deflecting magnet guiding them into the downstream transfer channel. This magnet is placed just outside the central aperture reserved for the circulating beam. One of its conductors acts as septum that separates the deflecting field from this aperture in order not to perturb the circulating beam. For clean extraction of a limited number of beam bunches, kicker rise and fall times had to be less than 80 ns, well below the 105 ns separation between two circulating bunches. During its first 60 years, the PS has known two different generations of kickers. The first, installed in the early 1960s, was of limited aperture and was retracted during injection when the beam size was large, and it was hydraulically positioned after beam acceleration. A stationary full aperture kicker system was proposed in 1964. It took nine years and two successive designs to obtain a satisfactory system with 12 separate modules. The magnets use ferrite for the yoke in the machine vacuum and form a 15 Ω delay-line. Figure 3.7 shows a schematic cross-section of such a kicker magnet (left) and one module (right).

The pulse forming networks (PFN) providing the driving pulse to the kicker are made of very low attenuation 15 Ω SF_6 pressurized polythene tape cables, charged to 80 kV in less than 4 ms by a resonant power supply. The discharge of the PFNs into the kicker is controlled by hydrogen thyratrons. The full-aperture kicker system was commissioned in 1974 and is still in use with high availability [14].

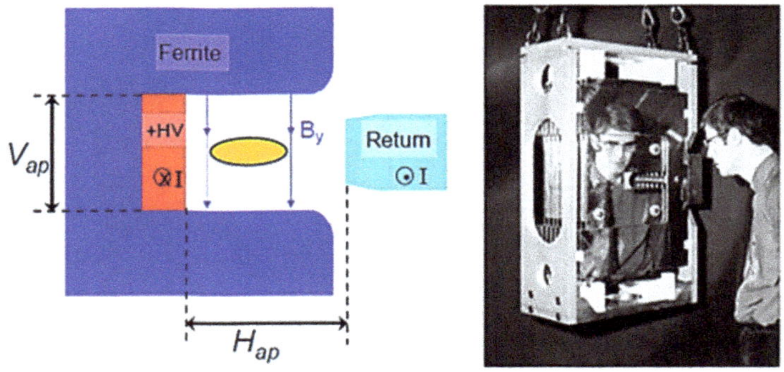

Fig. 3.7. Left: Schematic cross-section of the full-aperture kicker magnet showing the ferrite yoke, the one-turn coil (red/blue) and the beam to be deflected (yellow oval); Right: A module of this kicker with the gap for the beam seen in the middle of the shining front face.

Slow extraction

Slow extraction was introduced to provide primary beams to external targets for counter experiments. In this way external secondary beams can be produced with higher intensity and better quality than those from internal targets. During this extraction the focusing of the PS is adjusted such that the transverse oscillations of the particles grow due to a resonance driven by on-purpose installed magnetic lenses, which slowly increases the amplitudes of these oscillations until all particles have entered the deflecting field at the septum.

The first tests were conducted in 1963, reaching after improvements an extraction efficiency of 90%. In 1972 it became necessary to supply beams also to the new West hall. A new scheme was proposed based on a resonance occuring when the number of horizontal particle oscillation per turn, Q, is nudged to become a multiple of 1/3 (third-integer resonance). The resonance was driven by a standard quadrupole and a special semi-quadrupole combining quadrupolar and sextupolar fields. The scheme offered simultaneous beam sharing at will between internal targets and two external targets. The extraction efficiency reached 93% limited by the losses at the septa. The most recent scheme for proton slow-extraction was introduced in 1992 when the West hall extraction was suppressed. Taking into account that the internal targets had been eliminated by 1981 and the PS had to accelerate electrons and positrons for LEP, interleaved with proton cycles, its design used fewer and more standard magnetic elements for ease of maintenance, improved the ring vacuum, and provided component shielding against radiation from positrons or electrons. The first deflecting element of the extraction channel is equipped with a thin septum formed by an only 0.1 mm thick molybdenum foil to minimize beam loss (Fig. 3.8, left). It provides a deflecting electrostatic field of

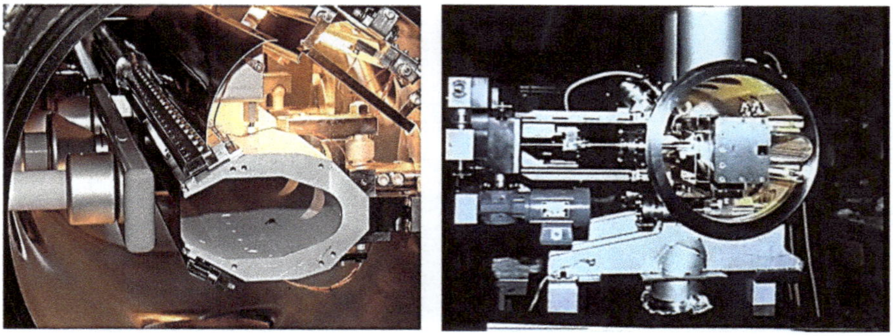

Fig. 3.8. Left: The electro-static deflector with a 0.1 mm septum in its vacuum tank; Right: Magnetic deflector with a 4 mm septum. The dark slot in the plate is the extraction channel while the aperture for the circulating beam is on the right of this slot.

10 MV/m (160 kV across a 17 mm gap) and is placed towards the inside of the ring to avoid exposure to radiation during lepton cycles [15, 16]. Further downstream, the deflected beam receives a second kick by a special magnet equipped with a thin (4 mm) septum before entering the transfer channel. This magnet also shares the ring vacuum and can be positioned remotely (Fig. 3.8, right). The extraction takes place at a particle momentum of 24 GeV/c providing 2×10^{11} protons per pulse over 400 ms (2×10^5 turns !) at the external target with an efficiency of 95%.

Multi-Turn Extraction

The development of this extraction method, called Continuous Transfer (CT), was driven by the need to fill the CERN SPS as uniformly as possible for fixed-target physics. Since the SPS circumference is eleven times that of the PS, the SPS could be filled by extracting the beam over several turns from the PS. The choice has been to fill the SPS with beam of two consecutive PS cycles, which terminate each time in an extraction over five turns at a particle momentum of 14 GeV/c. A gap of one 11th of the SPS circumference is left without beam to accommodate the fall-time of the pulsed SPS injection elements. The principle is based on a careful choice of the number of horizontal particle oscillations, on a variable-strength closed orbit bump, and on an electro-static deflector with a thin (0.1 mm) septum that peels the beam off each cycle turning it into a spill of five PS circumferences in length.

In the quest for an improved extraction mode to substantially reduce the unavoidable beam loss on the first septum, a novel approach was proposed, named Multi-Turn Extraction (MTE). In this process the beam is split in a beam four turns long ("long beam") and a central beam one turn long. This is achieved by generating, prior to extraction, stable islands in transverse phase space inside the circulating beam by crossing a resonance produced by appropriate non-linear magnetic fields. In this process the particles get trapped inside these islands thus generating around the central beam a new, well-separated stable beam which extends over four turns and closes in itself. By a controlled drift through the resonance these islands move towards larger transverse amplitudes until the horizontal separation between two beams exceeds the thickness of the septum. The resulting configuration in transverse phase space is a central beam around which the long beam closed in itself is wound. An example of the evolution of the beam distribution in horizontal phase space at a given azimuth along the circumference is shown in Fig. 3.9. Note that it is the same beam that appears in the four islands around the central beam.

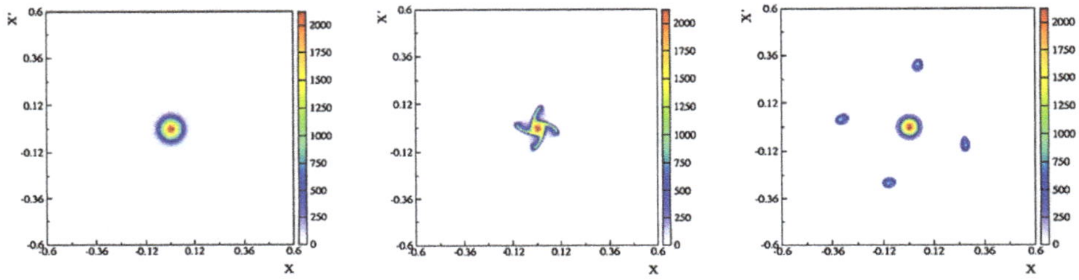

Fig. 3.9. Evolution of the horizontal beam distribution during resonance crossing in phase space (the ordinate is the horizontal momentum of the particles, while the abscissa is the horizontal position of the particles): The initial state (left); at resonance crossing particles are trapped inside the moving islands (centre); at the end of the process, the particles trapped in the islands are at larger amplitudes required for extraction (right).

Subsequently, a fast closed orbit bump created by two kickers, with a rise time short compared to the revolution time, is created with such an amplitude at the septum that the long beam enters the deflecting field region and is extracted over four turns while the central beam is not deflected. Finally, the central beam is extracted by a strong fast transverse kick creating a horizontal oscillation of sufficient amplitude making it enter the deflecting device. The timing is such that the central beam is contiguous to the long beam already transferred to the SPS.

The novelty of this method is that it replaces the brute force peeling by an intercepting device with a splitting performed by the non-linear resonance. This approach is superior as particle losses are limited to the fraction of the beam improperly deflected during the rise time of the fast orbit bump.

The proposed scheme was designed in 2006 and the first beam commissioning started in 2008. While the main concept of the beam splitting was quickly confirmed, the stability of the overall process has posed some problems. Moreover, the fraction of beam lost on the septum magnets during the rise time of the fast bump induced strong activation for which mitigation measures had to be devised [17]. This implied a major re-design of all fast extraction schemes, which were successfully commissioned by the end of 2014. This unique scheme has the welcome potential to reduce significantly the radiation dose to the equipment around the septum and, thereby, substantially increase component lifetime. As of September 2015 MTE has replaced CT as the operational extraction system from PS for SPS fixed-target physics. Thus, the PS moves closer to the ideal of loss-less extraction — the subject of a long-standing effort since its inception.

3.3 Acceleration and Bunch Gymnastics

Heiko Damerau and Steven Hancock

The basic function of the radio-frequency (RF) system is the acceleration of the particles grouped in bunches in synchronism with the rising magnet field which keeps them on orbit [Boxes 2.1 and 3.1]. In addition it is used to regroup the particles in user-defined bunch configurations along the orbit, different from the bunch sequence at injection, during the acceleration cycle, and to shape the final bunches before extraction according to the user requirement. Acceleration and bunch manipulations have to be performed avoiding beam instabilities and minimizing beam loss [18].

Key components of the RF systems are the cavities producing the accelerating electric RF field, the RF power supplies, ancillary equipment such as cavity tuners, devices damping undesired RF modes in the cavities, and control circuits to achieve the proper RF amplitude and phase. Two typical cavities are described.

The first example is the main RF system of the PS comprising 11 ferrite-loaded cavities each capable of delivering a peak voltage of 20 kV [19]. Originally it had been designed to operate on the 20th harmonic of the revolution frequency for proton acceleration from 50 MeV (kinetic) injection energy to 28 GeV. Since the protons had only 30% of the speed of light at injection but nearly the speed of light at top energy, the RF had to cover a swing from 3 to 9.5 MHz. Today protons from the PS Booster (PSB) are injected at 1.4 GeV to limit adverse space charge effects on today's high-intensity beams, reducing the relative frequency swing during acceleration to only 10%. Nevertheless, presently a range from 2.8 MHz to 10 MHz is employed together with the possibility to independently use the cavities in groups, allowing beams of protons and ions to be accelerated and manipulated using harmonics in the range from 6 to 24. Figure 3.10 shows schematically a cavity consisting of two resonators with separate accelerating gaps and the ferrites used for tuning surrounding the vacuum tube. The tuning is accomplished by modifying the effective inductance of the ferrites through changing their magnetisation with the help of the current loops.

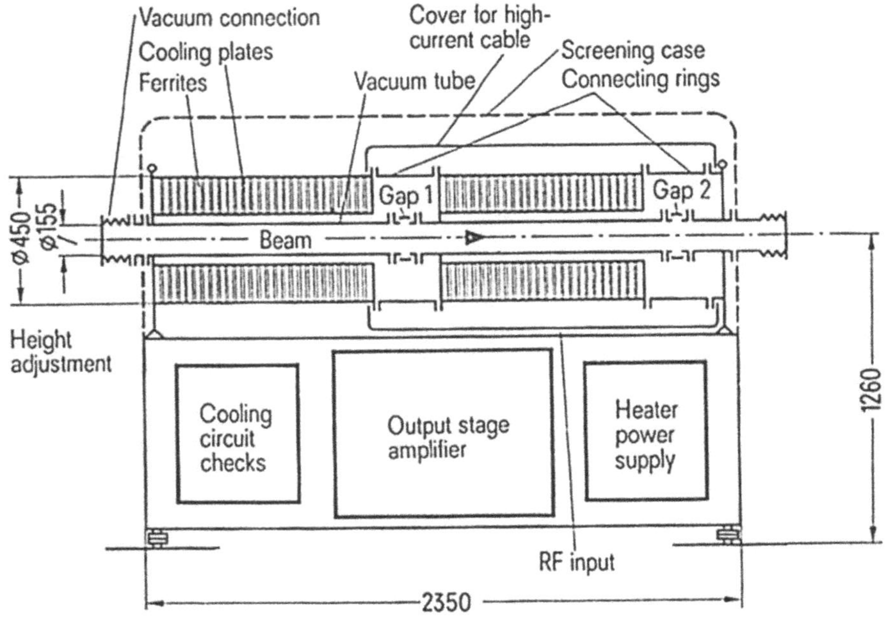

Fig. 3.10. Schematic drawing of one of the ferrite-loaded 3 to 10 MHz cavities; dimensions in mm.

The most complex task, the preparation of the beam for LHC, could only be mastered by adding a set of special RF systems operating at 20, 40 and 80 MHz to the main system [18]. From these systems the second example is chosen, a set of two special cavities operating at 80 MHz required for rapidly increasing the longitudinal focusing of the bunches destined to LHC just before extraction to the SPS [20]. These cavities produce a total of 600 kV and are pulsed for a brief moment prior to extraction. A damping antenna in the cavity suppresses higher-order electromagnetic modes which would adversely affect beam stability. TRIUMF contributed significantly to the design, construction and testing as an in-kind contribution of Canada to the LHC project. Figure 3.11 shows one of the cavities under test.

This battery of RF systems is used to produce a high-brightness proton beam for LHC [21]. The LHC requires short, intense bunches containing over 10^{11} protons of high transverse brightness, i.e. maximum transverse phase space density and a bunch to bunch separation of 25 ns or 50 ns. This beam is configured in the PS, the role of the SPS being restricted to acceleration from 25 GeV to 450 GeV, the LHC injection energy.

Fig. 3.11. Eighty-MHz RF cavity and 400 kW tetrode amplifier in the foreground in the test stand [20].

The main RF system cannot operate at 40 MHz corresponding to the 25 ns bunch separation. Hence, this bunch spacing is produced in steps during the acceleration cycle with the final steps just before extraction. Eight bunches are injected into the PS having six times the final intensity by using two cycles of the PSB producing each time one bunch in each of its four rings. After acceleration to 2.5 GeV, where the beam is less vulnerable to instabilities, the bunch train is compressed in time and two consecutive bunches are combined resulting in a bunch train of four bunches spaced by 300 ns, which is reduced to 100 ns in the subsequent triple splitting. All this is performed by the main RF system of three independent groups of three cavities each with the same voltage and phase. Figure 3.12 shows the beam during these manipulations, with the coloured lines marking the position of the bunches in time as a function of their relative position along the orbit. Using all cavities in all groups in unison, acceleration is pursued to 25 GeV where two consecutive bunch splittings take place resulting in the desired 25 ns spacing (for 50 ns only one splitting is performed). The first division of bunches is performed by the 20 MHz system, the second by the 40 MHz system while the main RF is off. In the final step of this virtuoso choreography, yielding 48 bunches of proper spacing, each bunch is longitudinally compressed to 4 ns by a rapid increase in the 40 MHz voltage and adding the full voltage of the 80 MHz system.

It is in this or similar schemes of bunch manipulation [18] that the PS RF systems display their full potential excelling in precise tuning, timing, phase and voltage control, testimony to their extraordinary flexibility and capacity, the result of continuous improvements over the years.

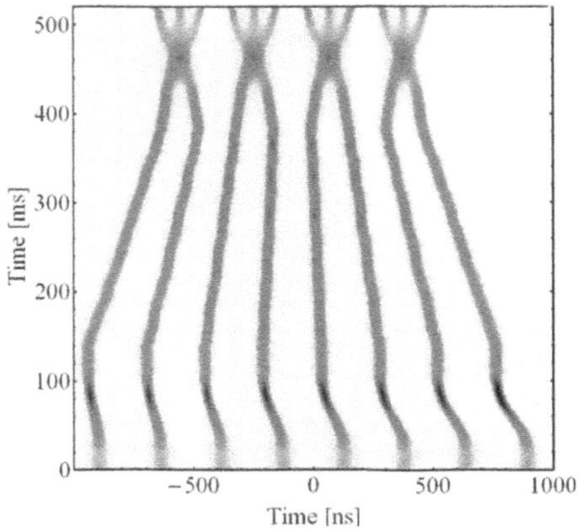

Fig. 3.12. Evolution of the bunch train between injection and at 2.5 GeV showing bunch train compression, combination of bunch pairs and triple splitting. Ordinate: time during the cycle; abscissa: longitudinal position of bunches along the orbit in terms of time [21].

3.4 Boosting PS Beam Intensity

Karlheinz Schindl

By the mid-1960s, CERN's 26 GeV Proton Synchrotron accelerated routinely $\sim 10^{12}$ protons per pulse, a performance largely exceeding its design. However, at this level there were signs of saturation, in particular at injection energy (50 MeV) where a phenomenon called "space charge" appeared to limit the intensity. With new potential clients — the ISR and the SPS — on the horizon, a study on how to increase the output beam intensity of the PS to $\sim 10^{13}$ protons per pulse (ppp) was launched. The space charge limit is brought about by the Coulomb force between the protons in a beam repelling each other and, therefore, working against the external focusing. The limit is strongly dependent on beam energy, and, during acceleration, the repulsive force gets weaker while the beam becomes stiffer. Hence, the limit is lowest at PS injection and it was overcome by raising the injection energy from 50 MeV to 800 MeV (kinetic) moving the limit up by a factor eight with an expected similar increase in intensity. A study of several alternatives led to the adoption of a very compact, slow-cycling 800 MeV synchrotron with four superimposed rings, the PS Booster (PSB) [22, 23], and re-use of the existing 50 MeV Linac. Operating each of the four rings (radius 25m, 1/4 of the PS) with five bunches at their individual space charge limit and transferring these four bunch trains sequentially to the PS would potentially deliver the desired 10^{13} protons to the PS.

The PSB was constructed between 1968 and 1972 after a very significant stumbling block was removed by diplomatic rather than technical skills: the PSB is the first accelerator ever built straddling an international border (Switzerland and France), paving the way for the much bigger accelerators also stretching across the border later. It was integrated in the PS injector chain in 1973, right away raising the PS output to 5×10^{12} ppp, just in time for the neutrino physics experiments with Gargamelle (Chapter 3.1).

It was the first particle accelerator with synchrotron rings stacked on top of each other. The four superimposed rings feature separate dipole (32) and quadrupole (48) magnets (as opposed to the PS where guide fields and gradients are generated by "combined-function" magnets) [24]. Each of them is a vertical stack of four magnets with a common yoke (Fig. 3.13). Their coils are connected in series and energized by one big power converter, while correction loops operating with rather small currents enable the guide and focusing fields to be adjusted for each ring. The stringent requirements on field quality and equality between rings gave rise to extensive studies of the prototype magnet stacks. The cross-section of both the magnet yokes and the coils were chosen large: the former mainly for better field quality, resulting in a relatively low magnetic field of ~ 0.6 T at 800 MeV, and the latter for minimizing integrated cost (construction + 10 years of operation). These latter features proved invaluable for later raising the PS Booster ejection energy to 1 GeV and 1.4 GeV. A further step to 2 GeV is planned.

A novel scheme [25] was also introduced to power the main magnets from a static power supply, rather than from a motor-generator set with a flywheel for energy storage, which draws constant active power from the mains as in the PS. Since the direct current varied between 200 and 3000 A during the acceleration cycle, this implied a power variation of about 7 MW. In order not to perturb the mains (voltage variation $< \pm 0.25\%$), the reactive power had to be compensated by a passive filter. A second passive filter was installed to ensure that the sinusoidal shape of the CERN mains voltage is maintained without parasitic oscillations at other points of the network (distortion $< 0.5\%$). This system works satisfactorily for repetition times as low as 1.2 s, lower repetition times showed resonance effects in the generators of the main Geneva hydro-electric power plant (Verbois).

Twelve low-cost kicker magnets for injection and ejection of novel design with short rise or fall times were built using thyratrons instead of the unreliable pressurized spark gaps and introducing ferrite-loaded pulse steepening lines which, depending on the task of the kicker, shortened either the rise or fall time of the deflecting field to less than 50 ns [26]. The design had to be modified later to provide the longer beam pulses required by the new client, ISOLDE, and the higher voltage for the new top energy [27].

Fig. 3.13. One (out of 16) magnet periods showing the four superimposed rings. From right to left: dipole (green); vacuum chambers in straight section (grey); another dipole; a quadrupole triplet (focusing-defocusing-focusing, orange); dipole; RF equipment for beam acceleration (grey).

The path to design performance and beyond

The path to design performance proved arduous. Obstacles were addressed one by one, profiting from a few key developments in accelerator physics and technology.

- The focusing in the accelerator was programmed to quickly reduce, during acceleration, the adverse effects of space charge (unavoidable at injection).
- The five bunches conspired to oscillate in a coherent manner and these coupled bunch oscillations leading to beam loss presented a hard limit. The PS Booster was not the first accelerator to suffer from such instabilities, but it was the one where a deeper understanding of this phenomenon was developed leading to the first electronic feedback system effectively damping these oscillations [28].
- A new Linac 2, still 50 MeV, but with up to 150 mA beam current, was built, replacing in 1978 Linac 1 which proved unable to reach 100 mA as specified; two further measures limited the adverse effects of space charge: new multipole correctors intended to compensate magnetic imperfections and an additional RF system operating at the second harmonic of the main system lowered the peak line density of bunches.

These developments and inventions, together with many other improvements, allowed the PS Booster to reach its design intensity of 10^{13} ppp by 1974, 3×10^{13} by 1985, and eventually 4.2×10^{13} by 2003 (Fig. 3.14).

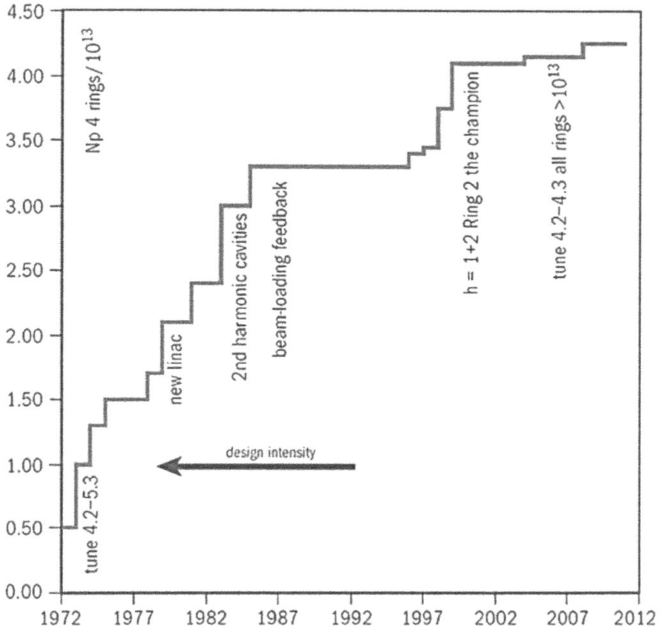

Fig. 3.14. The PSB peak intensity (sum of four rings) over the years. Major hardware additions and improvements are highlighted. (Adapted from [22].)

Also in the 1980s, the PSB got involved in ion acceleration, starting with deuterons and alphas, making its way through the Periodic Table to lead with the advent of Linac 3. This programme gave rise to numerous developments (RF, beam diagnostics) mainly to cope with ion intensities as low as 10^8 ions per pulse — five orders of magnitudes lower than proton beams within the same super-cycle. Since 1991 it replaced the aging CERN synchrocyclotron as proton source for ISOLDE.

In the LHC injector chain

In line with CERN's tradition of using existing accelerators as injectors for new machines, the LHC programme was no exception. It required a further increase of the top energy of PSB to 1.4 GeV and the replacement of the RF systems to deal with one instead of the originally used five bunches per ring [27]. In a further upgrade step the PSB ejection energy will be once more increased to 2 GeV and Linac 2 will be replaced by the new 160 MeV H^- Linac 4 as injector [29]. By this time, the PS Booster will be, in many parts, a new machine, while maintaining in the future its inherent versatility so successfully demonstrated in the past.

3.5 Capacitive Energy Storage Replaces Flywheel

Jean-Paul Burnet

The magnetic field in the 101 bending magnets of the proton synchrotron (PS) is cycled between 0.11 T at injection energy of the protons (1.4 GeV kinetic) to 1.25 T at top energy (26 GeV) with a repetition time of 2.4 s. About 6 to 8 million cycles are executed each year. The coils of this magnet string are in series representing an impedance of 0.32 Ω and 0.9 H and requiring a current of 5.5 kA at top energy. The active power for operating this string peaks at 40 MW at the end of the acceleration when the increase in stored magnetic energy reaches a maximum. After a constant period of about 0.2 to 0.3 s at top energy (flat top), the stored energy in the magnets has to be reabsorbed by the power supply, respectively by the energy storage system, during the ramp-down of the magnetic field implying a negative peak of 40 MW at the start of the ramp (Fig. 3.15). This power swing exceeds by far what could be tolerated by the mains in case of direct coupling though the average power is only 4 MW. The energy had therefore to be efficiently stored — a real engineering challenge. The original solution was energy storage in the rotational kinetic energy of the two rotors of a large motor-generator set acting as a flywheel. The set was supplied by industry and was replaced in 1968 in order to shorten the cycle time by a factor two to three. The new set shown in Fig. 3.16 consisted of a 6 MW alternating current (AC) motor and a 90 MW direct current (DC) generator providing a maximum of 6 kA at 12 kV [30].

The rotors had a total weight of 90 t capable of storing 233 MJ at 1000 rpm. The speed of the rotors decreased by 5% during ramp-up time while the motor absorbed 6 MW and, during ramp-down, the reflowing stored energy in the magnets peaking at 12 MJ reaccelerated the rotor to nominal speed.

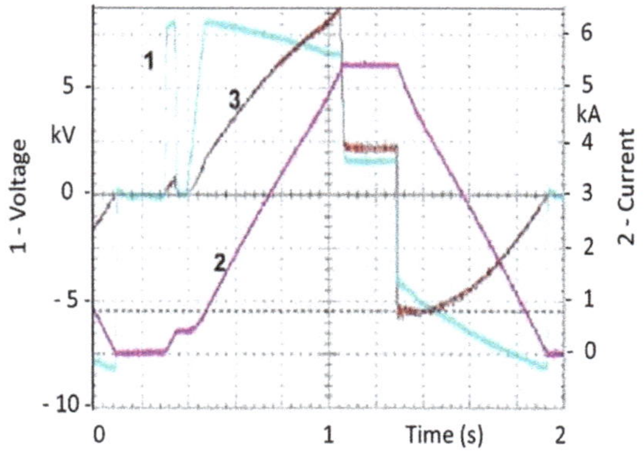

Fig. 3.15. The PS 26 GeV/c cycle: 1 voltage; 2 current; 3 active power 10 MW/div., vs. time [31].

Fig. 3.16. The 2nd generation PS main magnet power supply with the motor in the foreground.

Since maintenance of this set by industry came to an end and no supplier could be identified for a replacement, studies of alternative solutions were initiated in 2003. After discarding a solution with batteries due to their limited lifetime and energy storage in superconducting coils for lack of industrial products, the solution with storage in capacitors was preferred as they support nearly unlimited discharge cycles. Moreover, it could be constructed in a modular way based on industrial components. The new system, in operation since 2011, integrates two functions:

- Converting the AC current from the grid to DC current as required for the magnet powering;
- Storing the energy in six capacitor banks when not needed in the magnets [31, 32].

The capacitor banks are connected to the magnet string by six DC/DC converters which controls precisely current and voltage in the magnet circuit independently of the voltage of the capacitors. The voltage of the dry capacitors made from metalized self-healing Polypropylene decreases from 5 kV to 2kV during the ramping to top energy and increases again to 5 kV during ramp-down as shown in Fig. 3.17.

Only two AC/DC converters are connecting the system to the mains and supply the energy which is dissipated in the magnet coils. Since these losses peak at 10 MW during the flat top, the power rating of these converters is 5 MW. The DC/DC converters are based on medium-voltage power electronics produced by industry which allow a full control of the power flow in both directions depending on the requirement in the cycle. The key components are the 168 Injection Enhanced Gate Transistors (IEGT) which are used to switch currents up to 2.3 kA.

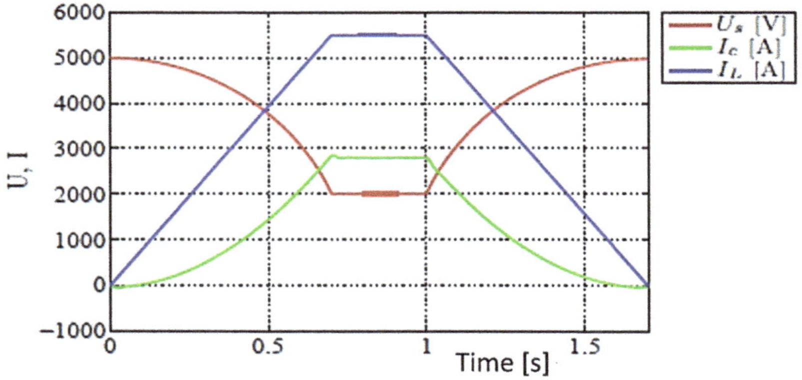

Fig. 3.17. Voltage (U_s) and current (I_c) of the capacitor bank with magnet current (I_l) as a function of time [32].

The system is complemented by output filters and a control system [33]. The total weight of capacitors is 60 t and they are placed inside six standard 40 foot shipping containers. Figure 3.18 gives an overall view of the new system. Although it was designed for high reliability, enough redundancy was included for functioning, albeit with reduced performance, in case of faults in one of the transformers, the AC/DC converter, the DC/DC converters, or in a capacitor bank. Its modular structure facilitates the management of spare components.

This novel CERN design has enabled the replacement of the aging large motor-generator set no longer supplied by industry with a fully solid-state system using the state of the art power electronics technologies. This system is a world first and, and being more efficient than the old motor-generator set, reduces the power demand from the mains to the strict minimum.

Fig. 3.18. The new power system for the PS main magnet.

3.6 Taking the Neutrinos by the Horns

Alan Ball

Four ingredients are needed to produce an intense muon neutrino (ν_μ) beam:

- An intense proton beam which is obtained by extracting all the circulating protons at top energy within one turn from the synchrotron [Highlight 3.2], the protons are focused onto a long thin target of low absorption materiel, often made from beryllium.
- A system to focus the secondary particles, produced in the target, charged pions (π) and kaons (K), which will eventually decay in flight into muons (μ) and muon neutrinos, the particles of interest. The challenge was to conceive a device that would focus a maximum of these "parent" particles, produced at different momenta and angles, towards the detector. The decay products, the neutrinos, are not focused by magnetic fields but are produced in roughly the same direction as the parent particles, ensuring maximum flux in the detector.
- A low density decay path, often in vacuum or helium gas, to ensure that parent particles do not interact with matter until they can decay in flight.
- An absorber to stop all hadrons remaining in the beam at the end of the decay path, so that only neutrinos and residual muons continue towards the detector.

The device invented by S. van der Meer [34] for the PS in the early 1960s was ingenious, revolutionary and decisive for obtaining neutrino beams of sufficient intensity. It is the *magnetic horn*. It works by focusing the secondary particles over a wide range of momenta and angles. The magnetic horn is a co-axial structure, typically 4–7 m long, where an electrical current flows along the inner conductor and returns along the outer conductor, producing a toroidal magnetic field in the space between them that is inversely proportional to the radius (Fig. 3.19). There is no field inside the inner or outside the outer conductors. By shaping the inner conductor, the integral magnetic field traversed by the particles can be designed to focus them towards the detector. The conductors are made as thin as possible to minimize absorption, yet sufficient to carry the large pulsed current.

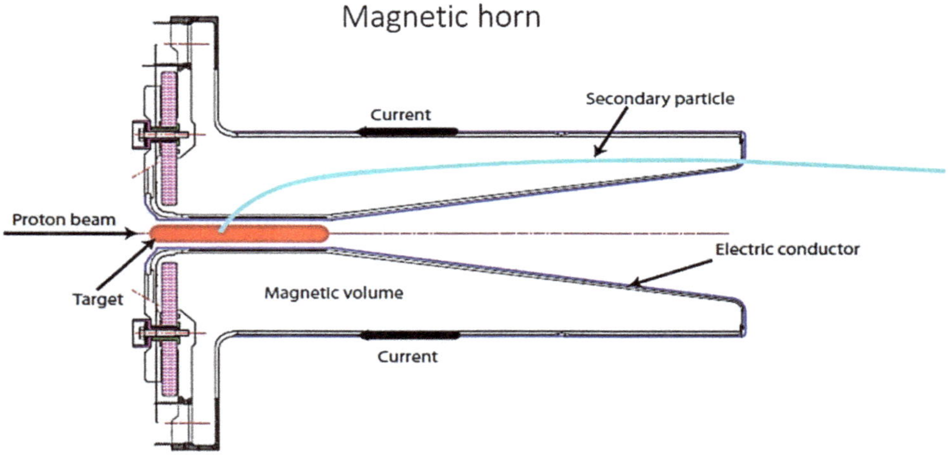

Fig. 3.19. Schematic cross-section of an early magnetic horn. The magnetic volume contains the azimuthal magnetic field.

Manufacturing the inner conductors of these co-axial lenses was no less of a challenge. The "neck" region of the horn was machined from a cylindrical bar of aluminium alloy, typically 70 cm long and 80 mm diameter. The machining could take nine months or more: the required "parabolic" form, of thickness of 1–5 mm, had to be cut and bored progressively with purpose-made tools, the swarf being removed at each stage to avoid piercing the thin-walled conductor. At larger diameters, the conductors were made from sheet metal rolled around a shaped mandrel and welded.

The insulated supports holding the inner conductor co-axially inside the outer conductor required precision adjustments. The inner conductor (Fig. 3.20) was cooled by water, sprayed from jets in the outer vessel and pulsed so there was no spray during the beam passage to add to the absorption. Insulation between conductors was complicated by power being supplied by capacitor discharge circuits giving a 200 μs pulse of up to 400 kA peak current at 12 kV. All this had to operate in an environment of high induced radioactivity.

The dimensions of this first dedicated neutrino facility at the PS [35] were quite modest; the decay tube was 15 m long and the detector, a heavy liquid bubble chamber (eventually Gargamelle) only a few metres downstream of the hadron stop. The total distance from target to detector was only 90 m.

Fig. 3.20. Inner conductor of magnetic horn for neutrino beam.

Figure 3.21 shows the layout of the most recent accelerator-based neutrino source operated at CERN, the CERN Neutrino beam to Gran Sasso (CNGS) [36], located at 732 km downstream of CERN. The 400 GeV proton beam originated from the SPS (Chapter 5). The target is followed by the first magnetic horn; a second horn or "reflector" is used to improve the collection efficiency. The secondary particles travel in a kilometre-long evacuated decay tube to minimize losses from absorption and scattering. All remaining hadrons are stopped in a massive carbon and steel "filter". The muons emerging from the hadron stop were used to measure the divergence of the beam and to steer it so that it was centred on the detector.

Magnetic horns have been used in almost all wide-band neutrino beams so far constructed and planned in Europe, Japan and US. No other focusing device has provided the wide energy and angular acceptance, sign selection of secondaries and overall reliability.

Comparison of early PS beams with CNGS shows the impressive development of neutrino sources over a span of 40 years. This would not have been possible without magnetic horns, and the inventiveness and engineering flair of the teams that first constructed and operated them.

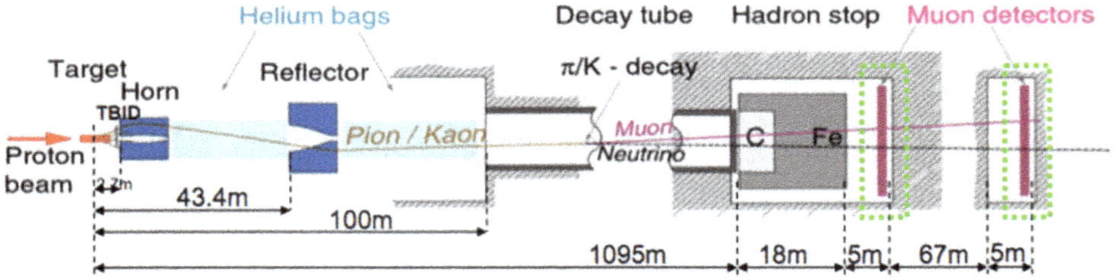

Fig. 3.21. Schematic layout of the neutrino source for CNGS [35].

3.7 OMEGA: Towards the Electronic Bubble Chamber

Werner Beusch and Emanuele Quercigh

Throughout the 1960s and 1970s bubble chambers were one of the principal instruments of particle physics research [Highlight 5.7]. These huge instruments were considered "facilities", which — similar to the accelerators providing the particle beams — were constructed and operated by teams of experts. These facilities provided the primary information allowing large collaborations to analyse a variety of new experimental data.

With increasing sophistication of the experiments, the use of electronic detectors, with the possibility of selecting particular phenomena and studying processes of small cross-sections, became obvious. The detector of choice was the spark chamber [Box 3.3], that could both visualise the events and be triggered to select a particular interaction. Placed in a magnetic field, the chambers provided position information of multiple particle tracks and hence their momenta. Spark chambers inside a magnet had been used at CERN since 1964. They provided data with statistics orders of magnitude larger than those of bubble chambers. The photographs of the sparks were measured with devices developed for bubble chambers. Later, these instruments, combined with novel algorithms for pattern recognition [37], provided particle trajectories without human intervention — major progress for analysis of large volumes of data.

With this background and as a part of the PS improvement programme a working group [38] was set up. The goal was to construct a large facility, the OMEGA spectrometer, also termed an "electronic bubble chamber", based on a strong magnet for good momentum resolution and large acceptance in order to investigate complicated interaction with many secondaries. A system of spark chambers with electronic readout and an efficient data handling was proposed.

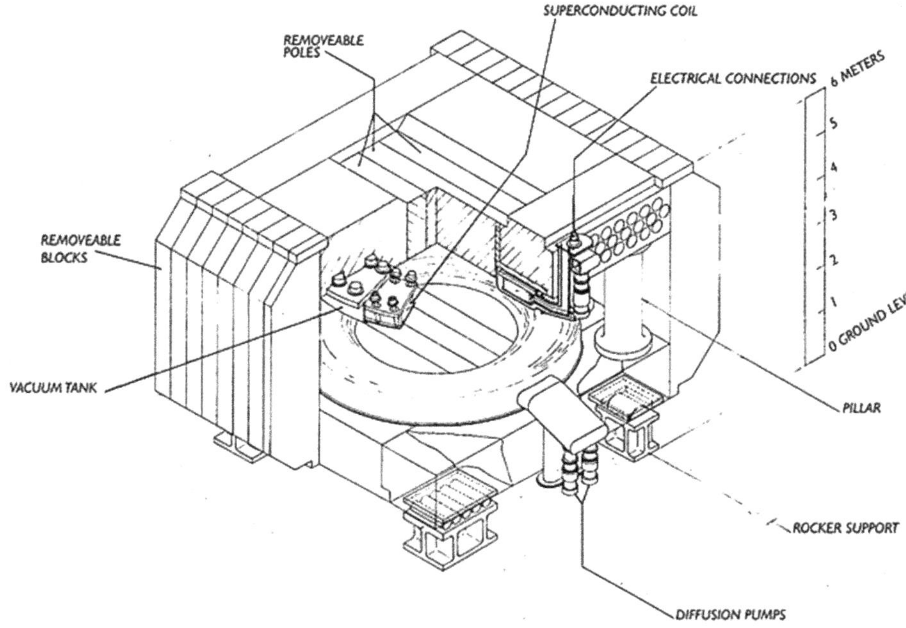

Fig. 3.22. Isometric view of the Omega magnet.

A courageous decision was taken: construct the largest superconducting magnet at the time (Fig. 3.22). The conductors were not to be the conventional ribbons in a bath of liquid helium, but tubes with the superconductor soldered to the surface, cooled by supercritical He at about 5 bar inside, a novel design that was intrinsically safe in case of a quench [Box 4.3]. The coils with an inner diameter of 3 m, producing a field of 1.8 Tesla, were clamped to the yoke. The supports had to withstand a force of about 2000 tons while transferring very little heat from the cold coils to the iron yoke. The whole system with helium compressors and liquefier was designed by a CERN team and realized by industry.

The readout of the spark chambers was another novelty. Tubes used in television cameras provided a stereo view of the chambers. The conversion of their video signal into coordinates useful for the physicists demanded some creative electronics design. Finding tracks from these data and reconstructing particle trajectories without human intervention was another new and difficult task.

Anticipating higher energy beams, increased precision of the momentum measurement was needed. As a first step two large (1.65 m × 3.2 m) drift chambers [Highlight 4.8] were constructed for a precise measurement of particles exiting the magnet downstream. Furthermore, large counter hodoscopes for triggering, a Cherenkov counter for particle identification and an electromagnetic calorimeter completed the spectrometer.

This hybrid solution (spark and drift chambers) worked for a while with beams up to 200 GeV/c, but the need of a fully electronic system without the shortcomings of spark chambers and their readout was becoming urgent. In 1977 a proposal for improving the particle detector system (Omega Prime) was approved [39]. The choice of proportional wire chambers allowed higher beam flux, more refined trigger selection by allowing more time for a decision, better two-track resolution and easier data handling. The use of the proportional wire chambers was also essential for the flexibility required by the various experiments. More than 40,000 signal wires were mounted in the chambers, each wire precisely tensioned to 0.5 Newton. The readout of the signals had to be encoded on the chambers because the conventional use of a twisted pair cable for each chamber wire would have been impossible for this density of detectors. To provide time for a trigger decision a novel signal delay was developed. It was easily adjustable for the specific experimental conditions, could accept any number of hits, was storing the signals in electronic registers advancing in 20 ns steps and read out selectively the information pertaining to the triggered event. It was realized making very extensive use of electronic chips available at the time. The electronics, mounted on cards attached to the chamber frames, was designed and produced by members of the OMEGA group.

These experiments relied also on the variety and quality of the particle beams directed to OMEGA which at the beginning originated from the PS and later from the SPS. The SPS delivered initially 200 GeV/c protons and later up to 450 GeV/c protons, which required displacing the 1500 ton facility from its original position towards the centre of the West hall. Hyperon beams (baryons including at least one strange quark) and beams of sulphur and lead nuclei were delivered in addition to the more conventional beams. Omega received also a RF separated beam, i.e. a momentum selected hadron beam, with two RF cavities accepting or rejecting particles, depending of their time of flight between the cavities, hence on their mass.

With the increasing complexity of the experiments, the user groups, together with the OMEGA staff, developed the required detectors. Apparatus for better particle identification [40], for measurement of high energy photons, high resolution semiconductor detectors [Highlight 5.9] and new devices for triggering were devised.

The theory of strong interactions, QCD [Box 4.2], was tested with photons and hadrons. The study of hadron structure functions led to the very first published

SPS result. The study of high momentum transfer processes confirmed QCD. Experiments searching for glueballs, exotic particles made only of gluons and allowed by QCD, gave intriguing hints on their possible existence.

Motivated by the discoveries of charm and beauty [Box 6.4], OMEGA experiments contributed to the knowledge of the particles containing these quarks. The photon beam led to early observation of charm photo-production.

With the beam of lead nuclei hitting a target of lead, a series of experiments were performed with increasing sophistication, e.g. the use of an array of silicon pixel micro-detectors as main tracking devices to cope with the high multiplicity of tracks — a first for particle physics. It allowed the observation of particles containing up to three strange quarks. Their enhanced production rate, later confirmed by other experiments, was OMEGA's evidence for a new state of matter produced in lead-lead collisions at SPS energies.

This advanced, very flexible and steadily evolving facility operated from 1972 to 1996. It provided the basis for a wide range of experiments, overall 48. It relied not only on a number of leading-edge technologies but also on their successful integration and led to a number of important physics results. For a complete report on OMEGA physics published up to 1996 see [41].

3.8 ISOLDE: Targeting a New Era in Nuclear Physics

Helge Ravn

Since the early days of CERN research was carried out on properties of radioactive nuclei far from stability. Such nuclei play an important role in many areas of science ranging from fundamental nuclear interactions and astrophysics to material science and medical applications. These nuclei were created by irradiating targets with beams from the SC and later from the PS-BOOSTER. After irradiation the targets were transferred to the nuclear chemistry laboratory by the "Student-Running-As-Fast-As-Possible" (SRAFAP) method. There, aqueous phase chemical separation methods were applied to isolate the produced isotopes of interest. Scandinavian groups in particular were very active in studying and mapping the formation cross-sections of the new range of nuclei made available by the high energy protons reactions at CERN.

A very complex mixture of many isotopes of each element is produced in these reactions, next to impossible to study with the techniques available then. Only a breakthrough in new experimental techniques would solve this dilemma. It came about when the nuclear chemists realized that an accelerator technique — electromagnetic isotope separation — could be developed to achieve the desired selection. The combination of this technique with chemistry provided high-purity samples of radioisotopes with a specific mass number A and atomic number Z.

One further obstacle had to be removed, requiring another invention and development. The inherent delay introduced by the "SRAFAP" off-line method limited the studies to nuclei with half-life > 10 minutes. This limitation was overcome by developing an on-line system, in which "chemically intelligent" targets were irradiated in an external proton beam. The chemical element of interest released selectively from the target was allowed to diffuse in a pipeline to the ion source of the mass separator from where mono-isotopic ion-beams emerged. To achieve this for all chemical elements a number of technological challenges had to be overcome, such as replacing the wet radiochemical techniques incompatible with the vacuum and high voltage requirements of an ion source.

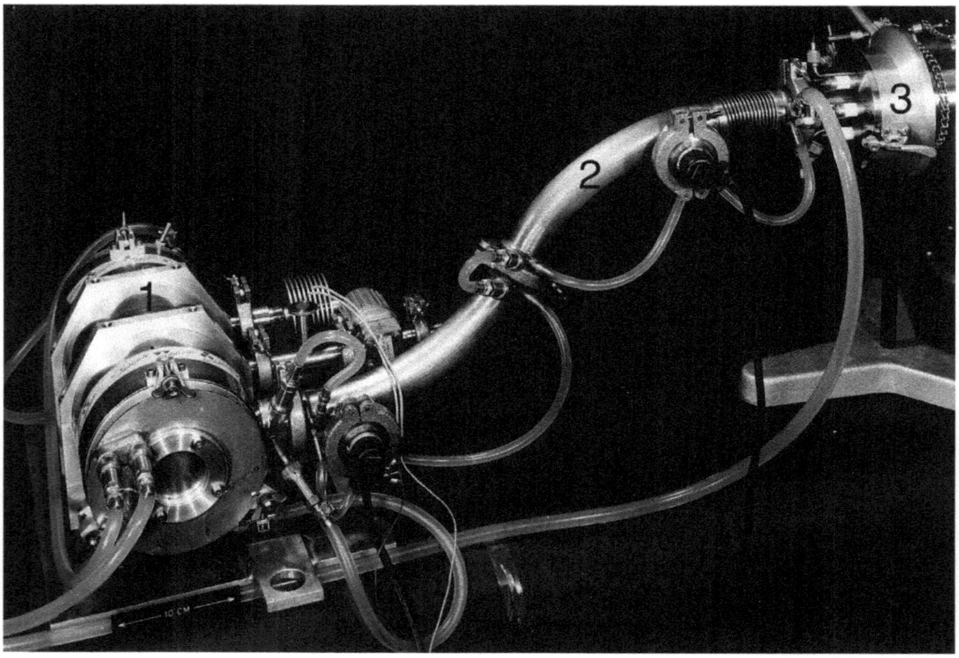

Fig. 3.23. The water cooled target vacuum-chamber (1) with its proton beam entrance window is pumped through the 1 m long vacuum tube (2) leading to the ion-source vacuum chamber (3). The target material (1) is molten lead or bismuth at 850 K. Inside (2) is the heated transfer line that allows the vapour of selectively released mercury isotopes to diffuse into the ion source.

Crucial for success was the understanding of the kinetics of this chemical mass-transport which allowed to separate the radioactive reaction products from the bulk target material and transferring them into the accelerator ion-source. This was solved by a vacuum-distillation process, often at temperatures above 2200 K. In the early system shown in (Fig. 3.23) the various steps in the process can be clearly identified [42]. It was thus that ISOLDE, the Isotope Separator On Line DEvice, was born. This system, invented at CERN by the ISOLDE group and the ISOLDE collaboration, allowed for the first time to continuously produce pure beams, e.g. of all the known mercury isotopes as well as discovering many new ones. Even the alchemist's dream of converting lead into gold was achieved by collecting the short-lived ^{197}Hg isotope from this first molten-metal accelerator-target and letting it decay into the stable gold isotope!

The speed of this separation process is governed by several factors: diffusion of the products out of the target material to its surface; desorption into the vacuum phase; adsorption and desorption steps encountered in collisions with the walls of the transfer-tube leading to the ion source and similar processes on the interior walls of the ion source. To assure the proper operation of the ion source, this process had to be performed in a vacuum of $< 10^{-4}$ Torr which also set the limit to the evaporation of the target matrix and other vapours accompanying the radioactive species at 10^{-8} mol/s [43]. For each element to be produced a separate unit had to be developed with its specific target and container material, temperature environment and ion-source type [44].

During the 50 years of CERN-ISOLDE and the ISOLDE collaboration target and ion-source units have been developed for about 60 of the chemical elements [45] providing more than 2000 individual ion-beams. For a typical example of one of these modern units see Fig. 3.24.

In addition to molten metal targets a number of high temperature — up to 2400 K — solid material targets were developed. With this method isotopes with half-lives as short as a few milliseconds could be studied.

In another development, high temperature chemical reactions inside the target were exploited for the first time to convert refractory elements into volatile chemical compounds, which then could be transported into the ion source. For example, the hafnium isotopes produced in a 2600 K tantalum-foil target can only be released from the Ta surface by letting them react with CF_4 gas continuously added to the target in order to form the volatile HfF_4.

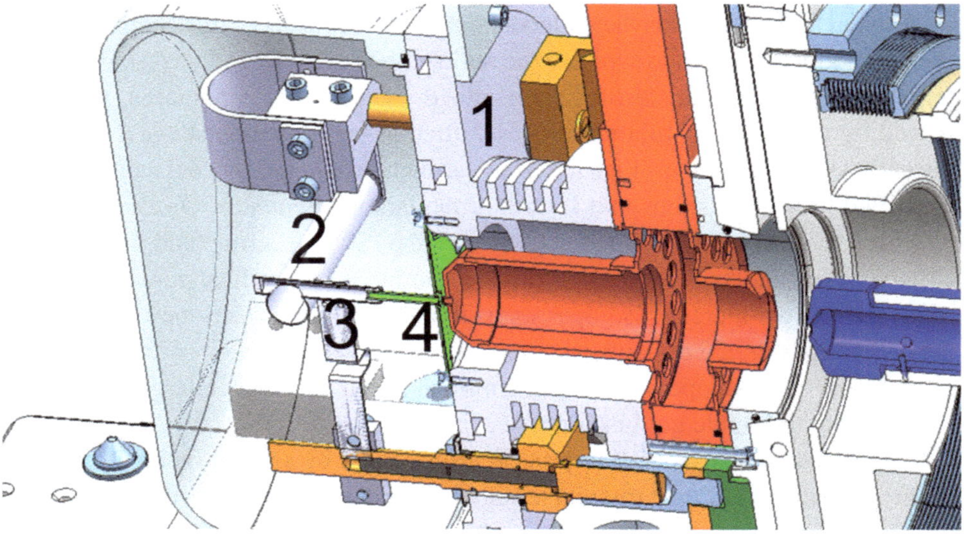

Fig. 3.24. Cut through the 30 by 30 cm universal vacuum chamber (1) containing the integral target, transfer-line and ion-source block which can be connected to the accelerator front end at 60 kV potential. In this example the 20 cm long Tantalum cylinder (2) contains a $^{235}UC_x$ target matrix kept at 2300 K. The released nuclei are conducted via a 5 cm long transfer line (3) to a 4 mm tubular ionizer (4) from which singly charged ions can be extracted into the accelerator to the right. By directing e.g. up to three laser beams opposite to the ion-beam direction a highly selective and efficient, stepwise resonant laser ionization is achieved. This target unit is equipped with a vacuum valve and quick connectors allowing rapid and remote exchange with an industrial robot.

Progress at ISOLDE in the beam production was paralleled by major progress in experimental methods developed to make best use of the radioactive ion-beams. Examples are the experiments based on methods adopted and adapted from atomic physics: optical pumping, laser spectroscopy atomic beam techniques, and mass spectrometry. Remarkably, due to a large degree of synergy between the accelerator and experiment communities, new accelerator techniques were developed based on some of these "imported" experimental methods.

A very successful invention of this type is the stepwise resonant laser ionization developed for many elements by CERN and the ISOLDE collaboration. While first used in the laser spectroscopy experiments in determining nuclear radii, it was adapted to become the primary ion-source as described in Fig. 3.24 [46].

It was the conjunction of these new technologies described that the vision of 1967 could finally be fulfilled [47]. The ISOLDE target and ion-source concept became the heart of today's efficient radioactive ion accelerator REX ISOLDE [48].

3.9 The CERN n_TOF Facility: Catching Neutrons on the Fly

Enrico Chiaveri

High precision neutron cross-section data are of importance for a wide variety of research fields in basic and applied nuclear physics: In nuclear astrophysics data on neutron–nucleus reactions are essential to understand the production of heavy elements in the Universe, which occurs mainly through neutron capture processes during the various phases of stellar evolution. In nuclear technology, renewed interest in nuclear energy production has triggered studies aimed at developing future generation systems that would address safety, proliferation and waste concerns. For these applications the available nuclear data for many nuclides are of insufficient accuracy or even not existing.

Based on these motivations and given that the PS can produce proton pulses of very high intensity, the neutron Time-Of-Flight facility n_TOF has been proposed and constructed at CERN [49, 50], Fig. 3.25. It is based on the insight that such an intense proton bunch extracted from the PS could produce an intense pulse of spallation neutrons, produced in a wide range of energies and a correspondingly large spread in velocities. Thus, the neutron arrival time at a detector, located far downstream from the target, gives the neutron velocity and hence, its kinetic energy. Measuring precisely the neutron-time-of-flight produces a beam of neutrons with excellent energy resolution.

Commissioning and operation started in 2001 with performances ultimately matching design after a substantial optimization of shielding. The PS provides up to 8×10^{12} protons per pulse every 1.2 seconds (or multiples thereof). These proton pulses of 20 GeV/c momentum impinge on a 1.3 ton cylindrical lead target 40 cm in length and 60 cm in diameter producing a bunch of 2×10^{15} neutrons of 6 ns width. The high neutron flux, the low repetition rates and the excellent relative energy resolution, reaching values as low as 3×10^{-4} for 1 eV to 10 KeV neutrons, open new possibilities for high precision cross section measurements from thermal to a few GeV energy on stable and, importantly, radioactive isotopes.

The n_TOF target [51] is cooled by a 1 cm water layer and with a subsequent layer of 4 cm of water or borated water (H_2O + 1.28%H_3BO_3, fraction in mass). Initially fast neutrons are moderated into the desired energy spectrum, which ranges down to thermal energies. The experimental area (EAR1) begins at 182 m from the spallation target and has a length of 7.9 m. Along the evacuated beam line a sweeping magnet (200 cm long, 44 cm gap and 3.6 Tm bending power) deflects and removes the charged particles in the beam. In a typical experiment, a sample is placed in the neutron beam, and the reaction products are detected with

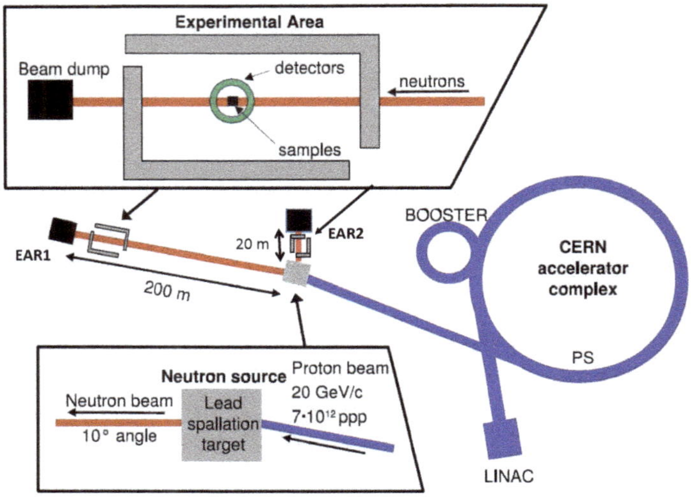

Fig. 3.25. Layout of the n_TOF facility [51].

specialized instruments. The reaction probability is measured as a function of the incident neutron flight time and hence its energy.

To extend the measurements on stable and short-lived isotopes, with very low cross sections or available only in extremely small quantities, the n_TOF Collaboration proposed the construction of a new experimental area (EAR2) at a shorter distance from the spallation target to exploit a much higher neutron flux [52]. It was convenient and advantageous to build the new experimental area on the surface, directly above the pit hosting the spallation target, which is located approximately 20 m below the surface. This layout reduces the time-of-flight between target and detectors by a factor 10 and increases the neutron flux by a factor of around 25 relative to EAR1. It allows measurements of correspondingly smaller samples, of smaller cross sections or in a shorter time. The factor 10 shorter flight time is also a crucial advantage for the study of radioactive substances. The spread in arrival time ΔT of neutrons in an energy interval ΔE and hence the necessary sensitive time of the detectors is reduced by the same factor 10. Reducing the measurement time reduces the dominant background from the decay of the radioactive isotope under study. The total gain in sensitivity of up to a factor of 250 relative to EAR1 means that isotopes with half-lives as short as a few months can now be studied [53].

The CERN n_TOF facility is worldwide unique due to its very wide energy spectrum and intensity of neutrons. It is home of a rich and in many ways unique scientific programme [52].

References

1. E.D. Courant, M.S. Livingston and H.S. Snyder, The strong-focusing synchrotron—a new high energy accelerator, *Phys. Rev.* **88**, 1190 (1952).
2. E. Regenstreif (ed.), *The CERN Proton Synchrotron, Vol. 1*, CERN-1959-029 (CERN, Geneva, 1959). http://dx.doi.org/10.5170/CERN-1959-029.
3. E. Regenstreif (ed.), *The CERN Proton Synchrotron, Vol. 2*, CERN-1960-026 (CERN, Geneva, 1960). http://dx.doi.org/10.5170/CERN-1960-026.
4. E. Regenstreif (ed.), *The CERN Proton Synchrotron, Vol. 3*, CERN-1962-003 (CERN, Geneva, 1962). http://dx.doi.org/10.5170/CERN-1962-003.
5. S. Gilardoni and D. Manglunki (eds.), *Fifty Years of the CERN Proton Synchrotron, Vol. I*, CERN-2011-004 (CERN, Geneva, 2011). http://dx.doi.org/10.5170/CERN-2011-005.
6. S. Gilardoni and D. Manglunki (eds.), *Fifty Years of the CERN Proton Synchrotron, Vol. II*, CERN-2013-005 (CERN, Geneva, 2013). http://dx.doi.org/10.5170/CERN-2013-005.
7. L. van Hove, M. Jacob, Highlights of 25 years of physics at CERN, *Phys. Rep.* **62**, 1 (1980).
8. S. Weinberg, Unified theory of elementary-particle interactions, *Sci. American*, July 1974.
9. P. Musset and J.P. Vialle, Neutrino physics with Gargamelle, *Phys. Rep.* **39**, 1 (1978).
10. R.K. Bock, Data processing for bubble chambers, *Nucl. Phys.* **36B**, 229 (1994).
11. M. Gell-Mann, *Phys.Lett.* **8**, 214 (1964) and
    G. Zweig, An SU(3) Model for Strong Interaction Symmetry and its Breaking,
    CERN preprints 8182/Th-401, and 8419/Th-412. http://cds.cern.ch/record/570209?ln=en.
12. R.P. Schutt (ed.), *Bubble and Spark Chambers* (Academic Press, New York, 1967);
    G. MacLeod and B. Maglic (ed.), *Informal Meeting on Film-less Spark-Chamber Techniques and Associated Computer Use*, CERN-1964-030 (CERN, Geneva, 1964).
    http://dx.doi.org/10.5170/CERN-1964-030.
13. M. Giovanozzi and C Steinbach, p. 91 in *Fifty Years of the CERN Proton Synchrotron, Vol. I*, S. Gilardoni and D. Manglunki (eds.), CERN-2011-004 (CERN, Geneva, 2011).
    http://dx.doi.org/10.5170/CERN-2011-004.91.
14. D. Fiander, Hardware for a full aperture kicker system for the CPS, *Proc.1971 Particle Accelerator Conference*, Chicago, USA (1971) 1022.
15. J. J. Bleeker, C. Germain, M. Thivent and R. Tinguely, Development of an electrostatic septum for the slow ejection of the CPS, *Proc. 8th Intern. Conf. on High Energy Accel.*, Switzerland, Geneva (1971) 113.
16. J. Borbugh, M. Hourican and M. Thivent, Consolidation Project of the electro-static Septa in the CERN PS Ring, *Proc. IEEE Particle Accel. Conf.*, USA, Chicago IL (2001) 1541.
17. C. Bertone *et al.*, Studies and implementation of the PS dummy septum to mitigate irradiation of magnetic septum in straight section 16, *Report CERN-ACC-2014-0043* (2014); http://cds.cern.ch/record/1697680?ln=en.
18. R. Garoby, RF gymnastics, pp. 69-89 in *Fifty Years of the CERN Proton Synchrotron, Vol. I*, S. Gilardoni and D. Manglunki (eds.), CERN-2011-004 (CERN, Geneva, 2011).
    http://dx.doi.org/10.5170/CERN-2011-004.69.
19. H.C. Grassmann, R. Jankovsky and W. Pirkl, New RF System for the 28 GeV Proton Synchrotron at CERN, *Siemens Review,* **XLIV** (4), 164-170 (1977).
20. D.G. Grier, E. Jensen, R. Losito, A.K. Mitra, The PS 80 MHz cavities, *Proc.6th Europ. Part. Acc. Conf. (EPAC'98),* Stockholm, Sweden (1998) pp. 1773-1775.

21. H. Damerau, A. Findlay, S. Gilardoni and S. Hancock, RF manipulations for higher brightness LHC-type beams. *Proc. Int. Part. Accel. Conf. (IPAC'13)*, Shanghai, China (2013) 2600-2602.
22. K. Hanke, M. Chanel and K. Schindl, The Booster hits 40, *CERN Courier* July/August 2012.
23. M. Chanel, The Proton Synchrotron Booster (PSB), in *Fifty Years of the CERN Proton Synchrotron, Vol. II*, S. Gilardoni and D. Manglunki (eds.), CERN-2013-005 (CERN, Geneva, 2013). http://dx.doi.org/ 10.5170/CERN-2013-005.
24. A.M. Ašner, G. Brianti, M. Giesch and K.D. Lohmann, The PS Booster Main Bending Magnets and Quadrupole Lenses, *Proc.3rd Int. Conf. on Magnet Technology,* Hamburg, Germany (1970) 418.
25. H. Fiebiger, Y. George, B. Godenzi, G. Legras and F.V. Völker, Thyristor static compensator for the CERN intermediate booster accelerator, *ACEC Rev.* **3**, 2 (1980).
26. A. Brückner, Kicking Protons, Fast and Cheap, *Proc. 4th IEEE Part. Accel. Conf.*, Chicago, USA (1971) 976.
27. M. Benedikt, P. Collier, V. Mertens, J. Poole, K. Schindl (eds.), *LHC Design Report, Vol. 3: the LHC Injector Chain,* edited CERN-2004-003 (CERN, Geneva, 2004). http://dx.doi.org/10.5170/CERN-2004-003-V-3.
28. F. Pedersen and F. Sacherer, Theory and performance of the longitudinal active damping system of the CERN PS Booster, *Proc.6th IEEE Part. Accel. Conf.,* Washington DC, USA, (1975) 1398.
29. K. Hanke *et al.*, Status of the Upgrade of the CERN PS Booster, *Proc.4th Int. Part. Accel. Conf.*, Shanghai, China (2013) 3939.
30. O. Bayard (ed.), *La Nouvelle Alimentation de l'Aimant du Synchrotron à Protons du CERN, Vol. 1: Description Générale*, CERN-1971-010 (CERN, Geneva, 1971). http://dx.doi.org/10.5170/CERN-1971-010.
31. C. Fahrni, A. Rufer, F. Bordry and JP. Burnet, A multilevel power converter with integrated storage for particle accelerators, in *Proc. Power Conversion Conference (PCC '07)*, Nagoya, Japan (2007) 1480.
32. C. Fahrni, Principe d'alimentation par convertisseurs multiniveaux à stockage intégré – Application aux accélérateurs de particules, CERN-Thesis-2008-154 (2008); http://cds.cern.ch/record/1375850?ln=en.
33. F. Boattini, JP. Burnet, G. Skawinski, POPS: The 60MW power converter for the PS accelerator: Control strategy and performances, *Proc.European Conference on Power Electronics and Applications, (EPE 2015),* Geneva, Switzerland (2015), *Report CERN-ACC-2015-0098* (2015); http://cds.cern.ch/record/2056733?ln=en.
34. S. van der Meer (ed.), *A Directive Device for Charged Particles and its Use in an Enhanced Neutrino Beam*, CERN-1961-007 (CERN, Geneva, 1961). http://dx.doi.org/10.5170/CERN-1961-007.
35. M. Giersch *et al.*, Status of magnetic horn and neutrino beam, *Nucl. Instr. & Meth.* **20**, 58- 65 (1963).
36. K. Elsener, E. Gschwendtner and M. Meddahi, Right on target: CNGS gets off an excellent start, CERN Courier, November 2006, http://cerncourier.com/cws/article/cern/29753.
37. P.M. Blackall, G.R. MacLeod and P. Zanella, The analysis of spark chamber pictures using a raster scan technique, *IEEE Trans. Nucl. Sci.* **NS13**, 34 (1966).
38. W.F. Baker, The OMEGA project: Proposal for a large magnet and spark chamber system, NP Internal Report 68-11 (1968); https://cds.**cern**.ch/record/299898.

39. W. Beusch, OMEGA PRIME: a project of improving the omega particle spectrometer, CERN/SPSC/77-70 (1977); cds.**cern**.ch/record/345021?ln=en.
40. R.J. Apsimon *et al.*, A ring imaging Cherenkov detector for the CERN Omega spectrometer – the design and recent performance, *Nucl. Instr. & Meth. A* **248**, 76-85 (1986).
41. M. Jacob and E. Quercigh (eds.), *Symposium on the CERN OMEGA Spectrometer*, CERN-1997-002 (CERN, Geneva, 1997). http://dx.doi.org/10.5170/CERN-1997-002.
42. A. Kjelberg and G. Rudstam (eds.), *The ISOLDE Isotope Separator Facility at CERN*, CERN-1970-003 (CERN, Geneva, 1970). http://dx.doi.org/10.5170/CERN-1970-003.
43. H.L. Ravn, Experiments with intense secondary beams of radioactive ions, *Phys. Rep.* **54**, 201 (1979).
44. H.L. Ravn and B. Allardyce, On-line mass spectrometers, p. 363 in *Treatise on Heavy-Ion Science* **8**, ed. D. Allan Bromley (Plenum Press, New York, 1989).
45. The ISOLDE Radioactive Ion Beam Facility; http://isolde.web.cern.ch.
46. V.I. Mishin *et al.*, A chemically selective laser ion source for on-line mass separation, *Nucl. Instr. & Meth. B* **73**, 550 (1993).
47. J. Bondorf, A short summary of the concluding discussion, p. 681 in *Proc. Int. Symposium on Why and How Should We Investigate Nuclides Far Off the Stability Line, Lysekil,* (1966), eds. W. Forsling, C.J. Herrlander and H. Ryde, *Ark. Fys.* **36** (1967).
48. REX ISOLDE, http://rex-isolde.web.cern.ch/
49. http://public.web.cern.ch/public/en/research/n_TOF-en.html.
50. C. Rubbia *et al.*, A high resolution spallation driven facility at the CERN-PS to measure neutron cross sections in the interval from 1 eV to 250 MeV, CERN/LHC/98-002-EET (1998); http://cds.cern.ch/record/357112/files/lhc-98-002.pdf.
51. C. Guerrero *et al.*, Performance of the neutron time-of-flight facility n TOF at CERN, *Eur. Phys. J. A* **49**, 27 (2013).
52. C. Weiss, E. Chiaveri, S. Girod, V. Vlachoudis and the n_TOF Collaboration, The new vertical neutron beam line at the CERN n_TOF Facility - design and outlook on the performance, *Nucl. Instr. & Meth. A* **799**, 90-98 (2015).
53. E. Chiaveri *et al.*, Proposal for n TOF Experimental Area 2 (EAR-2), CERN-INTC-2012-029, INTC-O-015 (2012); http://cds.cern.ch/record/1411635.

Chapter 4

The Intersecting Storage Rings (ISR): The First Hadron Collider

Christian Fabjan and Kurt Hübner

4.1 Introduction

In 1956, with the CERN Proton Synchrotron (CPS) still under construction, a group was formed at CERN to study new accelerators reaching higher energy and/or intensity. This initiative was to assure CERNs future development and strengthen its international standing [1].

The CPS was brought into operation in 1959. It worked very reliably, delivering protons at an energy of 25 GeV. This provided one more reason for the study group to propose adding two *storage rings* tangential to each other and to the CPS. Protons would circulate in opposite directions, stored in the rings for many hours, and would collide at one point providing collisions at a centre-of-mass energy inaccessible with conventional synchrotrons [Box 4.1]. It would require a synchrotron of 1300 GeV to reach the same centre-of-mass energy as these Intersecting Storage Rings (ISR). Clearly, this new technique offered a tremendous leap into new territory for particle physics [2].

By 1962 this idea was taken a step further by proposing concentric rings intersecting at eight points (Fig. 4.1), substantially reducing the foot-print of the facility and allowing a number of experimental groups to work independently and simultaneously.

Remarkably, the approach was driven not so much by the potential users, the particle physicists, but mainly by accelerator physicists and engineers who enthusiastically advocated the new technology. Having had leading positions in the CPS construction team, they brought their experience and expertise to the study group. The particle physicists were occupied with mastering experimentation at the CPS, and the few who reflected on the future rather thought of an extrapolation of the techniques they had been developing there.

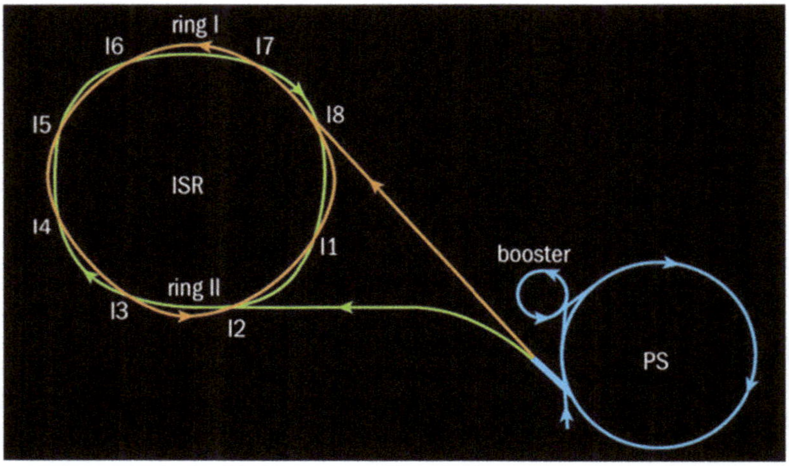

Fig. 4.1. Layout of the Intersecting Storage Rings with the PS injector.

The majority of particle physicists therefore lent support to an alternative to the ISR — a large proton synchrotron of much higher energy than the CPS. This project was preferred as it could provide a large variety of intense secondary beams, and would allow continuing the familiar style of experimentation. Eventually, a proton energy of 300 GeV was agreed upon by these scientists.

After some hesitation and considerable (at times heated) discussions between European particle physicists and CERN management, the latter decided in 1964 to opt for the ISR and to defer the more expensive 300 GeV accelerator.[a] The then Director-General, Victor Weisskopf, showed the vision and persuasiveness to convince both the particle physics community and the CERN Council, which approved the ISR project at the end of 1965 with Kjell Johnsen as project leader. It was an audacious bet to maintain CERN's competitiveness at relatively low cost but at the price of a yet unproven concept. The ISR project is an interesting example how reticent research is driven into a new domain by innovative technology.

The Machine

Already during the design phase two topics were identified as requiring major technological progress: (i) the average residual gas pressure in the vacuum chamber in which the proton beams circulated would have to be less than 10^{-7} Pa

[a]The 300 GeV study was intensively resumed in 1968 and led to the construction of the SPS operating from 1976 onwards (Chapter 5).

(about 10^{-12} times atmospheric pressure) to avoid that the beams would lose protons too quickly or increase their transverse dimensions due to interactions with residual gas. Four orders of magnitude had to be gained relative to vacuum achieved at the CPS. The required Ultra High Vacuum (UHV) technology would have to be developed with European industry. (ii) For the necessarily high interaction rate of the colliding beams, very intense beams had to be accumulated. This was to be obtained by stacking injected proton beam pulses from the CPS by a radio-frequency (RF) system imparting a small, step-wise acceleration to each injected pulse. This technique, first tested in the US, was vital for the new facility and would have to be mastered by CERN. To this end, a 2 MeV electron model of the ISR (CESAR) was constructed and successfully operated in 1964, proving the feasibility of technologies and at the same time qualifying the vacuum industry in a fine example of efficient technology transfer. The success of CESAR provided additional momentum to the project as it progressed through the approval phase.

In January 1971, barely five years after approval and slightly ahead of schedule, beams were injected and stored in both rings, producing the highest energy proton-proton collisions ever achieved on earth. This was duly celebrated at the inauguration ceremony in October (Fig. 4.2).

Fig. 4.2. ISR inauguration: The project leader, K. Johnsen, proudly hands the ISR key to the president of the CERN Council, E. Amaldi. Also on the platform (from left to right): former Director-General W. Weisskopf, the French Secretary of State M. Antonioz, Director-General W. Jentschke and W. Heisenberg, far right.

Colliders **Box 4.1**

Until the 1960s, the interaction of high energy particles with matter was studied in the laboratory by directing beams from accelerators onto targets, in which secondary particles are produced. The energy E_{cm} available in the centre-of-mass system of two colliding particles, one of which is at rest, increases with the square root of the energy E of the incident particle. It was recognized in the 1930s that colliding particles head-on would allow to fully exploit the kinetic energy of both particles, yielding $E_{cm} = 2E$ for identical particles, but it was thought to be too difficult, as the intensity of available beams was low. The principle was however successfully demonstrated in 1961 for e^+e^- collisions in a single storage ring where a bunch of e^+ and one of e^- circulated in opposite directions to meet twice per revolution, to be the first *e^+e^- circular collider*.

A *storage ring* consists of a synchrotron lattice [Box 2.1] and a beam tube under high vacuum to minimize losses and achieve a long beam lifetime. Increasing the number of circulating bunches per ring k_b leads to a proportionally higher collision rate, or luminosity [Box 6.1], as well as an increase in the number of experiments (up to 2 k_b) that can be accommodated. Sufficiently dense and intense beams can be accumulated by repeated injection. The synchrotron radiation emitted by the e^+ and e^- in the bends of the lattice reduces the spread in energy and the amplitude of transverse oscillations, which is important when merging the injected beam with that already stored. But the RF power needed to compensate losses due to synchrotron radiation, proportional to $(E/m_0{\cdot}c^2)^4/\rho$, limits the top energy E of e^+e^- circular colliders, even with a large bending radius ρ, because $e^{+/-}$ mass m_0 is small. For very high energy collisions ($E_{cm} \gtrsim 0.4$ TeV) *linear e^+e^- colliders* are favoured. These consist of opposing linacs with the beams tightly focused at the collision point to get a sufficient rate of interaction.

Proton-proton and p̄-p colliders, operate according to the same principle as e^+e^- colliders, with counter-rotating beams. The maximum beam energy is proportional to $\rho{\cdot}B$, where B is the magnetic field, so the design strives for large ρ and B. Two separate rings with opposite vertical magnetic field are required for *p-p colliders,* which provide collisions at intersection points. The magnets can be separate (ISR) or in common, each magnet having two bores (LHC). High beam intensity and/or dense beams are necessary to achieve a useful collision rate (luminosity). The accumulation mechanism of e^+e^- colliders is ineffective as synchrotron radiation is far less for heavy protons than for electrons. Beam intensity is built up quickly, either by injecting pulses side-by-side (ISR) or by sequentially stacking high-density bunches (LHC). This is very demanding for the injector system. If the required energy is higher than the injection energy the beam can be slowly accelerated in the storage ring.

The *p-p̄ collider* offers the advantage that counter-rotating beams can circulate in the same vacuum chamber in a single ring, as in e^+e^- colliders, provided the beam is correctly grouped in bunches. The price to pay is that the production of anti-protons requires an elaborate injector chain. A low-density p̄-beam is generated by firing a powerful primary p-beam onto a target. This is then accumulated and compressed by many orders of magnitude using stochastic cooling [Box 6.2]. Nevertheless, the number of antiprotons available remains limited, which impacts on p-p̄ luminosity.

Unsurprisingly, operation of this totally new machine revealed further technological challenges on the way to reaching, and later exceeding, the design specification, at the pressing request of the experimenters. It was the start of a more than decade-long close and fruitful collaboration between them and the ISR staff [3, 4].

The figure of merit of a particle collider is its luminosity L, which determines the collision rate of the beams [Box 6.1]. It is proportional to the product of the two beam currents divided by the effective height (h_{eff}) in the crossing point. All three parameters change as a function of storage time of the beams and are strongly affected by the quality of the vacuum.

The most pressing issue was the improvement of the UHV system to both increase beam lifetime and reduce the background to the experiments due to lost particles. A vigorous programme was launched to eliminate pressure bumps resulting from the gas release by ions impinging on the vacuum chambers and creating an uncontrolled pressure increase. These ions were created by the ionization of the residual gas traversed by the beam particles and were accelerated by the potential difference between beam and vacuum chamber. These pressure bumps were fought by a series of improvements eventually resulting in a totally upgraded vacuum system: extensive installation of titanium sublimation pumps in addition to the sputter-ion pumps, increase of the bake-out temperature to 300°C and glow-discharge cleaning of the vacuum components. During these upgrade interventions clearing electrodes were also systematically installed to remove the electrons trapped in the beam potential, causing beam instabilities. Ultimately, an average pressure of 4×10^{-10} Pa was reached, a unique achievement in such large a system (two rings of 943 m circumference each) resulting in a useful beam lifetime of up to 50 hours. Even more stringent were the requirements for the vacuum in the interaction regions, where a pressure lower than 10^{-10} Pa was required for acceptable background conditions. At the end of the 1960s it was thought that such a low pressure could only be obtained by condensation cryopumping [Highlight 4.2]. Such low pressures also required the development of vacuum gauges of novel design to measure down to the 10^{-11} Pa range [Highlight 4.3].

Effort was next concentrated on stable operation with high beam currents. This required: (i) using feedback to stabilize the incoming beam; (ii) a very low-noise RF-system for stacking; (iii) control of the electromagnetic forces due to beam current by counteracting them with intensity-dependent adjustment of the magnetic guide field during stacking; and (iv) using other feedback systems to counteract the electromagnetic interaction between the beam and the vacuum chamber.

Continuous improvement of beam diagnostics helped in the quest for ever increasing beam currents which eventually reached routinely 40 A, with a record of 57 A, corresponding to 1.14×10^{15} protons.

The ultimate increase in luminosity was achieved by further decreasing h_{eff} through stronger focusing of the beams at one of the interaction points with a "high luminosity insertion". It was initially implemented with focusing magnets having copper coils and later with more powerful superconducting magnets [Highlight 4.4]. These superconducting magnets, operated at 4.5 K, were installed in cryostats which were fed with liquid helium coolant via novel long screened and flexible coaxial transfer lines. Maximum use was made of the cooling potential of the He vapour from the baths to cool current leads, thermal shields and the lines [Highlight 4.5]. This technological advance was very valuable for superconducting equipment and is now extensively used at the LHC and worldwide.

For the precise determination of luminosity a new technique was invented to accurately measure h_{eff}, the so called "Van der Meer method" [Highlight 4.6], and the relative precision of the beam current measurement was pushed to nearly 10^{-8} with novel current transformers operating at up to 60 A.

During the final years of ISR operation the beam energy was pushed to a maximum of 31.4 GeV, beyond the 26.5 GeV provided by the CPS by the novel acceleration technique of RF phase displacement. It is a tribute to the flexibility of the combined CPS-ISR operation that besides protons the ISR would later store other particles, enabling the study of d–d, p–d, α–α, α–p, and p–antiproton collisions.

However, the single most outstanding ISR legacy to accelerator technology was the development and experimental proof of "Stochastic Beam Cooling", a technique to squeeze the particle beams and hence increase the luminosity. It takes advantage of the so-called Schottky noise, a statistical signal generated by the finite number of randomly distributed particles in a beam. The theory of Stochastic Cooling was initially developed by S. van der Meer in 1968, but appeared technologically too far-fetched. In 1972 with the advent of highly sensitive spectrum analysers it became possible for the first time to observe Schottky noise on the ISR beams. This observation motivated S. van der Meer to publish his concept of "Stochastic cooling" and led to its experimental proof in the ISR. It launched a development of what would later lead to the attribution of a Nobel Prize (Chapter 6 and Box 6.2). The observation of Schottky noise allowed also a revolutionary leap in beam diagnostic technology, resulting in a hitherto unknown means to non-intrusively monitor in real time many parameters of the beam. It has become a standard method of beam monitoring and diagnostics.

The ISR was a unique tool for particle physics which surpassed many design parameters, such as by a factor 35 the design luminosity, due to perseverant development of leading-edge technologies by a very devoted and stable staff complement. Most importantly, the ISR served as a test bed and laid the foundation for future Nobel-Prize winning CERN facilities, the Proton–Antiproton collider and the LHC.

Experimental Programme

While the very intense development phase for the ISR machine extended over many years, the committee guiding the ISR experimental programme (ISRC) started its work relatively late, just two years before the collider start-up planned for mid-1971. Two major lines of experimental programmes emerged: "survey" experiments to understand known physics in the new energy regime and "discovery" experiments searching for the Unknown.

In proton–proton collisions particles interact predominantly through the strong or hadronic interaction. In the late 1960s hadronic physics was couched in terms of phenomenological descriptions, lacking a deeper, fundamental understanding and providing little — sometimes even erroneous — guidance for experimental research. The elements of today's physics understanding, the Standard Model (SM) [Box 6.4] were just starting to take shape. The incipient revolution that was to establish the SM was paralleled by a revolution in experimentation. In 1968, Georges Charpak (Nobel Prize 1992) and collaborators had demonstrated the concept of a new particle position detector, the Multiwire Proportional Chamber (MWPC) [Highlight 4.8], propelling the community with a stroke of genius into the digital age. Nor should the sociological factor be forgotten: small groups, beam exposures of a few days to a few weeks, quick and easy access to the experimental apparatus, characterized the style of experimentation of the time.

The turmoil provoked by three simultaneous uprisings — the emerging SM confronting limited physics understanding, new tools sweeping away old experimental methods and a collaboration sociology struggling to adapt to new experimental imperatives — put its stamp on the early research programme. Particle physics was at the dawn of a "New Age". Experimentation at the ISR contributed to the "New Enlightenment" [5].

The first years of experimentation, 1971 to 1974

These brought a rich harvest of physics surprises. Among the lasting contributions were the startling observation of the rising of the total-cross section with the centre-of-mass energy, related to the effective size of the proton, and the measurements of elastic scattering, which leaves the protons intact [6]. It required to position particle detectors to within millimetres of the circulating proton beams, a feat accomplished with ingenious technology and excellent collaboration between the machine and experiment staff [Highlight 4.7]. The most sensational early discovery, however, was the observation of energetic particles frequently produced at very large angles relative to the direction of the proton beams [7]. It revealed also in the strong interactions the point-like constituents of the proton, as previously observed with the electromagnetic probe [8]. It confirmed the internal structure of the proton, containing the whimsically called "quarks". That protons contain quarks was the emerging consensus. Yet all experimental attempts to detect them as free particles failed, a major ISR legacy: "confinement" was recognized to be a fundamental, unique property of the strong interaction, profoundly shaping its understanding.

The Split-Field Magnet (SFM) was the first general-purpose ISR experimental facility. It was proposed in 1969 by Jack Steinberger, Nobel Prize 1988, as the strategy for exploring *terra incognita* at the ISR. Audaciously, it bet its existence on the novel MWPC-detector technology, invented just one year earlier. Within five years the facility was built and instrumented with 50,000 MWPC detector channels — an astronomically large number at the time. The simultaneous revolution in the electronics industry was a godsend: The invention of integrated electronic circuits (ICs) provided a cost-effective way of equipping the detector channels with signal processing electronics. The SFM, however, was conceived with the physics prejudices of the late 1960s: hadronic physics phenomena would reveal themselves in the direction about the incident particle beams, where most of the particles are produced. It was not optimal for the unexpected "real action", the new emerging physics with signatures predominantly at large angles.

While the production of energetic particles at large angles at surprisingly high rates was one of the early ISR physics "sensations", it was an equally unexpected, ferocious background to other new physics phenomena. It prevented ISR experimenters from discovering the J/ψ due to their limited experimental set ups. This particle was observed simultaneously at two American accelerator laboratories in 1974. It implied the existence of a further, fourth quark, the so-called charm quark, a crucial building block of the SM.

1974–1977: Learning the lessons

The discovery of the J/ψ brought sobering soul-searching to the ISR teams and painfully highlighted the lack of an experimental facility optimized for exploration of the new physics landscape [9]. Such a facility would be centred on a new major magnet. Several groups were studying a facility based on a superconducting solenoid, while another team explored a toroidal geometry. A working group, constituted by the ISR Experiments Committee, ISRC, received the remit to motivate and conceptualize a possible new magnetic facility. With exemplary speed — January to March 1976 — the working group documented the physics case and explored magnets and instrumentation, but even after extending discussions until August shied away from making a recommendation as to the magnet topology. A workshop in the autumn also failed to choose between solenoid and toroid, leaving it to the ISRC to clearly and decisively motivate its preference for a superconducting solenoid with large, openings in the return yoke for detector instrumentation. The merits of the toroidal geometry were recognized but considered too unproven for rapid realization: it would be another 30 years before a major toroidal magnet would be built for proton–proton collider physics — the ATLAS Muon Spectrometer Toroid [Highlight 8.12]. Finally, the CERN management also rejected the ISRC proposal of the large superconducting solenoid as being too costly and taking too long to build.

This working group nevertheless had a profound influence on CERN's research agenda. It provided an assessment of state-of-the-art collider experimentation and technologies. Many members of the group would use their work to shape the UA1 and UA2 facilities [Highlights 6.5 and 6.6] at the Proton–Antiproton collider, which were proposed at about the same time.

At the 1976 autumn workshop an innovative solenoidal magnet had been proposed, the Open Axial Field Magnet [Highlight 4.11]. Following the refusal to support a large superconducting facility this more modest spectrometer magnet was to become the centrepiece for the Axial Field Spectrometer (AFS). The AFS incorporated several state-of-the-art detector technologies. It was the first facility at a hadron collider providing very large solid angle coverage for the momentum and energy measurement of particles. It was designed to operate at collisions rates exceeding one million collisions per second, a totally new regime of experimentation. It required the development of a — for hadron colliders novel — "drift detector" [Highlight 4.8], which could register millions of particles per second and measure their trajectories with bubble-chamber type detail and with an accuracy of 0.1 mm. The energy measurement of all the particles was a further essential, new requirement. The technique is called "Calorimetry" [Highlight 4.10]: the particles are absorbed in specially selected dense materials, in which detectors are embedded which measure the cascade of particles resulting from the absorption process [Box 6.3]. The physics research demanded energy measurements at the percent level, prompting major R&D programmes. One technique used the ionization in liquid argon produced by the particles in the cascade, a concept, which later would be employed in many other experiments. Electrons and muons frequently reveal new physics. For the identification of electrons the effect of "Transition Radiation" [Highlight 4.9] was used for the first time and developed into a practical instrument: Ultra-relativistic particles, e.g. energetic electrons, produce soft X-rays in the passage through "radiators", a few hundred 20 micron-thin lithium or polyethylene foils, spaced some 200 microns apart. A special form of MWPCs was developed to detect these X-rays, covering many square metres.

These more evolved and novel experimental approaches brought a new level of complexity and longer lead-times from proposal to data-taking: the fruit of these efforts came a few years too late to make the potentially grand impact that was expected from, and deserved by the ISR. Despite this somewhat critical assessment of the experimental situation, a wealth of significant results were obtained, all of which contributed to shaping our understanding, as documented e.g. in [10].

Final years of ISR operation, 1977 to 1983

The experimenters focused on a variety of rare and energetic phenomena: leptons, photons, charmed particles, jets and search for new particles with masses beyond 30 proton masses. This strategy was vindicated by the discovery of the Upsilon (Υ) particle, about ten times more massive than the proton, albeit at the U.S. Fermi National Laboratory in 1977. This was a further crucial building block of the SM and yet another cruel blow for the ISR, especially as the first evidence for the Υ at the ISR was obtained just five months later by the R806 collaboration [11].

The evolving physics understanding put a stamp on the research programme and consequently on machine operation: go for the highest possible collider energy and collision rates. The collision energy was pushed to 63 GeV, and record-breaking collision rates in excess of one million per second were achieved — a successful rehearsal for the LHC. This operation allowed a multifaceted programme with emphasis on understanding the hadronic interactions through precise tests of the emerging theory, Quantum Chromodynamics, QCD [10] [Box 4.2]. One early major support of QCD was the discovery by the AFS collaboration (Fig. 4.3) of the prompt energetic photon production in p–p collisions, the QCD analogue of the electromagnetic Compton scattering. These photons are predominantly produced in the scatter of a quark on a gluon, providing evidence for the existence of gluons. The observation of "jets", the indirect manifestation of quarks and gluons, concurrent with the observations at the Proton–Antiproton collider, was further strong evidence for QCD [10, 11].

The ISR was a superb machine, a test bed for ground-breaking accelerator and detector technologies. It was a powerful and well-performing collider, but the detectors it would have deserved and required for discoveries within its reach came too late. One reason was that its experimental programme, starting in 1971, was in competition with that of the SPS, approved in the same year, resulting in tight resources, but the fact is that with regard to the experimental programme the physics community was simply not prepared for the enormous jump in energy and physics potential that the ISR provided.

Fig. 4.3. Partial view of Axial Field Spectrometer at I8. The vertical uranium/scintillator hadron calorimeter (just left of centre) is retracted to give access to the cylindrical central drift chamber. The yellow iron structure served as a filter to identify muons, with MWPCs and the array of Cherenkov counters to the right.

By the end of 1983 the CERN Proton-Antiproton collider had produced the first W- and Z- bosons, carriers of the electroweak interaction, a milestone towards the completion of the SM. The construction of LEP had started, which would vindicate the SM with near perfection. For the CERN management it was time to turn a page. At the closure ceremony of the ISR in June 1984, the former Director-General and staunch ISR supporter Viktor Weisskopf said he "had come to praise the ISR…The really important thing about the ISR is its success as an instrument, because *that fact did change the landscape of high energy physics.*"

Quantum chromodynamics (QCD) **Box 4.2**

In 1957 electron-proton scattering experiments showed that the proton has a finite size (about 10^{-13} cm radius), and in 1968 that it contains point-like scatterers, later identified as *quarks*. The quark model brought a dramatic simplification, explaining all known hadrons as bound states of these constituents: 3 quarks for baryons, a quark and antiquark for mesons. The spectroscopy of strange, charmed and beauty particles, and the discovery of the top quark established the 3 "families" [Box 6.4].

Colour, a new quantum number proper to quarks, was proposed in 1964 and *QCD*, starting from a model with 8 *gluons* as force carriers and a strong coupling constant α_S, was progressively developed to become the modern theory of the strong interaction. Experimental evidence for the gluon came in 1979. However, there remained a paradox: quarks and gluons appeared to be strictly *confined* within hadrons by a strong force, as if a "spring" prevented them being pulled apart until it broke, producing a jet of hadrons; but when struck by hard photons or W (i.e. lepton or neutrino deep inelastic scattering) quarks inside hadrons appeared to be free and point-like, not interacting with each other, but sharing the hadron momentum, a property responsible for *scaling*, i.e. having collision properties dependent only on dimensionless kinematical parameters, but not on an absolute energy scale.

The strong self-interaction (S-I) of gluons is the key to this mystery. In 1973 it was demonstrated that at high energies, i.e. at short distances between quarks, the colour coupling constant α_S tends to zero, referred to as *asymptotic freedom*. At long distance or low energy, the gluon S-I makes α_S grow, leading to *confinement*. Thus α_S "runs", i.e. changes with the energy scale or resolving power [Box 5.1].

QCD offers another amazing fact. Usually the mass of an object is the sum of masses of its constituents. However, already the atomic nucleus shows a tiny mass defect, which is the key to nuclear energy. For protons and neutrons, the mismatch becomes dramatic: they have a mass of about one GeV/c^2, but are made of "up" and "down" quarks of a few MeV/c^2 and of massless gluons. It is actually the kinetic energy of the frantically moving constituents which is responsible for most of their mass, hence of the mass of the visible universe, a condition referred to as "*mass without mass*".

In the expanding (cooling) universe the confinement of previously free quarks into hadrons, occurred during the "*quark-hadron transition*", a few μs after the Big Bang. Colliding energetic heavy ions allows us to re-create microscopically the temperature conditions reigning at that time. One observes effects interpreted as a brief inverse phase transition with deconfined quarks and gluons: the "*Quark-Gluon Plasma*".

Beyond QCD basic principles, its complex manifestations in multi-particle final states give rise to the main background in the search for new physics beyond the SM. Due to S-I, and large α_S, sorting through this is a monumental task. When LHC protons collide one has to know their composition in terms of elementary objects, i.e. the fractions and kinematic distributions of valence (leading) quarks, gluons and pairs of "sea" quarks and antiquarks of all flavours born from gluon splitting (*scaling violation*).

For further reading: F. Wilczek, QCD Made Simple, *Physics Today*, **53-N8** 22-28, (2000).

4.2 Vacuum Pumping by Freezing Molecules

Cristoforo Benvenuti

For the ISR intersections, where the p–p collisions take place, a pressure lower than 10^{-10} Pa[b] was required to minimize the disturbing proton–gas interactions. This pressure range corresponds to a molecular density of 10^3 to 10^4 molecules per cm^3. At the time of the ISR project proposal, condensation cryopumping was thought to be the only way to achieve this goal. This technique is based on the condensation of the residual gas molecules onto a metal surface having a temperature at which the vapour pressures (VP) are below the desired level.

At the boiling temperature of liquid helium (4.2 K) all condensable gases display VPs lower than 10^{-11} Pa, with the exception of hydrogen, H_2, for which it is higher by many orders of magnitude. However, by extrapolating its VP curve, a sufficiently low H_2 pressure should be within reach at 2 K. In order to verify experimentally the validity of this extrapolation, the equilibrium pressure of condensed H_2 was investigated. The initial results were surprising: below about 3 K this pressure departed from the VP curve, showing an irreducible limitation in the $10^{-7}/10^{-8}$ Pa range, no matter how low the condensation temperature. Further studies showed that this limit is proportional to the room temperature radiation absorbed on the metal surface where condensation occurs [12].

Phonons, the quanta of the atom's energy in a solid, are produced in the metal substrate by the absorbed radiation. It was natural to assume that these phonons could induce molecular desorption when reaching the surface of the condensed H_2. For this to be possible, they should cross the H_2 layer. Hydrogen is very particular in this respect; it is the only solidified gas able to transmit phonons of energy in excess of what is needed to desorb a molecule. A spectacular confirmation of this mechanism was obtained by interposing below the condensed H_2 a thin layer of another gas (e.g. N_2 or Ar) which could inhibit the transmission of the energetic, disturbing phonons. The interposed gas layer produced the expected effect and the H_2 pressure limit was decreased to below 10^{-10} Pa [12].

It would obviously be strange to inject heavy gases in an ultra-high vacuum system to improve the pumping of H_2, but fortunately a more reasonable alternative exists. The disturbing energetic phonons are produced by the thermal radiation from room temperature surfaces, so if the condensing surface is well protected by a low temperature shield, desorption can be decreased. In the CERN design (Fig. 4.4, left), only 5×10^{-4} of the radiation impinging on the shield

[b]The pascal, Pa, one N/m^2, is the official unit of pressure. It replaced the Torr, which is 133 Pa.

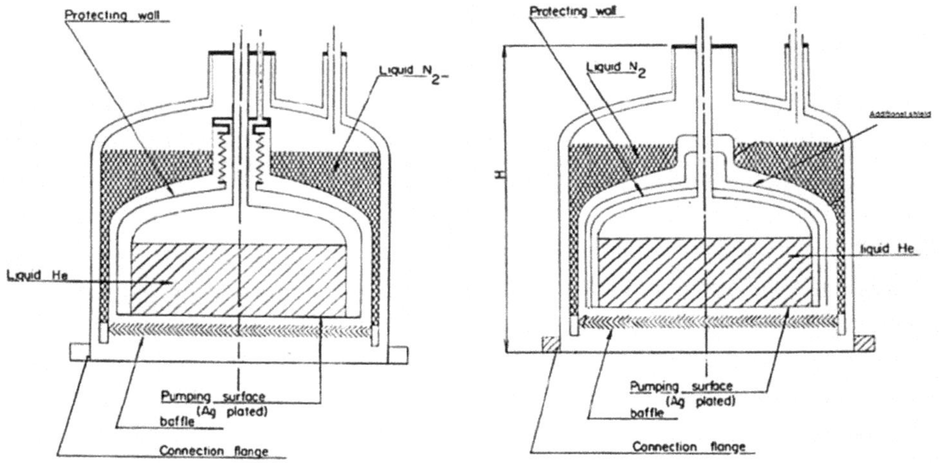

Fig. 4.4. (Left) first model — the function of the protecting wall is to avoid gas release due to the variation of He level. The double wall volume is under vacuum [13]. (Right) improved model — the double wall volume is filled with Ne, which provides good thermal contact during the initial cooling and thermal insulation at 4.2 K [14].

(a baffle cooled with liquid N_2) is transmitted to the condensation surface, resulting in a H_2 radiation induced pressure below 10^{-10} Pa [12]. The improved shield also led to a welcome reduction in the consumption of liquid helium.

One of the ISR experimental areas (point 6) was equipped with two large cryopumps, where a pressure lower than 10^{-10} Pa was achieved.

The initial model [13] was later improved by introducing an additional shield cooled by the evaporating He (Fig. 4.3, right). The liquid He consumption was then so low that it was possible to operate 200 days without refilling the 11 litre vessel [14]. The improved model was later used for the H_2 jet target experiment at ISR point 8, and for the Viksi cyclotron at the Hahn Meitner Institute in Berlin.

Cryopumping provides the advantages of a very high pumping speed and low ultimate pressure. On this ground it was considered the best pumping technique for about 10 years, at the end of the 1970s. But the advent of commercial large turbomolecular pumps and the use of Getter pumping (in the form of either Titanium sublimation or Non Evaporable Getters (NEG)) to produce extremely low pressures reduced the interest for this technique, which suffers the complication of having to provide and handle liquid helium.

Besides being a facility for physics research, the ISR was also a test bench for solutions which could be later adopted for a larger accelerator equipped with superconducting magnets. In these magnets the vacuum chamber would be cold and its behaviour with respect to the ISR pressure instabilities would be different.

To evaluate such "cold bore" behaviour, a cryostat containing a 1.3 m long vacuum chamber cooled at temperatures from 2 K to 200 K was installed in the ISR. Various quantities of different gases were condensed on the chamber surfaces while beams up to 40 A were circulating [15]. The vacuum was found to remain stable even in the most severe conditions thanks to the large pumping speed provided by cryopumping.

These positive results and the solid experience gained provided the confidence to propose cryopumping on a large scale in LHC where it is now used successfully in the cold sectors stretching over about 18 km (Chapter 8).

4.3 How to Measure Almost Nothing

Cristoforo Benvenuti

In the technology of Ultra High Vacuum, pressure is usually measured by means of ionization gauges. In such gauges the electrons emitted by a hot filament (cathode) are accelerated by the positive potential of a grid (anode) and ionise the residual gas molecules, which are then collected by a third electrode, the ion collector. The current reaching the collector is proportional to the pressure. However, at low pressure, the performance of this gauge is limited due to X-rays produced by electrons striking the grid. This radiation extracts electrons from the ion collector resulting in a current (I_x) that falsifies the result.

A major improvement was achieved by Alpert in 1950 who replaced the ion collector surrounding the grid with a thin wire at its centre (Fig. 4.5a). The large reduction of the collector area resulted in a decrease of I_x, and the low pressure limit of the gauge (called the Bayard-Alpert gauge, B.A.) was reduced by a few orders of magnitude, down to the 10^{-9} Pa range. Alpert himself remarked a few years later that it might not be possible to reduce I_x indefinitely by reducing the collector diameter because "there may be a critical size below which the probability of collecting ions reduces as rapidly as does the X-ray effect" [16]. This was because the tangential component of the velocity of the ions inside the grid would prevent a fraction of them from reaching the collector when first approaching it, and — assuming the electric field inside the grid to be perfectly radial — they would never be collected. Decreasing the collector diameter below about 0.1 mm would decrease I_x, but the ion current would also decrease in the same proportion and the performance of the gauge would not improve. This was confirmed by theoretical calculations and some experiments, and became vacuum technology dogma.

But such ideal conditions are always rare. The radial field inside the grid is perturbed both by the potential of the nearby filament and by space charge due to the accumulation of ions, which being fully enclosed by the grid are, in time, bound to be collected. This effect was confirmed by experiment at CERN showing that in fact ion collection efficiency remains constant for collector diameters down to 0.025 mm, extending the low pressure range of the B.A. gauge to 10^{-10} Pa [17].

This study had been motivated by the need to measure pressures much lower than that originally specified for the ISR (10^{-8} Pa), resulting from the vacuum system upgrade referred to in the introduction. Based on this work some 500 B.A. gauges of improved design were produced by industry and installed in the ISR, where they measured pressures typically in the low 10^{-10} Pa range. After the dismantling of the ISR these same gauges were used for LEP and later for LHC, some 40 years after their production.

For the ISR experimental areas, a few gauges were needed to measure pressures lower than 10^{-10} Pa, so a type was required that performed better than the improved B.A. gauge at such low pressures. Outside CERN and due to the above-mentioned dogma, all development of low pressure gauges was based on the same approach of moving the ion collector to outside the grid volume and properly shielding it from the detrimental X-rays. Various commercial gauges were tested, all of the

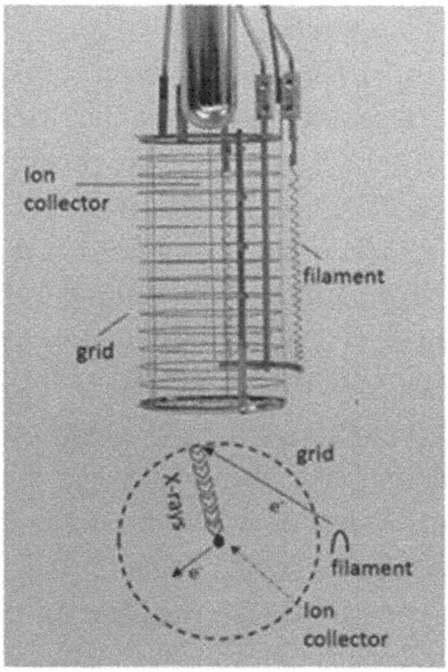

Fig. 4.5a: The Bayard-Alpert gauge, with an illustration of the X-ray effect.

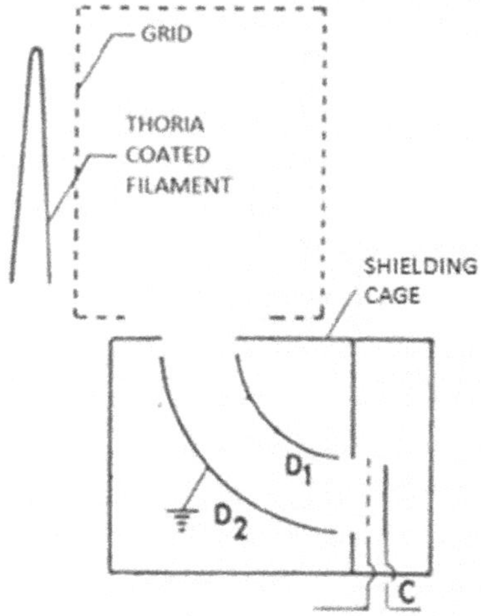

Fig. 4.5b: Schematic view of the Helmer gauge: D_1, D_2 electrostatic deflector, C ion collector [20].

"external collector" type, the most promising model being one that was marketed under the trade name of the Helmer gauge [18], with a power supply adapted for the 10^{-11} Pa range. The design of this gauge is shown in Fig. 4.5b. The ions produced inside the grid are extracted and driven by an electrostatic deflector to the ion collector, which is well shielded from X-rays. A thorough analysis of the gauge performance [19] showed that the atoms sublimating from the electron filament at ~ 2000 °C were producing a local parasitic pressure in the low 10^{-10} Pa range, in spite of the extremely low vapour pressure of tungsten, the material of the filament. By coating the filament with a better electron emitter (thoria) the filament temperature could be reduced to about 1000 °C, low enough to avoid atomic sublimation. By enlarging the grid diameter the length of the path of electrons inside the grid was increased, so extending the access of the gauge to the low 10^{-12} Pa range, which was and still is the lowest pressure ever measured inside a vacuum system [20].

4.4 Superconducting Magnets: Squeezing Beams to Extract More Collisions

Romeo Perin

The two main parameters of a colliding beam machine are the beam energy and luminosity [Box 2.1]. The energy is given by the size of the rings and the magnetic bending field: it cannot be altered, but there are ways of increasing luminosity. At a crossing point of the ISR the luminosity, inversely proportional to the height of the particle beams, can be increased by inserting a local focusing structure consisting of a pair of quadrupole magnets on either side of the interaction region (a quadrupole produces zero field on the axis and a linear gradient of field across the aperture). The concept was first validated with normal magnets, but stronger field gradients would be needed to increase luminosity by the desired factor of at least five. This would require superconducting quadrupoles — but of greater size, strength and quality than had been previously attained [Box 4.3]. In 1973 an R&D effort was initiated to verify the feasibility that culminated in the successful test of a prototype built at CERN [21, 22]. On this basis, specifications were written and orders placed in 1977–8 for critical components and for manufacture of eight quadrupoles. The high luminosity insertions (one per beam) started regular operation in 1981, as the first ever superconducting magnet system in an operating accelerator.

Cross-sectional views of the cryo-magnet are shown in Fig 4.9 [Highlight 4.5]. The main components of the quadrupole are: the room temperature bore, a stainless steel tube supporting 6-pole and 12-pole correction windings (which also

Superconductivity for magnets: A gift of nature **Box 4.3**

Superconductivity is observed in many metals and some ceramics. When such materials are cooled to temperatures ranging from about 100 kelvin (K) (about minus 173 °C), to near absolute zero (0 K), at a given *critical temperature* (T_c, a characteristic of the material) their electrical resistance suddenly vanishes. Besides the critical temperature, superconductors also feature a critical field ($\mu_0 H_c$), i.e. the magnetic field above which the material is not superconducting irrespective of the temperature. Examples of critical temperature and critical field are shown in the table. The current density that can be present in the superconductor thus depends on both the temperature and magnetic field, and for a material to be superconducting it must function beneath a *critical surface* as shown in the figure.

Material	T_c (K)	$\mu_0 H_c$ (T)
Tin-lead (Sn60-Pb40)	7.8	< 2
Niobium (Nb)	9.5	< 2
Nb-Ti (Nb53-Ti47)	10	15
Niobium tin (Nb_3Sn)	18	24
MgB_2	39	12
$YBa_2Cu_3O_7$ (HTS)	92	~ 100
$Bi_2Sr_2CaCu_2O_{8+x}$ (HTS)	108	~ 100

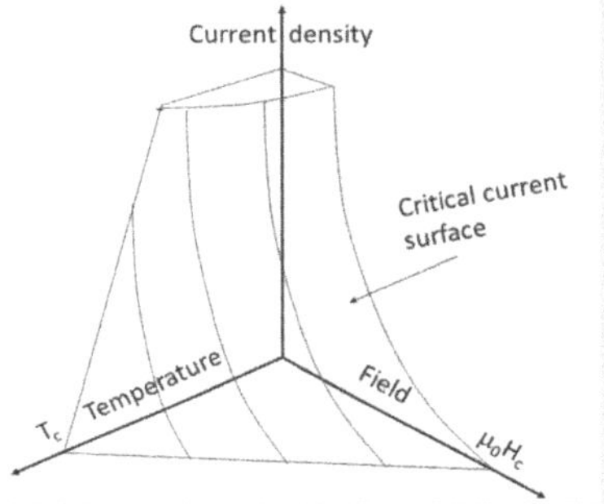

Having no electrical resistance (in DC conditions, see Box 7.4), superconductors can carry large electrical currents — wherein their interest for magnet builders. The power required for cooling is far less that than that which would be dissipated in resistive coils. Practical wires consisting of fine filaments of superconducting material embedded in copper (which stabilizes the conductor by providing a short-term parallel path for current) were first developed in the 1960s. They provide a very high current carrying capacity (~ 1000 A/mm^2), and led the way to an important increase in attainable magnetic field. The best understood and most used materials are the alloy Nb-Ti and the compound Nb_3Sn. Whereas Nb-Ti is strong and ductile, so wires and cables can be used directly for coil winding, Nb_3Sn is brittle, and coils with small radii of curvature, such as in accelerator magnets, must be wound prior to the heat treatment required to form the superconducting compound. Large-scale superconducting magnets were first built for high energy physics detectors (e.g. BEBC), and this tradition has been maintained with ever larger coils for collider experiments, e.g. ALEPH and DELPHI at LEP and ATLAS and CMS at the LHC. Coils for small and individually powered accelerator magnets can be wound using wire, but for the heavy currents adopted for series-connected magnets in large accelerators transposed cables are better [Highlight 8.2]. Flat cables that provide a good filing factor were developed for this purpose at the Rutherford Laboratory in the 1960s. Magnets for MRI constitute the most important commercial application of Nb-Ti; a large quantity of high performance wire was supplied for the LHC. Vigorous R&D on Nb_3Sn and High Temperature Superconductors (HTS) [Box 8.4] is underway for the next generation of accelerators [Highlight 12.3].

served as the inner wall of the helium vessel), the four main coils, a set of stainless steel spacers, a low-carbon steel yoke subdivided into four quadrants, five aluminium alloy rings shrink-fitted around the yoke, the outer shell of the helium vessel, the heat shield and the outer wall of the vacuum tank. The active part of the magnet, enclosed in the helium vessel, was cooled by natural convection in boiling helium at about 4.3 K and was suspended from the vacuum tank of the cryostat with four slender metallic rods of low thermal conductivity.

The main coils

In superconducting magnets the field is determined by the current in the windings. A pure quadrupole field is produced by current proportional to cos2θ distributed on a circle, θ being the polar angle. To be efficient the windings have to be as close as possible to the where the field is required. Taking into account the space for thermal insulation, correction windings and minimal clearance, the main coil had an inner diameter of 232 mm. The conductor was composed of about 1700 niobium-titanium filaments of 46 μm diameter embedded in pure copper, with cross-section 1.8 mm × 3.6 mm, and copper to superconductor ratio 1.5:1. The copper serves to support the fine filaments and to shunt the current in the event of a transition to the normal state (quench) for the time required to detect the fault and ramp the current down. A good approximation of a pure quadrupole field was found to be given by three blocks of conductors per half-quadrant, of constant thickness (38.5 mm) and uniform current density, separated by metal spacers as shown in Fig. 4.6. This configuration offers convenient surfaces for the transfer of forces between coils and clamping structure [23]. At operational field gradient, 45 T/m, the current was 1680 A and the peak field in the winding 6 T, approximately 6 times the strength of a typical superconducting MRI magnet used for medical imaging.

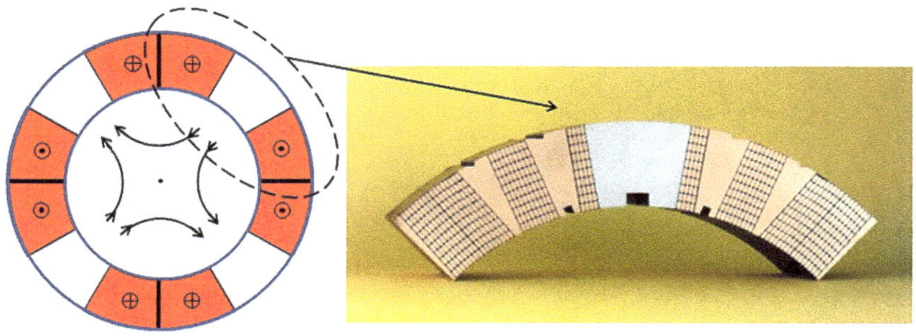

Fig. 4.6. Schematic cross-section of a quadrupole coil assembly and a cross-section of an actual quadrant showing conductors, stainless steel post and copper spacers.

The coil support structure

In operation, the main coils are subject to large electromagnetic forces (200 t/m length on each coil half). The coil support structure consists of the iron yoke and the aluminium shrinking rings. Besides providing a stiff support limiting deformation of the coils, the yoke limits the stray field and reduces the required excitation current.

In 1973, when the design of the superconducting quadrupole was started, the development of superconducting magnets of *circular* geometry (solenoids or similar) had already reached a stage where they could be designed with confidence. However, practically all magnet models *with straight-sided coils* necessary for handling high energy particle beams were plagued by degradation. In the light of this experience it was realized that a crucial difference between circular and straight-sided magnets lies in the way in which the electromagnetic forces are reacted by the structure. Sudden movements of even a few μm or microscopic cracks or slips occurring in the coil can generate enough heat locally to increase the temperature beyond the critical and cause the magnet to quench. The problem of mechanical stability was solved by the application of the principle of the Roman Arch [24]. The coil is compressed from outside, without support inside (Fig. 4.7), to prevent the appearance of tensile stress in operation. This concept, first applied intentionally to the ISR quadrupoles, has been generally adopted for superconducting magnets for accelerators.

A second objective of the project was to transfer superconducting magnet technology to European industry in anticipation of future projects. Effort was therefore put into preparing detailed specifications and manufacturing drawings, so that firms could estimate their costs and submit competitive tenders. Their responsibility would be to respect the choice of materials, dimensional tolerances,

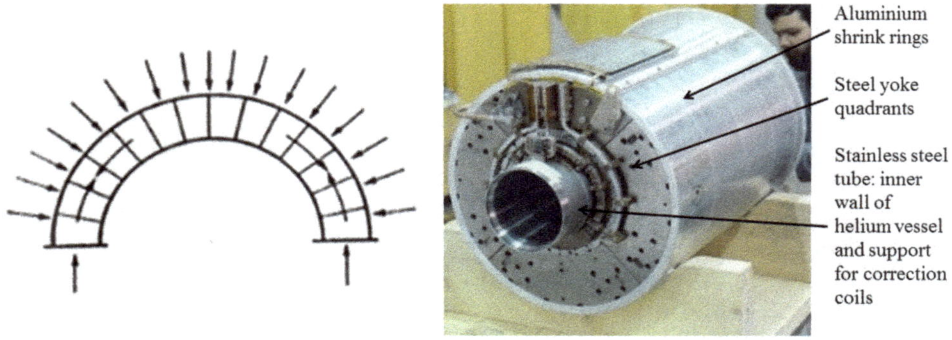

Fig. 4.7. Left: Roman Arch analogy — the elements of a Roman arch are maintained in compression. Right: an industrially produced quadrupole showing the major structural components.

and fabrication procedures, and to perform the checks and tests specified by CERN. CERN would take responsibility for performance. This project demonstrated by example CERN's strategy for the development of novel technology driven by cutting-edge research and for the procurement of components based on such technology: design, production and testing of prototypes at CERN until all the technical detail is understood, and then production in industry which has to meticulously follow the production steps and tolerances prescribed by CERN.

The manufacture of the 8 magnets took about two years. All magnets achieved the required performance [25]; there was no re-training or degradation after repeated thermal cycles from 293 K to 4.2 K. The insertions were installed at Point 8 of the ISR, to be used in combination with the Open Axial Field Magnet [Highlight 4.11], the centrepiece of a large detector [26]. Regular operation of the insertions started in 1981 and continued until the end of ISR operation in 1983. Luminosity was increased sevenfold, a record of 1.4×10^{32} cm^{-2} s^{-1} being reached for proton–proton operation. No quench of the magnets occurred during stable beam operation [27, 28]. Besides providing superb and unique experimental conditions for the researchers, the project led to an advance in magnet technology and allowed CERN to acquire expertise in the application of superconductivity and associated cryogenics that would prove to be invaluable in subsequent years.

4.5 Cryogenics for the Superconducting High Luminosity Insertion Magnets

Philippe Lebrun

The high luminosity insertion at Point 8 of the ISR [28] consisted of eight individually powered superconducting quadrupoles operating at currents of up to 2 kA in saturated helium baths at 4.5 K, housed in stand-alone cryostats [Highlight 4.4]. The cryostats also contained individually powered sextupole and 12-pole correction windings, also superconducting, and were equipped with corresponding current leads, as well as with room-temperature bores. The main challenge was to build, install and operate reliably — for the first time ever — a system of superconducting magnets with its ancillary cryogenics (see Fig. 4.8) in a functioning high energy accelerator, with minimum disturbance to the physics programme.

Fig. 4.8. The superconducting high luminosity insertion installed at Point 8 of the ISR.

Thermal optimization of the cryostats and cryogenic lines requires intercepting a large fraction of the heat in-leaks at higher temperature than 4.5 K, on thermal screens usually cooled by liquid nitrogen or cold gaseous helium. Rather than distributing the latter fluids in addition to the liquid helium filling the magnet baths, it was decided to feed the cryostats only with liquid helium, use locally the cold helium vapour boil-off to intercept heat on the thermal screens of the cryostat, cryogenic lines, and current leads, and return gaseous helium at room-temperature to the cryogenic plant, thus operating it as a liquefier rather than a refrigerator [29]. Although *a priori* less favourable from a thermodynamic point of view [30], this choice greatly simplified the cryogenic distribution, cryostat pipework and control system, thus reducing parasitic heat in-leaks, regaining overall efficiency and improving reliability. It also enables to decouple the "customer" cryostats from the cryogenic plant, which can then operate steadily liquefying helium into a dewar vessel acting as buffer.

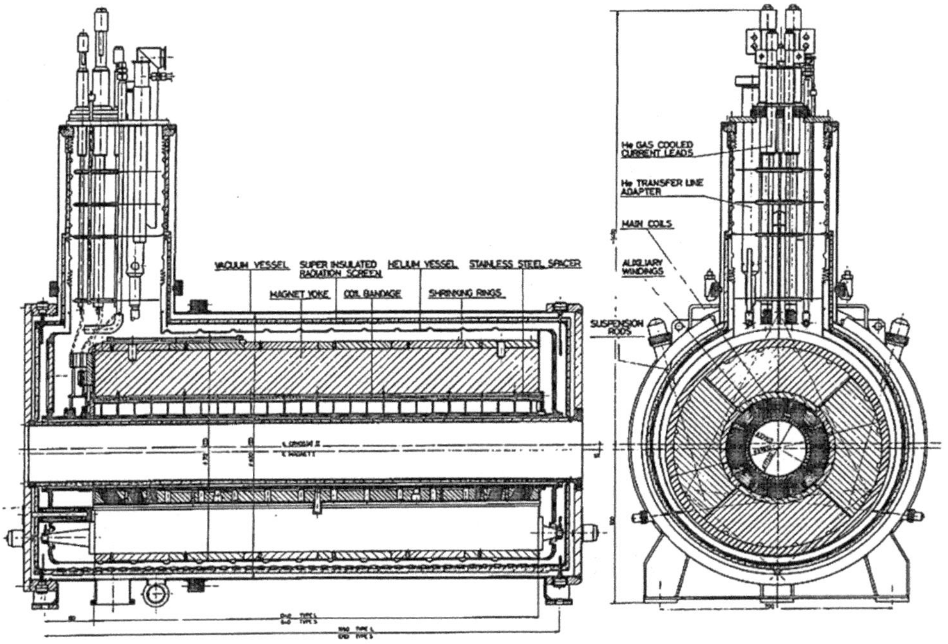

Fig. 4.9. Cross-sectional views of a superconducting magnet in its cryostat.

The cryostats (Fig. 4.9) were designed to fulfil their thermal, mechanical and magnetic functions, while being suitable for integration in the ISR. Technical solutions were chosen on the basis of their being adapted to manufacturing and assembly by industry, and their reliability in operation. The helium enclosure containing the magnet, made of AISI 304L austenitic stainless steel for the outer vessel and low-permeability AISI 304LN for the cold bore tube, was of an all-welded construction. It was surrounded by a radiation screen made of a hydro-formed, double-walled cylindrical shell of AISI 304L, wrapped with multilayer reflective insulation. The vacuum vessel outer shell was made of normal steel for shielding stray field, nickel-plated for surface protection, while the room-temperature bore tube used low-permeability AISI 316 LN. The cold mass was suspended from the vacuum vessel by four tie-rods made of Inconel 718, having four times lower thermal conductivity than standard stainless steel. The eight cryostats [31] were manufactured build-to-print and assembled by industry. Correct balancing of parallel helium vapour flows cooling the current leads was achieved by equipping them with compact, self-actuated thermostatic valves [32].

The cryogenic lines connecting the helium liquefier to the cryo-magnets in the ISR tunnel had to follow a contorted routing over a length of 50 m, imposed in particular by radiation shielding. Flexible multi-tube lines which could be entirely constructed and leak-tested in industry, shipped coiled on transport drums and rapidly installed in the field thus constituted an attractive solution. Unfortunately, such products were not industrially available in the lengths and configurations of interest, but the company Kabelmetal (now Nexans) had developed and was exploiting a process (Wellmantel®) for the continuous manufacture of corrugated tubes in a variety of engineering materials, to be used as metallic envelopes of electrical cables and pipelines for residential heating. Prototype lines made of four nested corrugated tubes of austenitic stainless steel (Fig. 4.10) were developed in a collaborative spirit between the company and CERN [33]. Following excellent test results, the solution was retained and eight 50 m long lines were supplied and installed by the company [34, 35], in a fraction of the time and effort that the conventional rigid alternative would have required. Together with the rest of the cryogenic system, the lines operated smoothly from 1980 until the ISR was closed in 1984. They could also easily be dismantled and were reused in other locations, either permanently or for specific tests [36]. After 35 years, one of these lines is still in daily use at a CERN cryogenic installation!

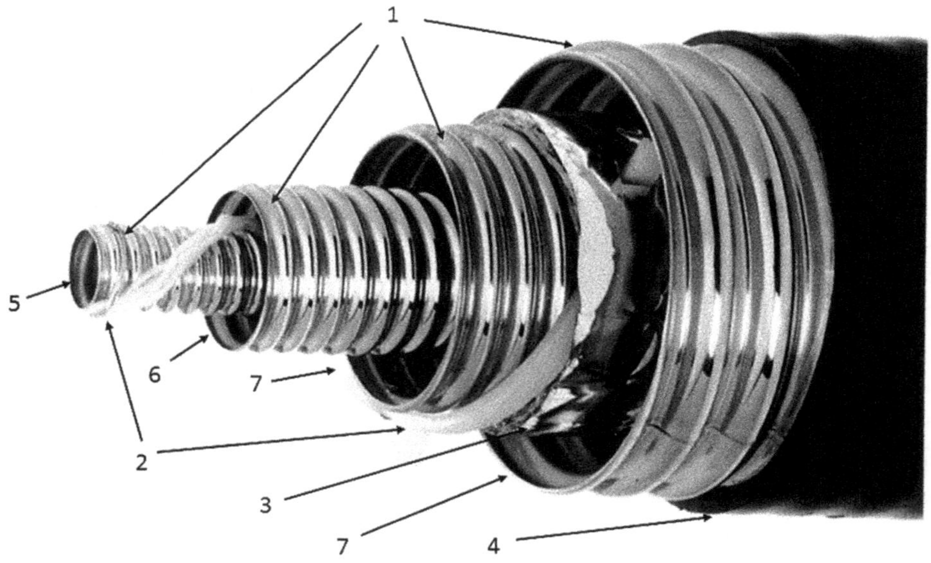

Fig. 4.10. Structure of the flexible helium transfer lines; 1: corrugated tubes; 2: helical-wound polyethylene spacers; 3: multilayer insulation; 4: polyethylene sheath; 5: liquid helium supply channel; 6: helium vapour return channel; 7: insulation vacuum space. (Photo: Nexans).

The cryogenic technology developed for the ISR superconducting high-luminosity insertions was later transposed to the LEP collider, thus enabling it to start operating from the onset with superconducting high luminosity insertions around its four large experiments [37], in spite of the very limited resources allocated to this part of the project. In the LEP experimental caverns, the superconducting quadrupoles were supplied in liquid helium by similar flexible lines of up to 92 m in length, spanning elevation differences of up to 25 m [38].

4.6 Van der Meer Scan: Proton Beam Tomography

Helmut Burkhardt

Knowledge of the luminosity L provided by a collider is fundamental. It quantifies the rate at which collisions occur. For a process of cross section σ, the collision rate N is given as the product $N = L\sigma$. In e^+e^- colliders, the luminosity can be determined to per-mille accuracy by measuring the rate of the Bhabha scattering process $e^+e^- \rightarrow e^+e^-$, for which the cross-section can be accurately calculated [39]. For hadron colliders the situation is more complex: there is no corresponding process with a well-known cross-section that can be used for calibration. A way out was pioneered at the ISR by Simon van der Meer (Nobel Prize 1984), who devised an ingenious method, now referred to as the "Van der Meer Scan" or "VdM" method [40].

The luminosity is determined by the flux of particles circulating in the collider, easily measurable as beam current, and the effective beam cross section, which is difficult to measure accurately. In the VdM method this cross section is measured by scanning one beam through the other (Fig. 4.11). The collision rates are recorded and plotted as a function of the relative displacement of the beams. The collision rate is at its maximum when the beams are colliding head-on, and decreases gradually when the transverse beam separation is increased. It can be shown that the effective beam cross section, irrespective of beam shape, is the area under the curve plotted divided by the maximum collision rate [40]. Varying the separation between the beams using magnetic steering is rather straightforward for machines in which particles of the same charge collide, such as the ISR, RHIC and LHC. A major strength of the VdM method is that beam profiles can be measured during physics operation, directly in the interaction region, which minimizes systematic uncertainties in the interpretation of experimental data.

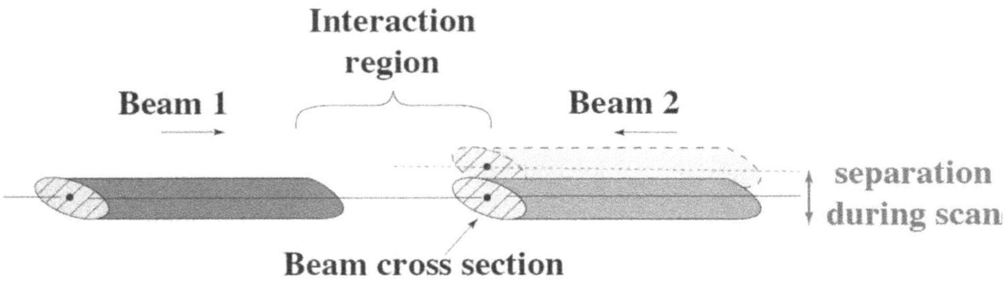

Fig. 4.11. Concept of the Van der Meer scan to determine the beam profile.

At the ISR the luminosity was determined with an accuracy of 1% [41]. This was helped by the fact that the ISR operated with continuous beams (not bunched), allowing absolute beam current measurements, and beam cross sections that were large enough for length scales to be calibrated precisely by mechanical means (using scrapers). For the Tevatron, which could not use the VdM method, the luminosity uncertainty was typically 15–20%.

The LHC renewed with the ISR tradition of VdM scans for determining luminosity [42] and optimizing the collisions [43]. After three years of operation, systematic uncertainties down to 1.5% have been achieved [42]. An automatic online procedure has also been developed for making VdM scans and directly sharing the results with the experiments [44]. The procedure includes a detailed analysis of beam dynamics uncertainties in VdM scans and has been adopted at both RHIC and the LHC. Automatic mini-VdM scans are now standard procedure in LHC operation and are used to optimize collisions for every physics run.

4.7 Roman Pots: Physics Next to the Accelerator Beam

Giorgio Matthiae

The "Roman pot" technique was invented at the ISR to study particles scattered at very small angles. These particles travel close to the circulating beams — in fact inside the vacuum chamber. They can only be detected by a special system able to place detectors a few millimetres from the beam. This system has come to be known as "Roman pots".

One may wonder why it is interesting to detect particles scattered at very small forward angles. The main motivation was to find out how the total cross-section of the scattering of a proton on a proton depends on the centre-of-mass energy. This fundamental quantity defines the overall probability of interaction of the two colliding particles. The total cross section is related through the "Optical Theorem" to the probability of elastic scattering of the protons in the forward direction. It is of course impossible to detect particles which are scattered precisely in the forward direction because they travel inside the beam; so what it is done is to measure particles scattered at very small angles and extrapolate to the forward direction. Using the Roman pots, the first experiments at the ISR made the startling discovery that the proton–proton total cross-section increases at the new energies probed by the ISR.

The basic theoretical ideas on the energy behaviour of the total cross section are discussed in [45] while the ISR measurements are described in [46, 47].

The Roman pots are special, movable sections of the vacuum chamber which contain small detectors. They are connected to the main vacuum chamber of the collider by bellows (Fig. 4.12, left panel), which are compressed as the pots are pushed towards the beam circulating in the collider. In their retracted position, the Roman pots do not obstruct the beam, thus leaving the full aperture of the vacuum chamber free for the beam during the injection process when the beam is very wide. Once the collider reaches its coasting energy with stable beams, the Roman pot is moved towards the beam with the aim of getting as close as safely possible.

Why Roman? This is because they were first used by the CERN-Rome group in the early 1970s to measure the total cross section at the ISR. And why pots? A picture of the first Roman pot used at the ISR is shown in Fig. 4.12, right panel. It was of a rather simple design, its rounded shape being at the origin of the name. When the physicists of the CERN-Rome group first proposed to use the pots, several people said that they were insane, because the inside of the vacuum chamber is a very inhospitable place, and it would be impossible to perform sensible experiments.

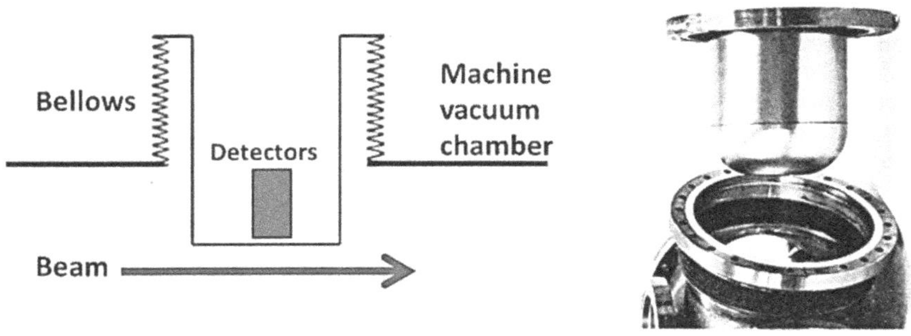

Fig. 4.12. Left: the concept of the Roman pot. Right: the first Roman pot used at the ISR.

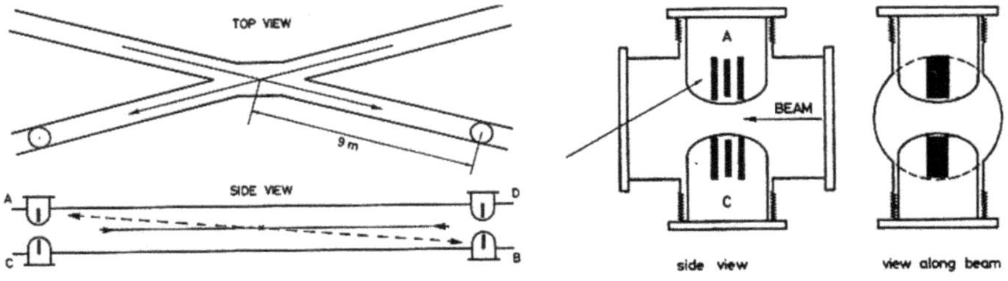

Fig. 4.13. Left: the two pairs of Roman pots of the CERN-Rome group were installed at the end of the ISR straight section, i.e. at about 10 m from the crossing. Right: sketch of the pots and of the detectors which could be approached to about 10 mm from the beam axis [47].

Connected with bellows to the vacuum chamber, a single Roman pot would be pushed strongly by the atmospheric pressure, increasing the difficulty of precise positioning with respect to the beam. This problem was solved by connecting mechanically the pots on either side of the chamber, bringing them into a nearly balanced state (Fig. 4.13).

Operation of the Roman pots required very close collaboration between the experimentalists and the machine physicists. It was crucial to have stable beams with minimal transverse haloes. Such very clean beams, resulting in very low background for the experiments, were obtained by "scraping", by inserting a thin absorber inside the machine vacuum chamber slowly and carefully — a simple but effective technique to remove the tails of the particle beams. By removing the tails it was possible to approach the beam, placing the pots within a few millimetres — in the best conditions at only 9 mm from the beam axis. This operation had to be repeated several times during a data-taking period because the tails of the beam built up again gradually after the "scraping" due to scattering of the circulating particles against molecules of the residual gas. In fact, a good, clean beam turned out to be the way for the detectors to work in the ferocious environment close by.

After the ISR, Roman pots were used at the CERN p–p̄ collider (Fig. 4.14), at the Fermilab Tevatron collider, at the DESY electron-proton collider, and are currently used by the TOTEM experiment at the CERN LHC proton collider.

The detector technique evolved with time, as required by the physics of elastic scattering at the new accelerators of higher energy [49]. The space resolution was of the order of 1 mm for the scintillator hodoscope at the ISR, of about 100 μm for the drift chambers at the SPS collider and eventually became as small as 10 μm for the Si microstrip detectors of TOTEM at the LHC [50].

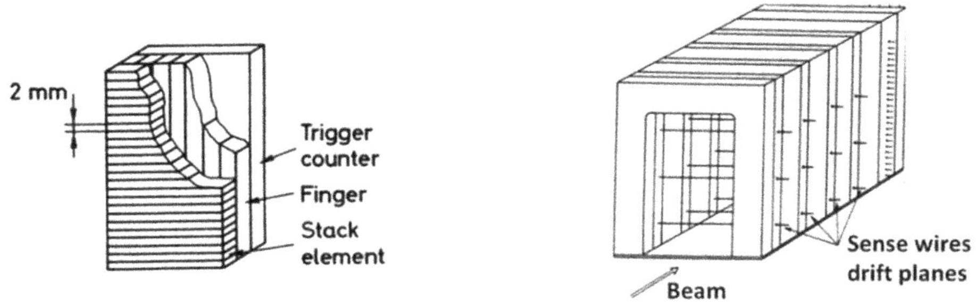

Fig. 4.14. Left: the small hodoscope of scintillation counters used by the CERN-Rome group at the ISR. Right: sketch of a special drift chamber used by the UA4 group at the CERN p–p̄ collider. The U-shaped frame of the chamber allowed the sensitive region of the chamber to approach the beams to within a few mm. An improved pot with a flat bottom plate was used for this experiment [48].

4.8 The Gas Detector (R)evolution

Fabio Sauli

The physics models current at the time of the conception of the ISR favoured the design of experimental setups optimized for detection of particles generated by proton-proton collisions in the forward direction. This required a detector design capable of bringing the sensitive area as close as possible to the vacuum chamber of the machine, and able to handle very high particle fluxes. None of the devices used in the sixties could meet these requirements. The invention in 1968 at CERN by Georges Charpak of the Multi-Wire Proportional Chamber (MWPC) [Box 4.4] completely changed the scenario [51]. Capable of detecting and electronically recording particle positions at a high rate, and permitting the coverage of large areas, they could be tailored to be sensitive a few cm from the vacuum chamber. It started a revolution in particle physics, recognized with the Nobel Prize to Charpak in 1992. The choice of this novel technology as main tracker of the Split-Field Magnet Detector (SFMD) was natural, albeit daring. A large-size MWPC prototype built by Charpak and collaborators is shown in Fig. 4.15 [52].

The original MWPC design, making use of heavy frames to tension and hold the stretched wires, was found to be ill-suited for installation within a magnet, where the ratio of sensitive to total detector area is a premium. An alternative assembly offered a much improved aspect ratio: light honeycomb plates tailored to cover the sensitive area, with the wires soldered to slender frames glued to the support plates [53]. This structure is very light and easy to handle, and has since been used for a large number of experiments.

Multi-wire proportional chambers **Box 4.4**

The multi-wire proportional chamber (MWPC), is a gas-filled radiation detector made with a grid of parallel and closely spaced thin metal wires (the anodes) stretched between two electrodes (the cathodes), see figure below. The typical wire spacing is one to few millimetres with a distance between cathodes of the order one cm. On application of a voltage between the electrodes, electrons released in the gas by ionizing radiation drift towards the anodes, where in the increasing electric field they collide with and ionize the gas molecules, resulting in charge multiplication, in a process called an electron-ion avalanche.

Even though multiplication factors of several tens of thousands can be reached in most gases, in order to detect small ionization yields it was necessary to use rather expensive electronics. This requirement made it problematic at the time to implement the large systems needed for particle physics. Discovered by Charpak and collaborators, filling with so-called "magic gas", a mixture of three or four component gases, permitted to reach gains above a million, and was paramount in the choice of the technology to equip the first large all-electronic detectors.

The MWPC localization accuracy, limited by the wire spacing of one or two millimetres, can be improved by an order of magnitude by measuring the charge distribution induced by the ion avalanches hitting the suitably segmented cathode planes.

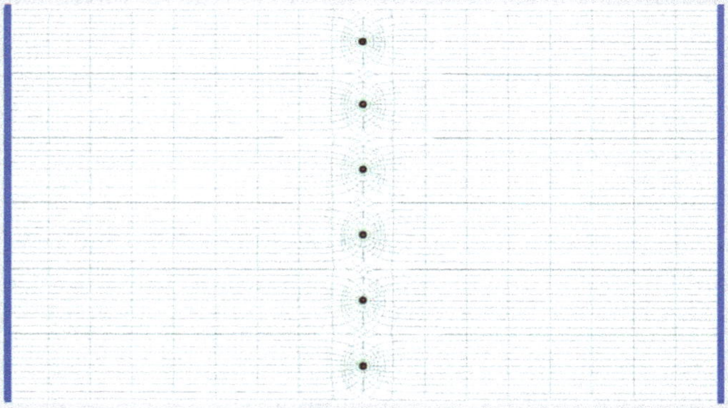

MWPC schematic showing electric field lines between anode (centre) and cathode wires (sides).

Fig. 4.15. Georges Charpak (left) and collaborators with one of the first large MWPCs.

A system of 40 MWPCs of this light-weight design with around 50,000 sense wires and the electronics to read the signals was completed in record time, and in 1972 the SFMD was ready to take data (Fig. 4.16). With one coordinate provided by the wire hit, and a coarser information given by pickup strips on the cathode planes, the detectors provided two-dimensional space points for the measurement of particle momenta from their bending radius in the magnetic field. The SFMD took data along the operating lifespan of the ISR, producing many results of relevance for fundamental physics.

The quest for free quarks, one of the physics motivations to build the ISR, was not successful, despite the demonstration that the single-electron sensitivity achieved using the "magic gas" filling permitted to efficiently detect the hypothetical particles. It became clear later that "confinement", a fundamental property of the strong interaction, prevents the existence of free quarks and hence their direct observation in the laboratory.

There were many derivatives of the revolutionary MWPC concept. In one frequently used form the drift time between the passage of a particle and the arrival time of the ionization charge on the anode wire is measured, providing much improved position information. These "Drift chambers" were widely used at the ISR [Box 4.5], the most advanced version being constructed for the Axial Field Spectrometer (AFS), shown in Fig. 4.17 [54]. Two half-cylinders around the central beam pipe were immersed in the field of the Axial Field Magnet [Highlight 4.11]. The full azimuth was subdivided into 4^0 cells, each one being instrumented radially with 42 anode wires, measuring the drift time. This very high subdivision was instrumental in imaging events with highly collimated groups of particles. These were some of the first observations of "jets", the incarnation of quarks and gluons expelled from confinement.

Fig. 4.16. The Split-Field Magnet Detector with the MWPCs of the novel light-weight design installed between the poles of the magnet.

Multi-wire drift, imaging and time projection chambers **Box 4.5**

A measurement of the time taken by the electrons released in the gas to reach the anodes, together with a precise knowledge of their drift velocity permits to infer the distance of the track from the wires with sub-mm accuracy. Developed in a wide variety of designs, from planar to cylindrical structures with cell sizes of several centimetres, multi-wire drift chambers provide localization accuracies of a few hundred microns or better with a relatively small number of measurement channels. Examples of large drift chamber systems are the cylindrical AFS detector used at the ISR (see main text), and the UA1 imaging drift chambers at the CERN proton-antiproton collider [Highlight 6.5].

In the time projection chamber, first used at SLAC and then in two LEP experiments and presently in ALICE at the LHC, the sensitive volume is a large gas vessel where ionisation trails are produced by the particles; the vessel is instrumented at one end with MWPCs. A measurement of the drift time gives the distance of the tracks from the wires, while a recording in short time intervals of the charge induction profiles on one cathode, stripped or padded, provides the other two orthogonal coordinates.

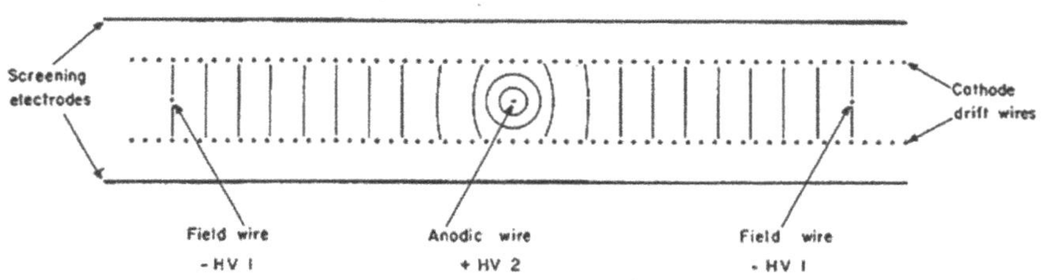

Schematic drawing of the high accuracy drift chamber.

Widely used in particle physics experiment and other applied research fields, MWPCs have several intrinsic drawbacks. The slow positive ions created in the multiplication process accumulate in the drift volume, causing field distortions and efficiency losses at high radiation fluxes; more serious, the formation in the avalanches of molecular aggregates of the main gas or of pollutants, coating the thin wires, results in permanent damages after long exposure to radiation (a process called "ageing"). A new family of gaseous devices, named micro-pattern gas detectors, (MPGD) solves many of these problems [55]. One frequently used detector in this family is the gas electron multiplier (GEM) [Box 4.6] introduced in 1997 [56]. Manufactured with a high-quality printed circuit technology developed at CERN, GEM electrodes can be tailored to the experimental requirements, and assembled using light yet sturdy honeycomb supporting frames.

Fig. 4.17. One half of the AFS cylindrical drift chamber. The anode and cathode wires were strung coaxially with the beam between the endplates, which carried the readout electronics.

The Gas Electron Multiplier (GEM) **Box 4.6**

Introduced in the late nineties, the gas electron multiplier (GEM) provides charge amplification in closely spaced narrow holes, typically fifty to hundred per square millimetre, etched on a metal-clad polymer foil with a photolithographic process. Electrons released in the upper region drift downwards into the high field of the holes, multiply and proceed towards the lower region; unlike other gaseous counters, the charge amplified by a GEM foil can be multiplied further in one or more cascaded electrodes, permitting to safely attain very large overall gains and reach single electron detection. Owing to their high granularity, GEM detectors are efficient at very high radiation fluxes, and achieve excellent spatial localization and multi-track resolutions.

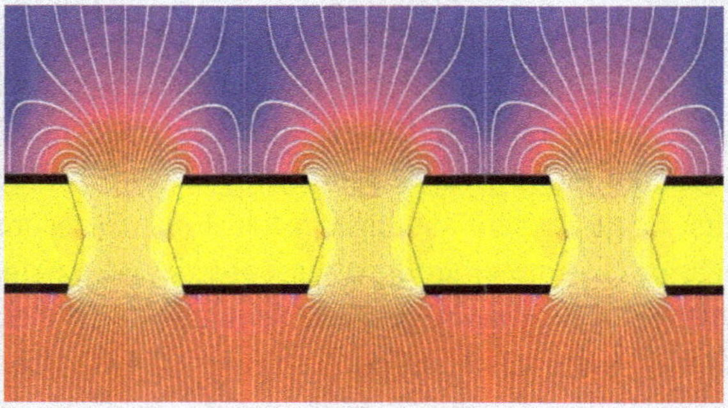

The electric field lines in the holes of a GEM electrode.

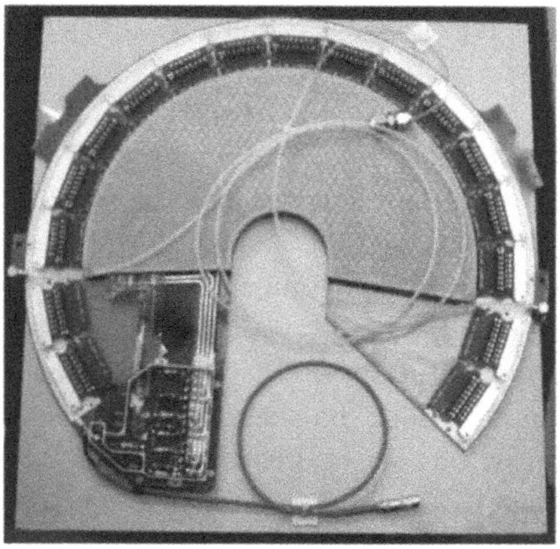

Fig. 4.18. A half-moon shaped GEM detector complete with its readout electronics.

Several hundred electrodes of this design have been produced at CERN for the construction of the tracker in the COMPASS spectrometer. Installed in 2001, the tracker is still in operation, demonstrating the reliability of this new detector technology. Figure 4.18 shows the half-moon shaped module developed for the TOTEM forward tracker.

Because of their superior performance, GEM detectors are foreseen to gradually replace wire-based devices in LHC experiments, e.g. the ALICE TPC end-cap MWPCs and the CMS muon detector, in order to better cope with increased interaction rates. Many other applications in medicine, biology, plasma diagnostics, neutron imaging are under development [56].

4.9 Transition Radiation: Imaging Relativistic Particles

Christian Fabjan

In the early years of research at the ISR the view emerged that «new» physics phenomena might reveal themselves through the observation of leptons, i.e. electrons and muons, not explainable with conventional physics. Such textbook examples are the discovery of the J/Ψ or the rare pion decay [Highlight 2.3]. But Nature likes to be cagy about revealing its secrets: These "new physics" leptons would be produced very rarely, if at all, and are difficult to detect. The ISR R209 collaboration, led by the co-discoverer of the J/Ψ and Nobelist S. Ting, developed a large Muon Spectrometer for this purpose. The R806 collaboration placed its bet on electrons. The stakes were high enough, daring the collaboration into developing two novel — most physicists thought exotic and unrealistic — detector concepts: Liquid Argon Calorimetry [Highlight 4.10] and Transition Radiation Detectors (TRDs).

Transition radiation (TR) is a close relative of Cherenkov radiation (CR) [Highlight 7.8]. When an ultra-relativistic particle transits from one medium to another, its own electric field and the atomic electrons in the new medium are affected. The particle reacts to this abrupt change of electric fields by "shaking off" photons, "Transition Radiation", which is emitted over a wide frequency spectrum, ranging from visible light to soft X-rays of some tens of keV. Like CR, TR is an electromagnetic effect. Its theoretical description is subtle and has occupied a generation of physicists, thanks to which it can now be precisely calculated. These changes of field extend over a characteristic length of some fractions of a millimetre along the particle trajectory. This must be taken into account in designing such TRDs. In contrast to CR, which is a function of the particle velocity v, TR is a function of $\gamma = \{1 - (v/c)^2\}^{-1/2}$, c being the velocity of light. Radiation in the X-ray domain starts to be emitted effectively only for particles with $\gamma > 1000$, i.e. $v > 0.999\,999\,5c$. At the ISR the only particles produced with such large γ-values are electrons, making TR the technique of choice for the identification of these messengers of "new physics". A detailed experimental verification of the theory was made at CERN, placing the development of the TR Detectors (TRD) for R806 on solid ground [57].

TR is quite a feeble process: the probability of emitting a photon of a few keV in passing through one such transition is calculated to be at the percent level. How can such a small effect be turned into a practical particle detector for large solid angle experiments? It requires the particle to traverse hundreds of interfaces between the two different media, still only producing a modest signal of a few X-ray photons. The photons are emitted close to the particle trajectory and must also traverse these material interfaces, before being measured in a suitable X-ray detector. None of these ingredients were available at the time. The theoretical optimization indicated that the required few hundred interfaces could be realized with foils, typically tens of microns thick, spaced by some hundreds of microns. To make use of all these foils the "Radiator" material must have a very high transmission probability for X-rays. Thin beryllium foils are the best, but are poisonous and difficult to handle. Next best is lithium, but this is hygroscopic and inflammable. Other materials are less performant. The collaboration considered the difficulties in making lithium radiators to be surmountable: it would require very dry air for the assembly, and an inert atmosphere (e.g. helium) for subsequent use in the experiment. A dry lab with relative humidity < 3% would not be difficult to set up — had it not been for the CERN Safety Group, concerned about the health risks associated with working in such an atmosphere over extended periods. Luckily, the US Air Force had previously studied the effect of dry air on pilots and found they performed better, without their health being affected! That settled the safety issue of radiator assembly, but the X-ray detectors had still to be developed. Given the large areas required (about 10 m^2), some form of a gas detector, such as MWPCs, was a plausible choice [Highlight 4.8], but this would require finding a gas which absorbs the X-rays efficiently. Xenon was found to be best: in such Xe-filled MWPCs the TR photons, when absorbed, produce one or more "photo-electrons", which deposit their energy through ionization in the MWPC. In this way all or most of the energy of the incident TR-photon is converted to ionization. However, in addition to this ionization the traversing relativistic electron also ionizes the gas in the chamber, and the signal recorded is the sum of the two processes. The much more abundant but heavier charged particles, e.g. pions, protons, do not produce TR photons so their associated signal is very much smaller than that of electrons. Tests were performed to confirm calculations which had indicated that in such an arrangement electrons could be clearly distinguished from the other more abundant particles.

In the experiment (Fig. 4.19) each of the particles produced in a collision traversed two such detectors to increase the quality of electron identification, with less than one percent of charged pions mistaken for electrons. Measurement of the

Fig. 4.19. Initial test set-up of the R806 collaboration to demonstrate the feasibility of Transition Radiation and Liquid-Argon calorimetry.

energy in an associated calorimeter provided further discrimination lowering the level of miss-identification of pions to 10^{-4}. This was sufficient for the physics research, but came six months too late: the ISR had missed the discovery of the Upsilon meson, made at Fermilab in the summer of 1977.

Following the R806 demonstration of TRDs, this technique became widely used whenever electron identification was at a premium [58]. The UA2 and UA6 collaborations at the p–p̄ collider built such instruments; the NA31 collaboration used this technique in its pioneering CP violation experiment [Box 3.3]; the NA34 collaboration at the CERN SPS took the concept one step further, combining eight sequential, rather compact TRDs to form an electron identifier with tracking capability. A novel geometry of detector chambers allowed to distinguish electronically between the continuous ionization along the track and localized energy deposit of the X-rays, further improving the electron-pion discrimination. The NOMAD collaboration developed a large system for electron identification to study neutrino interactions. The technique found its way into the HERA collider at DESY, the p–p̄ collider, the Tevatron, at Fermilab and the e^+e^- collider, Tristan, at KEK in Japan. And the LHC experiments employ this technique on a grand scale. The ATLAS collaboration built the "Transition Radiation Tracker", TRT, using an ingenious geometry to track charged particles with bubble-chamber like detail and to simultaneously identify electrons.

Left: one of the 18 modules of the ALICE TRD.

Below: the radiator consists of mats of polypropylene.

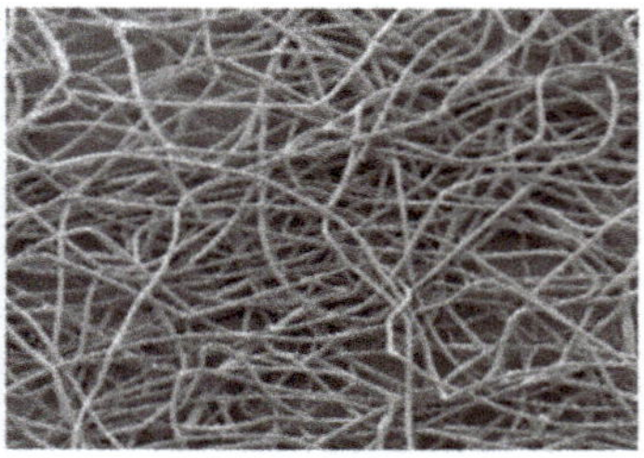

Fig. 4.20. One of the 18 modules of the ALICE TRD during final tests prior to installation. Six radiator-detector layers are contained in such a module. The last layer seen in the photo shows the signal and real-time trigger electronics, directly behind the X-ray detector. The radiator (insert) is made for technical reasons from mats of polypropylene fibres. (Source: University of Heidelberg).

The ALICE collaboration developed the NA34 concept further, building the world's largest TRD (Fig. 4.20) and using novel signal processing techniques to enhance the TR-signal and to identify electrons in real-time for event selection [Highlight 8.11] in a daring — yet successful — tour-de-force [59].

These detectors are now also making an impact in astrophysics, the most spectacular application being the choice of a powerful TRD as a key component of the Advanced Magnetic particle Spectrometer, AMS, based on the International Space Station, ISS. This experiment studies with superb detail features of cosmic rays, with an emphasis on antimatter.

4.10 Precision Calorimetry: Honing an Essential Tool

Christian Fabjan

Calorimetry, the energy measurement of particles, is ubiquitous in modern particle physics experiments, as explained in the Box 6.3. Conceptually, a calorimeter is a block of material, in which an incident particle is absorbed by transferring its energy through sequential interactions to a "shower" of lower energy particles. Measuring the sum of the energy of the shower particles gives the energy of the incident particle. In the early 1950s this technique was pioneered for the detection of *electrons* in the Nobel-prize winning determination of the size of the proton in the famous electron scattering experiments by Hofstadter at SLAC. In the early 60's the technique was extended to detect *hadrons* in cosmic rays. Experiments with calorimeters at the CERN SC and PS, and in other laboratories followed. Some of the first generation ISR experiments required large arrays of photon or electron/positron calorimeters [9].

Calorimetry became a focal point of detector R&D at the ISR with the emergence of experimental concepts requiring a high quality energy measurement of particles [60]. Part of the motivation came from the nascent ideas of the Standard Model of particle physics with its force carrier bosons, revealed in decays into electrons, muons and neutrinos. Neutrinos can of course not be measured directly, but their energy can be inferred. At an electron-positron collider the sum of the particle momenta is zero before and, by conservation, also after the collisions. At a proton–proton (antiproton) collider the components of momenta and energy transverse to the direction of the beams are zero. A sufficiently precise measurement of the energy balance, e.g. summing the energy components in the plane transvers to the collision, should add up to zero. A neutrino will manifest itself as an unbalance in this transverse energy distribution.

In the quest for precision calorimetry two lines of R&D were pursued: understanding the underlying detector physics determining the intrinsic performance, and developing instrumentation which would reach this performance while allowing the construction of apparatus that englobes the collision point quasi-"hermetically".

The physics of the absorption process depends on the particle type. Electrons (positrons, photons) interact via the electromagnetic, hadrons through the strong interaction (and also electromagnetically, if charged). These processes are quite distinct, leading to different optimizations and instruments for electrons and hadrons. Both breeds of calorimeters come in a large variety, depending on the way the absorbers are instrumented to extract the energy information.

In materials of high atomic number Z the electromagnetic (e-m) shower is very compact compared to the hadronic one: a 100 GeV-electron can be absorbed in some 20 cm of material, while a 100 GeV hadron needs an absorber 1 m to 2 m thick. Therefore, high-quality electromagnetic calorimeters (EMCs) can be built, where the absorber also functions as the particle detector. One example, extensively used at the ISR, is glass enriched with lead, "lead-glass". More recently, certain crystals have been specifically engineered to meet the experimental requirements [Highlight 8.8].

Besides this special class of homogeneous crystal EMCs calorimeters used at accelerators can be of a "sandwich" construction: suitably chosen absorber plates alternate with layers of an active detector medium, which provides a signal related to the absorbed energy (Fig. 4.21). In these devices only a fraction of the total energy is absorbed and measured in the detector medium, potentially an overriding limitation. Plastic scintillator plates were the detectors of choice: they are relatively cheap, easy to handle and provide a sufficiently strong light output to be

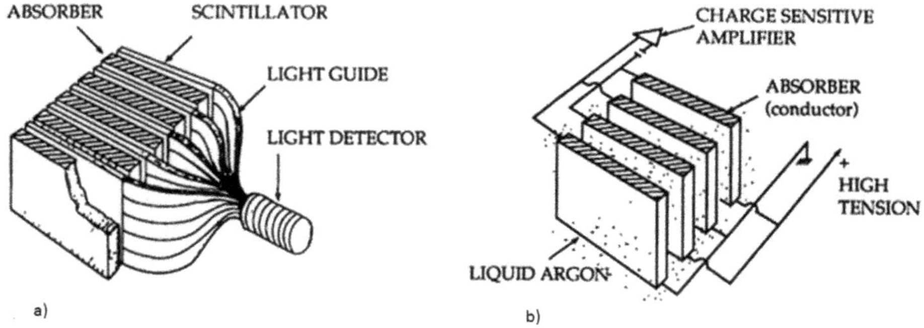

Fig. 4.21. Schematic of frequently used calorimeter readout techniques. (a) Plates of scintillator optically coupled to a photomultiplier. In modern instruments more compact readout methods of the scintillator light are used; (b) Ionization charge produced in an electron-transporting medium (e.g., liquefied argon) is collected at electrodes that also function as the absorber plates.

registered with photon detectors (Fig. 4.21a). The downside is that the photons are easily "lost", being absorbed in the scintillator or not recorded. The relation of the photon signal relative to the energy of the particle depends on many parameters, as does the signal of photomultipliers, requiring extensive calibration procedures to relate the signal to the energy.

The instrumental disadvantages of scintillators led to the unconventional suggestion to replace the scintillator with liquefied argon (LAr), an inherently stable medium. In LAr the charged particles of the cascade produce an ionization charge sufficiently large to be well measured (Fig. 4.21b) [60]. As Argon is used on an industrial scale it is amazingly cheap: a litre of LAr is cheaper than a litre of beer! This technique also offered a further, decisive advantage: the individual contributions to the energy resolution due to the construction and readout and due to the intrinsic physics effects could be identified and separately measured [61]. These advantages come at a price: LAr needs to be cooled to about 88 K, using liquid nitrogen, requiring a cryogenic installation. A view of the initial test device is shown in Fig. 4.22. This technique has found its way into many particle experiments. It has become the detector of choice for neutrino physics: a 700 ton LAr detector is operational; plans for 100 000 ton neutrino detectors are on the drawing board. Sometimes more exotic versions of this technique, using liquid krypton [Highlight 5.6] or liquid xenon are justified on performance grounds.

Fig. 4.22. Preparing for the first major liquid argon calorimeter test: a technician is preparing the cryostat prior to insertion of the calorimeter module on the right, consisting of 2 mm thick metal plates alternating with 2 mm gaps for the liquid argon. Liquid nitrogen is used to liquefy argon via a heat exchanger (stainless steel spiral).

Understanding the physics of the hadronic cascade and the *intrinsic* performance of hadronic calorimeters (HCs) turned out to be the real challenge [61]. It required measurements to isolate different contributions affecting the performance. Crucial input was given by novel computer simulations of the complex physics of the hadronic cascade at the individual particle and collision level. In these showers a vast range of particles is produced. High energy hadrons generate energetic secondary hadrons, but also break up (fission) the nuclei producing many low energy photons and neutrons. Energetic neutral pions are produced, which decay into photons, initiating an e-m cascade. These varied particles tend to be recorded with different signal strength in the active detector medium. This detailed understanding opened the door towards precision hadron calorimetry: it is crucial to equalize the calorimeter response to the electromagnetic and hadronic parts of the cascade.

Among the proposals to achieve this equality ("compensation") the most unconventional one was surely the suggestion to use depleted uranium 238 as absorber plates, left over from the production of fuel elements for nuclear reactors. Calculations, confirmed by measurements, showed that the particle-induced fission in these plates by the hadronic cascade produces additional particles, whose energy is detectable in the detector medium. Remarkably, this additional signal can be tuned to contribute just at the right level to achieve the desired compensation. The improvement in intrinsic energy resolution is a factor of 2, which for ISR experimentation was important. A 300 ton uranium scintillator calorimeter with full azimuthal coverage was constructed for the Axial Field Spectrometer [Highlight 4.11] [62]. These U-HCs measure a 10 GeV pion with a relative accuracy of some 6%, a 100 GeV hadron with a respectable 2%. It permitted the first observation at the ISR — concurrent with experiments at the CERN p–p̄ collider — of "hadronic Jets": these are collimated sprays of particles, the manifestation of highly energetic collisions between the constituents of the protons, and crucial confirmation for the newly-developed theory of strong interactions. This concept was taken a step further at the electron-proton collider HERA at DESY, where the physics required the best HCs possible.

Imaginative applications of the calorimetric techniques have opened new fields in astronomy, studying the highest energy phenomena produced in cosmic accelerators [61]. In the search for the putative "Dark Matter" particles calorimeters are optimized for very low energy deposits. In medical diagnostics tumours are detected with Positron Emission Tomography, using crystals, similar to those used in the Higgs-Boson discovery [Highlight 10.4].

4.11 The Open Axial Field Magnet: Barrier-Free Access

Thomas Taylor

During the discussions in 1976 on a new facility for the ISR much emphasis was put on the importance of "openness" — both for minimising material in the path of particles produced by the collisions, and to allow good access for installation and maintenance of detectors during the short time allocated between runs. This led to the novel proposal of a normal conducting magnet, providing better access than either the superconducting solenoid or the toroid that were being discussed.

The concept was seized upon to figure in the proposal for an experiment presented in January 1977. By then basic field calculations and cost estimates for the Open Axial Field Magnet (OAFM) had been prepared, and it was confirmed to be an interesting alternative to the superconducting option. The field was lower, but it was shaped to be more efficient for momentum measurements than in a regular solenoid, the access was far better, and the cost would be less by a factor of 5. The design was optimised and the experiment was approved in March 1977. The magnet with its dimensions and field shape is shown in Fig. 4.23. The field is cylindrically symmetric out to a radius of 1.5 m, simplifying analysis of particle tracks. Small dipole and skew quadrupole correction magnets were installed close by to render the OAFM transparent for the circulating beams.

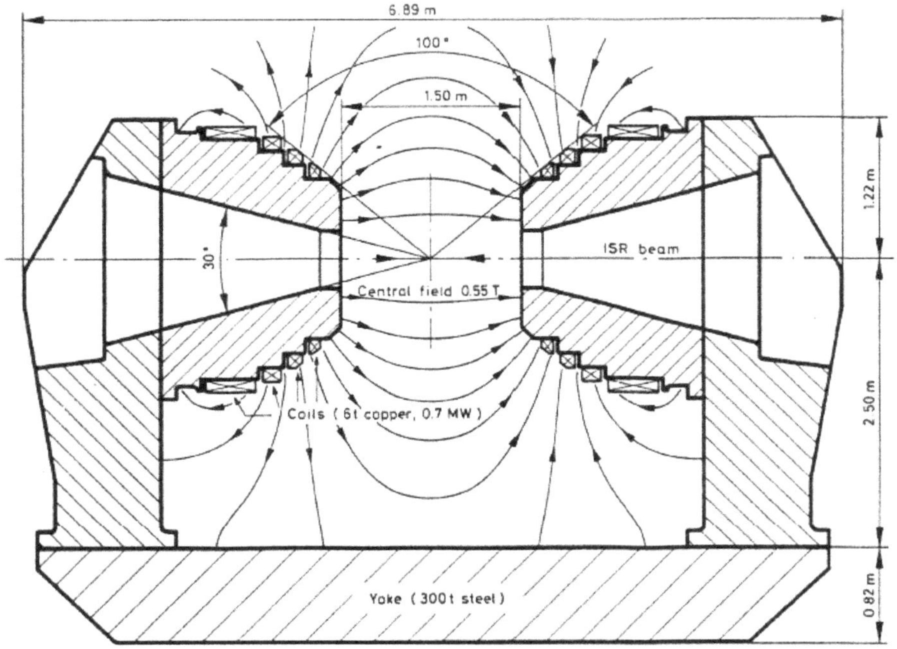

Fig. 4.23. Longitudinal cross-section of the OAFM, showing dimensions and field lines.

The field, 0.55 T at the centre, is produced by circular water-cooled copper coils clamped to hollow conical steel poles; it provides free access through azimuthal angles 0° to 15°, 40° to 140°, and 165° to 180°. The pole separation is 1.5 m. The magnetic flux is returned from the poles via cast low-carbon steel upright pieces, shaped to optimise the distribution of magnetic flux and bolted to a rectangular steel base. The base consists of two 60 ton castings, the uprights are also 60 tons each, and the total mass of the magnet is about 300 tons. The yoke and coils were bought from industry to CERN specifications. Power consumption was 700 kW.

The magnet was assembled and tested, and the field mapped at ISR point 8 during the 1978–9 winter shutdown (Fig. 4.24), to be used by The Axial Field Spectrometer Collaboration (R807) [26], in conjunction with the superconducting high luminosity insertion [Highlight 4.4] until the ISR was closed in 1983. It was later installed at LEAR as the magnetic spectrometer for the OBELIX experiment. The field-shaping concept was further developed [63], was adopted by the PHENIX collaboration for the "Relativistic Heavy Ion Collider" RHIC at Brookhaven [64], and featured in proposals for experiments at high energy hadron colliders (SSC and LHC) [63, 65].

Fig. 4.24. OAFM during the field mapping campaign.

References

1. D. Pestre, The second generation of accelerators for CERN, 1956–1965: the decision process, pp. 679–778 in *History of CERN, Vol. II, Building and running the Laboratory*, by A. Hermann, J. Krige, U. Mersits and D. Pestre (North Holland, 1990).
2. A. Russo, The ISR. The construction and operation of CERN's second large machine and a survey of its experimental programme, pp. 100–170 in *History of CERN, Vol. III*, ed. J. Krige, (North Holland, 1996).
3. K. Johnsen, The ISR and accelerator physics, *Particle Accelerators*, **18**, 167-182 (1986).
4. K. Hübner and T. Taylor, The birth and development of the first hadron collider, the ISR, in PAS *Symposium Subnuclear Physics*, 119 (2011).
   https://cds.cern.ch/record/1626809/files/CERN-ACC-2013-0248.pdf;
   P.J. Bryant, p. 15 in *40th Anniversary of the First Proton-Proton Collisions in the CERN Intersecting Storage Rings (ISR)*, U. Amaldi, P.J. Bryant, P. Darriulat and K. Hübner (eds.), CERN 2012- 004 (CERN, Geneva, 2012), http://dx.doi.org/10.5170/CERN-2012-004.15.
5. M. Jacob, *CERN: 25 years of physics* (North-Holland, 1981).
6. U. Amaldi, Small-angle physics at the ISR forty years later, *Report CERN 2012-004* (2012).
7. P. Darriulat, p. 63 in *40th Anniversary of the First Proton-Proton Collisions in the CERN Intersecting Storage Rings (ISR)*, U. Amaldi, P.J. Bryant, P. Darriulat and K. Hübner (eds.), CERN 2012- 004 (CERN, Geneva, 2012), http://dx.doi.org/10.5170/CERN-2012-004.63.
8. Jerome I. Friedman, Henry W. Kendall, Richard E. Taylor, Nobel Prize in Physics 1990; https://www.nobelprize.org/nobel_prizes/physics/laureates/1990.
9. C.W. Fabjan, Evolution and revolution: detectors at the ISR, *CERN COURIER*, January 2011.
10. G. Giacomelli and M. Jacob, Physics at the CERN-ISR, *Phys. Reports* **55**, 1 (1979).
11. C.W. Fabjan and N. McCubbin, Physics at the ISR 1978–1983, *Physics Reports* **403-404**, 165–175 (2004).
12. C. Benvenuti and R.S. Calder, The desorption of condensed hydrogen from various substrates by infrared thermal radiation, *Physics letters* **35A**, No. 4, 291 (1971).
13. C. Benvenuti, Study of a cryopump for possible use in the ISR, *Le Vide* **168**, 235 (1973).
14. C. Benvenuti and M. Firth, Improved version of the CERN condensation cryopump, *Vacuum*, **29**, No. 11/12, 427 (1979).
15. C. Benvenuti and N. Hilleret, Cold bore experiments at CERN ISR, *IEEE trans. Nucl. Sci.* **NS-26**, No. 3, 4086 (1979).
16. D. Alpert (1958), p. 624 in *Encyclopaedia of Physics* **12,** ed. S. Flügge (Springer, Berlin, 1958).
17. C. Benvenuti and M. Hauer, Low pressure limit of the Bayard-Alpert gauge, *Nucl. Instr. & Meth.* **140**, 453 (1977).
18. J.-C. Helmer and W.H. Hayward, Ion gauge for vacuum pressure measurements below 10^{-10} Torr, *Rev. Sci. Instr.* **37**, 1652 (1966).
19. B. Angerth and Z. Hulek, The tungsten evaporation limit of hot-cathode ionization gauges, *J. Vac. Sci. Technol.* **11**, No 1, 481 (1974).
20. C. Benvenuti and M. Hauer, Improved Helmer gauge for measuring pressures down to 10^{-12} Pa, *Le Vide* **201**, (Suppl.) 199 (1980).
21. J. Billan, R. Perin, L. Resegotti, T. Tortschanoff and R. Wolf (eds.) *Construction of a Prototype Superconducting Quadrupole Magnet for a High Luminosity Insertion at the CERN Intersecting Storage Rings*, CERN-1976-016 (CERN, Geneva, 1976).
   http://dx.doi.org/10.5170/CERN-1976-016.

22. L. Resegotti, Superconducting low-beta insertion and high luminosity prospect at ISR, *Report CERN ISR-BOM/77-58* (1977).
23. R. Perin, T. Tortschanoff and R. Wolf, Magnetic design of the superconducting quadrupole magnets for the ISR high luminosity insertion, Report CERN ISR-BOM/79-2 (1979); https://cds.**cern**.ch/record/119973/files/**1979**04141.pdf.
24. R. Perin, Mechanical stability of superconducting quadrupole coils, *Proc. 5th Int. Conf. on Magnet Technology*, Rome (1975) p. 559.
25. J. Billan *et al.*, The eight superconducting quadrupoles for the ISR high luminosity insertion, *Proc. XI Int. Conf. on High Energy Accelerators,* CERN, Geneva (1980) p. 848.
26. H. Gordon *et al.*, The Axial Field Spectrometer at the CERN ISR, *Nucl. Instr. & Meth.* **196**, 303–313 (1982).
27. J. Billan *et al.*, Operational experience with the superconducting high-luminosity insertion in the CERN Intersecting Storage Ring, *IEEE Trans. Nucl. Sci.* **NS-30**, No. 4 (1983).
28. J. Billan *et al.*, A superconducting high-luminosity insertion in the Intersecting Storage Rings (ISR), *IEEE Trans. Nucl. Sci.* **26**, 3179-3181 (1979).
29. H. Laeger and Ph. Lebrun, The helium cryogenic system for the superconducting high-luminosity insertion at the CERN-ISR, *Adv. Cryo. Eng.* **29**, 359–367 (1984).
30. P. Lebrun, pp. 41-86 in *Proc. of the CERN Accelerator School: Superconductivity in Particle Accelerators, DESY, Hamburg, Germany, 30 May - 3 Jun 1988*, S. Turner (ed.), CERN-1989-004 (CERN, Geneva, 1989), http://dx.doi.org/10.5170/CERN-1989-004.41.
31. H. Laeger, Ph. Lebrun and P. Rohmig, Eight liquid helium cryostats for the superconducting magnets of the ISR high-luminosity insertion, *Proc. ICEC 8,* Genoa (1980) pp. 124-129.
32. H. Blessing, H. Laeger and Ph. Lebrun, Modular thermostatic vapour-cooled current leads for cryogenic service, *Adv. Cryo. Eng.* **29**, 199–206 (1984).
33. H. Laeger, Ph. Lebrun and P. Rohner, Long flexible transfer lines for gaseous and liquid helium, *Cryogenics* **18**, 659–662 (1978).
34. H. Blessing *et al.*, Four hundred meters of flexible cryogenic helium transfer lines, *Proc. ICEC 8,* Genoa (1980) pp. 261–266.
35. H. Blessing *et al.*, High performance flexible cryogenic helium transfer lines, *Adv. Cryo. Eng.* **27**, 761–768 (1982).
36. H. Blessing *et al.*, Controlled downward transfer of saturated liquid helium across large differences in elevation, *Proc. ICEC 12, Southampton,* (1988) 222–226.
37. J.-P. Dauvergne *et al.*, Helium cryogenics at the LEP experimental areas, *Adv. Cryo. Eng.* **35B**, 901–908 (1990).
38. H. Blessing *et al.*, Very low-loss liquid helium transfer with long flexible lines, *Adv. Cryo. Eng.* **35B**, 909–916 (1990).
39. ALEPH Collaboration, Precision electroweak measurements on the Z Resonance, *Phys. Rep.* **427**, 257 (2006).
40. S. van der Meer, Calibration of the effective beam height in the ISR, *Report CERN/ISR–PO/68–31* (1968).
41. G. Carboni *et al.*, Precise measurements of proton-antiproton and proton-proton total cross sections at the CERN Intersection Storage Rings, *Nucl. Phys.* **B 254**, 697 (1985).
42. P. Grafstrom and W. Kozanecki, Luminosity determination in proton colliders, *Prog. Part. Phys.* **81**, 97-148 (2015).
43. H. Burkhardt, Optimisation of LHC beam conditions, *Proc. 35th Int. Conf. on High Energy Physics*, Paris (2010).

44. S. White, Determination of the absolute luminosity at the LHC, CERN-Thesis-2010-0139, (2010).
45. M. Froissart, Asymptotic behavior and subtractions in the Mandelstam representation, *Phys. Rev.* **123**, 1053 (1961).
46. U. Amaldi, Proton interactions at high energy, *Scientific American*, November 1973, http://inspirehep.net/record/214689?ln=en.
47. U. Amaldi, An ISR discovery: The Rise of the Proton-Proton Cross-Section, in *60 years of CERN experiments and discoveries*, eds. L. Di Lella and H. Schopper (World Scientific, 2015).
48. R. Battiston *et al.*, The 'Roman Pot' spectrometer and the vertex detector of experiment UA4 at the CERN SPS collider, *Nucl. Inst. & Meth. A* **238**, 35 (1985).
49. G. Matthiae, in *Scattering; scattering and inverse scattering in pure and applied science*, eds. R. Pike and P. Sabatier (Academic Press, 2002).
50. TOTEM Collaboration, Performance of the TOTEM detectors at the LHC, *Int. J. Mod. Phys.* **A28**, 1330046, (2013).
51. G. Charpak, *et al.*, The Use of Multiwire Proportional Counters to Select and Localize Charged Particles, *Nucl. Instr. & Meth.* **62**, 262 (1968).
    G. Charpak and F. Sauli, High-resolution electronic particle detectors, *Ann. Rev. Nucl. Part. Sci.* **34**, 285 (1984).
52. G. Charpak, *et al.*, Some features of large multiwire proportional chambers, *Nucl. Instr. & Meth.* **97**, 377 (1971).
53. R. Bouclier, *et al.*, Proportional Chambers for a 50 000-wire detector, *Nucl. Instr. & Meth.* **115**, 235, (1974).
54. D. Cockerill, *et al.*, Operation of a Drift Chamber Vertex Detector at the ISR, *Nucl. Instr. and Meth.* **176**, 159 (1980).
55. F. Sauli and A. Sharma, Micro-Pattern Gaseous Detectors, *Ann. Rev. Nucl. Part. Sci.* **49,** 341 (1999).
56. F. Sauli, Gas Electron Multiplier (GEM) detectors: Principles of Operation and Applications, *Comprehensive Biomedical Physics* **8**, 367 (2014).
57. C. W. Fabjan and W. Struczinski, Coherent emission of transition radiation on periodic radiators, *Phys. Lett.* **57B**, 483 (1975).
58. B. Dolgoshein, Transition radiation detectors, *Nucl. Instr. & Meth. A* **326**, 434 (1993).
59. A. Andronic and J. P. Wessels, Transition radiation detectors, *Nucl. Instr. & Meth. A* **666**, 130 (2012).
60. W. J. Willis and V. Radeka, Liquid-argon ionization chambers as total-absorption detectors, *Nucl. Instr. & Meth.* **120**, 221 (1974).
61. C.W. Fabjan and F. Gianotti, Calorimetry for particle physics, *Rev. Mod. Phys.* **75**, 3 (2003).
62. T. Akesson *et al.*, Properties of a fine sampling uranium-copper scintillator hadron calorimeter, *Nucl. Instr. & Meth. A* **241**, 17 (1985).
63. E. Dolgosheina, T. Taylor and W.J. Willis, Field shaping by iron for muon measurement at hadron colliders, *Nucl. Instr. & Meth. A* **301**, 451-453 (1991).
64. K. Adox *et al.*, PHENIX detector overview, *Nucl. Instr. & Meth. A* **499**, 469 (2003).
65. A. Dudragne *et al.*, A shaped solenoid for muon spectroscopy at high energy hadron colliders, *Nucl. Instr. & Meth. A* **324**, 93 (1993).

Chapter 5

The Super Proton Synchrotron (SPS): A Tale of Two Lives

Niels Doble, Lau Gatignon, Kurt Hübner and Edmund Wilson

5.1 Introduction

In the early 1960s groups of accelerator designers on both sides of the Atlantic made plans for large proton synchrotrons based on the recently completed CPS and AGS. Each study group aimed for 200 GeV to 300 GeV, roughly ten times the energy of the existing machines, and both proposals were scaled up versions of the combined-function magnet lattice used for the AGS and CPS. Combined function refers to the ring of magnets which guide particles in a circle. The magnets combine in one type both focusing and bending of the circulating beam [Box 2.1].

A new US proposal for a 200 GeV synchrotron finally chose a different lattice in which the functions of bending and focusing the beam were assigned to separate dipole and quadrupole magnets. This choice permits a higher field in the dipoles resulting in a more compact design allowing a higher energy in a circle of a given radius. CERN considered this design but found the cost of the alternatives would differ very little and, as a "green-field site" provided by one of the CERN Member States was being considered, a 15% smaller size was not crucial and the combined-function design was retained.

The original plan of CERN had been to build both the ISR (Chapter 4) and a 300 GeV machine at the same time, but the latter was delayed by discussions among the Member States over the choice of the site and by a request to consider superconducting magnets [1].

In 1968 John Adams was invited by the CERN Council to oversee CERN's by now well and truly stalled 300 GeV Synchrotron Project. He made two important decisions to save the project. First, the combined-function design was abandoned and the separated-function design was chosen by the new team as this reduced the size of the accelerator for the same top beam energy and allowed for a "missing magnet" option: to install all the quadrupoles but only half of the dipoles, i.e. every second dipole magnet. Although this would lead to one half of the top energy, the initial budget would be reduced and the accelerator would be operational earlier; the missing dipole positions could be filled later — possibly with superconducting magnets, if they became available [2].

The second decision settled the thorny site problem, after years of protracted discussions. The proposal to site the new accelerator next to the existing laboratory using the PS as injector and the existing infrastructure had been put forward already in 1961. However, at that time, it was felt that (i) the new large facility should be constructed elsewhere in Europe to avoid concentration at one site; (ii) building the accelerator on the surface by the cut-and-fill method would be too major a disturbance of the local countryside; (iii) siting it in a tunnel would require expensive underground experimental areas as internal targets for the production of the secondary beams were the norm at that time. By 1968, however, tunnelling technology had advanced, the beam extraction techniques had been refined and the project team argued that experimental halls on the surface could very well be served by an underground accelerator, a novel concept and decisive for CERN's future (LEP, LHC). On top of that, the existing large West Hall including its bubble chambers would be perfect as the initial experimental area. The idea to build the new facility near to the CERN site promised substantial savings, an earlier start — not to mention an interesting perspective for the existing laboratory.

In May 1970 CERN proposed to Council to build the 300 GeV accelerator adjacent to CERN I (Meyrin) site, and to install at first only half of the magnets, in an underground tunnel of 7 km in circumference, see Fig. 5.1. Such a machine would reach 400 GeV with normal magnets installed in all the gaps. The missing magnet scheme would even permit higher energies if filled with superconducting magnets. This satisfied critics that 400 GeV was not high enough and encouraged the advocates of superconducting magnets to continue developing this technology.

Finally, the Member States, after much hesitation, accepted the new ideas and approved the machine in February 1971. During the construction it became clear that superconducting technology was developing more slowly than expected, while conventional dipole magnets proved cheaper than budgeted, and in 1973 there was a decision to build a 300 GeV machine with all normal-conducting magnets — to the great relief of the SPS construction team.

The tunnel boring in the amenable molasse (local sandstone) rock was completed on 31 July 1974, when the machine excavating the SPS tunnel returned to its starting point. The joy of achieving the important milestone of completing the boring of the 7 km long tunnel is captured in Fig. 5.2. The tunnel is situated at an average depth of 40 m below the surface and straddles the Franco-Swiss border, making the SPS the first major cross-border accelerator.

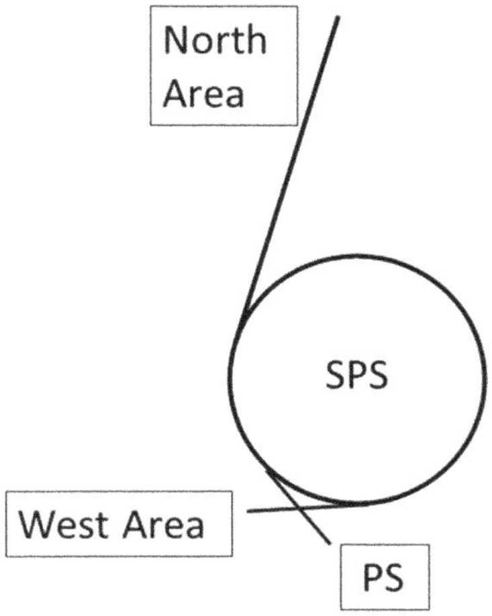

Fig. 5.1. Layout of the SPS with its experimental areas and its injector the PS. The latter is on the original Swiss site of CERN.

Fig. 5.2. Break-through moment: completion of the SPS tunnel.

The construction proceeded rather uneventfully except for the discovery in 1975 that, after about 250 dipoles had been manufactured and about 100 installed, some showed a low resistance between the insulated coils and ground which could lead to breakdowns. The SPS magnet team investigated and found the cause to be the use by the manufacturer of an inappropriate acid to clean the tag ends of magnet coils before brazing the terminals, which damaged the insulation. All suspect magnets were repaired without delaying the project. During construction the team also quietly found a way to equip the machine to be capable of acceleration up to 400 GeV. A section of the fully equipped SPS tunnel is shown in Fig. 5.3.

Five years after approval the SPS machine was operational. In the morning of the June 1976 Council meeting Adams announced the successful acceleration to 300 GeV and formally asked permission to re-programme the machine for 400 GeV operation, although Council was doubtless aware of the capability. Acceleration to 400 GeV was announced a few hours later, just before dinner.

Later in the year, the West Hall received the first extracted beams, initially limited to 200 GeV not to exceed the legal radiation background levels [3,4]. The experiments were already installed and the experimental programme could start without delay. Thanks to the careful design, the top energy could later be increased to 450 GeV at the request of particular experiments which later turned out to be a good investment for the LHC, which would benefit from a higher injection energy.

Fig. 5.3. The SPS magnet string with dipoles (red) and quadrupoles (blue). The vacuum chamber is visible before entering a sextupole downstream of the dipole.

During the construction known technologies had to be further developed and novel approaches introduced to cope with the sheer size of the facility and the high beam energy. New alignment methods had to be developed as the tunnel could only be accessed by six shafts. First, survey pillars were erected covering the site for accurately determining the position of the shafts and the position relative to the PS. The tunnelling machine was laser-guided which took its reference position from a gyro-theodolite set-up every 32 m resulting in an average error of only 20 mm at the end of a 1.2 km long sector.

Although most of the main components of the SPS were rather conventional, a high standard was applied which turned out to be a real bonus for reliability and, later, enabled the conversion of the SPS into a collider (Chapter 6). An example is the vacuum system, which though much less demanding than that of the ISR, was constructed with the same care, to maximize reliability over the 6.9 km length. It later permitted the average pressure to be reduced from 3×10^{-5} to 3×10^{-7} Pa (nitrogen equivalent) for the collider operation by the simple addition of vacuum pumps.

A major effort was devoted to develop a flexible extraction system providing both fast extraction (burst duration 1 μs to 23 μs) and slow extraction with a spill duration from 70 μs to 4.8 s. Certain modes of extraction offered the option to stop the spill after extraction of part of the beam and re-accelerate the remaining beam for extraction at higher energy [5]. After an upgrade of the West area, simultaneous slow extraction to the West and North areas at up to 450 GeV became possible. To master these tasks, technology known from the PS had to be developed to accommodate the much higher beam energy. Electrostatic deflectors of 12 m length were developed to provide a deflecting field of typically 12.5 MV/m over a 20 mm gap using as septum a row of vertical 0.1 mm diameter wires of tungsten-rhenium alloy [6]. The pulsed magnetic deflectors downstream of the electrostatic deflectors provided a field of 0.47 T in a 10 cm gap with a copper septum of 4.2 mm thickness carrying 7.5 kA [7]. As all these devices operated in vacuum, they needed to be baked-out in situ to reduce outgassing. The alignment precision in the vacuum tanks of ± 0.15 mm and the fierce radiation environment imposed a strict selection of materials.

For accelerating the beam special travelling-wave structures were developed operating at a frequency of 200 MHz [8]. Lower energy synchrotrons such as the PS accelerate protons through velocities ranging from a fraction to near the speed of light. The standard way had always been to tune the accelerating cavities through a comparable range of resonant frequencies with blocks of ferrite within the cavity. These were tuned by changing the bias current in a winding around the ferrite to alter its permeability and hence the resonant frequencies of the accelerating cavities [Highlight 3.3]. At the SPS, in which particles are already injected very close to the speed of light, a much smaller tuning range (0.5%) was needed. This led to the clever idea of using the more efficient travelling wave structure, as used in linear accelerators. Such a structure accepts a band of driving frequencies which easily covers the narrow range of velocities in the SPS. Complex and lossy ferrite tuners were not required.

When the SPS was designed, computers were used in accelerators in a stand-alone way generating, e.g., time-dependent operational functions, or for operator consoles, but were not linked. The sheer dimension of the facility and technological advance in industrial products suggested a new approach to the control system. The first truly distributed multi-computer control system was developed introducing a number of novel features, which served as a springboard for application to other accelerator projects at CERN [Highlight 5.2]. This facilitated the interaction of the operators with the system, e.g. using the operator consoles equipped with the now ubiquitous touchscreens for easy monitoring and control. A general purpose "knob" was also made available to the operator for programmable assignment of several functions to a certain task.

Another major undertaking was the construction of the long tunnels and beam lines to the experimental halls (Fig. 5.1). The North area received the first beam in early 1978 to produce a high-intensity muon beam with energy of up to 280 GeV. This muon beamline was modified a few times to match the requirements of the various experiments and is still in use today [Highlight 5.4]. A further example of the many innovative beam lines, which were and still are being constantly optimized for a particular experiment, is the line producing simultaneous and nearly-collinear K^0 meson beams. This is the most complex beam line ever constructed at CERN [Highlight 5.5].

The most dramatic change, however, to the SPS was its second incarnation as a proton-antiproton collider, which enabled CERN's Carlo Rubbia and Simon van der Meer to earn their Nobel Prize for the discovery of the W and Z (Chapter 6).

In the late 1980s, at the end of collider operation, the SPS was equipped to accelerate electrons and positrons from 3.5 GeV to 20 GeV whenever LEP (Chapter 7) required filling, interleaved with fixed-target proton or ion operation. To this end, an additional RF system was installed consisting of 32 newly developed standing-wave accelerating cavities. These were later gradually replaced by superconducting cavities, the first units being prototypes of the novel niobium-clad copper standing-wave cavities for LEP, before mass production [Highlight 7.4]. Substantial effort was devoted to shield the magnet coils in the ring against the highly directional synchrotron radiation of the circulating electrons or positrons [9].

The most recent major change occurred in a long shutdown in 2005, when this versatile accelerator was transformed into an efficient injector of protons and lead ions for the LHC. This required installing an extraction system for 450 GeV protons and 176 GeV/nucleon lead ions for the LHC anti-clockwise ring, the upgrade of the 200 MHz and 800 MHz RF systems, and rebuilding the injection kickers in order to shorten the rise-time and to reduce substantially the ripple of their pulse [10].

And so, while there were engineering innovations in the SPS which had far-reaching impact, the machine was most remarkable for its high standard of construction, which has allowed this amazingly versatile accelerator to respond to the constantly evolving demands of a dynamic physics research community.

The SPS physics programme

The SPS physics programme was discussed and detailed in workshops and meetings involving large audiences of the HEP community and providing input concerning the variety (hadronic, leptonic, charged and neutral) and characteristics (intensity, momentum resolution) of the possible beams from the SPS [11]. The locations of beams and of some of the experiments (WA for West Area, NA for North Area) are shown in Figs. 5.4 and 5.5. The flexibility, versatility and high quality provided by the SPS machine yielded a wealth of remarkable physics results [12]. In many cases the science output was only made possible by technological innovations both in the beam lines and in the detectors. A brief overview of the two experimental areas and research highlights illustrates the interplay between technology and research.

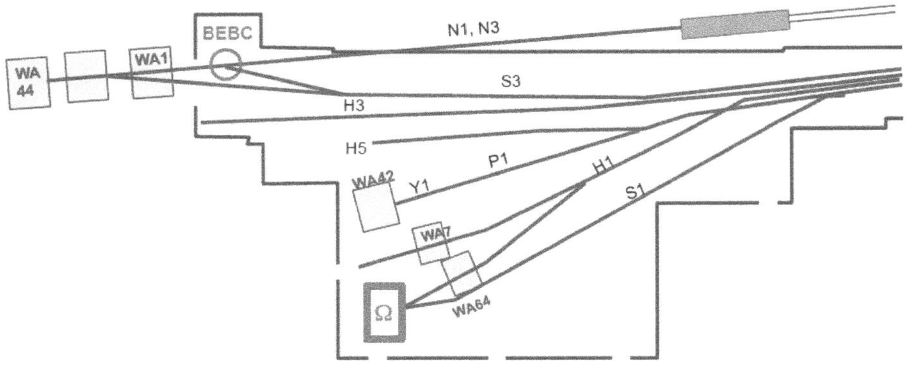

Fig. 5.4. Initial layout of beams in the SPS West Experimental Area, WA.

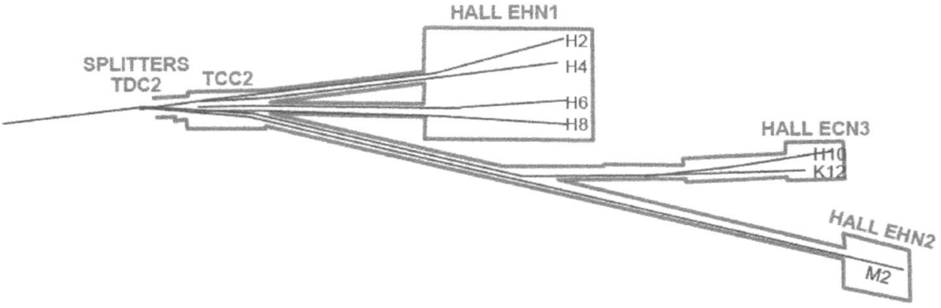

Fig. 5.5. Schematic layout of the North Experimental Area in 1980.

The West Area (WA), initially used as a PS area, was first to receive beams from the SPS. Operation began in 1976 with secondary beams derived from 200 GeV/c (later 250 GeV/c) primary protons. The charged hadron beam S1 led to the first SPS physics publication and was later used as a RF separated beam [Highlight 3.7]. The initial layout of the West Area is shown in Fig. 5.4.

Later the WA was transformed to receive two high energy secondary beams (H1 and H3), produced by 400 (450) GeV/c protons in an underground target station, modelled after the beams to EHN1 in the North Area. Additionally, two 'long' beams (N1, N3, S3) were derived from 400 GeV/c fast extracted protons targeted in underground caverns, near the extraction point from the SPS. These beams were directed towards the Big European Bubble Chamber (BEBC) and electronic detectors downstream of BEBC.

The North Area (NA), a new facility, tailor-made to receive the high energy beams, incorporated several innovative aspects. Its layout in the year 1980 is shown in Fig. 5.5. The NA features a separate underground target cavern, TCC2, and surface experimental halls, EHN1 and EHN2, joined by tunnels housing the beam lines. The layout resulted from the requirement to provide beams of well-defined momentum ($\Delta p/p \approx \pm 2 \times 10^{-4}$) up to the highest momenta. This beam layout also ensures that the main muon background is absorbed in the ground underneath the experimental hall (Fig. 5.6).

Upstream of TCC2, the slow-extracted, primary proton beam of 400 GeV/c (later 450 GeV/c) from the SPS was split three-fold, with two series of septum magnets, so as to impinge on two general-purpose target stations and one dedicated to the production of a high energy muon beam. Each of the two general-purpose

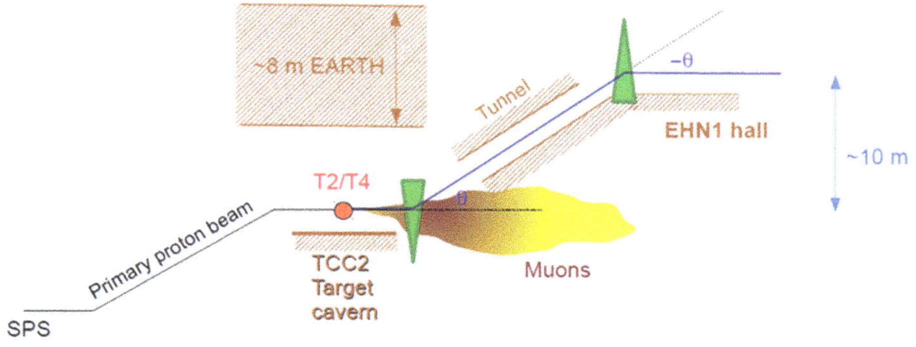

Fig. 5.6. Schematic vertical section through beam lines in the North Experimental Area.

target stations comprised horizontally-deflecting magnets, allowing the angle and position of the proton beam to be varied ('wobbled') onto the target. Thereby the momenta of two secondary beams emitted in fixed directions could be decoupled, whilst the majority of muons produced were not accepted by the subsequent vertical bends of the beams. A wide choice of particles of different momenta could be transported by the four beam lines: primary protons, attenuated in intensity, charged secondary particles and relatively pure tertiary particles, derived from neutral particles, such as polarized protons from $\Lambda^0 \rightarrow p\,\pi$, low-energy pions from $K^0_S \rightarrow \pi^- \pi^+$, $e^\pm$ from $\gamma \rightarrow e^- e^+$. These beams could also transport various ions species from the SPS.

Later, an underground experimental cavern, ECN3, was added as part of the North Area High Intensity Facility (NAHIF), designed for experiments requiring the highest-intensity primary and secondary beams.

The many facets of the research landscape

Exploration of the nucleon structure

One major research activity was the exploration of the nucleon structure [Box 4.2], using the weak probe (i.e. neutrino scattering), the electromagnetic probe (i.e. muon scattering), as well as hadron scattering, producing photons and lepton pairs born from the energetic or "hard" scatter of their constituents. The use of polarized beams and targets was an important asset of these programmes [13].

Neutrino scattering [14] was performed in the WA with the WANF neutrino facility. Both wide- and narrow-band neutrino beams (N1, N3) were available, employing magnetic horns [Highlight 3.6] to focus the parent pions and kaons into an evacuated decay tunnel, followed by the hadron absorber made from iron and concrete. These beams served BEBC and a series of 'counter' neutrino experiments: WA1/CDHS, WA18/CHARM, later NOMAD, CHORUS, installed on the same line, downstream of BEBC and pushing detectors to new dimensions. The heavy liquid bubble chamber GARGAMELLE (chapter 3) was installed in the neutrino beam line, downstream of BEBC. BEBC was complemented with the External Particle Identifier (EPI), which measured the energy loss dE/dx of the charged tracks leaving the bubble chamber [Box 5.2] and with the external muon identifier (EMI).

Later on, the same beam was used to search for τ neutrino interactions. No ν_τ event was observed. This was because the scale of ν oscillations (that were to be discovered later elsewhere) was too large to be detected by the contemporary experiments at CERN.

Muon scattering [15], performed in the north area, used a most remarkable muon beam. The programme started with the experiment of the European Muon Collaboration (NA2/EMC), a powerful open spectrometer, and NA4. The latter used a 40 m long liquid hydrogen target surrounded by toroidal magnetized iron discs interleaved with detectors to measure the properties of scattered μ's. EMC was followed by an upgraded version, the New Muon Collaboration (NMC). The muon section of the beam line was rebuilt for the Spin Muon Collaboration (SMC) experiment, exploiting the high degree (~80%) of polarization of the muons in the beam and using a high-performance polarized target [Highlight 5.8]. The beam was again modified for the NA58/COMPASS experiment, using the M2 beam line both for a muon beam and as a high energy hadron beam. This beam line is a fine example of productive longevity!

Hard hadron scattering [16] experiments were performed with several beams and in both areas. The signal under study was either lepton pairs in, for example, WA11, WA12 (OMEGA), NA3, NA10, NA34 or energetic photons in, for example, WA70 (OMEGA). Information was obtained on the structure of both the target nucleon and the incident hadron (pion, kaon, antiproton).

The identification of beam particles was essential for this programme. This was achieved by using an RF separated beam, such as the WA S1 beam or with special differential Cherenkov counters [Box 3.2]. These 'CEDAR' detectors employed a mirror at the downstream end of a pressure tank filled with helium gas, and an achromatic ring-focusing lens to concentrate the Cherenkov light through an annular diaphragm onto 8 photo-multipliers at the upstream end (Fig. 5.7).

Loops **Box 5.1**

Elementary particles live in a quantum world and the properties of the quantum vacuum lead to non-intuitive effects.

A particle propagating may "borrow" energy from the vacuum and turn into a pair of other particles (creating a "loop"), provided the relevant quantum numbers (electric charge, etc.) are conserved. The larger the loan, the faster it is reimbursed, in agreement with Heisenberg's energy-time uncertainty relation. The loop disappears again and these "virtual" particles do not come to "real life". Their existence has nevertheless an effect and modifies the rate of a process in which they occur. In the SM [Box 6.4] this effect depends on the mass and nature of the particle appearing as virtual and is precisely calculable.

In e^+e^- collisions on the Z^0 resonance (see Chapter 7), the propagating Z^0 may fluctuate e.g. into a top-antitop pair (Fig. 1a), even though the top is too heavy to be really produced. Measuring accurately the Z^0 final states at LEP1, one obtained an idea of the top mass. The effect is substantial and had to be well mastered before being sensitive to the even tinier effect of the virtual BEH ("Higgs") boson [Box 8.2], Fig. 1b.

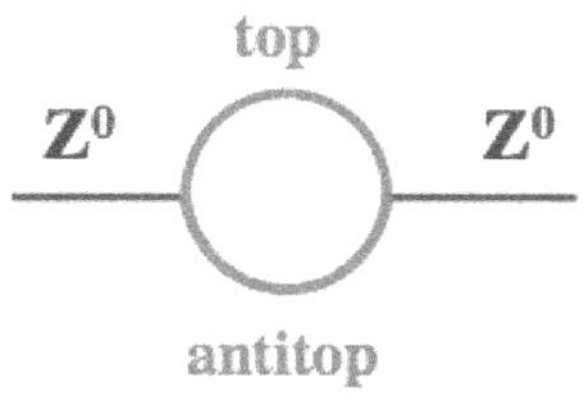

Fig. 1a. Top-antitop loop.

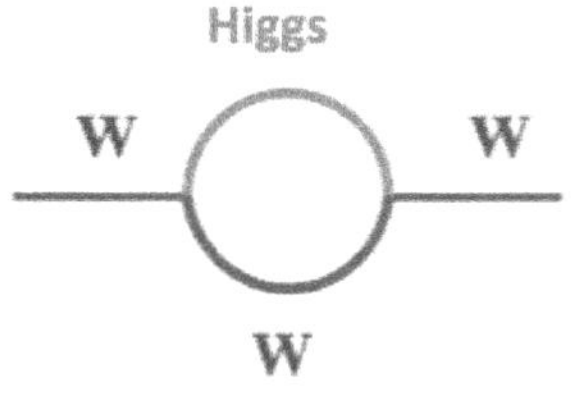

Fig. 1b. W-Higgs loop.

This effect concerns any kind of virtual particle, in particular hypothetical ones, e.g. supersymmetric particles [Box 7.2] and provides a way to search for them. This is one goal of beauty physics performed at B factories and LHC (Chapter 8).

An "indirect" hint of a virtual particle is a precious guide but "discovery" of a new particle requires producing and observing it. The agreement between the direct and indirect measurement of its mass tests the SM at the "loop" level. This confirmation of the SM was one of the major achievements of the LEP programme.

Another effect of virtual particles is the "running of coupling constants" which evolve with the energy scale of the process. The one associated with the weak force is almost constant; with the electromagnetic force it increases at smaller distance, i.e. higher energy. For strong coupling, due to the self-interaction of gluons, it does the opposite [Box 4.2]. Triple convergence of the coupling constants at very high energy is quasi-perfect with Supersymmetry.

Probing particle reactions at higher energies improves the resolving power and exhibits more details, such as higher sensitivity to fluctuations into virtual particles. These studies (violation of "scale invariance") were one of the main topics of Deep Inelastic Scattering at the SPS (Chapter 5).

In the special underground area ECN3 the primary proton beam served to produce high intensity pion and photon beams for a series of experiments: NA10 studying muon pair production, NA14 performing direct elastic scattering of real photons on quarks, called QED Compton scattering, and NA34 searching for unconventional sources of lepton production.

There were important results on nucleon structure. Comparing structure functions measured in neutrino and lepton scattering, as well as QED Compton scattering, brought the final word about the fractional electric charge of quarks.

Another important result was the evidence that the measured nucleon structure actually depends on the resolving power of the observation, a basic prediction of quantum chromodynamics (QCD) [Boxes 4.2 and 5.1]. This is the so-called "scaling violation", observed in particular in neutrino scattering, first by the bubble chambers BEBC and Gargamelle and subsequently, with higher accuracy, by the electronic detectors. Muon scattering revealed that the nucleon structure depends on the nucleus in which the nucleon is contained (EMC effect) and that the motion of quarks in nucleons inside the nucleus decreases for heavier nuclei. One series of muon scattering experiments used a polarized target, where the spins [Box 2.2] of the target protons and therefore also the spins of the constituent quarks were aligned along the direction of a magnetic field. Conservation of angular momentum makes muon scattering sensitive to the spin alignment and significant differences in scattering cross sections are expected depending on the direction of alignment. This was not born out experimentally and revealed that the nucleon spin is only partly carried by its valence quarks, and that sea quarks and gluons also contribute. This "spin crisis" changed our understanding of the nucleon.

For energetic hadron scattering, the observed production rates of lepton pairs or prompt photons were found to be higher than estimated at that time, by some multiplicative "K-factor", which called for more elaborate theoretical estimates, in the framework of the then emerging QCD to which the SPS experiments contributed substantially, in particular with results from collisions of heavy ions.

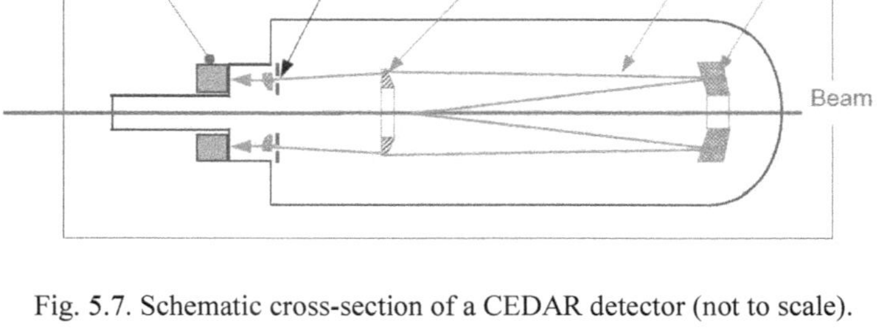

Fig. 5.7. Schematic cross-section of a CEDAR detector (not to scale).

Particle Identification (PID) via energy loss (*dE/dx*) **Box 5.2**

A charged particle traversing a medium interacts with the electrons of the atoms, which are excited or ionized. At each collision some energy (*E*) is transferred. The *E* loss per unit distance, *dE/dx*, depends on the particle velocity $v = \beta c$. The average $\langle dE/dx \rangle$ is given by the Bethe-Bloch (BB) curves (Fig. 1), which show:

- a decrease ~ β^{-2} (electric force of atomic e^- acts more briefly on faster particles);
- after a minimum, a rise ~ $\ln(\beta\gamma)$ due to the relativistic increase of the particle's transverse electric field (TEF), $\propto \ln \gamma = \ln(E/mc^2)$, increasing interaction cross-section;
- saturation at high $\beta\gamma$ due to medium polarization, shielding TEF far from the track.

E loss over a path length is the sum of *E* transferred in many collisions, with *E* fluctuating statistically about $\langle dE/dx \rangle$. Rarer large *E* transfers in single collisions skew the energy loss distribution to high energies (Landau distribution). Sampling *dE/dx* many times along the track gives an estimate of $\langle dE/dx \rangle$ with high statistical significance. Due to the skewness of the distribution, the average obtained by excluding 15–50% highest *E* losses (Truncated Mean, TM) is a better estimator for the energy loss than the mean. The TM distribution of many tracks of same *v* is centred about a value TM(*v*) and is close to a Gaussian with variance σ_{TM}^2. It can be related to $\langle dE/dx \rangle$ as given by the BB curves and hence to the $\beta\gamma$ (or *v*) of the particle. For PID, momentum *p* and TM are measured. Two species of particles with the same *p* but different *m*, hence different *v*, will produce different TM distributions. The resolution R ($=\sigma/E$) is the relative accuracy of the TM measurement. Gaseous detectors can achieve $R < 5\%$ for a careful choice of gas mixture [1]. PID works well at low β, fails around the minimum, and is (statistically) useful in the relativistic rise region as shown in Figs. 2 and 3 [2]. Various devices have been made to provide PID. The External Particle Identifier (EPI) was installed behind BEBC [Highlight 5.7]; the large pictorial drift chamber ISIS for the European Hybrid Spectrometer achieved a TM resolution of 3.5% in the 2–80 GeV/c range. TPCs (ALEPH, DELPHI at LEP, ALICE at LHC) provide good sampling along tracks, and $R < 5\%$.

[1] I. Lehraus, *Nucl. Instr. & Meth.* **217**, 43 (1983). [2] C. Lippmann, arXiv:1101.3276.

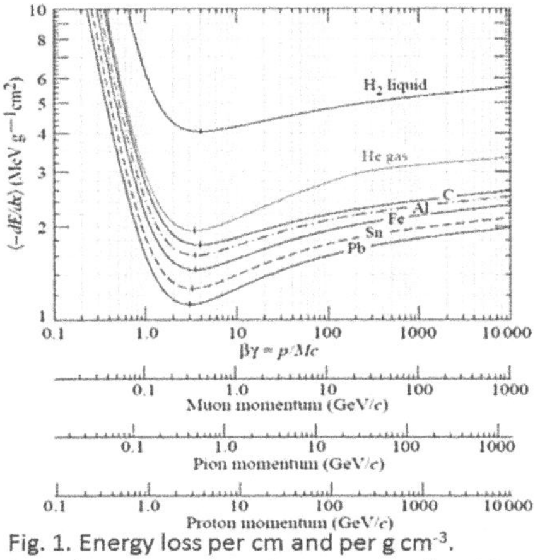

Fig. 1. Energy loss per cm and per g cm^{-3}.
[K. Nakamura *et al.*, *J. Phys G* **37** 0287 (2010)]

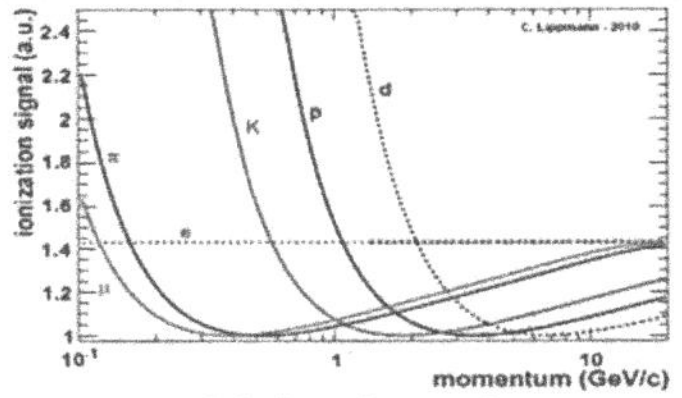

Fig. 2. Typical dE/dx of particles.

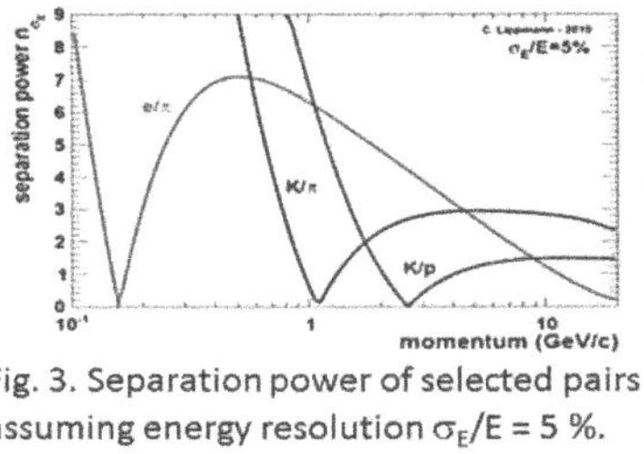

Fig. 3. Separation power of selected pairs assuming energy resolution $\sigma_E/E = 5$ %.

Heavy flavours [17]
A second major research area was the study of the heavy quarks or flavours: charm, discovered in 1974, and beauty discovered in 1977. While these discoveries came from US laboratories, CERN was very active in the study of heavy flavour spectroscopy, benefiting from the variety and quality of hadron and photon beams and pioneering the use of very accurate micro-strips silicon detectors [Highlight 5.9] exploiting the finite lifetime of charm and beauty particles. Later, pixel detectors provided even higher spatial resolution.

Micro-strip detectors were first used in the WA75 experiment. The OMEGA spectrometer performed a long series of experiments, using hadron, electron/photon, hyperon and ion beams, which included, e.g. the observation of several particles containing beauty quarks. Pixel detectors also found their place in some OMEGA experiments, such as WA89 and WA97. In the NA, silicon devices, active targets, strip detectors, pixel and CCD detectors made essential contributions in several experiments, such as NA1, 10, 11, 14, 32, 34, and 57.

The European Hybrid Spectrometer (EHS), in the H2 beam line, combined the virtues of small to medium-sized bubble chambers with those of a large magnetic spectrometer. It also housed an early *dE/dx* detector (ISIS) to study charm decays.

Heavy ions [18]
The SPS pioneered the study of ultra-relativistic heavy ion collisions. This programme aimed at the study of nuclear matter under extreme conditions of energy and temperature. It opened a new field of physics, initially explored with several survey experiments and a range of dedicated experiments looking for a putative "quark-gluon plasma", characterized by unconfined quarks and leptons, whose existence is one of the "golden" predictions" of QCD.

Initially, oxygen and sulphur ions were accelerated, followed in a later phase by heavier ions, up to lead. The programme was performed in EHN1 (NA34, NA45, NA49, NA57 and now NA61) and ECN3 (NA38, NA50, NA51, NA60).

In 2000 CERN announced the discovery of a 'new phase of matter', consistent with the onset of the Quark Gluon Plasma.

CP violation [19]
One outstanding achievement was the accurate measurement of CP-violating processes in neutral kaon physics [Box 3.4], most notably the evidence for direct CP-violation, using elaborate and unique beam and experimental techniques.

For experiment NA31, 450 GeV/c primary protons were targeted to produce a beam rich in long-lived neutral kaons, K^0_L, which entered and decayed in a large vacuum tank. Alternately, a K^0_S beam was generated from a 360 GeV/c secondary proton beam, incident on a separate target. This target was mounted on a train that could move into 41 longitudinal positions inside the evacuated decay volume. With this beam NA31 obtained the first positive indication of direct CP-violation, while a competing US experiment obtained a result compatible with no signal [20,21]. A second generation experiment, NA48, was designed to establish unambiguously and measure the direct CP-violation parameter ε'/ε in the K^0 system. The novel and decisive concept of the experiment was based on a pair of simultaneous, nearly collinear beams of K^0_L and K^0_S and used a state-of-the-art liquid krypton calorimeter [Highlight 5.6]. These efforts paid off: after 30 years of measurements, both the CERN NA48 and the US KTeV experiments finally proved the existence of direct CP-violation, a major facet of the Standard Model. The beam line and detectors have now been transformed for NA62 that aims to measure the ultra-rare decays of K^+ into $\pi^+\nu\bar{\nu}$.

Recent developments

A very special muon-neutrino beam has been constructed for a tau-neutrino experiment (Fig. 5.8). The neutrino beam derived from secondary pions and kaons decaying in an evacuated decay tube approximately 1 km long, points through the earth to about 730 km away in Italy, where the OPERA detector in the Gran Sasso Underground Laboratory searched for events indicating the appearance of τ-leptons, produced by τ-neutrinos interacting in the detector. These τ-neutrinos were to appear from the transformation the original muon-neutrinos. Five events were detected from 1.8×10^{20} protons on the neutrino production target. The neutrino beam has now been dismantled and the facility is being transformed into a test bed for proton driven plasma wake field acceleration, the AWAKE project. (Chapter 12). This may in turn become a future technological highlight!

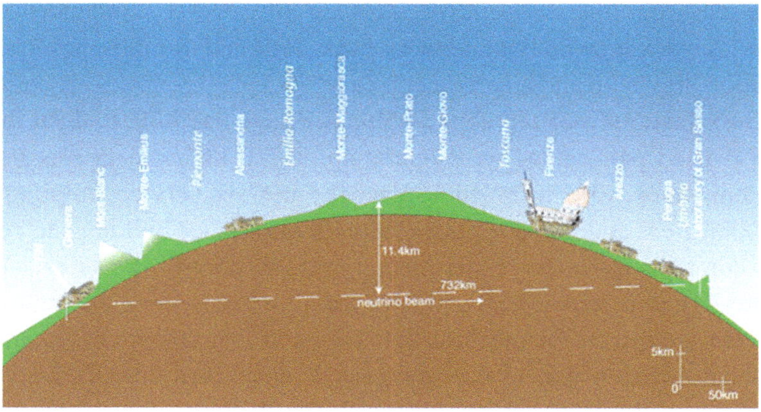

Fig. 5.8. Schematic cross section showing the CERN Neutrino to Grand Sasso (CNGS) beam line.

5.2 SPS Distributed Control System: The Emergence of Local Area Networks

Pier Giorgio Innocenti

The control system for the SPS incorporated a host of innovations. These included the first truly distributed multi-computer control system, where any action could be carried out from any part of the system, the first multi-computer operating system, the first large scale use of an interpretive language for the applications programs, the introduction of the Data Module for isolating the applications programmer from the hardware details, and the provision of simple-to-program colour graphical and other interfaces to the operator [22,23].

In the early 1970s, when the SPS design was being discussed, computers had been applied to issues relating to control in a number of accelerator laboratories, but mostly in an *ad hoc* fashion. When more than one computer was involved, they either performed independent control on different parts of the accelerator, or were organized in a master/slave relationship. For the SPS, given the large size of the ring, with its six access shafts and equipment buildings, and the need for local commissioning, it was mandatory to have the computer system distributed around the accelerator. Each of these distributed computers had to be able to operate autonomously. Instead of having them report to a single large central computer, as was normal practice for such a system, they had to be connected to a network that would allow any two computers to communicate independently of the rest. Such a network is now called a Local Area Network (LAN) [Highlight 9.3], but no such system existed at the time. Even Ethernet, for which it was early days (not yet proposed as a standard), would not have satisfied all the requirements, e.g. long range operation with cable runs of up to 6 km. Moreover, high level network protocols ensuring the cooperative working of several computers were not available. A packet-switching network was designed for this application, the specification drawn up and a contract placed with industry for its implementation. The network featured star topology, with a message handling node at the centre. Links of twisted pair cables were operating at 750 kbit/s, with repeaters every 2 km. Computers were installed in each of the six service buildings, in the experimental areas and in the control centre: in the service buildings to control the machine components, in the experimental areas to control the secondary beams and to interface with experiments, in the control centre to drive operator consoles, provide data storage for the machine data base and for program development.

A crucial decision was to use interpreted rather than compiled applications programs. It was clear that developing the applications programs for the operation of the SPS would require a very large effort in a relatively short time. The typical solution would have been for the equipment experts write detailed specifications and to engage a large number of programmers to turn them into code, debug and modify them as the experts changed their specification. This was not the route followed, because it was thought preferable to provide a user-friendly and safe computing environment, such that the accelerator experts could write most of the applications programs themselves. The key was the use of an interpreter, a program which executes instructions in a high-level language directly, rather than compiling them and then executing the resultant machine code. Since no interpreter for a multi-computer system was available, a new one was developed, called NODAL [24]. This was the first true multi-computer operating system. Commands could be sent from any computer to a target machine for executing locally a subroutine (procedure) and returning the results to the caller: it performed what was to become standard practice — Remote Procedure Calls (RPC).

Precise timing and sequencing required for accelerator operation could not be executed by the computer network: a separate timing network distributed clock and coded "events" with microsecond precision. The clock and events were used to trigger accelerator components directly and to start programs in computers.

Another innovation was the 'Data Module'. Until then, in most accelerator control systems using computers, the applications programmer was expected to incorporate direct references to individual hardware interface devices in his programs and carry out any necessary conversions. However, as subsequent changes to the interface might involve changes to many programs, the risk of introducing errors was large. A data module dealt with all the low level interaction needed for the control of a given type of equipment, e.g. power supplies, taking into account variations between different types. The application programs could then call the data module, only having to give the name of the power supply, the action to be carried out and values in engineering units. A change to an interface, or a modification to a power supply, only required the data module to be modified, leaving the application programs unchanged. The Data Module combined data and associated actions (methods) into a logical structure (object): it was the first example of what is now called Object-Oriented Programming (OOP).

Innovative devices were developed for presenting information to, and receiving instruction from the operator in the control room: Programmable 'touch buttons' on a display [Highlight 5.3] and a track ball pointing to regions of the display itself, as done today with a mouse, touchpad or joystick. In addition, a computer programmable knob could be dynamically assigned to functions more easily performed by turning such a knob. All computers in the network could be equipped with a colour graphics system which allowed computer generated displays to be incorporated in a common manner.

5.3 SPS Controls: A Part of Touch Screen History

Bent Stumpe

The choice of a distributed computer network for controlling the SPS had profound consequences for the interaction of the accelerator operator with the equipment. No direct (hard) wiring would connect the control centre to the equipment to change settings and observe signals. Every control action or piece of information would transit through the computer network via a limited number of operator consoles. The issue was to develop an "intelligent" system based on minicomputers, which would replace the thousands of buttons, switches and oscilloscopes a conventional control system would need for a machine of the size and complexity of the SPS [Highlight 5.2]. Three programmable devices were either newly designed or improved for the machine controls project: a programmable knob which could mimic a variety of hardware knobs and two devices to point at an image on a screen: a trackball, and a transparent touch screen [25]. The operator would need only a few devices to interact with the accelerator and would observe its status and performance on at most half a dozen displays. The devices and displays could be redefined quickly to select accelerator subsystems for control and monitoring, and choose from hundreds of analogue signals those to be displayed at any given time.

The trackball was invented in the 1940s for a British military research project and remained classified for a long time. The CERN design of the track ball had incremental optical encoders, working on the same principle as the ball mouse that became popular in the 1980s. Buttons with labels programmable by a computer [26] could easily be generated on a cathode-ray tube (CRT) display. How the computer would detect which button was being selected, was the question to be answered. The existing rather complicated mechanical designs were unsuitable for the SPS control system. This prompted the development at CERN of one of the world's earliest operational capacitive transparent touch screens, and the first to be used to control a machine of the complexity of the SPS.

The transparent Touch Screen [25,27] consisted of a set of capacitors etched into a film of copper on a sheet of glass, each capacitor being constructed so that a touching finger would change the capacitance by a significant amount. In the final device, a simple lacquer coating prevented the fingers from actually touching the capacitors. The two conductors of the capacity were two interleaved flat copper combs (Fig. 5.9). The fineness of the lines, which constituted the teeth of the combs, and their pitch required a great deal of care in producing the screen, but could be achieved with techniques used to make printed circuit boards. At first, using vacuum deposition of the copper layer on the glass did not result in films

Fig. 5.9. Schematic layout of the transparent double-comb capacitor sputtered on the glass, matching the location of a button.

with reliable adherence. Ion sputtering gave better results. Ensuring that the glass was scrupulously clean and depositing the copper slowly provided films with adherence strong enough to solder connections to the sputtered layers.

The capacitance of each button was about 200 pF, increasing by about 10% under the touch of a finger. The change in capacitance was detected with a phase-locked oscillator circuit, which had become available as a single integrated circuit chip. One circuit acted as a reference oscillator, while each button had a similar circuit. The oscillator attached to a button locked into the frequency of the reference oscillator (120 kHz), so that a change in capacity altered the phase but not the frequency. The phase shift was converted into a voltage shift, indicating that the button had been touched. The circuit was very immune to noise and transients. Changing conditions would be common to all 16 "touch circuits", so that good thermal stability could be obtained with commercial components.

A simple computer algorithm detected a "touch": the computer could handle 16 simultaneous touches. The touch screen developed in 1972, nowadays called a self-capacitance screen, was therefore a "Multi Touch" screen and was able to react to simple gestures. The "Touch" information was then used by the software to execute the desired operation. By 1974 industry was capable of producing robust 16-button glass screens with low surface reflections. In a further development an X/Y touch screen (nowadays called a mutual capacitance screen) was constructed in 1978. When the SPS started operation in 1976 its control room was fully equipped with touch screens. Touch screens later took their place in modernized control systems for the PS, in the Antiproton Accumulator (AA) control room (Fig. 5.10), and subsequently for the much bigger LEP collider. Some of these screens continued to operate until the new CERN Control Centre took over operations in 2006. CERN touch screens were also adopted by other laboratories around the world. The commercial breakthrough for touch screens came some 25 years later, thanks to the technological progress in electronics which led to powerful microchips replacing the electronic circuits of the 1970s.

Fig. 5.10. The "Touch Terminal", equipped with a CRT and driven by a 16-bit microprocessor.

The present CERN Control Centre (CCC), which oversees the control of CERN's entire accelerator complex, including the PS, SPS and LHC, has no touch screens for accelerator control: the use of the ubiquitous mouse as a pointing device provides the same function. However, touch screens still play a role in the CCC, as the operators frequently communicate with colleagues by mobile phones with their capacitive touch screens and, if needed, can even control the accelerator!

The details of the practical Touch Screen, as applied in 1972 to accelerator control at CERN, were published, and following CERN policy, no patent was taken. The concept has since conquered the world with a multitude of applications, from vending machines to rail and airline ticket machines, and after further development to the ubiquitous multifunction smartphone.

5.4 The SPS Muon Beam: Energy, Intensity and Precision

Niels Doble and Lau Gatignon

The high energy muon beam M2 was designed for the NA2 (EMC) and NA4 experiments, located in experimental hall EHN2 in the SPS North Area (Fig. 5.5). These experiments required a high intensity muon flux at beam momenta up to 280 GeV/c. Muons are dominantly produced by two-body in-flight decays of pions and kaons. The lifetime of pions in the laboratory, proportional to their momentum, is about 55 m per GeV/c,[a] so that pions in the 100 to 300 GeV/c range have average decay lengths of from 5 to16 km. This led to the concept of the M2 beam design [28], which consists of a pion and a muon section (Fig. 5.11).

---

[a] Kaons have a shorter average decay length of about 7.5 m per GeV/c.

The hadron decay section

A very high pion flux is a prerequisite for a high muon intensity. A high-intensity primary proton beam of 400 GeV/c or 450 GeV/c impinges on a Beryllium target of up to 500 mm length. Pions and kaons produced in the interactions are captured by a series of six high-gradient quadrupoles immediately following the production target. Three different beam optics were designed to achieve maximum angular acceptance in three momentum ranges. The accepted pions and kaons are then deflected by two groups of horizontal dipoles through angles of 18 and 8 mrad, respectively. In between these dipoles a collimator selects a hadron momentum band. A vertical dipole next to this collimator bends the beam upwards with a 10 mrad slope. From there the beam is matched into a ~ 600 m long decay channel, equipped with regularly spaced quadrupoles of alternating polarity, tuned to transmit the parent pions and kaons and their decay muons with high angular and very wide momentum acceptance. At the end of this section the muons are focused on a hadron absorber, which consists of nine 1.1 m long blocks of beryllium, 55 mm in diameter, encapsulated in aluminium. This absorber stops all the hadrons with minimum perturbance to the muons, which lose less than 3 GeV in energy.

The muon section

The muons traversing the hadron absorber must be momentum selected, using a vertical set of dipoles and a collimation system. The available decay length for the pions and kaons was maximized by installing the hadron absorber modules inside the first three vertical dipoles. As muons lose very little energy when traversing material [Box 3.2], classical collimators are not at all efficient, and 5 m long

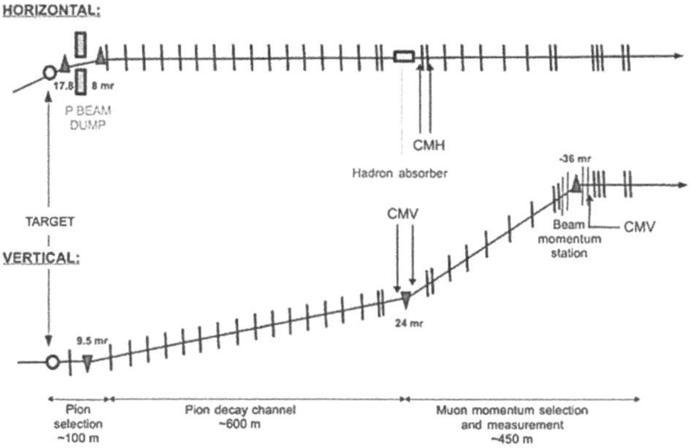

Fig. 5.11. Conceptual layout of the M2 muon beam.

magnetized iron toroids ('scrapers') with adjustable gap are used as collimators. These also define the geometrical acceptance of the muon beam. After collimation and cleaning with additional toroids, the muons are transported to the surface and deflected to be horizontal for its passage through the experiments. Four planes of hodoscopes surround this vertical bend for fast and precise momentum measurements of the individual muons. The final section of the beam serves to focus the beam at the experiments.

The beam performance

At present, the beam delivers up to 5×10^8 muons at 280 GeV/c per SPS cycle of 2.4×10^{13} protons impinging on the production target, but the intensity drops rapidly with increasing momentum. The size of the beam spot at the final focus is 8 mm RMS in both planes with a RMS divergence of 0.8×0.4 mrad. The maximum muon momentum spread is 3% RMS. The momentum can be varied between 60 and 280 GeV/c. In normal operation the ratio between the central muon and hadron momenta is chosen to be of the order of (92–94)%, i.e. selecting forward muons in the decay rest frame. This ratio gives the best muon flux per incident proton, and in addition, due to parity violation in the pion and kaon decays (the spin of the μ^+ is antiparallel to its direction in the pion rest frame), gives a very high degree of beam polarization, close to −80% for μ^+ and +80% for μ^- (the spin direction flips due to charge conjugation) [Box 2.2]. The availability of polarized muon beams was (and still is) essential for studies of spin dependence of structure functions [Box 4.2], most recently by the COMPASS Experiment (Fig. 5.12) [29].

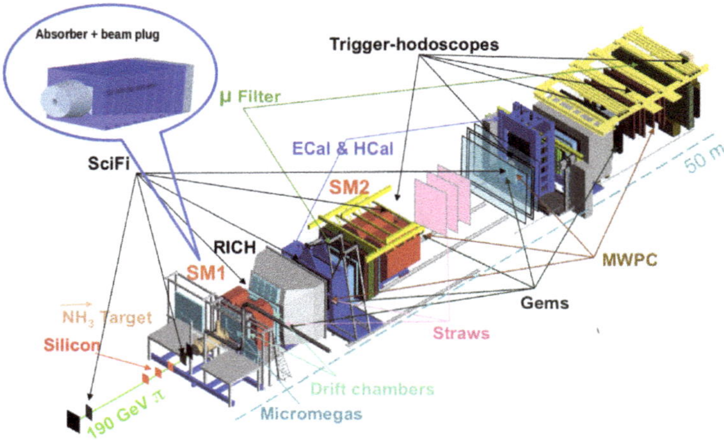

Fig. 5.12. The COMPASS experiment, showing a layout typical for fixed-target experiments. A panoply of experimental techniques is deployed for wide-ranging research. (Courtesy COMPASS).

Channelling **Box 5.3**

When positively charged particles impinge on a mono-crystal at a sufficiently small angle with respect to the orientation of the crystal planes, they experience the collective effect of the electrical fields of the ions in the crystal lattice, as if the charges were smeared over the plane. The particles see therefore a potential well, in which they are trapped in case the transverse energy of the particles with respect to the plane is smaller than the depth of the potential well. In that case the particles are kept away from the nuclei and the particle is said to be 'channelled'. Depending on the particle momentum, the transverse energy limit translates into a critical angle, which for the (110) plane of a Silicon crystal is about 7 μrad for 450 GeV/c protons. The critical angle varies with $1/\sqrt{p}$. Particles that directly hit a crystal plane, will most likely interact with one of the ions and be lost. This leads to a maximum transmission of particles into the bulk of the crystal that limits the ultimate efficiency for channelling.

In case the crystal is slightly bent, the transverse potential is modified by the interaction with the charged particle and the potential barrier is lowered at the outside plane, as a function of the radius of curvature. Particles that have a transverse energy lower than the reduced well depth are still channelled and will thus follow the bend. Those that have a larger transverse energy will be de-channelled. Below a critical radius of curvature R_c, no particle can be channelled any more. For a radius of curvature larger than R_c, the channelling efficiency increases with the radius of curvature. For a given crystal length and bending angle, one may compute the equivalent magnetic field that would be required to give the same deflection over the same length. In the figure below the measured deflection efficiency for 450 GeV/c protons is shown for protons along the (110) plane in a Germanium crystal. The dashed line describes the expectation from a model. A curvature of 10 mrad over the 50 mm length of the crystal in this case would correspond to a field of 300 Tesla!

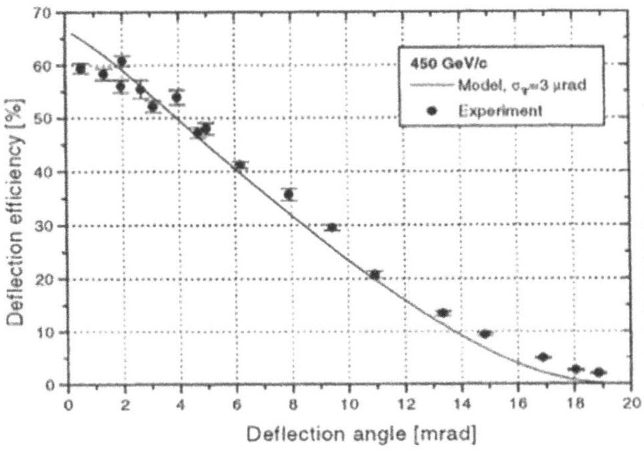

One of the first applications of channelling in a bent crystal in an operational beam line was its use in the K° beam for the NA48 experiment [Highlight 5.5]. Recently the use of bent crystals as collimators for circulating beams has been tested in the SPS with a view to possible application in higher energy accelerators. Bent crystals are also being considered to extract fixed target beams from high energy proton colliders.

5.5 Two Very Special K^0 Beams: Discovery of Direct CP Violation

Niels Doble and Lau Gatignon

A pair of simultaneous and nearly collinear beams of long- and short-lived neutral kaons, K^0_L and K^0_S, formed an integral part of the NA48 experiment, aimed at studying direct CP-violation [Box 3.4] at the SPS. This study required the four decay rates, $K^0_L \rightarrow \pi^0 \pi^0$, $K^0_L \rightarrow \pi^+ \pi^-$, $K^0_S \rightarrow \pi^0 \pi^0$, $K^0_S \rightarrow \pi^+ \pi^-$, to be measured with extremely high relative accuracy in a common set of detectors [30], among which a liquid krypton calorimeter [Highlight 5.6] played an essential role. The two beams were obtained by a novel application of a bent silicon crystal.

Design Concept

Key to the success of NA48 was its concept to minimize painstakingly systematic errors in the relative decay rate measurements. The K^0_L and K^0_S beams were therefore designed to enter the same fiducial region simultaneously and along adjacent lines, converging at a small angle towards the NA 48 detectors.

A primary proton beam of high momentum (450 GeV/c) and high flux ($\sim 10^{12}$ protons/s) was required to produce sufficient numbers of the rare, CP-violating $K^0_L \rightarrow 2\pi$ decays. A much lower proton flux (by a factor $\sim 10^{-5}$) sufficed to produce comparable numbers of K^0_S decays. The high-intensity K^0_L beam was obtained from a branch of the slow-extracted high-intensity primary proton beam from the SPS. A small fraction of these protons was shaved off using the phenomenon of 'channelling' in a bent crystal to derive the proton beam for the K^0_S production [Box 5.3]. The channelling method offered three advantages: the selected protons were reduced in flux, deflected and defined to have a suitable small emittance. Moreover, the relatively low flux of protons could be detected and tagged individually, so as to attribute them to the correct K^0 beam.

Two target stations, one for high-rate K^0_L production, the other one for low-rate K^0_S production were used. They had to be located at quite different distances from the detectors, given the very different decay lengths of K^0_L and K^0_S ($\lambda_L = 3480$ m, $\lambda_S = 6.0$ m, at a mean momentum of 110 GeV/c).

The acceptances for detecting $\pi^0 \pi^0$ and $\pi^+ \pi^-$ decays were each functions of the kaon momentum (p_K) and longitudinal position of the decay vertex (z). It was therefore important that the momentum spectra of the K^0_L and K^0_S decays were as similar as possible over the useful range of momenta ($70 < p_K < 170$ GeV/c) and of z. To this end, the z-range was chosen taking into account the K^0_S decay length (λ_S) and weighting factors were then applied to equalize the z-distributions and hence the detector acceptances for K^0_S and K^0_L decays.

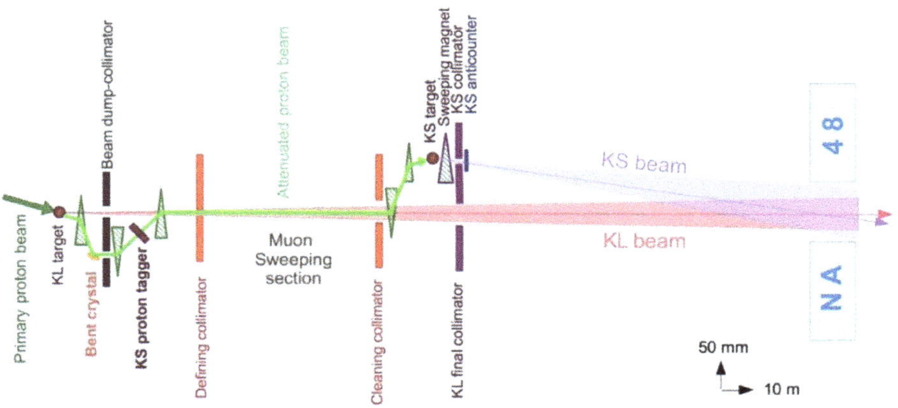

Fig. 5.13. Schematic layout of the K^0 beams.

Layout

The beams were derived from protons striking two separate targets, located at 126 m and 6 m, before the beginning of the decay region, upstream of NA 48. The layout of the beams is shown in Fig. 5.13.

The K_L^0 beam

The K_L^0 beam (red cone in Fig. 5.13) was produced by the primary beam from the upstream target at a small angle to reduce the neutron flux. The beam, after emerging from the opening in a beam dump-collimator, passed through three successive collimators, designed to maximize the acceptance of the beam and to minimize backgrounds outside the central apertures of the detectors.

A vertically-deflecting dipole magnet after the K_L^0 target swept away charged particles, whereby the primary protons, which had not interacted, were deflected downwards to impinge on the channelling crystal, situated below the K_L^0 beam axis. The crystal was designed to split off the wanted, small fraction of the protons and to deflect them upward, back to the horizontal. The majority of protons and other particles, which were not deflected by the crystal, were absorbed in the beam dump-collimator.

The K_S^0 beam

The low-intensity proton beam required for the K_S^0 beam was generated with the channelling crystal (Fig. 5.14) [31]. A silicon mono-crystal was used, cut to dimensions $60 \times 18 \times 1.5$ mm^3, parallel to the 110 plane. It was bent through an angle θ_0 (greater than the required beam deflection angle, $\theta = 9.6$ mrad) over 56 mm of its length, by pressing it with rollers against the cylindrical surface of an aluminium block, precisely machined to have a radius of curvature, $R = 3$ m. The

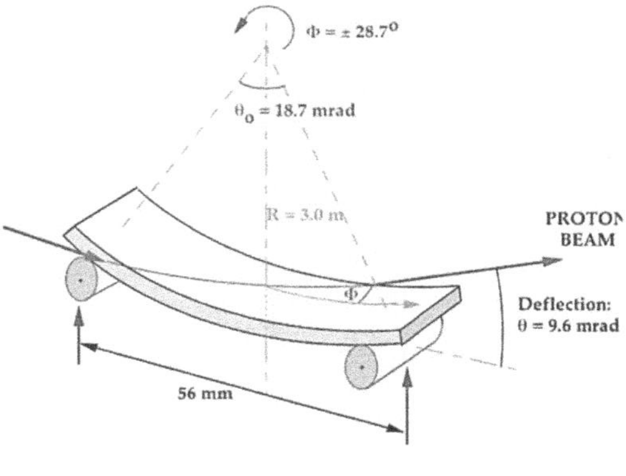

Fig. 5.14. Schematic arrangement of the bent crystal [31].

holder was mounted on a motorized goniometer, which allowed the crystal to be aligned with two transverse displacements and two rotations about axes perpendicular to the incident beam. When the crystal was rotated through an angle Φ about the vertical axis, the beam traversed the crystal from one edge to the other. The effective distance traversed in the crystal, and hence the vertical angle of the beam deflection, was controlled by adjusting Φ. With this set-up the orientation of the crystal could be chosen to adjust the deflection of the beam and its emittance.

The beam dump-collimator following the crystal contained a passage for the channelled protons, below that for the K^0_L beam. The protons (green line in Fig. 5.13) were brought back onto the K^0_L beam axis, passing through a 'tagging station' on the way. With a coincidence measurement between this station and the principal detectors of the experiment, an observed K^0 decay could be attributed or not to the K^0_S beam. The protons were then refocused and steered onto the K^0_S target.

The production angle of the K^0_S beam (blue cone in Fig. 5.13) was chosen to make the ratio of K^0_S and K^0_L decays in the fiducial region as similar as possible over the momentum range used (70 GeV/c–170 GeV/c). After a sweeping magnet and appropriate collimation with passages for the two neutral beams, the K^0_L and K^0_S beams emerged into the common decay volume and converged to the centre of the detectors at a distance of 120 m from the K^0_S target.

To count the $K^0_S \rightarrow \pi^+\pi^-$ and $K^0_S \rightarrow \pi^0\,\pi^0$ decays with the same acceptance, the beginning of the decay fiducial region had to be defined precisely. For the charged decay mode, this was done directly using a thin scintillation counter at the exit of the K^0_S collimator to detect the $\pi^+\pi^-$. For the neutral decay mode, the photons from the quasi-instantaneous decay of the π^0's were converted to e^+e^- in a disc of high-Z material placed immediately in front of the counter. The thickness and scattering effect on the beam of this material was minimized by using another, carefully-aligned, crystal (iridium), which coherently enhanced the conversion probability.

This pair of beams served to establish direct CP-violation in the decay of neutral kaons by experiment NA48 from 1997 to 2002 [32]. The line was subsequently transformed to provide another pair of simultaneous beams, this time of oppositely-charged kaons (K^+ and K^-), into the same set of detectors [33].

5.6 Liquid Krypton Calorimetry: Elucidating Nature's Subtle Asymmetries

Italo Mannelli

The discovery of the violation of Charge–Parity (CP) symmetry shook the foundations of physics [Box 3.4]. Considerable experimental efforts followed, spanning more than 20 years, seemingly indicating that this effect was confined to K^0 – anti-K^0 mixing. In parallel, however, with the development of the Standard Model incorporating three families of quarks [Box 6.4], the presence of a further mechanism of CP Violation (Direct CP Violation) was predicted in kaon decays. This would manifest itself as a difference in the ratio of K_L decays into charged to neutral pion-pairs relative to this ratio in K_S decays. The effect was predicted to be subtle. Experimental searches turned out to be not sensitive enough, mostly due to the lack of adequate techniques to measure the kaon decays to two neutral pions.

First evidence for Direct CP violation was obtained at the CERN SPS by the NA31 experiment in 1987. An analogous experiment at FNAL did not support the CERN result. The need was recognized by both collaborations for new experiments, taking advantage of novel technical developments, which could achieve an order of magnitude higher statistical and systematic accuracy.

Key to the success was the quality of the measurement of the neutral pion decay into two photons with energies up to 100 GeV. This required the e-m calorimeter to be homogeneously sensitive for best resolution. It also had to locate the position of the photon showers with millimetre accuracy to reconstruct the neutral pions. The newly-formed CERN-based NA48 collaboration bet on an audacious, novel concept: the development of an ionization calorimeter with liquid krypton (LKr) as the active absorber.

Why an ionization calorimeter? Previous incarnations, using liquid argon (LAr), had demonstrated the superior control of systematic effects and long-term stability [Highlight 4.10]. Why LKr? It has three times the absorption power compared to LAr. A 100 GeV photon shower is contained in a length of 125 cm; the transverse shower size is reduced by about a factor two compared to LAr, improving the localization of the shower. The calorimeter had a 5.3 m^2 octagonal cross-section and a depth of 127 cm. The total krypton mass was 25 t (Fig. 5.15).

The ionization charge has to be collected with electrodes. Again, ingenuity and inventions were required: the electrodes had to be nearly massless not to disturb the energy measurement, and placed with sub-millimeter precision in the 6 m^3 volume. A system of longitudinal copper-beryllium ribbons, each 1.8 cm wide and 40 μm thick, precisely spaced to form 13248 signal cells, 2 cm × 2 cm in the transverse direction, measured the initial current induced by the drift of the ionization charge. Every 20 cm in depth, the ribbons, with applied tension of 3 N, were held in position in a 50 mrad zig-zag geometry via slots (Fig. 5.16), accurately machined in glass-fibre-epoxy diaphragms. A slight increase in the perforation pitch vs depth made the axes of the cells converge to the centre of the observation volume of the kaon decay, 110 m in front of the calorimeter. The 1 cm spacing between ground and charge collecting electrodes had to be accurate to within 50 μm, as the initial current is inversely proportional to the gap width [34].

Fig. 5.15. The calorimeter, prior to insertion into the cryostat. On the right side, after the second mechanical support disk fibre bundles exit towards the photomultipliers.

The preamplifiers, the HV biasing and the charge injection for calibration were placed inside the cryostat immediately after the end of the ribbons to avoid signal distortion due to the inductance of long connections. The signals of each of the 13248 cells were brought out from the cryostat via 50 Ω coaxial cables through feedthroughs to the signal processing electronics, which included a 40 MHz flash ADC. The choice of placing the first electronic stages inside the LKr cryostat, practically precluding maintenance interventions, was more than compensated by the achieved time resolution and by the quality and stability of the electronic response inherent to the cryogenic operation at 120 K.

At a depth of 40 cm from the front of the calorimeter a vertical layer of scintillating fibres was placed in bundles which were brought to a set of photomultipliers operating in the LKr. Their signals were used to select the interesting kaon decays to neutral pions.

The novel calorimeter worked flawlessly, matching the design parameters. It routinely achieved an energy resolution $\sigma(E)/E = 3.2\%/\sqrt{E} + 90\ \text{MeV}/E + 0.42\%$, still at world record level, 20 years later. A 100 GeV photon is measured with 0.5% accuracy. The position resolution in the transverse directions is $\sigma(x,y) = 5.5/\sqrt{E} + 0.3$ mm. The time resolution is a remarkable 2.6×10^{-10} sec.

The bet paid off. The NA48 collaboration could unambiguously establish the existence of Direct CP Violation in kaon to two pion decays, as did their American competitors [32]. With just a few improvements to the external electronic chain to adapt it to higher data recording rates, the NA48 calorimeter has been operated without interruption since 1996, with no significant change in the energy calibration and no loss of krypton.

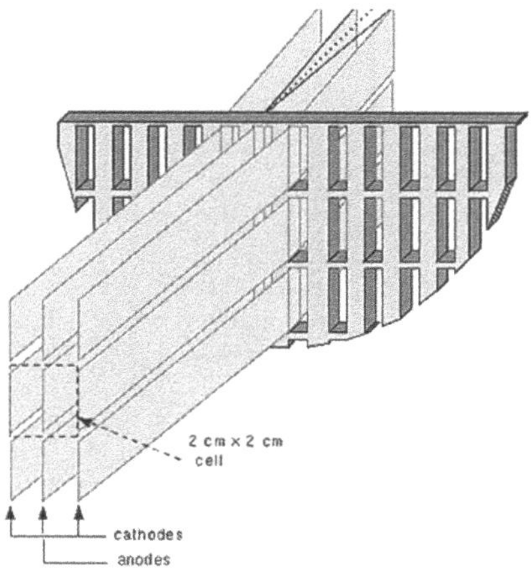

Fig. 5.16. Schematic drawing of the ribbon cathodes forming 2 cm × 2 cm cells.

5.7 Bubble Chambers at the SPS: A Technique at its Apogee

Horst Wenninger

The story goes that in 1953 Donald Glaser observed bubble forming in his beer glass triggering him to study bubble formation in liquids traversed by cosmic rays. Glaser himself confirmed this during the discussion session of the last International Bubble Chamber Conference, at CERN in July 1993.

The era of bubble chambers at CERN lasted 30 years, and finished in 1984 with the closure of BEBC — the Big European Bubble Chamber — and of GARGAMELLE — the large French heavy liquid bubble chamber, famous for the discovery by the neutrino physics collaboration of so-called neutral current interactions, in striking support of the emerging theory of electroweak interactions. Analysis of data from experiments with bubble chambers in Europe ended ten years later with the conference cited above.

Bubble chambers reveal the trajectories of charged particle as they pass through a superheated liquid (Fig. 5.17). A string of gas bubbles produced along the path of ionizing particles form tracks that can be recorded on stereo view photographs. A magnet creating a uniform field allows to determine the momentum of particles by measuring the curvature of the tracks (Fig. 5.18). Direct visible evidence of particle interaction patterns on millions of photos were thus analysed in physics institutes around the world [35].

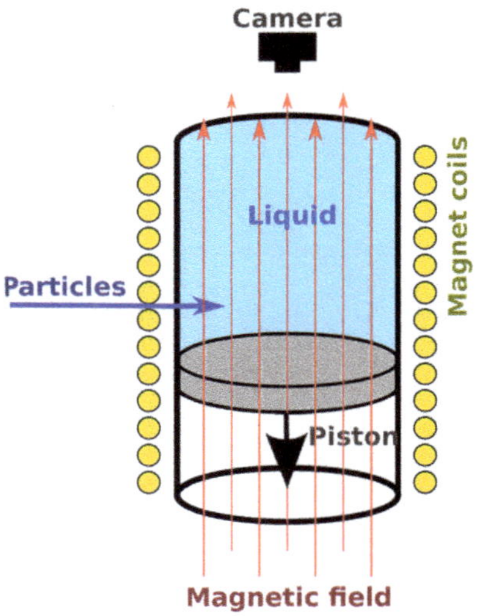

Fig. 5.17. Schematic showing the principle of bubble chamber operation.

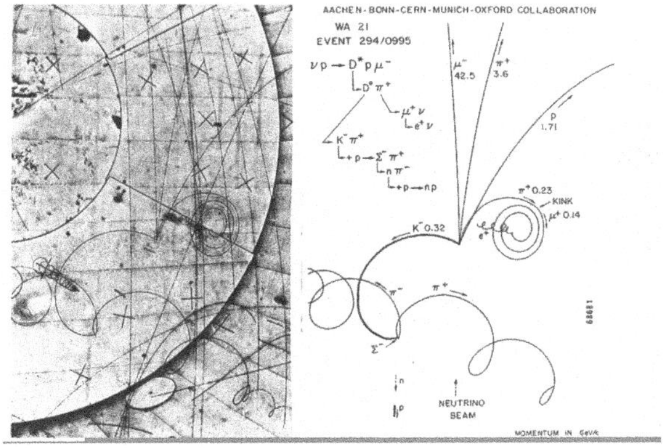

Fig. 5.18. Photo of neutrino interaction in BEBC, and reconstruction of a neutrino "event".

CERN engineers and technicians developed the technologies required for small and large, for track-sensitive, for holographic, and for rapid cycling bubble chambers, as well as for sophisticated automatic film measuring devices. The technologies required to construct and operate bubble chambers included cryogenics,[b] hydraulics, vacuum, superconductivity,[c] fish-eye optics, lasers, holography, and controls; the film analysis required pattern recognition algorithms and world-standard reconstruction software. Knowhow and competences acquired by CERN staff during this period provided the basis for later success of the laboratory in support for electronic particle physics experiments.

CERN-specific developments and technological highlights during the Bubble Chamber era focused on (i) increased photon conversion efficiency (e.g. photons from π^0 decays) in hydrogen bubble chambers, improved particle identification using a wire chamber Internal Picket Fence for muons (IPF) plus an External Particle Identification by *dE/dx* (EPI) (Fig. 5.19), and (ii) fast cycling chambers combined with tagging of selected rare events — e.g. charm decay events studied by the collaboration of the CERN Hybrid Spectrometer. Trials with holographic bubble chamber were also started in parallel.

In order to overcome the low conversion efficiency of photons in hydrogen bubble chambers the US laboratories installed iron converter plates. In BEBC, a track sensitive liquid hydrogen-filled LEXAN box was used to observe primary

---

[b] Cryogenics and superconductivity have since assumed an important role at CERN.

[c] A large superconducting magnet, providing a field of 3.5 tesla over the sensitive volume of 35 m^3, the largest such magnet in the world in terms of stored energy until 1996, was built for BEBC at CERN. In addition, an External Muon Identifier (EMI), the Internal Picket Fence (IPF), the External Particle Identifier (EPI) and a Track-Sensitive Target (TST) converted BEBC into a hybrid detector.

Fig. 5.19. BEBC surrounded by External Muon Identifier EMI and External Particle Identifier EPI.

beam particle interaction, surrounded by a heavy liquid neon-hydrogen mixture, developed in collaboration with the Rutherford Laboratory, for improved photon detection [36]. The trick of this "bubble chamber inside a bubble chamber" was to find a temperature and pressure regime such that the pressure cycle transmitted from the heavy liquid to the hydrogen via LEXAN[d] walls allowed to attain simultaneous track sensitivity inside and outside the transparent box, and visualize the interactions and tracks in both media of the chamber on the same photo.

The development and operation in BEBC of the LEXAN track sensitive target resulted in a second CERN invention with the construction of a high resolution fast cycling LEXAN bubble chamber (LEBC/HOLEBC) which was used for charm physics and allowed one of the first definitive studies of charm production at a proton facility [35].

Besides Neutral Currents the large bubble chamber led to other most interesting results. By comparing the structure functions of the nucleon derived from neutrino and muon scattering, it was proved at Gargamelle that quarks have indeed a fractional electric charge. The Gargamelle experiments discovered that the neutrino–nucleon cross-section rises linearly with energy. Finally, by combining their results, the BEBC and Gargamelle teams obtained the first evidence for scaling violation, namely the proof that the nucleon structure functions actually evolve with the resolution power of the observation [37].

---

[d] A specific development was launched with an US company for an optically transparent multi-layer plastic (trade name LEXAN) keeping sufficient flexibility at 20 K to transmit the pressure from surrounding liquid neon-H_2 to H_2 inside the 3 m^3 LEXAN box [36].

5.8 Polarized Targets: Pointing to New Directions

Tapio Niinikoski

In Polarized Targets (PT) nuclear spins are polarized dynamically in a solid material containing a high density of the nucleons of interest. Such targets are required for some scattering experiments where the spins of the beam and/or target particles have to be aligned with respect to a static magnetic field. To develop, build and operate solid PTs draws on several fields of physics and technology. In this highlight we address advances where CERN has made a major contribution.

Quantum statistics to describe Dynamic Nuclear Polarization (DNP)

Particles with spin have an associated magnetic moment that couples with external and internal magnetic fields. Spin polarization is observable by magnetic resonance methods, whereby a small oscillating magnetic field, transverse to a strong static field, is absorbed by the system of magnetic moments.

Spins in thermal equilibrium with the lattice have a small natural polarization because the magnetic energy levels of the moments have tiny difference in a static external magnetic field. For proton spins this is only a few tenths of a percent at 1 K temperature and 2.5 T magnetic field. Electron spins, on the other hand, under the same circumstances, have about 99% polarization, because their magnetic moment is about 1000 times higher. This large polarization can be transferred to the nuclear spins by certain types of magnetic resonance: the double resonance of electron and nuclear spin pairs, or by cooling of the whole spin system using off-resonance microwave irradiation. The former requires a strong magnetic component of the oscillating microwave field, at 70 GHz frequency, that is parallel to the steady field. The latter requires a much weaker transverse microwave field. The concept of "spin temperature" helps to understand this mechanism.

The spin temperature T_s can be seen in the populations of the magnetic energy levels E_m that are proportional to the Boltzmann factors $\exp(-E_m/k_B T_s)$, after relaxation transitions have established thermal equilibrium among the levels m. This is the classical static view. With a transverse field oscillating at a frequency close to Larmor precession, the spin system reaches a new equilibrium spin temperature that is positive, zero or negative, respectively, and can be much lower than the temperature of the lattice. We may then speak of "cooling of spin system".

The effects are measurable and dramatic: the new "dynamic" spin temperature is transmitted to all other non-resonant spin systems by spin-spin relaxation transitions, assisted by the oscillating field. Quantum statistics must then be used to predict the thermodynamic parameters such as spin temperature, polarization and entropy. At high temperatures this is relatively easy, because the quantized

Boltzmann factors can be linearized. At low temperatures this approach does not work. It was resolved with the formalism of "Spin Temperature Model of DNP", which gives numerically solvable formulae [38].

The development of this model was one of the major original contributions of CERN to polarized target physics It explains the previously obtained, but poorly understood, results in hydrogen-rich materials by extending the early quantum statistical treatment of magnetic resonance saturation to electron spin systems at low temperatures where the electron polarization is high. The quantum statistical formalism was developed to properly describe and evaluate spin temperature and polarization for the regime of PT applications, where the simplifications of the so-called "high temperature" approximation are not valid. The Polarized Target team at CERN also validated experimentally [39] the key prediction of the model, namely that all nuclear spin species cool towards a common temperature. This common temperature is that of the electronic spin system under a saturating microwave field. It can be as low as one thousandth of the lattice temperature, which is typically 0.2 K to 0.7 K. This new formalism pointed the way to obtain proton polarizations over 99% and deuteron polarizations of about 90% in large targets with corresponding spin temperatures below 0.001 K.

The double resonance method was limited to small targets, owing to the high microwave power required, and to exotic single crystal materials that contain relatively few protons, the target nuclei of interest. In the dynamic cooling method, the microwave power can be reduced so much that very little heat is generated in the material, thus allowing the target to be cooled with ^{3}He evaporation refrigerators at 0.5 K temperature, and even lower with ^{3}He–^{4}He dilution refrigerators, with large gains in the polarization.

The DNP described by the new model benefits from lower coolant and lattice temperatures, and can be extended to a large class of materials that are rich in hydrogen or deuterium. Importantly, the much lower temperatures available by ^{3}He–^{4}He dilution refrigeration (Fig. 5.20) enable "freezing" the polarization. A lower and less homogeneous "holding" field of a spectrometer is sufficient to maintain the polarization of the target after DNP in an external, strong and homogeneous field. Such target technology was developed at CERN where the first frozen spin target was constructed and operated in 1974 [40].

Dilution refrigerators were first introduced to the polarized targets at CERN in the 1970s, with horizontal [40,41] and vertical [42] designs. For these targets CERN developed sintered counterflow heat exchangers [41], which have had a substantial impact on current industrial design of machines for reaching down to temperatures of 0.002 K.

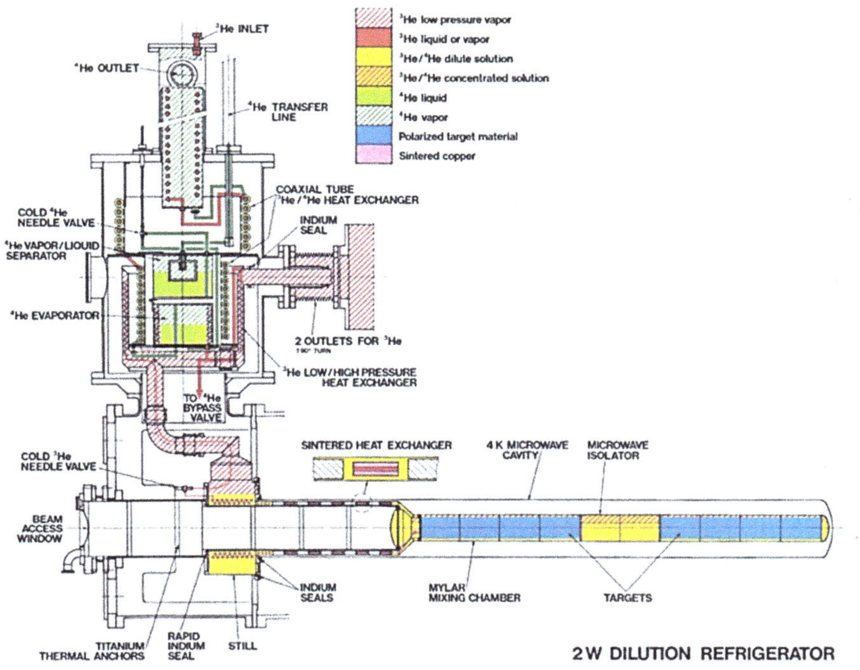

Fig. 5.20. Cross-section of EMC PT cooled with a dilution refrigerator absorbing 2 W power generated by DNP. The target is split in halves polarised in opposite directions, and has 1 m total length. The superconducting solenoid magnet is not shown, for clarity.

Development of materials and magnet technology for polarized targets

DNP in hydrogen-rich chemically doped materials requires the uniform dispersion of suitable paramagnetic impurities. Such materials were initially developed by trial-and-error until the CERN team discovered that they share the common property of being excellent glass formers [43]. The glass "solidifies" by becoming increasingly viscous already at high temperature, which prevents, upon fast cooling, the phase separation of the dissolved dopant as microcrystals.

Another way of doping PT materials uniformly is to create free radicals by suitable irradiation. A technique was developed at CERN to prepare and irradiate solid ammonia, which is richer in H_2 than any hydrocarbon (except methane). Over 90% polarization was reached in the first samples tested by the CERN team [44].

Large polarized targets were designed for the deep-inelastic muon scattering experiments of EMC (NA2), SMC (NA47) and COMPASS (NA48). The NA2 PT was made of irradiated ammonia, while the NA47 target [45] used doped deuterated butanol. The low lattice temperature enabled reaching > 60% deuteron polarization in a 2.5 T magnetic field. Even higher deuteron polarizations have been achieved in materials doped by new trityl free radicals [46], in irradiated deuterated butanol [46, 47], and in irradiated deuterated ammonia [48]. Common to these and to irradiated ^{6}LiD is a narrow "Electron Spin Resonance" (ESR) line that allows one to reach a low deuteron spin temperature by dynamic cooling of the spin–spin interactions of the unpaired electrons, as had been predicted in earlier studies by the PT team at CERN [49].

For the large targets an axial field geometry was required, with a field uniformity in the range of 10^{-5}. Superconducting solenoid magnets were designed with improved coil winding techniques, using rectangular wire to reduce the number of compensation and trimming coils. The rectangular wire fabrication technique was developed in collaboration with industry. The technology for achieving a uniform field in superconducting solenoids has found other applications, including magnetic resonance imaging (MRI).

While this highlight focused on technical development at CERN, clearly other laboratories also contributed to the field and CERN profited from exchanges with them. This development has now enabled us to reach over 90% polarization, positive and negative, of almost any nuclear spin. Targets can be operated in frozen spin mode in a relatively low field of any orientation, and in DNP mode for high intensity beams. Beyond experiments in nuclear and particle physics, applications of DNP are emerging in macromolecular chemistry and in MRI.

5.9 The Silicon Age: Micrometre Precision Millions of Times a Second

Erik Heijne

Silicon sensors and custom-designed integrated circuits ("chips") are essential components in today's particle physics experiments. These silicon devices replaced earlier tracking instruments such as spark chambers and bubble chambers which recorded the particle trajectories photographically. While gaseous wire chambers remain preferred for covering large surfaces [Highlight 4.8], they are no longer competitive in spatial resolution and signal speed for the smaller areas close to the interaction region. The characteristic features that contributed to the breakthrough of silicon (Si) detectors between 1980 and 2000 were: (i) the small, stable and well-defined dimensions of the sensor cells of typically around 100 μm; (ii) the potential to resolve several particles, incident simultaneously on mm-size area; (iii) the short signals of less than 20 nanoseconds (ns); and (iv) the progressive miniaturization of the associated electronics for the processing of the particle signals: by 2010, 65 000 signal amplifiers, digitizers and memory could be placed on a single substrate of Si (15 mm × 15 mm), a miniaturization which was, and continues to be, driven by industry. The introduction of Si devices designed for multiple particle detection was spearheaded by scientists at CERN who exploited the advances in microelectronics technology. Si microstrip detectors were soon widely adopted in experiments that studied short-lived charm and beauty particles. Their decay-path, less than about a mm, could now be distinguished separately from traces that originate from the primary interaction vertex. Early steps in this innovation are illustrated here.

Semiconductor materials such as Si, germanium (Ge) or diamond (C) contain only few free charge carriers, contrary to metals in which a large density of free electrons easily allows an electrical current to flow. Nevertheless, a small current signal can be generated in a semiconductor by the passage of a fast, charged particle, which loses some of its energy and liberates charges. An electric field is applied between metallic contacts (Fig. 5.21) under which the charges drift to the electrodes and induce a signal, which is sensed by a connected amplifier. In Si it is necessary to provide rectifying contacts (the p+ contact in Fig. 5.21) in order to prevent a 'leakage current' which would obscure these small signal currents. Such a structure is called a 'diode', and technology has been developed to imprint a large matrix, e.g. 256 × 256 of adjacent diodes on the same Si substrate. In most cases, the particle generates signal currents simultaneously in two or more sensor cells and, by interpolation, it is possible to determine its position in space with a few μm precision, considerably smaller than the dimension of the sensor cell itself.

A complete reconstruction of the trajectory of a particle requires the measurement of several points in space and 'telescopic' arrangements of three to four successive Si detector planes usually surround the interaction region in an experiment.

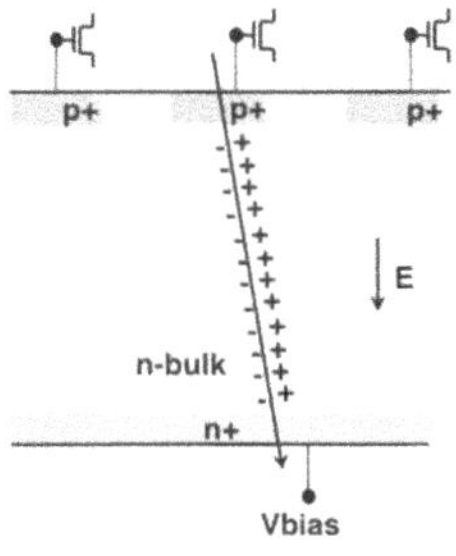

Fig. 5.21. Schematic drawing of a segmented Si detector, with an incident ionizing particle. Segmented, rectifying contacts p+ are made by ion-implantation of boron, the continuous n+ contact is made by implantation of phosphorous or arsenic ions. A biasing voltage Vbias is applied across the detector, generating an electrical field E throughout the n-type bulk Si, which separates the charges created by an incident particle. The resulting signal current is sensed by the amplifiers connected to the rectifying contacts (Drawing not to scale).

In 1980 vertex reconstruction was achieved for the first time using a Si microstrip detector [50], in collaboration with the NA11 experiment at the SPS, using the setup shown in Fig. 5.22. This detector was a one-dimensional matrix of 100 metal-Si diodes, 200 μm wide gold-evaporated bands on a 20 mm × 30 mm Si substrate. The design was inspired by the earlier, so-called 'Si checker-board' segmented detectors used in the 'BOL' experiment [51] at the Nuclear Physics Institute IKO in Amsterdam, operating at their cyclotron since 1967. This IKO device had perpendicular segmentation at the front and back, hence providing 2-dimensional particle positions, while the CERN detectors had strips only on one side, and several planes were required for 3-dimensional spatial coordinates. Soon thereafter, the MPI/TU Munich and CERN collaboration perfected the exploitation of Si microstrip detectors for charm physics research. In parallel, experiment WA75 using the Omega spectrometer [Highlight 3.7] also adopted Si microstrip detectors to determine the decay point of charm and beauty particles in an upstream plate of sensitive emulsion. WA75 published the first direct observation of beauty particles decaying into charmed particles in 1985 [52].

Research into new phenomena in elementary particle physics experiments frequently needs increased energy, typically producing more particles in an interaction. However, the new phenomena tend to occur more rarely than the already known processes, so higher beam intensities and interaction rates are required, motivating development of innovative instrumentation. In Si detectors high signal rates (well over 40 MHz) can be handled, as the freed electrons move at speeds of greater than 10000 km/h: the electrical charge in a 0.3 mm thick Si

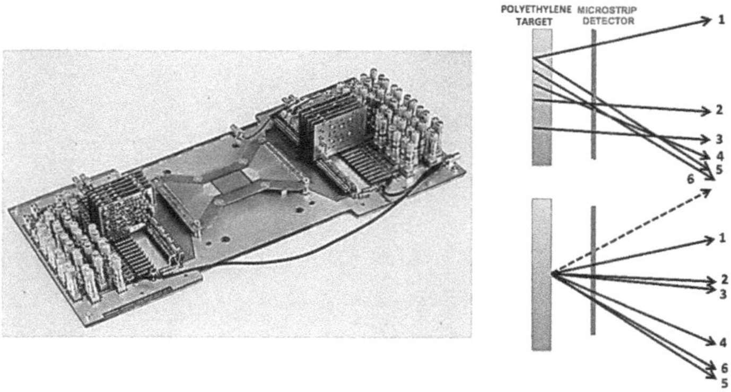

Fig. 5.22. Left: the first operational setup of a segmented, 100-element Si microstrip detector (the rectangle in the centre of the photo), on a support containing a few of the readout amplifier cards. Right: (top) reconstruction of an interaction using only the wire chamber information; (bottom) vertex reconstruction after including the signals from the Si microstrip detector. An additional particle is found as well, the broken line. (Courtesy NA11).

layer can be collected in less than 20 ns. Immediately afterwards, the diode detector is ready for recording another particle. Several particles can be measured simultaneously, even if they are as close as 0.5 mm. For these event topologies, a true 2-dimensional matrix of sensor cells is preferable to combinations of 1-dimensional microstrip detectors. In 1989 it became possible, with the help of the microelectronics team at the EPFL, to produce a first 9 × 12 matrix of a so-called Si pixel detector. In this case a 2-D Si sensor matrix is connected pixel by pixel to a corresponding matrix of readout cells on an integrated circuit, made in CMOS (Complementary Metal-Oxide Silicon) microelectronics technology. Each of these cells comprises a full signal processing circuit with amplifier and digitizer. In recent implementations a memory in each cell can even store the time of incidence and amplitude of a signal from one or several successive particles. In 1992 the second iteration of the Si pixel detector could resolve hundreds of particle tracks on a 6 × 6 cm^2 area and was successfully used for WA94, an ion experiment in the Omega spectrometer. This demonstration helped to dispel much of the prevailing scepticism surrounding the practicality of pixel detectors. Figure 5.23, left shows the two half planes that together constituted one of the seven arrays of the Si pixel telescope. A reconstruction of an ion-event using this instrument in WA97, the successor to WA94, is shown in Fig. 5.23, right.

With the introduction of segmented Si microstrip and pixel detectors, the number of sensing elements in the experiments increased to tens of thousands or

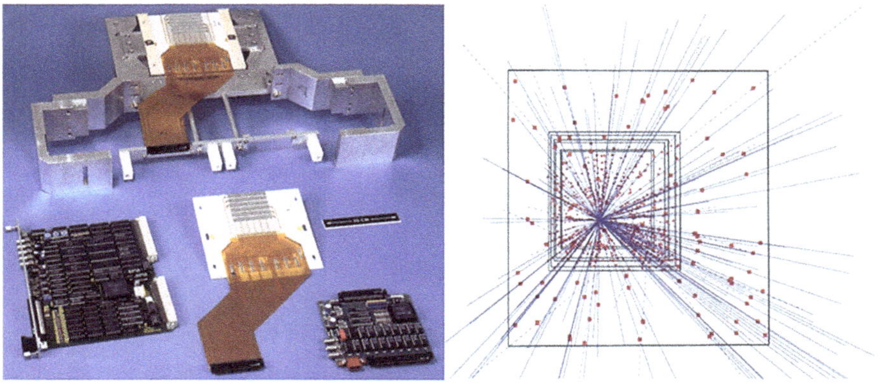

Fig. 5.23. Left: two half-planes from the first CERN silicon pixel telescope for the WA94 Omega spectrometer ion experiment. When mounted together, face-to-face, the six horizontal Si ladders on one, white ceramic support cover the spaces between those on the other support. Also shown are the receiver board (left) and the driver module placed directly near the telescope. Right: reconstruction of tracks from an interacting ion in the WA97 experiment, using the pixel telescope with 7 double planes (Source: WA97).

even hundreds of millions. It is obvious that miniaturized, integrated electronic circuits are required for the processing of such a multitude of signals. From 1985 onward, a Si integrated circuit design expertise was established at CERN, with the 16-channel AMPLEX chip as the first outcome, used in the UA2 Si pad detector [Highlight 6.6]. Intensive contacts with experienced groups in universities such as the EPFL in Lausanne and IMEC in Leuven helped to achieve a professional level in chip design. The custom-designed circuits were then manufactured by CMOS chip processing industry, while evaluation and testing proceeded at CERN. Other particle physics institutes soon participated in concerted developments of Si circuits and detector systems. The early steps in the development of the Si pixel detectors have been described in [53, 54]. More recent achievements on electronic circuits and the development of their radiation resistance are reported elsewhere [Highlights 8.6 and 8.7]. The use of silicon detectors in particle physics experiments as well as in space-borne research has grown exponentially over three decades, as illustrated in Fig. 5.24, showing the Si detector area in an experiment as a function of the year of its first operation.

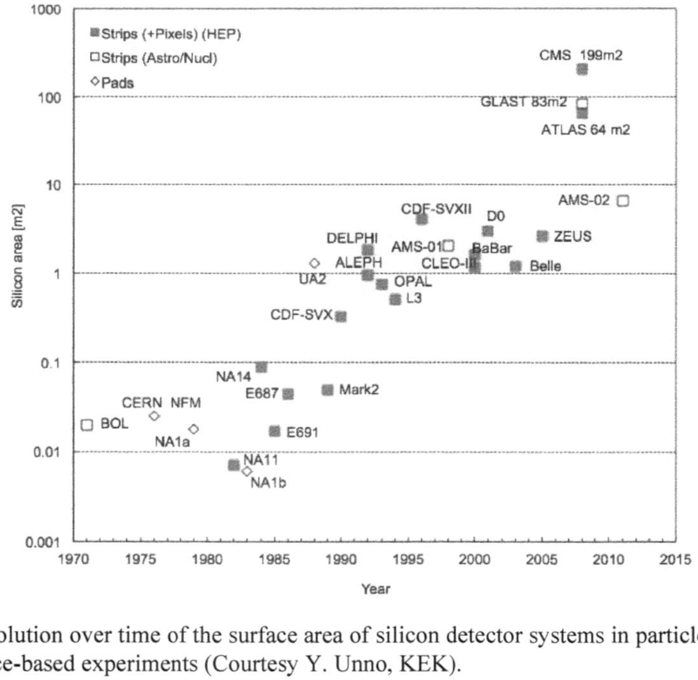

Fig. 5.24. Evolution over time of the surface area of silicon detector systems in particle physics and in recent space-based experiments (Courtesy Y. Unno, KEK).

References

1. D. Pestre, The difficult decision, taken in the 1960s, to construct a 3-400 GeV synchrotron in Europe, in *History of CERN*, Vol. III, ed. J. Kriege (Elsevier, 1996) p. 65.
2. J.B. Adams and E.J.N. Wilson (eds.), *Design Studies for a Large Proton Synchrotron and its Laboratory*, CERN-1970-006 (CERN, Geneva, 1970). http://dx.doi.org/10.5170/CERN-1970-006.
3. J.B. Adams, The CERN 400 GeV Proton Synchrotron (CERN SPS), *Proc. 10th Int. Conf. on High-Energy Accelerators, vol. 1*, Protvino, USSR (1977) p. 17.
4. M. Goldsmith and E. Shaw, *Europe's Giant Accelerator*, (Taylor & Francis, London, 1977).
5. Y. Baconnier, P.E. Faugeras, K.H. Kissler, B. de Raad and W. Scandale, Extraction from the CERN SPS, *Proc. 7th Part. Accel. Conf.*, Chicago, USA (1977) p. 1434.
6. R.L. Keizer, L. Reinder, K.H. Kissler and L.R. Oberli, Improvement of the extraction channels of the CERN-SPS accelerator, *IEEE Part. Accel. Conf.*, Washington D.C., USA (1987) p. 1836.
7. Y. Baconnier *et al.*, Septum magnets for the extraction channels of the 400 GeV SPS, *6th Int. Conf. on Magnet Technology,* Bratislava, Slovakia (1977) p. 504.
8. W. Schnell, A wide-band travelling-wave accelerating method for a 300 GeV proton synchrotron, *5th Int. Conf. on High-Energy Accelerators*, Frascati, Italy (1965) p. 38.
9. K. Cornelis and R. Schmidt, Commissioning of the SPS as LEP injector, *Proc. 1st European Part. Accel. Conf. (EPAC),* Rome, Italy, (1988) p. 329.
10. M. Benedikt, P. Collier, V. Mertens, J. Poole, K. Schindl (eds.), *LHC Design Report, v. 3, The LHC Injector Chain*, CERN-2004-003 (CERN, Geneva, 2004). http://dxdoi.org/10.5170/CERN-2004-003-V-3.
11. ECFA 300 GeV Working Group: Volume I - Status Reports of the Working Parties, July 1972; Volume II - Proceedings of the Second Tirrenia Study Week, September 1972; Volume III - Final report of the Executive Committee, CERN/ECFA/72/4 (1973).
12. J.R. Ellis *et al.* (eds.) CERN: the second 25 years, *Phys. Rep.* **403-404** (2004) pp. 1-504.
13. M. Fidecaro, Polarization experiments at the SPS, *Proc. AIP Conf.* **35** (1976) pp. 158-160.
14. F. Dydak, Neutrino Physics, in [12] pp. 57-67.
15. R. Voss, Deep inelastic scattering with muons, in [12] pp. 3-18.
16. B. French, E. Quercigh, Physics with hadron and photon beams at the SPS, in [12] pp. 69-90.
17. C. Lourenco, H.K. Worhi, Heavy flavour hadro-production from fixed-target to collider energies, *Phys. Rep.* **433** 127-180 (2006) hep-ph/0609101.
18. E. Quercigh, K. Safarik, Ion physics at the SPS: the experimental challenge in [12] pp. 51-56.
19. H. Wahl, First observation and precision measurement of direct CP violation: the experiments NA31 and NA48, in [12] pp. 19-25.
20. H. Burkhardt *et al.*, The beam and detector for a high-precision measurement of CP violation in neutral kaon decays, *Nucl. Instr. & Meth. A* **268**, 116-143 (1988).
21. H. Burkhardt *et al.*, First evidence for direct CP violation, *Phys. Let. B* **206**, 169-176 (1988).
22. M.C. Crowley-Milling (ed.), *The Design of the Control System for the SPS*, CERN-1975-020 (CERN, Geneva, 1975). http://dx.doi.org/0.5170/CERN-1975-020.
23. M.C. Crowley-Milling (ed.), *Experience with the Control System for the SPS*, CERN-1978-009 (CERN, Geneva, 1978). http://dx.doi.org/10.5170/CERN-1978-009.
24. M.C. Crowley-Milling and G.C. Shering (eds.), *The NODAL System for the SPS*, CERN-1978-007 (CERN, Geneva, 1978). http://dx.doi.org/10.5170/CERN-1978-007.

25. F. Beck and B. Stumpe (eds.), *Two Devices for Operator Interaction in the Control of the New CERN 300 GeV Accelerator*, CERN-1973-006. (CERN, Geneva, 1973). http://dx.doi.org/10.5170/CERN-1973-006.
26. E.A. Johnson, Touch display – a novel input/output device for computers, *Electronics Letters* **1**(8), 219 (1965); https://dx.doi.org/10.1049%2Fel%3A19650200.
27. B. Stumpe and C. Sutton, The first capacitive touchscreens at CERN, *CERN Courier* **50** 3 (2010); http://cds.cern.ch/record/1734498.
28. N. Doble *et al.*, The upgraded muon beam at the SPS, *Nucl. Instr. & Meth. A* **343**, 351-362 (1994).
29. G. Mallot *et al.*, The COMPASS experiment at CERN, *Nucl. Instr. & Meth. A* **577**, 455-518 (2007) and *Report CERN-PH-EP-2007-001*, arXiv:hep-ex/0703049.
30. V. Fanti *et al.*, The beam and detector for the NA48 neutral kaon CP violation experiment at CERN, *Nucl. Instr. & Meth. A* **574**, 433-471 (2007).
31. N. Doble, L. Gatignon and P. Grafstroem, A novel application of bent crystal channelling to the production of simultaneous particle beams, *Nucl. Instr. & Meth. B* **119**, 181-191 (1996).
32. J.R. Batley *et al.*, A precision measurement of direct CP violation in the decay of neutral kaons into two pions, *Phys. Lett. B* **544**, 97-112 (2002).
33. J.R. Batley *et al.*, Search for direct CP violating charge asymmetries in $K^{\pm}\rightarrow\pi^{\pm}\pi^{+}\pi^{-}$ and $K^{\pm}\rightarrow\pi^{\pm}\pi^{0}\pi^{0}$ decays, *Eur. Phys. J. C* **52**, 875-891 (2007).
34. G.D. Barr *et al.*, Performance of an electromagnetic liquid krypton calorimeter based on a ribbon electrode tower structure, *Nucl. Instr. & Meth. A* **370**, 413-424 (1996).
35. H. Wenninger, In the tracks of the bubble chamber, *CERN Courier*, July 2004 (2004), G.G. Harigel, Bubble chambers, technology and impact on high energy physics; http://www.bo.infn.it/antares/bolle-libro/harigel.pdf.
36. D. Haidt, H. Leutz and H. Wenninger, Optimization of size and shape of a track sensitive H_2/D_2 target for neutrino interactions in BEBC, *Proc. 5th Int. Conf. on Instrumentation for High Energy Physics*, Frascati, Italy (1973) pp. 71-78.
37. P.C. Bosetti *et al.*, Analysis of Nucleon Structure Functions in CERN Bubble Chamber Neutrino Experiments, *Nucl. Phys. B* **142**, 1-28 (1978).
38. M. Borghini, Spin temperature model of DNP, *Phys. Rev. Lett.* **20**, 419 (1968).
39. W. de Boer, M. Borghini, K. Morimoto, T.O. Niinikoski and F. Udo, Dynamic polarization of protons, deuterons and carbon-l3 nuclei: thermal contact between nuclear spins and electron spin-spin interaction reservoir, *J. Low Temp. Phys.* **15**, 249 (1974).
40. T. Niinikoski and F. Udo, "Frozen spin" polarized target, *Nucl. Instr. & Meth* **134**, 219 (1976).
41. T. Niinikoski, Dilution Refrigeration: New Concepts, p. 102 in *Proc. 6th Int. Cryo. Eng. Conf., Grenoble 1976*, Ed. K. Mendelssohn (IPC Science and Technology Press, Guilford 1976).
42. R. Bernard *et al.*, A Frozen Spin Target with Three Orthogonal Polarization Directions, *Nucl. Instr. & Meth. A* **249**, 176 (1986).
43. T. Niinikoski, Viscometric Study of Binary Mixtures of Butanol and Pentanol with Water and Pinacol, in *Proc. of the 2nd Workshop on Polarized Target Materials*, eds. G. R. Court, S. F. J. Cox, D. A. Cragg and T. O. Niinikoski (SRC, Rutherford Laboratory, 1980).
44. T. Niinikoski and J.-M. Rieubland, Dynamic nuclear polarization in irradiated ammonia below 0.5 K, *Phys. Lett. A* **72**, 141 (1979).

45. D. Adams *et al*., A New Measurement of the spin-dependent structure function $g_1(x)$ of the Deuteron, *Phys. Lett. B* **357**, 248 (1995), and
D. Adams *et al*., The polarized double-cell target of the SMC, *Nucl. Instr. & Meth. A* **437**, 23 (1999).
46. S.T. Goertz *et al*., Highest polarizations in deuterated compounds, *Nucl. Instr. & Meth. A* **526**, 43 (2004).
47. D.G. Crabb, Polarization in radiation-doped butanol and CD_2, *Nucl. Instr. & Meth. A* **526**, 56 (2004).
48. J. Heckmann *et al*., Electron spin resonance and its implication on the maximum nuclear polarization of deuterated solid target materials, *Phys. Rev. B* **74** , 134418 (2006).
49. M. Borghini (ed.), *Choice of Materials for Polarized Proton Targets*, CERN-1966-003 (CERN, Geneva, 1966). http://dx.doi.org/10.5170/CERN-1966-003.
50. E.H.M. Heijne *et al*., A silicon surface barrier microstrip detector designed for high energy physics, Nucl. Instr. & Meth. **178**, pp. 331-343, (1980).
51. L.A.C. Koerts *et al*., The 'BOL' nuclear research project, *Nucl. Instr. & Meth.* **92**, 157-160 (1971).
52. J.P. Albanese *et al*., Direct observation of the decay of beauty particles into charm particles, *Phys. Lett. B* **158**, 186-192 (1985).
53. F. Anghinolfi *et al*., A 1006 element hybrid silicon pixel detector with strobed binary output, *IEEE Trans. Nucl. Sci.* **NS-39**, 654-661 (1992).
54. E.H.M. Heijne, Semiconductor micropattern pixel detectors: a review of the beginnings, *Nucl. Instr. & Meth. A* **465**, 1-26 (2001).

Chapter 6

The CERN Antiproton Programme: Imagination and Audacity Rewarded

Vinod Chohan and Pierre Darriulat

6.1 Introduction

The accelerator developments

As early as 1966, C. Rubbia, the Physics Nobel laureate in 1984, had realized the potential of the projected "300 GeV Machine" later built at CERN as the Super Proton Synchrotron (SPS) and proposed to use it as a proton–antiproton collider [1]. Two years later, in 1968, Simon van der Meer, also Physics Nobel laureate in 1984, invented a method to increase the collision rate ("luminosity") [Box 6.1] of the ISR. This was the concept of stochastic cooling [Box 6.2], a brilliant idea. It was published only in 1972 [2], and later proven experimentally at the ISR [3]. This was the necessary ingredient for the bold and imaginative proposal in 1976 [4] of using the CERN SPS as a single ring p–$\bar{p}$ collider aimed at the discovery of the putative W and Z bosons [Box 6.4].

In 1976 CERN decided to launch two actions in parallel. One was to construct rapidly a small ring, (Initial Cooling Experiment — ICE) to study beam cooling; the other was to set up a design group for a p–$\bar{p}$ facility using the SPS as a storage ring and collider. Meanwhile, many experimental tests in the ISR, further theoretical developments, and most importantly, the successful test of a faster and more efficient method of longitudinal cooling in ICE confirming cooling in all planes [5], gave the confidence that with stochastic cooling and stacking an overall increase in antiproton density from the production target to the final accumulated beam core of over 10^9 particles could be achieved. Although the cooling times obtained (10 s) were still about 30 times longer than required, it was decided to go ahead with the construction of the Antiproton Accumulator (AA) based on stochastic cooling. The AA was the key element of the scheme, a fixed magnetic field, single storage ring, designed to accumulate antiprotons at 3.5 GeV/c, generated from 26 GeV (PS) protons impinging on a metal target.

The stochastic accumulation (stacking) process, an essential feature in the AA scheme, could not be tested in ICE. This process involved simultaneous cooling in both transverse planes and increasing the longitudinal density by four orders of magnitude while moving the particles into the dense core. This was the biggest gamble in the launch of the AA because it could only be studied in detail by theoretical calculation. Fortunately, in this respect the AA performed as expected.

The AA project was launched in 1978, with beam commissioning in July 1980. The first SPS proton–antiproton collisions at 270 GeV occurred in July 1981 and the first real period of physics runs in 1982. After the W and Z discovery, Nobel awards in 1984 and the ever-increasing appetite for higher luminosities, the Antiproton Collector (AC) ring was built and commissioned in 1987, to bring a ten-fold increase in the accumulated antiprotons. The SPS ran as a collider as well as for short fixed-target runs during these years. After the last collider run in 1991 [6], the SPS returned to its mode as a fixed-target physics accelerator.

The systems and processes described here briefly represent major technological innovations and breakthroughs for the antiproton programme at CERN; similar, improved systems were adapted later for the antiproton source at FNAL/USA, leading to the top quark discovery there in the 1990s.

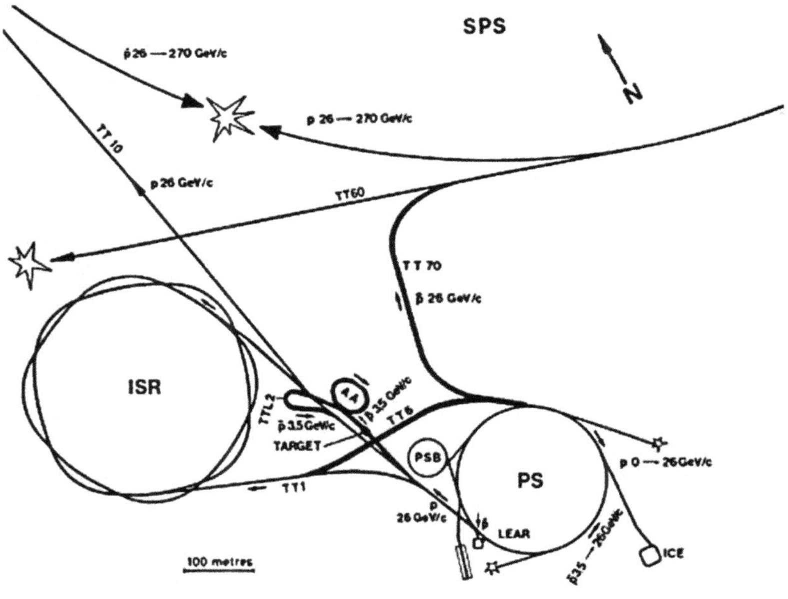

Fig. 6.1. The layout of the CERN accelerators in 1981 with the existing and new tunnels (thick lines) constructed for beam transfer from the AA to the PS, ISR & SPS [1].

Luminosity **Box 6.1**

Luminosity, L, is a figure of merit of the instantaneous rate of events at a particular interaction point of a collider [Box 4.1]. For a given process having cross-section σ_i, the event rate is $\sigma_i \cdot L$. The cross-section is measured in barns, where 1 b = 10^{-24} cm^2. Event rate is measured in Hz, so luminosity is measured in cm^{-2} s^{-1} (or b^{-1} s^{-1}). Except in the special case of continuous beams (ISR), luminosity is proportional to the number of particles per bunch N_b squared and the local rate of bunch encounters, and it is inversely proportional to the transverse area of the beam at the interaction point $s_x \cdot s_y$ where s_x and s_y are local horizontal and vertical beam sizes. Thus, N_b and the bunch encounter rate have to be maximized while $s_x \cdot s_y$ has to be a minimum (i.e. the beams have to be strongly focused). *In a circular collider*, the rate of bunch encounters is given by the product of the number of bunches k_b circulating and the revolution frequency. *N_b is limited* by the performance of the injector, by instabilities arising from bunch-vacuum tube interaction, and by effects between the opposite beams in the interaction points (beam–beam effect). To limit adverse beam–beam effects the beams are separated at crossing points without an experiment. The *total beam current* $k_b \cdot N_b$ is constrained — in e^+e^- colliders by the available RF power replacing the energy lost by the beam due to synchrotron radiation or by cryogenic power required to cool the superconducting RF cavities; in p–p colliders, by particle losses tolerated by the superconducting magnets; in p–p̄ colliders, by the performance of the p̄ injector. The *beam sizes s_x, s_y* to be minimized are determined by the square root of the product $\beta_i \cdot \varepsilon_i$ where β_i is a local magnet lattice parameter, inversely proportional to the focusing strength, and ε_i, is the beam emittance, i.e. the area of transverse phase space occupied by the beam. Small β requires strong local focusing: this is produced by a sequence of quadrupoles on either side of the interaction point forming a so-called *low-β insertion*. In e^+e^- colliders, synchrotron radiation effects lead to an equilibrium value of ε_i that grows with the square of beam energy. In hadron colliders, the injector determines the invariant $\beta_b \cdot \gamma_b \cdot \varepsilon_i$ where β_b and γ_b are the relativistic parameters of the beam, so ε_i shrinks during acceleration — an advantage for high energy colliders. The beam intensity and the beam size also vary with time: the beam intensity decreases due to collisions with residual gas and with the opposing beam. While the beam sizes are nearly constant in e^+e^- colliders, in hadron colliders they increase for the same reasons as beam loss occurs. For the experiments, the relevant parameter is *L integrated over time* and is measured in terms of barn^{-1}. It is optimized by a careful programming of the operation cycle of injection, beam acceleration, luminosity production, and beam dump. In a *linear e^+e^- collider*, the rate of bunch encounters is given by the number of bunches in the bunch train times the repetition frequency. The latter is only about 100 Hz, so high-intensity bunches are imperative, and these must be very tightly focused at the interaction point. This adds to the focusing but also creates synchrotron radiation — which causes a spread in centre-of-mass energy and background in the detector. Longitudinally polarized beams were achieved at the SLC and are planned for future linear e^+e^- colliders. This feature which enhances the effective luminosity and enables improved discrimination between weak-interaction processes by choice of spin directions [Box 2.2] of the interacting particles.

Supplying antiprotons to the SPS as a collider

The overall scheme (Fig. 6.1) involved some major modifications to the PS and the SPS and the construction of new beam transfer lines for the antiprotons. The scheme evolved over the years improving continuously the performance.

The PS injectors, the 50 MeV Linac, the 800 MeV booster-synchrotron and the 26 GeV PS were pushed to their limits to deliver an intense proton beam on the production target with an intensity of up to 1.4×10^{13} protons per pulse, as much as the target could withstand, yielding up to 7×10^{7} injected antiprotons. The transverse emittances had to be small to permit focusing to a small size over the whole length of the target. Finally, the proton burst length had to correspond to the AA circumference, exactly one quarter of the PS.

Initially, the AA was used as a collector as well as an accumulator; in 1987 the ACOL project was launched with the construction of a large acceptance Antiproton Collector (AC) ring encircling the AA, for collecting the antiprotons from the target; this permitted a 10-fold increase in the accumulated particles. A powerful RF system (1.5 MV, 9.5 MHz) in the AC reduced the momentum spread by a factor four before cooling was applied. Stochastic cooling in the AC reduced the six-dimensional phase space density by up to 4×10^{9} before the beam was bunched by an RF system and extracted to the AA, where the beam was accumulated over hours or days.

In the AA, complementary stochastic cooling systems acted continuously on the antiproton stack. Over a day, a stack with a dense core of several 10^{11} antiprotons was accumulated. An antiproton bunch was picked from the stack and moved to the ejection orbit. From there it was sent through a "loop" (TTL2, Fig. 6.1) to the PS. In the PS, this bunch was accelerated to 26 GeV/c and sent to the SPS.

Prior to a transfer, checks were performed, concluded by the dispatch of a small "pilot bunch" of $\sim 10^{9}$ antiprotons all the way from the AA to the SPS. This assured that the big shots (up to 6 bunches), containing a day's worth of accumulated antiprotons, would safely find its way. The antiproton beam was accelerated in the SPS to the top energy of 270 GeV simultaneously with the counter-rotating proton beam which had been injected just before the antiproton transfer. After some adjustments the beams were brought into collision. The acceleration time of ~ 3.4 s was short compared to the time the beams were kept colliding with typically up to 16 h of useful beam lifetime.

The AA and AC storage rings

The AA had a circumference of 157 m. Its magnetic field was constant, for a beam momentum of 3.5 GeV/c on the central orbit. It was installed in a new hall and covered with concrete shielding blocks. Figure 6.2 (left) shows the AA and the AC rings, before being covered with concrete shielding. Figure 6.2 (right) shows the layout of the two rings with the "dog-leg" incorporated in the antiproton injection line to diminish the flux of secondary electrons and pions reaching the hall.

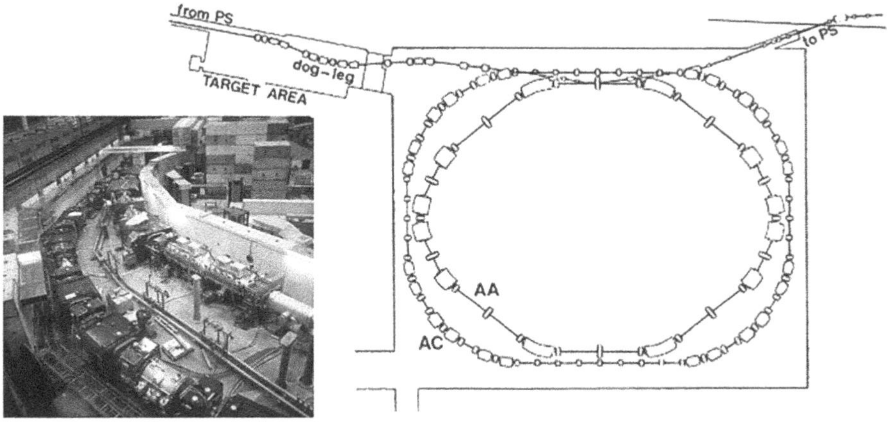

Fig. 6.2. The AC ring built around the AA ring, and a partial view of the installation.

The circumference of the AC was 187 m. It was installed around the AA in a period of just eleven months in 1986–87. Its primary task was to capture an order of magnitude more antiprotons than the AA, for which it had much larger acceptances, both, in transverse and in momentum space.

The AA, and later the AC had to have a large acceptance in order to capture a maximum number of antiprotons coming from the target. The phase volume occupied by the beam had therefore to be reduced and the particle density in phase space increased. In fact, the density in 6-dimensional phase space (the particles have position and momentum in transverse and longitudinal directions) had to be boosted by nearly a factor 10^9 in the AA. The only way to achieve this was to fully exploit the method of stochastic cooling.

To this end, the AA was the first accelerator in the world to be equipped with a full set of stochastic cooling systems; there were 7 systems (frequency range 150 MHz to 2 GHz): a precooling system for the momentum plane and 3 systems, one for each phase plane for stack-tail and core regions [7]. With the arrival of the AC ring, the AA systems were converted to 5 systems, operating in the higher frequency range of 1–8 GHz [8]. The AC had 9 systems in 3 bands (1–3 GHz) applying cooling in all three phase planes in each band [8]. All these systems had their specific functionalities and significant variations in pickup-kicker technology, amplifiers, gains, bandwidths and optimization. For example, in the AC, both pickup and kicker transverse apertures followed a programmable mechanical movement in time to keep in step with the transverse beam dimensions as the cooling progressed during the 4.8 sec that the beam stayed in the AC before ejection towards the AA. The components at low RF power (pickups, terminations, preamplifiers) were cryogenically cooled (20 K at the terminating resistors) to reduce thermal noise. Common amplifiers were used to switch from transverse to longitudinal plane systems at appropriate times in the AC cycle.

Overall, the AC & AA stochastic cooling systems constituted as high-power, low-noise, cryogenic multi-band microwave systems, with multiple transmission lines, feed-throughs and fast-moving electrodes in ultra-high vacuum.

The SPS and its modifications as a proton–antiproton Collider

The SPS started collider operation in 1981 at 270 GeV (centre-of-mass, CM, energy 540 GeV), for the same magnet power dissipation as the 400 GeV accelerator cycle. In 1984, increased water cooling of the magnets allowed the CM energy to be pushed to 630 GeV. For short periods of operation, a pulsed mode, cycling between 100 GeV and 450 GeV, extended the collider reach to 900 GeV CM energy. Usually, there was one "fill" per day, and the beams were dumped when the luminosity had become too low for the experiments to take useful data.

The fixed-target machine SPS had to be significantly modified to function as a collider from 1978 onwards in parallel to the AA construction. The machine needed increased beam focusing at the two collision points. Two huge underground experimental areas were constructed: The UA1 and UA2 detectors had to be retractable to avoid high levels of radiation during the period of the yearly fixed-target physics run. Functioning as a storage ring required a considerable improvement in the vacuum system. The RF system needed many modifications to simultaneously accelerate protons and antiprotons with precise synchronisation between bunches of the counter-rotating beams and to bring them to collision at the two experiments. Earlier studies of using SPS as a collider had indicated that RF noise could be a limiting factor for beam life-time; the various sources of noise in the closed loop RF system were analysed and their influence reduced sufficiently. The SPS had been built for an injection momentum of 14 GeV/c. The injection system had to be modified to permit 26 GeV/c proton injection via the TT10 transfer line, which also had to be upgraded. For the counter-clockwise circulating antiprotons a new transfer line, TT70, had to be built from the PS to the SPS and a new injection system installed for 26 GeV/c antiprotons. With the tenfold increase in antiproton production from 1987 onwards and collisions with

6 bunches in each beam (up to 10^{11} particles/bunch), a peak luminosity of 3×10^{30} $cm^{-2} s^{-1}$ was eventually reached, having increased by a factor of 60 since 1982.[a]

The low energy programme spawned by the p–$\bar{p}$ collider: LEAR and AD

Low-Energy Antiproton Ring (LEAR)

Initially, the scheme [9] was proposed as a 'parasitic' user, using only a small fraction of the accumulated antiprotons in AA, the main client still being the SPS Collider. Antiprotons at 3.5 GeV/c were returned from the AA to the PS for *deceleration* from 3.5 GeV/c to 0.6 GeV/c and transferred to LEAR, where they were further decelerated to below 100 MeV (or accelerated up to about 2 GeV); the antiproton beam was extracted into the PS South Hall for different experiments at dedicated beam lines. One of the major highlights and technical innovation was the development of a technique for ultra-slow extraction; the LEAR beam of modest quantity would remain circulating for many hours and an extremely feeble quantity of antiprotons (equivalent and down to about one antiproton per turn) would be extracted using this special technique [Highlight 6.4].

Antiproton Decelerator (AD)

LEAR operated until end 1996. However, with the approval of the LHC in 1994, the scheme conceived originally for the SPS Collider was considered too expensive to continue providing low-energy antiprotons for the small but vibrant physics community at LEAR. This spawned the idea converting the AC into a multi-function Collector and Decelerator of antiprotons with extraction for physics into the middle of the ring, within the existing Hall. In this scheme, the target area for antiproton production remained unchanged but the AC ring was modified to permit ramping from the injection momentum of 3.5 GeV/c down to 100 MeV/c and an extraction channel for experiments was installed in the AD Hall. In 1997, the AA was dismantled, LEAR closed down and the AC was converted into the AD [9].

While the AD extracted beam energy of 5.3 MeV was adequate, it was still a far cry from the needs of the antiproton trapping experiments which required use of degrader foils to further decelerate, losing however almost 99% of the beam in the process. This led to the innovative idea to use a radio-frequency quadrupole to decelerate the antiprotons even further, down to 10 keV, for a particular physics experiment ASACUSA [Highlight 6.3]; The Extra Low-ENergy Antiproton ring (ELENA) under construction at CERN today envisages AD extraction into this lower stage decelerator from a momentum of 100 MeV/c to 13.7 MeV/c (kinetic energy 100 keV).

---

[a] The editorial team thanks Lyn Evans for his contribution to this chapter.

The experimental programme

In 1976, when Carlo Rubbia proposed to produce the putative weak bosons in proton-antiproton collisions in the SPS, the prestige that such an important discovery would mean for European physics in general and for CERN in particular convinced the physics community and the CERN management that the risk was worth taking: the short time table and the relatively low cost of the project matched well the window of opportunity open in the pre-LEP period. However, as many in the accelerator community were frightened by the "beam acrobatics" implied by the proposed scheme, and as many in the Fixed Target community were upset to see part of the SPS time slip out of their hands, approval of the project was subjected to three conditions: positive outcome of a feasibility study that was immediately initiated; success of a feasibility test of stochastic cooling (ICE) that was concluded in Summer 1978, providing evidence for fast cooling in both transverse and longitudinal phase space; setting up of a detector study group, under Rubbia's chairmanship, aimed at showing the feasibility of experimentation in general and of detecting the W and Z bosons in particular. The work of this study group served as the basis for the design of the UA1 detector [10].

In June 1978, the SPS committee approved the UA1 experiment and encouraged the submission of a second proposal. UA2 was approved in December 1978. The first proton-antiproton collisions were observed in June 1981. The W and Z bosons were detected soon after, and Carlo Rubbia and Simon van der Meer (Fig. 6.3), thanks to whom the discovery was made, were awarded the 1984 Nobel prize for "their decisive contributions to the large project which led to the discovery of the field particles W and Z, communicators of weak interactions" [10, 11].

Fig. 6.3. Carlo Rubbia and Simon van der Meer following the announcement of the Nobel Prize.

Detectors for the W- and Z-boson discovery

The detection of the weak bosons was less of a challenge than their production. Their large invariant mass made it easy to eliminate promptly, at trigger level, the low transverse momentum particles produced in the vast majority of proton-antiproton collisions: the problem was their identification among the selected sample of high mass, centrally produced final states, dominated by pairs of hadron jets. It was solved by selecting their leptonic decays, namely W → eν or μν and Z → ee or μμ, expected to give a clear signature, in spite of their low branching fractions (20% and 6.5%, respectively). To the extent that the misidentification of a hadron jet faking a lepton could be kept below one part in ten thousand, clear W and Z signals were expected. It was important that the detector could measure the sign of the lepton charge, allowing for an elegant probe of the nature of the weak interaction: in W decays, the positive lepton is preferentially emitted along the antiproton beam, the negative lepton along the proton beam.

Experience with the operation and exploitation of so-called 4π detectors, namely detectors surrounding the beam interaction region as completely as possible, was still embryonic when UA1 and UA2 had to be designed. However, at that time, other detectors were already being conceived for upcoming electron-positron colliders, using the experience of MARK I at SPEAR. In the case of hadron colliders, the constraints on detector design were stronger but one knew pretty well, from experimentation at the ISR, how such detectors must look like. The importance of efficient and accurate calorimetry was well understood, in particular for the detection of the transverse momentum imbalance caused by the presence of a non-interacting neutrino in W decays. Insensitive zones would be very damaging in this respect; detectors had to be, as one said, "hermetic". A major unknown was how "hostile" the collider environment would be, and how much one would have to protect the detectors from the background produced by beam losses. The collider turned out to be remarkably "quiet", producing much less background particles than anyone of us had anticipated, facilitating the experimentation.

The UA1 detector [12] had been designed as a multipurpose detector with no right to fail or to be blind to any significant manifestation of the new physics to be explored. It had to cover as large as possible a solid angle, to detect individual hadrons, hadron jets, electrons as well as muons. Its design precisely met these objectives. This spectrum of detection possibilities, however, had been obtained at the price of compromises on the performance of individual components, in particular the quality of the calorimetry and the strength and configuration of the magnetic field. The technically safe choice of a warm shoe-box-like dipole, rather than a superconducting solenoid, limited the field to 0.7 T and imposed the presence of dead regions in the field direction and in the gap between the two half-yokes. Moreover, it severely limited the segmentation of the hadron calorimeter, fitted in the magnet yokes, and of the electron calorimeter, lacking azimuthal segmentation. The main strength of the UA1 detector was a large-volume, high-resolution central tracking detector of an original and high-performing design. Its remarkable pioneering achievements make it without any doubt a major technological highlight of particle detection at the proton–antiproton collider [Highlight 6.5].

UA2, the second detector [13], had been approved under the condition of being cheaper than UA1 (by about a factor three, as it turned out) (Fig. 6.4). Moreover, it had to be constructed and assembled in less than three years in order to catch up with UA1. Such constraints imposed restrictions on its design: it would detect electrons but not muons, it would focus on the central rapidity region and would not measure particle charges except in two cones along the beams where the W decay charge asymmetry was expected to be maximal. The central high resolution calorimeter, of a very careful and precise design, was innovative in several of its features, in particular its projective geometry and the use of special thin plastic plates ("wavelength shifters") to collect separately the light from different sections of the lead-scintillator lamination [Box 6.3]. It proved to be a superb tool for the

Fig. 6.4. UA2 detector during construction.

detection and energy measurement of electrons and hadron jets. Toroidal magnetic fields in the forward and backward cones did their job in revealing the expected charge asymmetry in W decays but otherwise severely limited the solid angle coverage and the hermeticity of the detector. When the installation of ACOL gave the opportunity to upgrade the detector, they were replaced by calorimeters extending the central structure to smaller angles.

It was also at that time that the central vertex detector, originally a set of cylindrical drift chambers that had to fit in a very small volume around the beam, was replaced by scintillating fibres and silicon pads. While the latter are described in more detail below [Highlight 6.6], the former was the first detector of this type to be used in an experiment; it featured several novel technologies: scintillating fibres as the active detector element, custom image intensifier with integrated CCD readout to collect the signals and custom digitizers to process the images. It included layers of fibre triplets, one parallel and two at opposite small angles relative to the beam, allowing for space reconstruction of the particle tracks and incorporated a layer of lead for initiating electron and photon showers.

The UA1 and UA2 detectors turned out to nicely complement each other. UA1 could do more than UA2: it could detect electrons, muons and tau leptons; observe the branching ratios for the weak bosons decaying into different leptons, a crucial test of the theory; detect muons in the neighbourhood of a hadron jet as a possible signal for new physics, give early evidence that mesons containing a beauty quark could change, "oscillate", between their matter and antimatter forms and make an early and elegant measurement of the W spin. But what UA2 could do, it did better than UA1; it provided the most accurate measurements of the W and Z masses and gave important contributions to the exploration of the strong interaction sector, using its excellent jet detection capability.

The remarkably smooth operation of the CERN proton–antiproton collider has been an opportunity for other experiments than UA1 and UA2 to open new chapters of physics in an unexplored energy domain. The novelty of the experimental environment that it offered and the challenge it presented in terms of experimentation were an incentive to innovate in the design of the detectors.

A particularly successful example is the UA5 experiment [14], which had only a few days to collect its data before the UA2 detector was rolled into the tunnel and yet succeeded in giving to the physics of hadron collisions some very important results. It used two very large streamer chambers, the largest ever used in an experiment, photographed through image intensifiers and triggered by scintillation counters [Box 3.3]. Lead glass converters inside the chamber volume were used to convert photons. Charged particles were detected down to less than a degree, a remarkable achievement.

In the same way as the ISR had been a test bench for the development of detectors used at the proton-antiproton collider, the latter, together with the Fermilab Tevatron, has been a test bench for the development of LHC detectors. Among many such examples, the case of the Roman pots is an interesting illustration. Developed at the ISR to measure the total cross-section [Highlight 4.7], they were revived at the proton-antiproton collider, instrumented with fine resolution detectors which could be placed to the beam within a few millimetres; later they were used to test a silicon strip detector, with six planes totalling 46 000 channels, which became the prototype of an important element of the LHCb detector.

Exploring antimatter

Not to be outdone by the collider detectors, those at LEAR, using nearly stopped antiprotons, displayed much ingenuity in their design. Particularly innovative were two 4π detectors, Crystal Barrel and CPLEAR. The former included a projective assembly of 1380 CsI crystals and a segmented drift chamber operated in a 1.5 T magnetic field; it discovered a new particle considered as the lowest mass scalar glueball. CPLEAR developed a technique to identify the quark (or "flavour") content of the produced neutral kaons, allowing detailed measurements of CP violation [Box 3.4] and placing stringent limits on CPT conservation in weak interactions. The experimental method that was used for identifying the flavour can be considered as a pioneer for some of the methods used today in LHCb.

A major success story of the LEAR experimental programme has been a succession of experiments aiming at precision measurements of the properties of antiparticles, first antiprotons and later anti-hydrogen. The first in the line was the Penning trap experiment, which succeeded in trapping antiprotons and keeping them trapped for up to 2 months. Compared with trapping protons, which had been already done in a few laboratories, the difficulty was to reach a much better vacuum in order to prevent the trapped antiprotons from interacting and annihilating with the normal matter of the residual gas. The experiment verified the equality of the antiproton and proton charge-to-mass ratios with an accuracy of 1 part in 10^{10}. In parallel with these advances in trapping techniques, formation of anti-hydrogen atoms was observed in flight in 1996 for the first time [Highlight 6.7]. This double success paved the way to the trapping of anti-hydrogen that was achieved at LEAR in 2002 [15] by the two successor experiments, ATHENA and ATRAP. Other remarkable contributions of LEAR to atomic physics include studies of antiprotonic helium and the X-ray spectroscopy of protonium, the exotic atom of a proton and an antiproton orbiting each other. These successes were at the root of the recent decision to continue fundamental research on antimatter at CERN with the addition of the ELENA ring.

Looking back at the early days of the proton–antiproton collider and the many contributions it gave to fundamental particle, nuclear and atomic physics, one cannot help but being impressed by the role it played in technological advances, which lead to the construction and exploitation of the LHC. Progress on the two fronts feed each other, accelerating the rate of innovative achievements in each of them. The degree of sophistication, in both accelerator and detector technology, which prevailed at the collider was inconceivable at ISR time, as that which now prevails at LHC was unconceivable by then. Yet, all this happened in less than fifty years. It is the pride of CERN, together with the Fermilab Tevatron, to have hosted and fostered such progress so successfully.

6.2 Stochastic Cooling: Technology to Compress the Beams

Fritz Caspers and Lars Thorndahl

Stochastic cooling of particle beams, pioneered by CERN, was a condition *sine qua non* for the feasibility and success of the proton–antiproton collider. The required novel technology was developed by a small team, whose nucleus had been formed for the cooling tests at the ISR and included S. van der Meer, the inventor of stochastic cooling. The development concentrated in particular on electrodes, picking up with high selectivity the signal from the beam, the "pick-ups" (PU), and on the pulsed elements (kickers) imparting a kick to the beam particles derived from the signal of the PUs [Box 6.2]. The signal on its way from the pick-up electrode (PU) to the kicker had to be conditioned, without introducing significant delays, such that the signal generated by a beam slice at the PU would arrive in time to impart a corrective kick on the same beam slice (Figure in Box 6.2). Cooling of transverse deviations requires a transverse kick, longitudinal cooling (i.e. in the beam direction), reducing the energy spread and increasing the longitudinal density, requires acceleration or deceleration of the slice. The signal treatment between the PU and kicker needs amplification in well-defined frequency bands and filtering. The systems must have a wide microwave bandwidth as the cooling rate is proportional to the bandwidth of the system. For technical reasons the bandwidth of a system is typically limited to an octave, so the required total bandwidth is often covered with several systems operating in adjacent ranges of frequency. The coupling to the beam should be as strong as possible to keep the power of the amplifier inserted between PU and kickers relatively low [16]. A few prominent examples of these technologies, some of which were at the time at the limit of feasibility, are given below.

Stochastic cooling: Domesticating beams **Box 6.2**

The purpose of stochastic cooling [1, 2] is to increase the density of a beam of charged particles. During this process, the particles are "compressed" into a denser beam with less angular divergence and less energy spread: empty space between the particles is squeezed out. The phase space occupied by the beam is reduced so that a new beam can be added during beam accumulation in the limited phase space of the accelerator, and the final beam can be made denser, imperative for achieving good luminosity in antiproton colliders. Cooling is also used to prevent the blow-up of a stored beam.

The principle of stochastic cooling is illustrated for the case of reducing the wide horizontal dimension of a beam caused by oscillations of the particles around the nominal orbit. Consider first a single particle. A sensor (pick-up) acquires an electrical signal proportional to the displacement of the particle (see Figure) [1]. The signal takes a short-cut across the ring so that the particle receives the kick required to put it on the nominal orbit. Fast signal transmission from pick-up to kicker is vital for the cooling system, as the particles move close to the speed of light. In reality, pick-up and kicker act on a short slice of the beam which contains a small fraction of the total number particles N. The slice/sample duration is determined by the bandwidth W of the electronics. The damping of the oscillation of the particle is disturbed by the signals from the other randomly distributed particles in the slice. Given the finite number of particles and the sample containing different particles at each pass due to a spread in revolution frequencies (perfect mixing), the displacement of the slice will be different at each passage. The effect of the companion particles in the sample averages out to first order but they produce a second order adverse blow-up of the beam depending on the gain of electronic system. For a properly chosen gain the correction of many samples leads to a slow increase in beam density, i.e. cooling, at a rate $\propto W/N$. So high bandwidth is imperative, and the system works well for low N, typical of antiproton beams. Cooling rate is also reduced by imperfect mixing; thermal noise in the electronics, the power limit of the large-band amplifiers, and pick-up to kicker time-of-flight errors. Longitudinal cooling to reduce the energy spread in a beam and increase longitudinal density works according to the same principle. To sense the energy deviation, a pick-up is used at a point on the orbit where an energy error leads to a large transverse displacement. The signal feeds a kicker producing a longitudinal kick. Transverse and longitudinal cooling must be combined to achieve high beam density.

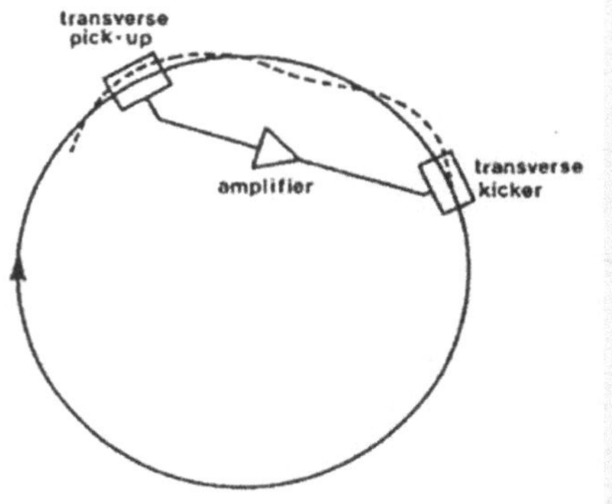

[1] S. van der Meer, *Rev. Mod. Phys.* **57**, 689-697, (1985).

[2] D. Möhl, *Stochastic Cooling of Particle Beams* (Springer-Verlag, Berlin, 2013).

For the cooling of the longitudinal momenta of the beam particles a method based on *filters with periodic frequency behaviour* was invented [17]. Longitudinal cooling must be sensitive to the different longitudinal momenta of the particles, making use of their slightly different revolution frequencies. The spread in longitudinal momentum is reduced by comparing the revolution frequency of the particles to the nominal value and by a subsequent correction of the deviation with an electric kicker field in a downstream gap, either accelerating or decelerating. The comparison is accomplished by a filter between the PU and kicker which generates the appropriate phase-shift of the signal, depending on the deviation detected leading to acceleration or deceleration. Such filters, having extremely high selectivity, were based on transmission lines with a length corresponding to the nominal revolution frequency. The signal of the beam slice is generated by a PU and amplified before filtering, yielding an excellent signal-to-noise ratio after amplification. This advantage is particularly relevant if the beam has a very narrow energy spread.

Novel *slot-type structures* [18] were developed for both PUs and kickers, for frequencies above 1 GHz, a frequency range not covered by the technology available at the time (coupling loops). They were superior to the loops, which are not wide-band and had no means for the suppression of undesired waveguide modes. The same type of structure acts as both PU and kicker for stochastic cooling of both transverse and longitudinal momentum spread. In order to improve the signal-to-noise ratio at the PU many of these elements were used in a sequential array in the ring. This type of structure was first tested and put into use in the ISR and later adopted for the Antiproton Accumulator (AA). Furthermore, these structures have the attractive features of constructional simplicity and requiring only a small number of vacuum feedthroughs. Figure 6.5 shows such a structure with the coupling slots between the beam chamber and pairs of (TEM) transmission lines on top and bottom. In a PU, these lines transmit the signal from the beam to the amplifier, and when used as kicker transmit the signal from the

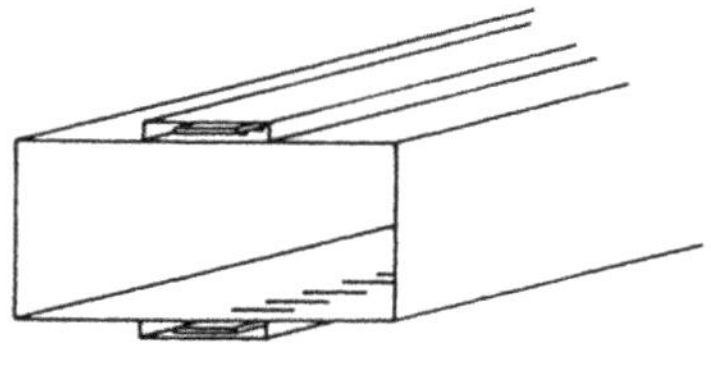

Fig. 6.5. Slot-type PU or kicker. The beam circulates at the centre of the chamber [16].

amplifier to the beam. signals from the individual slots add up to give strong coupling, provided that (i) the slots are small compared to the wavelength of the signals and (ii) the phase velocity in the transmission lines transporting the signal equals the particle velocity which is close to the velocity of light.

In order to satisfy the ever-increasing appetite of antiproton users, the Antiproton Collector (AC) was built around the AA in 1986. From 1987 on, it boosted the accumulation rate, eventually by an order of magnitude. As a maximum of antiprotons had to be accumulated, the beam had large transverse dimensions and the AC provided a matching large transverse acceptance. However, this beam had to shrink very quickly, within 4.5 s, to a size fitting into the acceptance of the AA. This was achieved by using novel *plunging pick-ups and kicker electrodes*, which maximized the coupling to the beam. They had moving electrodes, which followed the shrinking of the beam as it was cooled [19]. Even this was not enough: the PUs, pre-amplifiers, terminating resistors and combiner boards had to operate at cryogenic temperatures — as low as 20 K — to minimize the thermal noise in the RF circuits, which substantially complicated the mechanical design. Figure 6.6 shows such a PU in its vacuum tank.

Two aluminium support structures (to the right and left) would move the many pickup loops (seen here as triangles) by 45 mm each towards the particle beam axis. The many horizontal channels (only their cross-sections are visible) were parts of the signal combiner for the multiple pickup loops. The upper and lower fixed aluminium support structures of 2.2 m length were holders of ferrite tiles absorbing undesirable microwave signals. The silver-plated undulated foils (adapting to the movement) were connections to fixed cold aluminium profiles. Similar flexible foils served as signal connections between the moving support structures and the vacuum feedthroughs. Fixed copper braids established the final thermal leads to the cryogenic sources. Six complete systems covering three adjacent bands in the 1 GHz to 3 GHz range were in operation in the AC. The two systems in each band were for horizontal and vertical cooling respectively; the sum signal of the two PUs served for longitudinal cooling.

Fig. 6.6. Plunging PU in the AC, seen along the beam. The outer diameter of the vessel is 0.5 m.

The AC also required the development of 100 W *power amplifiers* for the bands 1–1.6, 1.6–2.4 and 2.4–3 GHz. These compact amplifiers were based on four power Field Effect Transistors (FET) per module and four modules per amplifier. Signal input splitting and output combination was performed with four-way elements absorbing electrical mismatches. This design was superior to the competing commercially available traveling wave tubes for several reasons: no high voltages, no cathode heating, better linearity, small phase change with amplitude and better life time [20]. The manufacture of the large series was entrusted to industry after a successful transfer of know-how.

6.3 Radio Frequency Quadrupole: Slowing Down Antimatter

Werner Pirkl

Many fundamental studies with antiprotons require extremely slow antiprotons, with velocities far below the kinetic energy of 5.31 MeV of the antiproton beam, extracted from the AD synchrotron [Highlight 6.7]. To this end a novel "Decelerator" was developed, a variant of the Radio Frequency Quadrupole (RFQ), which decelerated the antiprotons to 55 keV, followed by an integrated superimposed electrostatic energy correction to adjust the output energy between ~10 keV and 120 keV [21].

The RFQ is essentially a modified electric quadrupole consisting of two electrode pairs of opposite polarity, positioned at the diagonals of a square with the beam at the centre. This provides transverse focusing in one plane but causes defocusing at the orthogonal plane. Nevertheless, overall focusing can be achieved by alternating static focusing and defocusing along the beam line. Alternation *in space* can be replaced by alternation *in time*. Feeding longitudinally continuous electrode pairs with properly chosen Radio Frequency (RF) provides also overall focusing due to the changes in polarity, as W. Paul, Nobel laureate 1989, had earlier demonstrated with linear "traps".

The decisive ingredient towards the RFQ was added in the seventies by "modulating" the radial distance of the electrode tips by opposite peaks and valleys in longitudinal direction [22]. This adds a longitudinal component to the transverse field pattern. Hence, the RFQ combines transverse focusing with longitudinal acceleration in a very compact geometry. It is therefore ideally suited for pre-accelerators up to the MeV range and supplanted quickly the traditional accelerator front-ends with their huge high-voltage Faraday cages.

The suitability of the RFQ for beam *de*celeration (RFQD) was quickly recognized [23, 24]. The basic electrode structure is the same as for the RFQ. A top view of four modulated electrodes is given in Fig. 6.7. Half of two electrodes is cut away for clarity. The distance between two electrode peaks, called a double cell, has to correspond to the time the particles travel during one RF cycle, being 4.94 ns in the case of the RFQD, determined by the operating frequency of 202.56 MHz.

As the velocity of the particles decreases gradually along the RFQD that distance decreases in proportion, here by a factor of about nine, from 157 mm at the input to 17.5 mm at the output. Contrary to accelerating RFQs it is no longer possible to use 30 to 50 double cells for "soft" beam capture and shaping as those are now on the high energy side, where their much longer cell length would lead to prohibitive overall length. Instead, bunching of the beam in a single step outside and upstream of the RFQD has to be used with less than perfect efficiency. Only about half of the beam falls in the acceptance of the RFQD.

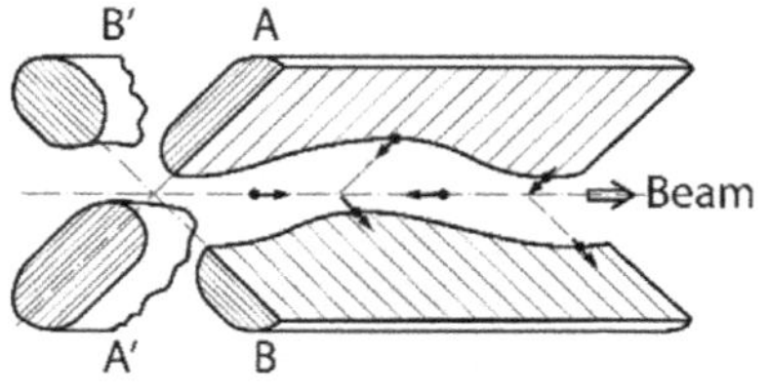

Fig. 6.7. The modulated electrodes of an RFQ; arrows show the electric field at a given moment.

Fig. 6.8. The RFQD tank with the inner structure extracted for inspection.

The inner structure of the RFQD forms a "ladder" of length 3.4 m (Fig. 6.8). Continuous rails on top and bottom are connected by 35 rungs with the electrodes along the axis. The electrode pair A–A′ is connected to even-numbered rungs, pair B–B′ to the odd-numbered. All ladder parts are made of copper. The resulting 34 double-periodic cells are individually tuned to the operating frequency of 202.56 MHz. An RF amplifier chain (not shown) delivers the necessary RF power of more than 2 MW in 400 ns pulses at a repetition period around 300 s.

Figure 6.9 shows a cut through the RFQD in front of an even-numbered rung. Electrode pair A–A′ is firmly connected whereas pair B–B′ runs through an opening in the rung, at a distance of 10 mm to hold the RF operating voltage of 167 kV peak.

The ladder has to be mounted electrically "floating". It is held in place by two high-voltage (HV) insulators on the top and three on the bottom, where the central one acts also as a HV feedthrough for the energy-correcting voltage of +/− 65 kV DC, see Fig. 6.9. Curved HV shields on the vertical faces, made of stainless steel, prevent discharges. The inner face of the tank with 380 mm inner diameter is copper plated to reduce the RF losses.

Initial deceleration tests were carried out with a proton beam at the University of Aarhus, Denmark [25]. Current operation at CERN yields about one million decelerated antiprotons per shot. Compared to passive deceleration by degrader foils it provides a one to two orders of magnitude higher transmission together with improved beam quality which is close to the theoretical maximum for non-cooled deceleration.

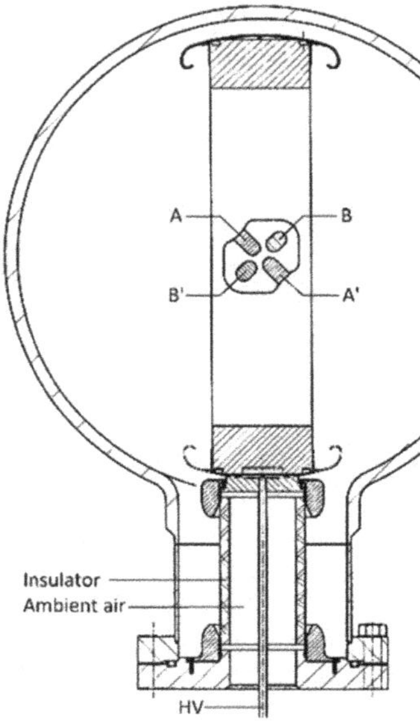

Fig. 6.9. Cross section of the RFQD.

6.4 The LEAR Ultra-Slow Beam Extraction: Trickling Antiprotons

Michel Chanel and Kurt Hübner

During the early discussions of the CERN proton-antiproton programme it was realized that antiprotons from the Antiproton Accumulator (AA) could be decelerated to lower energies providing pure, low-energy antiproton beams of unprecedented intensity for experiments [26]. This led to the construction next to the PS of a synchrotron ring of 78 m circumference, the Low-Energy Antiproton Ring (LEAR) [27] that was operating from 1983 to 1996. The beam, after injection from the PS at around 200 MeV (kinetic) energy, could be either accelerated up to 1270 MeV or decelerated to 5.3 MeV. At the chosen energy, the beam was left coasting either interacting with an internal target surrounded by a detector or extracted slowly and transferred to the experiments in the PS South Hall. The experimental programme embraced the study of p–p̄ and p̄–neutron interactions, of fundamental symmetries and the production of antihydrogen for matter–antimatter symmetry tests, and this called for new methods to manipulate beams.

Typically, a few 10^9 antiprotons, close to the minimum the PS could handle, were injected into LEAR. These particles were taken from the AA stack with an average consumption of 10^6 particles/s or about 10% of the AA accumulation rate and, hence, with little impact on the collider programme operated in parallel with

LEAR until 1991. Stochastic cooling and electron cooling of the beam was eventually used to control and to reduce the beam dimensions, particularly important at low energy. Electron cooling, invented in Novosibirsk [28] and tested in ICE, is based on the corrective action of "friction" experienced by particles deviating from the nominal momenta in a dense electron beam moving with the antiprotons over a small part of the circumference. It is particularly efficient at low energy.

In order to avoid saturating the particle detectors of the experiments, the users of the low-energy beams required an extraction rate, the spill rate, to be around 10^6 particles/s. Hence, for 10^9 antiprotons or more circulating the spill had to last about 10^3 s or longer, far beyond what had been achieved previously, e.g. typically 0.3 s in the PS. Compared to a one-time spill, the spill length of the beam arriving at the experiment had to be effectively stretched by a factor up to $3 \cdot 10^9$, implying that one would not skim off more than about one particle per turn from the circulating beam.

This problem was solved by stochastic slow extraction [29, 30] combined with a powerful feed-back system [31]. Slow-extraction is based on a mechanism which makes the amplitude of the horizontal particle oscillations grow on a time scale long compared to the revolution time. Once the amplitude is increased beyond a certain point the particles enter an electrostatic septum deflecting them into an external beam transfer channel guiding them to the experiments. The electrostatic septum provides the required electric field for this deflection at the edge of the aperture but shields the circulating beam from this field. The amplitude growth is generated by driving the particles into a resonance with a magnetic perturbation of proper azimuthal distribution introduced by design in the ring. Since there is a spread in the oscillation frequencies, the beam is gradually extracted. The oscillation frequency of the particles is expressed in terms of Q, the number of oscillations of the antiprotons per turn. The resonance occurs at a certain oscillation frequency Q_{res} outside the Q distribution of the beam. Since Q is a function of the revolution frequency, a modulation of the revolution frequency transforms into a modulation of Q. Hence, applying band-width limited white RF noise in a band at around a harmonic of the revolution frequency superimposes a noise spectrum on Q which can be positioned between Q_{res} and the Q distribution of the beam (Fig. 6.10); the particles diffuse in this noise band from the beam distribution to Q_{res}. The noise at a harmonic of the revolution frequency is generated by a kicker of the type used for longitudinal cooling which accelerates or decelerates the particles. The strength of the noise in Q is adjusted such that the particles diffuse towards the resonance at a speed determined by the required spill rate.

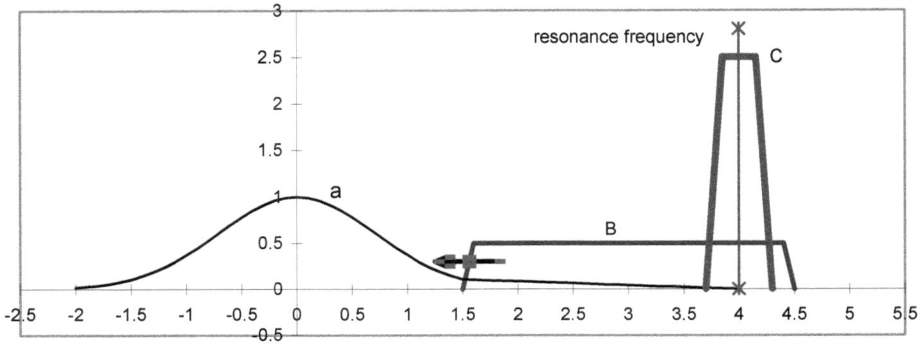

Fig. 6.10. Density distribution of antiprotons (a) and noise power spectra (B) and (C) versus $Q - Q_0$ where Q_0 refers to the centre of the beam distribution. Arbitrary units.

Particles slowly moving towards the resonance by diffusion in the noise band B are exposed to a much stronger noise C around the resonance. The low-frequency edge of band B is moved slowly into the stack eating into the distribution at the required rate. The band C increases strongly the diffusion rate once the particles are close to the resonance making the diffusion faster to reduce the adverse effects of the wobbling of the resonance which is brought about by the ripple in the powering of the magnets.

Two further steps were taken to smooth the spill in time. First, an air-core quadrupole was installed in the ring which was driven in antiphase of the mains ripple to strongly reduce the Q ripple and in turn the wobbling of the resonance. Secondly, the advance of the noise band into the stack was controlled by a sophisticated feed-back system, its input being an error signal derived from comparing the measured spill-rate at the experiment with the nominal rate.

Thanks to this set of measures spills lasting up to 14 hours eventually became routine, with spill-rates sufficiently low to satisfy the requirements of the experiments. Figure 6.11 shows an example of a spill lasting 10 hours where the beam was shared between two experiments by splitting the extracted beam.

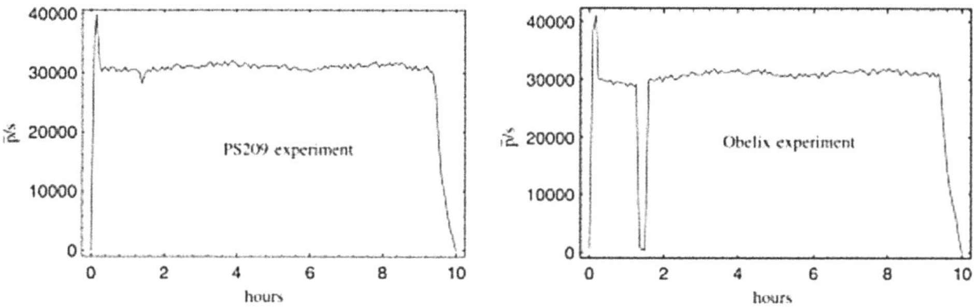

Fig. 6.11. Spill of antiprotons to two experiments. Each point on the curve is the number of particles recorded per s averaged over 10 s. The brief dip in the spill-rate of OBELIX is a pause for calibration.

6.5 The UA1 Tracker: An Electronic Bubble Chamber

Bernard Sadoulet

Physics goals and detector design philosophy

The priority for the UA experiments was the detection of the putative W and Z bosons of the Standard Model [Box 6.4] with emphasis on their "golden" signatures of their leptonic decay modes W → eν or μν and Z → ee or μμ, producing high transverse momentum charged leptons. Another major goal was the search for physics beyond the Standard Model, revealed by e.g. the simultaneous production of electrons and muons with specific charge configurations.

The accent was therefore put on the precise measurement of leptons, implying new performance requirements on the charged particle detector measuring the curvature of their tracks inside the magnetic field. The required excellent identification of the electrons would be obtained by comparing their momenta measured in the central tracker with their energies measured in the electromagnetic calorimeter. Muons would be identified as tracks penetrating through a metre of the iron return yoke of the magnet and by comparing the momentum measured in the central detector with that estimated from their deflection in the magnetized iron, obtained from the direction of the track emerging behind the yoke.

The measurement of high transverse momentum "jets" of particles was another design requirement of the UA1 detector. These jets could signal the emission of quarks or gluons, originating e.g. from the dominant quark–antiquark decay modes of the W and Z. Jets had been hinted at by experiments at the ISR (and later observed by AFS collaboration [Highlight 4.11], but at the ISR energy they were broad and difficult to identify. At the energy of the p–p̄ collider, however, they were expected to be fairly easily detectable as a group of collimated particles.

Electromagnetic and hadronic calorimetry would measure their total energy, including the neutral particles (π^0's and neutrons).

The collaboration recognized rapidly that a detector was needed covering as much as possible of the solid angle around 90° from the beams. This is a prerequisite for the detection of the W or of any new physics based on "missing transverse energy", signalling the escape of a neutrino or possibly other neutral non-interacting particles such as supersymmetric particles [Box 7.2]. The design therefore aimed at achieving as close a "hermetic", i.e. 4π solid angle coverage, as possible, minimizing regions where particles could not be detected.

The geometry of the magnet was a major topic of discussion. The choice was between a solenoid coaxial with the beams or a dipole with its field perpendicular to them. While considerations of symmetry favoured a solenoid, a dipole was finally chosen. This was motivated in large part by the requirement of a detector with a fairly large aspect ratio (6 m long with a diameter of 2.5 m), in order to match the topology of the final states produced in the collisions. A solenoid would have to be instrumented with a self-supporting drift chamber with 6 m long wires, for which the technology was not available, and the CERN engineering team came forward with a very elegant solution for a tracking detector optimized for the dipole [Box 4.4]. To meet the short construction time a conventional warm magnet was adopted with a 0.7 T field and a magnetic volume of 80 m^3.

The UA1 tracker

With the spectrometer magnet chosen, the collaboration converged rapidly on the technology of the central detector. This instrument had to offer very good spatial resolution, high efficiency for reconstructing tracks in events of very high particle multiplicities and a bubble-chamber like clarity of the track pattern to clearly expose the precious leptons. The "Time Projection Chamber", invented in 1974 [32], shows this kind of performance and this concept was adapted to the UA1 magnet geometry. Instead of measuring the track curvature in projection, the measurement of the drift time was used due to its intrinsic precision; this led to having the sense wires parallel to the magnetic field. The design goal materialized in a drift chamber system with 6000 sense wires [Box 4.5]. The novel electronic read out could register several independent hits on a drift wire per collision, simultaneously digitizing the drift time and the charge information, which allowed a pictorial track reconstruction in three dimensions. At the time, this tracker (Fig. 6.12) was at the frontier of drift chamber development [Highlight 4.8].

Detecting invisible particles **Box 6.3**

Particles properties are revealed in particle detectors, devices in which the particle "interacts" and leaves a telltale signal [Box 3.2]. No single detector can measure all their properties and several specialized ones are combined in an experiment. Through electromagnetic (e-m) interaction, charged particles ionize or excite a specially chosen material in the detector. It may be a gas [Highlight 4.8], a liquid [Highlight 5.6] or a solid [Highlight 5.9]. The level of ionization or excitation depends characteristically on the particle velocity v [Box 5.2]. A variety of detectors have been developed to measure position, velocity and energy of the particles.

Tracking detectors measure the charged particle position or trajectory. They have one important and distinct feature: a very gentle interaction, hardly influencing the properties of the particle, suffices to produce a useful signal.

When these detectors are immersed in a magnetic field B, the radius of curvature ρ of the charged track, proportional to p/qB (where q is the charge), gives its *momentum p*. The figure of merit of such a "spectrometer" is its accuracy, expressed as $\Delta p/p, \approx p\,\sigma/\,B\,L^2$, where σ is the precision of the position measurement and L the distance over which the particle track is analysed. This explains the quest for improved σ and higher B, and the increasing size (L) of the spectrometers to measure ever increasing p [Highlight 8.12].

Other instruments measure the *energy E*. Interacting via e-m and/or strong interaction the particle develops a *cascade ("shower")* of particles absorbed in a massive detector. Nearly all E of the incident particle is finally transferred to excitation of the absorber molecules, increasing its temperature, whence the name *"calorimeter"*. The absorber is usually instrumented with detectors sensitive to the cascade of particles [Highlight 4.10], but can also be sensitive in its full volume [Highlight 7.9]. The number of cascade particles being proportional to E, the statistical accuracy of the E measurement is $\approx 1/\,E^{1/2}$, a bonus for high E. Calorimeters work for both charged and neutral particles.

To *identify a particle* via its mass, one needs p and its velocity v. Besides time of flight measurement for slow particles [Highlight 8.9], one uses also the v dependence of e-m processes: ionization, Cherenkov or transition radiation [Highlights 4.9, 7.8]. Charm or beauty particles live of the order of 10^{-12} s. Being produced with v close to the speed of light, the relativistic time dilation lets them fly of the order of 1 mm before decaying, a decay pattern accessible to Pixel detectors [Highlight 8.6]. Assembling such specialized detectors leads to the typical onion-like sequence at colliders. "Fixed target" experiments (such as COMPASS, Chapter 5) have a quite different structure.

These detectors, first developed for particle physics, are now frequently found in medical diagnostic and imaging techniques [Highlights 10.3, 10.4].

Fig. 6.12. Partial view of UA1. The detectors surrounding the long, cylindrical central tracker are retracted to provide access to the Tracker. The thin beam pipe traverses the centre of the tracker.

The mechanical design was very innovative. As a self-supporting cylinder, made of six independent half-cylinder chambers, each with a length of 2 m and a diameter of 2.2 m (Fig. 6.13a), it provided close to 4π coverage of the collision products and achieved the bubble chamber-like "view" of the events (Fig. 6.14). One striking feature of the system was the cylindrical shell of the chambers supporting the wires, made with a pre-stressed honeycomb structure, sandwiched between glass fibre-epoxy layers. The amount of applied pre-stress was calculated to precisely balance the tension applied to the wires.

The 6000 sense wires (10 mm spacing, up to 2.2 m length) and the 17 000 field-shaping wires were parallel to the magnetic field. They were organized in horizontal planes in the four forward modules, and in vertical planes in the two central ones. This arrangement is a consequence of the horizontal magnetic field. With this orientation of the wire planes approximately the same number of measurements along the tracks over the detector volume was obtained, given the expected topology of the interactions and the dipole magnetic field. A typical drift cell is shown in Fig. 6.13b. The electrostatic forces were controlled with an intermediate field plane at voltage V_c, while V_f controlled the gas amplification. The position of the sense wires in a plane was known within 50 microns, the coordinate of a "typical" plane to about 200 microns, a tribute to the engineering quality of the supporting structure.

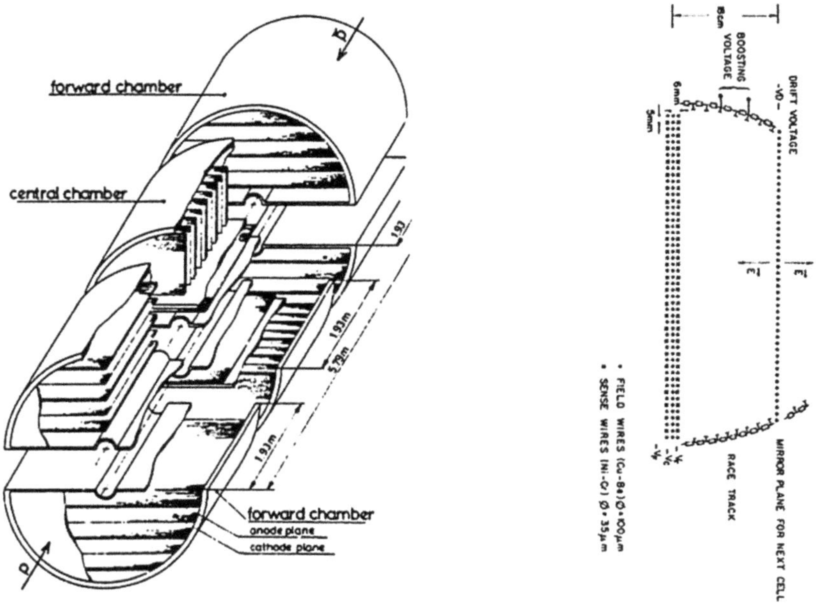

Fig. 6.13. Left: exploded view of the six half-cylinders constituting the central tracker; right: cross-section through a half-cylinder, showing the arrangement of signal wires and wire electrodes shaping the drift field and the signal amplification region.

The longest drift distance was 18 cm, such that at a 1.5 kV/cm drift field the drift time was less than the interval between bunches (3.8 μs). This drift length also limited the distortion of the electric field by space charge from backgrounds from the machine. The state-of-the-art signal processing used 6-bit "Flash Analogue-to-Digital Converters (FADCs), two per sense wire, digitizing the induced signal charge. The sense wire had kΩ level resistance; the ratio of the charges recorded at each end gave a measure of the longitudinal position of the charge along the wire to about 2% of the wire length. The drift time was measured with Time-to- Digital Converters (TDCs) with 4 ns accuracy. The digital outputs of TDCs and FADCs were stored in a buffer memory, recording the event history of the preceding 4 μs. The data volume (1.6 MB) was impressive at the time and had to be synchronously reduced by processors.

The analogue signal processing electronics required careful calibration. The linearity was calibrated by injecting a precisely known and variable charge into the preamplifiers. For each drift cell, the relevant voltages and currents were precisely measured and monitored. An alarm system provided fast indications of any malfunctioning.

To achieve the intrinsic performance of the tracker a number of parameters had to be measured and controlled, such as temperature, gas mixture variations and distortions of the electric fields. Two independent methods were used. Artificial straight tracks were generated with precisely positioned ionizing X-ray and laser beams. Secondly, the built-in "auto-calibration" of the detector was used, made possible by the special drift cell structure: as two adjacent drift spaces had opposite E-fields, tracks traversing them had to "line up" in a correctly calibrated detector. These efforts paid off. The globally achieved resolution of the drift distance was a most remarkable 290 microns for the reconstructed tracks.

Figure 6.14 shows an historical event, the first W decay recorded in the UA1 experiment. With its combination of optimal detection geometry, matching well the expected interesting collision topologies, an advanced technology for the support structures, innovations in electronics, calibration and read-out, this tracker was at the cutting edge of technology and the first true "electronic bubble chamber". Its quality and reliability were a key to the Nobel Prize-winning success of UA1.

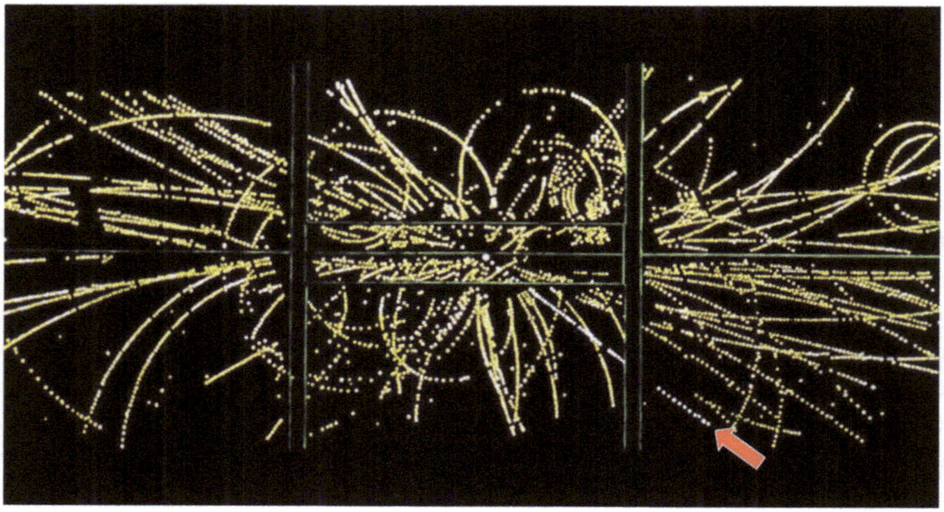

Fig. 6.14. The first observed W-decay, as recorded in the central tracker.

The Standard Model (SM) **Box 6.4**

The SM is a remarkably successful theoretical description of the elementary particles and their interactions, from which our visible world is built. It ignores gravity, described by the General Theory of Relativity. Figure 1.2 shows the SM constituents (fermions of spin 1/2). The carriers of the 3 types of forces between fermions, bosons of spin 1, are shown in Fig. 1.3. To these elementary objects, one must add the spin 0 Higgs boson.

The SM describes mathematically the particles as field quanta. These exist not only in space-time, but also "live" in abstract, so-called *internal spaces* (Fig. 1.2). E.g., a quark lives in the 3-D space of *colour*, in which it can be thought of as a vector. The subscripts L (Left) and R (Right) refer to their handedness H [Box 2.2] and point at a basic feature of the SM that the behaviour depends on H: in the *weak isospin* (WI) *space*, LH ones live in couples, RH ones as singles. In the SM the W couples only to LH fermions.

In such spaces, one can think of performing "rotations": for instance, a rotation in colour space changes a blue quark into a red quark. Or, in the WI space, an e^- rotates into a neutrino, or a strange quark into a charm quark. The SM gives the processes: emission of a gluon in the former case, of a W boson in the latter. There is thus a unique mode of interaction between fermions, whatever the force: the exchange between them of the relevant boson, as suggested by the diagrams of Fig. 1.3.

The SM provides the mathematical framework describing these rotations. But it does much more: its guiding principle posits the freedom to perform any arbitrary such rotation *locally*, i.e. in any point of space-time, without changing the physics, which cannot depend on an arbitrary choice of local coordinates. The equations of the SM must thus be invariant (symmetrical) under the relevant rotation. This symmetry requirement is a great help in defining the actual mathematical formulation. Such a rotation of the field configuration is called a "gauge" transformation (GT). The lack of change (invariance) of measurable quantities under a GT is called *gauge invariance*.

In quantum electrodynamics, QED [Box 2.3], describing e.g. the e^- and γ, the GT changes the phase of the e^- wave function, through γ emission by the e^-. The γ feels the e^- electric charge but does not carry it. Being neutral it does not self-interact (SI).

An important success of the SM is a unified description of QED and the weak interaction, the Electroweak Interaction. Its "abstract" bosons W^0, B^0 (not shown in Fig. 1.3) "mix" to give the physical γ and Z^0, through the mixing angle θ_W [Box 7.1]. The initial unified formulation (in 1961) only worked however for massless particles, and could not explain fully our perception of reality. The BEH mechanism [Box 8.2] completed the picture, enabling the SM to much better describe the visible world.

In the weak and QCD sectors, W/Z and gluons feel and carry the weak and colour charge, respectively, and *self-interact*. Gluon SI is responsible for the specific properties of the strong interaction [Box 4.2].

6.6 A Novel Particle Detector for UA2: The Power of Silicon

Claus Gößling and Pierre Jarron

In 1983 — the UA1 and UA2 collaborations had just discovered the W and Z bosons — the UA2 collaboration decided to upgrade its tracking detector by replacing part of the existing detector with a silicon detector [33]. It represented a major step forward in silicon detector technology, replacing the conventional strip-geometry of the detector elements [Highlight 5.9] with a novel checkerboard or "pad" configuration, providing improved track reconstruction. This detector worked very well. It encouraged the collaboration in 1986 [34] to push this concept further and to develop a second Silicon Pad Detector (SPD) with finer pad segmentation to be placed directly around the collision interaction beam pipe.

The layout of the new inner SPD matched the detector geometry of the outer silicon detector in order to optimize track reconstruction. The detector was built as a cylinder surrounding closely the beam pipe in the available radial thickness of 9 mm, with almost no dead space. The size of the pads was 17.3×33.5 mm^2; they provided 3072 channels. It was the first incarnation of a silicon tracker with a detector geometry adapted to a collider experiment, an ancestor of the present silicon vertex detectors.

The project presented several challenges: the barrel detector had to fit into the available radial space of less than 1 cm, including silicon sensors, electronics and associated circuit boards. Miniaturization of the detector was mandatory and was achieved with at that time two brand new technologies, the silicon sensor [35] and an Application Specific Integrated Circuit (ASIC), representing totally new approaches in particle physics experimentation. At that time tracking technology was based on gaseous detectors (MWPC) [Highlight 4.8] and discrete electronics comprising miniaturized components with hybrid electronic technology.

Fig. 6.15. ASIC AMPLEX, a 16 channel readout system developed for the Inner Silicon Pad Detector.

Developing the readout electronic system scaled to the miniaturized dimensions of the SPD represented probably the biggest challenge. Existing electronics was too bulky — incompatible with the required thickness and the channel density of the UA2 Silicon Pad Detector. It required the development of a novel ASIC tailored to the UA2 silicon sensors and experimental requirements. Figure 6.15 shows the device mounted in a 10×10 mm^2 package. All 16 electronics channels, providing the readout of 16 pads, were contained in a silicon die of 4×4 mm^2.

The ASIC, called AMPLEX [36], was fabricated in a 3 μm gate length CMOS technology. Stringent requirements were placed on the electronic noise, because a relatively tiny charge of about 24 000 electrons (4 fC) is produced by a minimum ionizing particle in 300 μm thick silicon. The typical noise figure obtained with the AMPLEX input amplifier stage was the equivalent of 1200 e^- r.m.s. resulting in an excellent signal-to-noise ratio of 20. Power consumption per channel was 1.5 mW.

The second feature of the AMPLEX circuit was the readout technique. The CERN proton–antiproton collider had a machine cycle of 3.8 μs. Therefore, the signal had to be available within 2–3 μs for a possible "readout decision" within the machine cycle prior to the serialized readout of the analogue signals of all channels. An appropriate signal shaping circuit was implemented, a proven electronic design for nuclear instrumentation, followed by a sample-hold circuit.

One further critical step in the construction was the assembly of the silicon detectors and electronics. A stave-like long multilayer-board, 3.5 mm thick, carried on the outer side all the AMPLEX-ASICs (plus a few capacitors) and on

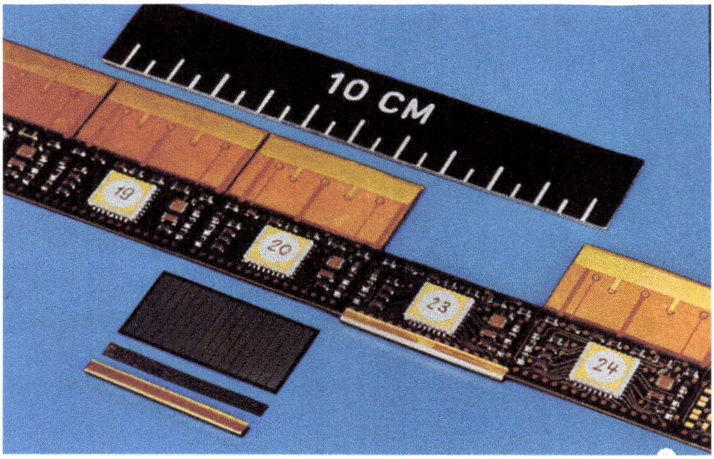

Fig. 6.16. Assembly of the silicon pad sensor and AMPLEX ASIC's on the stave.

the inner side the silicon sensors. The electric contact from the sensor to the multilayer-board was established by a strip of vertically conductive rubber under mechanical pressure of an elastic clip. A similar approach has been later used for the construction of the silicon tracker and vertex of ATLAS and CMS [Highlight 8.6].

The assembled and tested detector boards were positioned on a carbon fibre support, prior to insertion into the UA2 detector (Fig. 6.17). The insertion of the detector was assumed to be a trivial task, until it was realized that the detector diameter was 0.3 mm bigger than the available space of 9 mm. As always, the devil is in the details. The problem was solved by replacing certain capacitors which were out of specification and the Inner UA2 Silicon Pad Detector was successfully installed.

Both UA2 Silicon Pad Detectors performed very reliably during the data-taking from 1987 up to the end of operation at the p–p̄ collider in 1991. The unambiguous two-dimensional tracking information was instrumental for the pattern recognition of the many secondary tracks produced in these p–p̄ collisions. The AMPLEX chip was the first complex readout chip operating inside the sensitive volume of an inner detector and in the direct vicinity of the interaction point.

Fig. 6.17. Insertion of the inner silicon pad detector into the UA2 detector.

6.7 Antimatter's Disappearing Act

Michael Doser

Many attempts have been made to explain the apparent complete absence of antimatter in the Universe, which runs counter to the expectation that right after the Big Bang, both matter and antimatter should have been present in equal amounts. Although some of the proposed explanations have been experimentally investigated and identified, they do not suffice. Further mechanisms remain to be discovered. Searching for minute differences between the properties of particles and their antiparticles lies at the heart of many of these attempts, and still holds out the best chance at sniffing out the culprit. The challenges are daunting: on one hand, fundamental symmetries and conservation laws set severe limits on any possible differences between particles and antiparticles, and finding any such differences would trigger a scientific revolution; on the other hand, carrying out precision experiments on antimatter systems requires developing or adapting techniques that go far beyond the state of the art. Progress in this field has thus been slow, if relentless, and is closely tied to invention and implementation of new technologies.

Most amenable to being precisely studied are individual stable antiparticles, or systems composed of them: antiprotons, positrons or antihydrogen atoms. While it is relatively straightforward to obtain positrons (from the radioactive decay of radioisotopes such as ^{22}Na), or to form antiprotons (which requires particle accelerators, and was first done in 1955 at the BEVALAC at LBL, and subsequently carried out at CERN on a quasi-industrial scale from 1980 onwards), precision studies require antiparticles at energies that lie many orders of magnitude below what can be achieved at accelerators, μeV instead of GeV. Furthermore, with minute numbers of antiparticles to work with, detection of their signals will always be daunting: either their detection will be non-destructive, in which case electronic noise represents the largest background, or their detection will be destructive, in which case background from cosmic rays (which can occasionally mimic annihilation topologies) represents the largest challenge. But before tackling these hurdles, the first crucial technology that needed to be developed was trapping of antiparticles.

Antiprotons are produced at CERN by colliding 24 GeV protons from the Proton Synchrotron with a fixed target (made of copper); the tiny fraction of antiprotons produced in these collisions (one antiproton–proton pair per million collisions) are filtered out by mass and charge. They are subsequently decelerated from their initial energy of several GeV down to 5.3 MeV at CERN's Antiproton Decelerator (AD). Trapping these requires a further reduction in energy. The TRAP collaboration conceived and successfully implemented the first such scheme in 1986 [37]: like all charged particles, antiprotons lose energy by interacting with the electrons of the atoms of any material that they traverse. Furthermore, charged particles with energies below a keV can be easily manipulated electromagnetically (in vacuum) within so-called Penning traps (stacks of ring-shaped electrodes held at different electric potentials, and situated inside a strong magnetic field). The TRAP collaboration thus placed a thin foil (whose thickness was chosen to maximize the tiny fraction of antiprotons coming out with energies below a few keV) upstream of a Penning trap whose furthest electrode was held at several kV. Before the "slow" reflected antiprotons (whose velocity is of the order of 1000 km/s, or 1 m/μs) could return, they rapidly switched the first ring (immediately after the foil) from ground potential to the same high voltage. The energy of these now trapped antiprotons can then be further reduced (via "electron cooling") from keV down to meV, where their properties are studied precisely by the ATRAP and BASE experiments [38]. It is these cold antiprotons that, combined with cold positrons, can form antihydrogen atoms, as demonstrated for the first time in 2002.

The first method for producing cold antihydrogen was developed by the ATHENA and ATRAP collaborations and subsequently improved by the ALPHA collaboration, all working at CERN (Fig. 6.18). This method consists of setting up a specific configuration of electric fields inside a Penning trap, in order to create an overlap between a cloud of cold antiprotons and a cloud of positrons. Formation of an antihydrogen atom requires three partners: an antiproton, a positron, and an additional positron which carries away the binding energy of the newly formed antiatom. Demonstrating that this process is taking place means detecting these antihydrogen atoms. Once formed, they are neutral, and are free to drift through the electric and magnetic fields of the Penning trap. The scheme developed by ATHENA relies on imaging the annihilation products that appear when the antihydrogen atoms reach the inner surface of the Penning trap rings. A multilayer silicon microstrip detector, coupled with a CsI-based photon detector, all operating at cryogenic temperatures, allowed reconstructing the trajectories of the charged particles produced in the annihilation of the antiprotons (and thus the position of the annihilation vertex), as well as detecting the direction and energy of the photons produced in the annihilation of the positrons (Fig. 6.18). Such imaging detectors were crucial not only in demonstrating the production of antihydrogen [39], but also in elucidating the complex plasma dynamics taking place in its formation, and enabled the ALPHA collaboration to take the crucial next step, that of trapping the coldest formed atoms.

In order to study the formed antihydrogen atoms, they must either be shaped into a beam (as done in the ASACUSA experiment) or trapped. Although electrically neutral, antihydrogen atoms can be trapped in a magnetic multipole

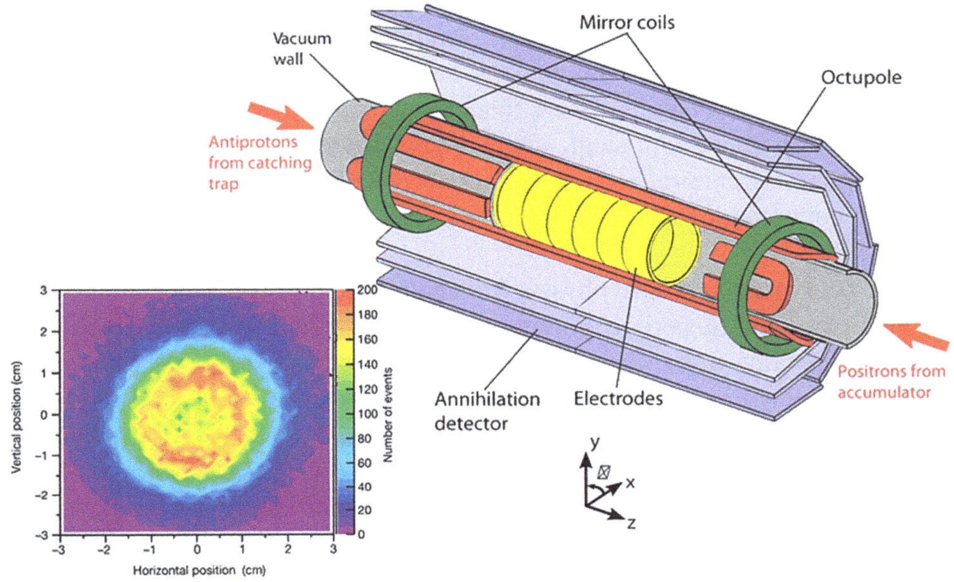

Fig. 6.18. Left: Distribution of reconstructed antiproton annihilation vertices in the ATHENA experiment, reflecting the formation and subsequent annihilation (on the inner Penning trap surfaces) of antihydrogen atoms. The colour code (violet to red) corresponds to number of detected annihilations [39]. Right: A schematic, cut-away diagram of the antihydrogen production and trapping region of the ALPHA apparatus, showing the relative positions of the electrodes, the minimum-B trap octupole (red) and axial pinch (green) magnets and the 3-layer silicon microstrip antihydrogen annihilation detector [41].

trap (with a field minimum at its centre) through the interaction between the antihydrogen atoms' internal energy states and the magnetic field. Technical limitations on the magnets however entail that only the coldest atoms, with a temperature of less than 0.5 K, can be trapped this way, while the bulk of the formed atoms have temperatures far in excess of this value.

Detecting the tiny fraction of (possibly) trapped atoms required another technical breakthrough: rapid switching of magnetic multipole traps. To prove that they had formed and trapped antihydrogen atoms, the ALPHA collaboration opened their magnetic multipole trap, allowing antihydrogen to escape and annihilate. A major confounding background signal comes from cosmic rays interacting in their apparatus, which occasionally mimic signals associated with antiproton annihilations. Reducing this background is thus of tantamount importance, and one way to do this is to ramp down their multipole trap more rapidly than the 10 seconds usually required to reduce the current of the superconducting trap. The ALPHA design [39] relies on a unique technique developed at BNL [40] that provides great freedom in defining the winding patterns of the magnet, as well as great mechanical strength, and that allows a current decay time of 9 ms. Using these technologies, the ALPHA experiment could establish trapping of single antihydrogen atoms in 2010 [41]. Other exotic atoms containing antiprotons are also possible: ASACUSA uses precision lasers to study antiprotonic helium, in which an electron in helium is replaced by an antiproton.

The field of antihydrogen physics will continue to place great demands on new technologies, as the experiments move deeper into the field of atomic physics. The current generation of experiments is working towards first precise comparative studies of antihydrogen and hydrogen. The next few years should see optical and microwave spectroscopy of antihydrogen, as well as first tests of the gravitational behaviour of antihydrogen atoms by the AEgIS, GBAR and ALPHA-g experiments. Reaching and going beyond these initial steps, into the domain of high precision studies, will however require cooling of trapped antihydrogen atoms to their recoil limit of ~ 2 mK, development of new cooling techniques to enter the realm of sub-mK temperatures, and enhancement of the formation rates of ultra-cold antihydrogen atoms by orders of magnitude. An increasingly large and interdisciplinary community is focused on developing the necessary technologies that will allow the experimental sensitivities to gradually reach the range where fundamental discoveries could become possible.

References

1. H. Koziol and D. Möhl, The CERN antiproton collider programme: accelerators and accumulation rings, *Physics Reports* **403-404**, 91 (2004).
2. S. van der Meer, Stochastic damping of betatron oscillations in the ISR, CERN-ISR-PO/72-31 (1972), https://cds.cern.ch/record/312939/files/197200067.pdf.
3. P. Bramham, G. Carron, H.G. Hereward, K. Hübner, W. Schnell and L. Thorndahl, Stochastic cooling of a stored beam, *Nucl. Instr. & Meth.* **125**, 201(1975).
4. C. Rubbia, P. McIntyre and D. Cline, Producing massive intermediate vector bosons with existing accelerators, *Proc. International Neutrino Conference*, Aachen, Germany (June 1976) pp. 683–687.
5. G. Carron *et al.*, Stochastic cooling tests in ICE, *Phys. Lett. B* 77, 353-354 (1978).
6. V. Chohan *et a*l., The CERN Antiproton Accumulator Complex (AAC): current status and operation for the nineties, in *Proc. 15th International Conference on High-energy Accelerators*, Hamburg, Germany (July 1992) pp. 106-108.
7. S. van der Meer, Stochastic cooling in the CERN antiproton accumulator, *IEEE Trans. Nucl. Sci.* **28**, 1994-1998 (1981).

8. E. Jones, ACOL, CERN's Upgrade of the antiproton accelerator complex, *Proc. 6th Topical Workshop on Proton-Antiproton Collider Physics*, Aachen, Germany (June 1986) pp. 691-704.
9. H. Koziol and D. Möhl, The CERN low-energy antiproton programme: the synchrotrons, *Physics Reports* **403-404**, 271-280 (2004).
10. C. Rubbia, The discovery of the W and Z bosons, *Phys. Rep.* **239**, 241 (1994);
P. Darriulat, History of original ideas and basic discoveries in particle physics, *NATO ASI Series B. Phys.* **352**, 757 (1996), and *Eur. Phys. J. C* **34**, 33 (2004).
11. D. Denegri, When CERN saw the end of the alphabet, *CERN Courier*, April 30, 2003;
L. Di Lella, *Proc. XXXII Int. Meet. on Fund. Phys.*, Alicante, ed. CIEMAT, Madrid (2004) 291.
12. M. Barranco Luque *et al.*, The construction of the central detector for an experiment at the CERN anti-p p collider – UA1, *Nucl. Instr. & Meth.* **176**, 175 (1980);
M. Calvetti *et al.*, *Nucl. Instr. & Meth.* **176**, 255 (1980);
K. Eggert *et al.*, *Nucl. Instr. & Meth.* **176**, 217 (1980);
A. Astbury, *Physica Scripta* 23, 397 (1981).
13. A.G. Clark *et al.*, *Proc. Int. Conf. on Inst. for Coll. Beam Phys.*, SLAC-250 (1982); p. 67 in *2nd Int. Conf. on Phys. in Colliders*, Stockholm (Plenum, New York, 1983);
B. Mansoulié, p. 609 in *Proc. 3rd Moriond Workshop on $\bar{p}p$ Physics*, (Editions Frontières, 1983);
A. Beer *et al.*, The central calorimeter of the UA2 experiment at the CERN anti-p p collider, *Nucl. Instr. & Meth.* **224**, 360 (1984);
K. Borer *et al.*, *Nucl. Instr. & Meth. A* **257** (3), 591 (1987);
R.E. Ansorge *et al.*, *Nucl. Instr. & Meth. A* **273** (2), 826 (1988).
14. J.G. Rushbrooke *et al.*, *Physica Scripta* 23, 642-648 (1981);
G.J. Alner *et al.*, UA5: A general study of proton-antiproton physics at $\sqrt{s}$ = 546 GeV, *Physics Reports* **154**, 247-383 (1987).
15. M. Amoretti *et al.*, Production and detection of cold antihydrogen atoms, *Nature* **419**, 456-459 (2002);
G. Gabrielse *et al.*, Background-free observation of Cold Antihydrogen with field-ionization analysis of its states, *Phys. Rev. Lett.* **89**, 213401 1-4 (2002).
16. S. van der Meer, Stochastic Cooling and the Accumulation of Antiprotons, *Rev. Mod. Phys.* **57**, 689-697 (1985).
17. G. Carron and L. Thorndahl, Stochastic Cooling by filter techniques, *CERN Internal Report* CERN-ISR-RF-78-12, (1978), http://cds.cern/record/119952/files/19704143.pdf.
18. L. Faltin, Slot-type pick-up and kicker for stochastic beam cooling, *Nucl. Inst. & Meth.* **148 (**3), 449-455 (1978).
19. B. Autin, G. Carron, F. Caspers, S. Milner, L. Thorndahl and S. van der Meer, ACOL stochastic cooling systems, *Proc. Part. Accel. Conf. (PAC 1987)*, Washington, USA (1987) pp. 1547-1551.
20. G. Carron, F. Caspers and L. Thorndahl (eds.), *Development of Power Amplifier Modules for the ACOL Stochastic Cooling Systems*, CERN-1985-001, (CERN, Geneva, 1985). http://dx.doi.org/10.5170/CERN-1985-001.
21. Y. Bylinsky, A.M. Lombardi and W. Pirkl. RFQD: a decelerating RFQ for the CERN antiproton facility. *Proc. 20th Int. Linac Conf.*, Monterey, USA (Aug. 2000) pp. 554-556. http://accelconf.web.cern.ch/AccelConf/l00/papers/TUD05.pdf.
22. I.M. Kapchinskij and V.A. Teplyakov. Ion linear accelerator with spatially homogeneous strong focusing. *Prib. i. Tekhn. Eksp. (Instr. and Exp. Techn.)* **2**, 19-22 (1970).

23. C. Biscari and F. Iazzourene, Post-deceleration of the LEAR beam by a radiofrequency quadrupole. *Proc. 3rd LEAR Workshop,* Tignes, France, (Jan., 1985) pp. 115-120, http://cds.cern.ch/record/159830/files/cer-000071161.pdf.
24. J. Bosser *et al.* (ed. W. Pirkl). Feasibility study of a decelerating Radio Frequency Quadrupole system for the Antiproton Decelerator AD, *CERN internal report CERN/PS/HP Note* 97-36, (1998), http://cds.cern.record/558230/files/open-2002-040.pdf.
25. A.M. Lombardi, W. Pirkl and Y. Bylinsky. First operating experience with the CERN decelerating RFQ for antiprotons. *Proc. 2001 Part. Accel. Conf.*, Chicago, USA (2001) pp. 585-587, http://accelconf.web.cern.ch/accelconf/p01/PAPERS/ROPA009.pdf.
26. K. Kilian, U. Gastaldi and D. Möhl, Deceleration of antiprotons for physics experiments with low-energy antiprotons, *Proc. 10th Int. Conf. on High-Energy Accelerators*, Protvino, USSR, (1977) pp. 179-184.
27. https://home.**cern**/about/accelerators/low-energy-antiproton-ring.
28. G.I. Budker, Status report of the work on storage rings at Novosibirsk. *Proc. Symp. Int. sur les Anneaux de Collisions à Electrons et Positrons*, p.II-1-1, Saclay, France (1966).
29. S. van der Meer, Stochastic Extraction- A low ripple version of resonant extraction, *CERN Int. Report CERN/PS/AA* 78-6 (1978), http://cds.cern/record/118881/files/cer-000029365.pdf.
30. R. Cappi, R. Giannini and W. Hardt, Ultra-slow extraction – Status report, *CERN int. report CERN-PS-LEA-82-3*, presented at LEAR Workshop, Erice, Italy, (June 1982). http://cds.cern/record/1414413/files/CERN-PS-LEA-82-3.pdf.
31. S. Jacobsen, G. Molinari and H. Mulder, A feed-back system to control the flux during ultra-slow extraction at LEAR, *Proc. 6th European Particle Accelerator Conference (EPAC96)*, Sitges, Spain (June 1996) pp. 1845-1847.
32. D.R. Nygren, Proposal to Investigate the Feasibility of a Novel Concept in Particle Detection, *LBL internal report*, February 1974;
J.N. Marx and D.R. Nygren, *Physics Today* **31**(10), 46 (1978).
33. K. Borer *et al.*, Construction and performance of a 1 m^2 silicon detector in UA2, *Nucl. Instr. & Meth. A*, **257**, 591–593 (1987).
34. K. Borer *et al.*, The implementation and performance of a 1 m^2 silicon hodoscope array in UA2 and proposed construction of a second array, *Nucl. Instr. & Meth. A* **273**, 605-610 (1988).
35. E.H.M. Heijne and P. Jarron, The impact of microelectronics on particle detection, *Nucl. Instr. & Meth. A*, **226**, 12-15, (1984).
36. E. Beuville *et al.*, Amplex, a low-noise, low-power analog CMOS signal processor for multi-element silicon particle detectors, *Nucl. Instr. & Meth. A* **288**, 157 (1990).
37. G. Gabrielse *et al.*, First capture of antiprotons in a Penning trap: A kiloelectronvolt source, *Phys. Rev. Lett.* **57**, 2504 (1986).
38. G. Gabrielse *et al.*, Cooling and slowing of trapped antiprotons below 100 meV, *Phys. Rev. Lett.* **63**, 360 (1989).
39. W. Betsche *et al.* (ALPHA collaboration), A magnetic trap for antihydrogen confinement, *Nucl. Instr. & Meth. A* **566**, 746 (2006).
40. http://www.bnl.gov/magnets/BEPCII/BEPCII.asp.
41. G.B. Andresen *et al.*, (ALPHA collaboration), Trapped antihydrogen, *Nature* **468**, 673 (2010).

Chapter 7

The Large Electron Positron Collider (LEP): Probing the Standard Model

Thomas Taylor and Daniel Treille

7.1 Introduction

LEP: Its origins

B. Richter, co-discoverer of the J/ψ particle and Physics Nobel Laureate of 1976, was visiting CERN in 1974, where he launched the idea of a very Large Electron Positron (LEP) Collider [1]. It needed to be large to provide enough energy for the discovery of the postulated Z and W bosons [2]. Their masses were estimated to be close to a hundred times that of the proton, based on experimental constraints and growing confidence in the Standard Model (SM). The physics community was rapidly convinced that this should be the next big European facility. Building on the successes of circular colliders such as SPEAR, with PETRA and PEP on the way, and anticipating a circular tunnel that could eventually host a very high energy proton–proton collider, the choice of a circular machine appeared logical.

In 1978 a first important milestone was reached in a workshop held in the alpine village of Les Houches [3]. The physics programme was evaluated, machine parameters studied and conceptual experimental installations developed. The Director-General at the time, J. Adams, presented a possible planning, which in the event turned out to be quite close to the mark. The mood of the community was upbeat: "LEP's do it" was the motto adopted by the workshop participants. Under the aegis of ECFA a working group was set up where accelerator and experiment communities worked together, creating strong motivation for the new project [4], and other conferences on experimentation were organized in quick succession [5].

The second milestone was the proposal in 1980 to use the existing synchrotrons, the PS and the SPS, as injector for LEP, once again demonstrating the wisdom of the strategy of keeping all accelerators together on one site [6]. This preempted all discussions on the possible site of LEP elsewhere in Europe and avoided the highly political debate on siting which had overshadowed and delayed the SPS project.

In 1982 the LEP project was authorized by all Member States. Meanwhile the LEP team had continued to work on the optimization of the design and on the exact location of the ring. The LEP Management Board, with E. Picasso as Chairman, the LEP Experiments Committee and the LEP Machine Advisory Committee were put in place. In 1983 the civil engineering started. The same year the Z^0 was discovered with a mass close to 92 GeV/c^2, as predicted by the SM, well within LEP's reach. In February 1988 the LEP tunnel excavation was completed. Only 17 months later the first Z^0s were enthusiastically registered in the four experiments. The competition of the Stanford Linear Collider, SLC, in Stanford, USA, with much less luminosity but with longitudinal polarization of the beam [Box 7.1], further increased the motivation. After 7 years devoted to Z^0 physics (LEP1), the beam energy E_b was progressively increased (LEP2).

Pre-LEP physics landscape

In 1986 a seminal Workshop at CERN focused on the LEP physics programme [7]. The foundations of precision tests of the electroweak (EW) theory [Box 6.4] were laid and the experimental methods defined in detail.

For the strong interactions, experimental evidence pointed increasingly to the validity of QCD [Box 4.2]. Asymptotic freedom, the vanishing of the strong coupling at small distance — the most unexpected property of the strong force — had been theoretically predicted in 1973. Experimental evidence for the gluon, mediator of this force, was found at PETRA at DESY in 1979. But the predicted energy-dependence of strong coupling still had to be demonstrated and jet studies (started at the ISR, p–$\bar{p}$ collider, PETRA and TRISTAN) had to be pursued.

Of the three generations of quarks and leptons, only the top quark remained elusive. There was little doubt as to its existence, and its mass was suspected to be around 40 GeV. If so, LEP would be ideal to perform new subtle "onium" spectroscopy of top–antitop bound states, just as charm and beauty had been revealed through gamma transitions measured with appropriate high resolution calorimetry [Highlight 7.9]. But "top" turned out to be much heavier than expected and this physics did not appear: the possibility of more than three generations was still open. The physics of the tau lepton and heavy flavours became an obvious goal.

Supersymmetry (SUSY) had been postulated in the early seventies, with decisive contributions of J. Wess and B. Zumino, then visitors at CERN, but its phenomenology was only starting to be explored. The fact that it required a light SM-like BEH boson was recognized, but an upper limit on its mass (< 130 GeV) had to wait for more theoretical work and a precise measurement of the top mass.

The LEP machine

The Large Electron–Positron storage ring (LEP) collider was located beneath the border between Switzerland and France, near Geneva. It was to be by far the largest accelerator in the world, with the main ring tunnel having a circumference of the order of 30 km. Although studies and plans for LEP machines of various diameters started as early as 1976 [8], it was not until October 1981 that the project was approved [9]. The tunnel would pass under villages, so the environmental impact had to be evaluated and presented to the municipalities [10].

The maximum beam energy for which the major systems of the 26.7 km circumference machine were to be designed was 125 GeV [11].

While the tunnelling under the plain went well, the 3 km section passing through the Jura limestone was difficult due to water infiltration [12–14]. By dint of a crash programme half of the resulting delay of about a year would be made up during the installation of the machine components. The first beams circulated in July 1989, with physics runs starting a month later [15].

LEP was designed to bring very high energy electrons and positrons into collision [14–16]. Initially, the beams were accelerated to provide a centre-of-mass energy at the collision point of 91 GeV, the rest mass of the Z^0 boson, thereby opening a new world of investigation into the neutral current interactions of the electroweak force. This first phase of LEP, called LEP 1, was a huge success, with over 900,000 Z^0 bosons being produced in the first year. Data continued to be accumulated at the same energy until 1995. Then the machine was upgraded to become LEP 2 [17]. Collision energy was gradually increased by adding accelerating cavities as they became available, which allowed the experiments to search for new particles as well as to study in depth $W^{\pm}$ and Z pair production. The maximum centre-of-mass achieved was 209 GeV, with the straight sections at the even points filled with cavities, when in 2001 the machine was closed due to pressure for building the next accelerator, the LHC.

Before injection into the main ring the particles had to be accelerated in a number of other, smaller accelerators most of which had been designed and operated previously for other projects. Indeed, it was the existence of these injectors that helped to convince the community that LEP should be built at CERN. Figure 7.1 shows the layout of the complete accelerator complex. The electrons were produced by thermionic emission and accelerated along a linear electron accelerator, the LEP Injector Linac (LIL). Some of the electrons were diverted onto a tungsten target to produce the positrons; the electrons, along with the positrons, were then passed into the Electron Positron Accumulator ring (EPA), where they were accumulated and stored. The particles were then sent to the Proton Synchrotron (PS) where they were accelerated in bunches up to a few GeV.

Following this they were transferred to the Super Proton Synchrotron (SPS) where further acceleration took place prior to injection into the LEP ring [18].

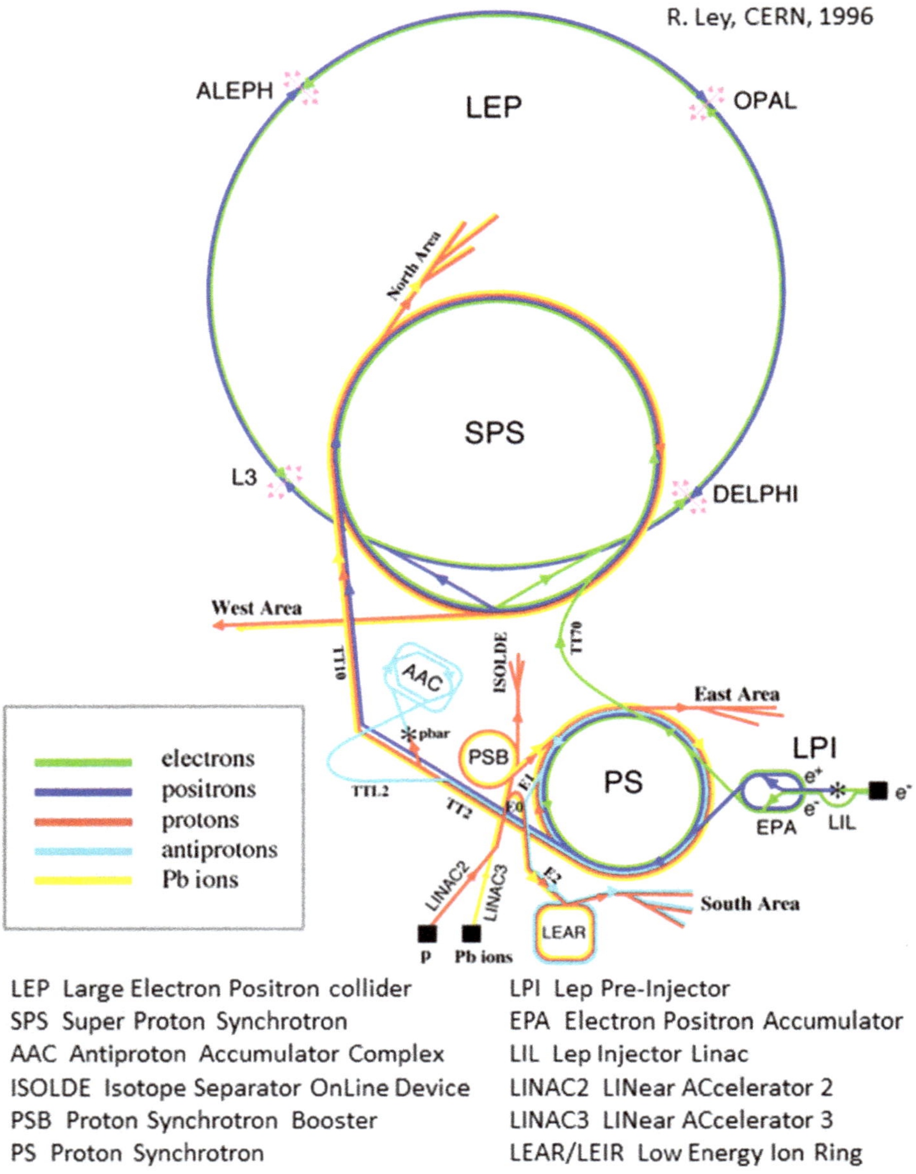

Fig. 7.1. Diagram showing the CERN accelerators at the time of LEP and how they were used to produce the high energy beams circulating in the collider.

The 26.7 km LEP tunnel is situated at between 40 and 150 m below the surface. The plane of the ring is inclined by 1.4%. This was to ensure that no shaft be deeper than 150 m, and that the underground caverns and tunnel be located to the largest extent possible in solid molasse (local sandstone). The ring consists of eight arcs and four straight sections. The arcs were equipped with magnetic cells to guide the beams around the ring. Each such cell was composed of a sequence of bending (dipole), focusing (quadrupole) magnets [Box 2.1] and orbit correctors. The total length of a cell was 79.11 m, with each arc containing 31 cells. Families of sextupole, octupole and skew quadrupole magnets were installed to allow the fine tuning of beam parameters. The beams were accelerated in the straight sections on either side of points 2 and 6 for LEP1 and at all even numbered points for LEP2. The four experiments, L3, ALEPH, OPAL and DELPHI were located at points 2, 4, 6 and 8, where the beams were brought into collision.

This exceptionally large accelerator called for an elaborate geodesy: it was even necessary to take into account the non-uniformity of local gravity due to the proximity of the Jura. This required the adoption of the latest techniques in survey: satellite bearings; two-frequency laser metrology; verification of measurements with stars, etc. The tunnel was excavated in two directions and the closing error was less than 2 cm. The slope of the plane of the machine meant that every piece of equipment that was installed required its own survey parameters [Highlight 7.5]. It was also found that the position of the moon had a measurable effect on the diameter of the machine; this had to be taken into account to determine precisely the energy of the beams and thus the Z boson mass [Highlight 7.6].

Synchrotron radiation leads to an energy loss per turn proportional to E^4/R, i.e. increasing with the fourth power of the beam energy E, and gives rise to upper limits for both current and energy. The radius of curvature of the ring R was therefore maximized, respecting other constraints. The main loss of particles is through beam–gas interactions, so good vacuum is essential. Synchrotron radiation also causes heating of the vacuum chambers, and so the chamber walls were made of aluminium, featuring channels through which cooling water flowed, and were covered with lead shielding to protect surrounding equipment. The vacuum was ensured by the incorporation of an innovative system of linearly distributed Non-Evaporable Getter (NEG) pumps [Highlight 7.3].

The designers of the LEP magnet system also had to face up to unusual challenges. The dipole bending magnets had to deliver a maximum field of 0.14 T, about a factor of ten less than usual. To obtain good field quality over the working range it was decided to space the laminations that make up the yokes. In order to have an inexpensive rigid block the space was filled with concrete, pre-stressed with tie-bars, rather than using a structure of girders. An unexpected problem was posed by the change in the magnetic properties of the iron laminations caused by shrinkage of the mortar [Highlight 7.2]. Another innovative feature of the magnet design was the use of water-cooled aluminium bars, welded end to end, instead of coils wound around the poles, to provide the excitation with reduced overall cost.

On powering the first sector of the machine, unexpected beam parameters (the wavelength and coupling of particle oscillations in the beam) were observed, and traced to magnetization of the nickel flash required to solder the lead shielding to the aluminium vacuum chamber. A method was quickly devised to demagnetize the nickel in-situ, and a crash program was launched to rectify the situation without impacting on the overall schedule.

Due to the size of the machine and the pressure to minimize expenditure on the project, it was important to take into account the cost of cables when optimizing the magnet system. This led to the choice of exceptionally low excitation currents of 2.5 A and 5 A for the corrector magnets. The machine was also equipped with powerful wiggler magnets, to control damping, emittance and polarization.

In order to optimize total system cost it is usual to minimize the aperture in the arcs of a synchrotron, but in the case of LEP, due to the low field, the aperture could be relatively generous without excessive expense. However, this posed a problem for the design of the final focus system as the projection of the arc aperture into the interaction region would have led to very large and expensive quadrupole magnets. This problem was addressed by installing retractable aperture limiting collimators in the arcs, and allowed the use of arc quadrupoles in the insertions. To maximize luminosity [Box 6.1], the final focus magnet system had to be close to the crossing point, so an iron-free superconducting quadrupole was developed that could be cantilevered into the detectors through the poles of the spectrometer magnets [19]. It was also crucially important to control the background in the LEP experiments, so systems of primary, secondary and tertiary collimators were installed on either side of the crossing points.

Copper RF accelerating cavities were used for LEP1. Such cavities are not very efficient, so the RF group developed a novel system of coupled low-loss resonators to store some of the energy during the absence of beam and thereby save power. The 352 MHz copper cavities delivered about 1.5 MV/m and 128 such cavities were used to accelerate beams to 45 GeV. The power consumption problem was also addressed via the development, in collaboration with industry, of efficient klystrons. Despite these efforts, it was clear that to achieve the beam energy for LEP2, superconducting cavities would be essential. The first such cavities were made out of sheet niobium, delivering 5–6 MV/m. It was also decided to explore a new route, namely that of sputtering of niobium onto the surface of a copper cavity — copper has better thermal conductivity than niobium, leading to improved stability [20]. The technique was perfected and most of the cavities for the upgrade to LEP2 were made in this way. The performance of the cavities gradually improved, and the accelerating gradient could be pushed from 6 MV/m to over 7 MV/m [Highlight 7.4]. The maximum energy achieved, in May 2000, was 104.5 GeV per beam, with 272 niobium-plated cavities, 16 niobium cavities and 56 copper cavities.

The many superconducting cavities of LEP2 called for an abundant and reliable supply of liquid helium, so the CERN cryogenic specialists got together with the two major suppliers of cryogenic equipment to design powerful and efficient He liquefiers. Two plants, each capable of delivering 18 kW of cooling at 4.5 K, were installed at each of two points (2 and 6), one plant feeding the left side and one the right, interconnected so that if a fault occurred in one plant, the other plant could supply both sides and operation be continued (if necessary at a lower energy). The same plants are now used for cooling the LHC magnets [Highlight 8.3].

The project leadership of Emilio Picasso and the enthusiastic support of the Director-General, Herwig Schopper, were vital for the success of LEP (Fig. 7.2).

Fig. 7.2. LEP Project Leader Emilio Picasso and CERN Director-General Herwig Schopper celebrate the completion of the tunnelling of the 27 km ring.

The experimental facilities

For detectors [Box 6.3] the main revolution, which changed the course of particle physics, had occurred at CERN in 1968 with the invention of the Multi-Wire Proportional Chamber (MWPC) by Georges Charpak, for which he was awarded the Nobel Prize in 1992. The MWPC and its derivatives (e.g. drift chambers) had already been developed and used on a large scale at the ISR, SPS and elsewhere [Highlight 4.8]. The Time Projection Chamber, TPC, invented in Berkeley/US had appeared in the late seventies. Liquid argon, scintillating crystal, refined heavy absorber/scintillator calorimeters were already in use (see ISR and SPS). The large collaborations at the SPS and the appearance of the World Wide Web [Highlight 9.7] had paved the way to efficient collaborations on a worldwide scale.

At LEP four state-of-the-art multipurpose experimental facilities had been approved, to be operated simultaneously. Keeping a balance between playing safe and introducing innovative features was a constant concern during their conception and approval phase. All four were ready at the start of LEP and their performances turned out to be similar and, overall, excellent. Several reasons motivated the choice of four simultaneous facilities: to enable a cross-check of experimental results; to stimulate competition between teams; to spread the risk of using novel measurement techniques and to allow the large and interested physics community to participate fully in the research.

While ALEPH [21] and DELPHI [22] incorporated large superconducting solenoids, a bold enterprise at the time, OPAL [23] was equipped with a room temperature solenoid using water cooled aluminium conductor. ALEPH and DELPHI chose as main trackers big TPCs, offering excellent multi-track capability and a good particle identification by *dE*/*dx* [Box 5.2].

L3 [24] made use of a large scintillating crystal calorimeter of BGO [Highlight 7.9] and a very large precise muon spectrometer [Highlight 7.10], which required in turn a very large magnet and new methods of alignment and calibration for the spectrometer chambers, paving the way to similar and still larger facilities at the LHC [Highlight 8.12].

The Ring Imaging Cherenkov (RICH) technique was used in DELPHI for particle identification [Highlight 7.8]. Its realization was difficult, but it worked well and stimulated a variety of important developments.

Having realized the importance of charm and beauty identification, all detectors incorporated silicon micro-vertex detectors [Highlight 7.7]. This technique was further developed with the introduction of multi-layer double-sided micro-strip detectors of up to 1 m^2 in area. In the spirit of the UA2 pad detectors [Highlight 6.6], DELPHI introduced pixel detectors, a first at a collider.

Small "luminometers" in the very forward region of the detectors gave the absolute scale for cross section measurements. This started a spectacular competition between theorists computing e^+e^- (Bhabha) scattering and experimentalists measuring it ever more accurately. The race led finally to an accuracy 30 times better than expected (see below).

A very important step was the introduction by ALEPH and DELPHI of their own "PC farms", using inexpensive Personal Computers interlinked on a large scale, to revolutionize the way computing and data processing was performed.

LEP experimentation also saw the advent of very elaborate quality control procedures for the measurements and analysis and the use of powerful visualization, taking advantage of the "pattern recognition" capability of the human eye, "seeing" the features of events and checking the results of algorithms [Highlight 9.6]. Last but not least, to extract a maximum of information from the LEP measurements the collaborations introduced the practice of combining the results (the ALEPH, DELPHI, L3, OPAL ("ADLO") collaborations) [25, 26].

The LEP legacy

From its conception and realization up to the completion of its rich physics programme, LEP stands as a very successful project and a landmark in establishing the validity of the Standard Model in all its subtlety. Figure 7.3 shows the scenery of LEP physics in a nutshell, showing the Z resonance and the boson pair cross sections at higher energy, and how the peak luminosity increased over time, to consistently exceed design specifications [16].

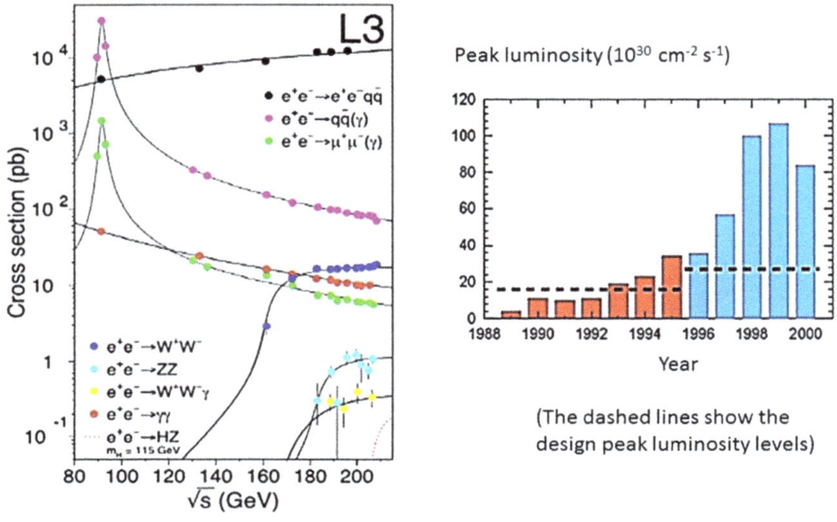

Fig. 7.3. LEP physics in a nutshell (see text).

LEP1

The Z^0 mass, one of the three basic entries of the SM, has been measured at LEP1 with 23 ppm accuracy to be 91.1875 ± 0.0021 GeV/c^2, by using a resonant depolarization method [Box 7.1]. This value of the mass now appears in the textbooks of particle physics.

The accuracy of measurements of many electroweak quantities improved dramatically during the LEP era. This quest of accuracy was neither an obsession nor art for art's sake, but fundamental to probing the SM at the "loop level", the mission of LEP, as explained in Box 5.1. The accuracy of the weak angle, a measure of the quality of SM testing, was improved by two orders of magnitude.

Similar progress occurred concerning the number N of neutrino flavours. LEP1 rapidly settled the question, showing that there are three and only three species of light active neutrinos. In fact, N was found to be slightly below 3 — potentially interesting, as it could signal the existence of "sterile" neutrinos, interacting only very weakly with SM particles. Such particles could contribute to Dark Matter and are still actively sought.

Through radiative effects, LEP1, without producing the top particle, showed rapidly that it was quite heavy. It was discovered at the Tevatron in 1996 with a measured mass of 172 ± 6 GeV, while the indirect LEP measurements gave 157^{+16}_{-12} GeV. Nowadays, more precise values are still in marvellous agreement, another triumph of the SM (Fig. 7.4). An accurate top mass is also a key input of modern considerations in cosmology.

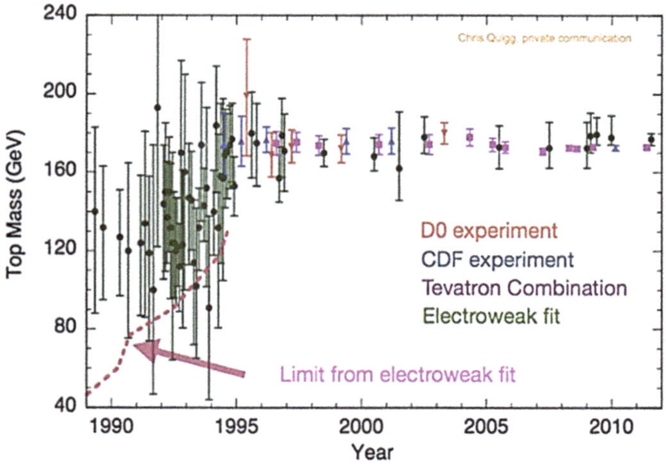

Fig. 7.4. Increasing accuracy of the measurement of top mass over time.

Asymmetries **Box 7.1**

In e^+– e^- collisions, a Z boson can be produced by a left-handed (LH) or right-handed (RH) e^-, the e^+ being then of opposite handedness. The figure shows a LH e^- colliding with a RH e^+. Given the different couplings of the Z to LH and RH e^-, flipping the handedness of the incoming e^-, as was done at SLAC SLC, allows to measure different Z production rates and the *left–right asymmetry* A_{LR}, and to extract the mixing angle [Box 6.4].

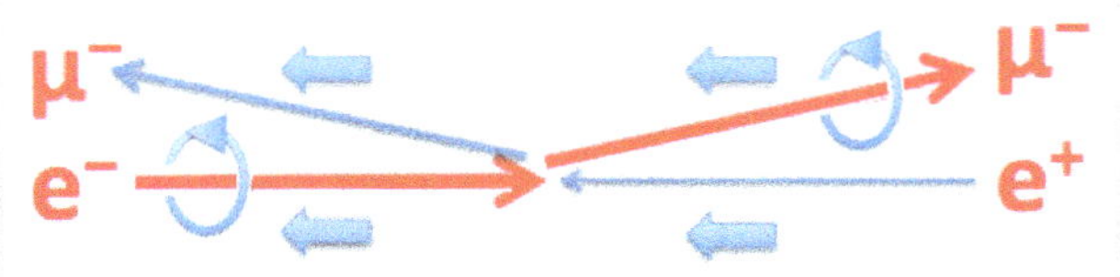

If the incident e^- beam is not polarized, as in LEP, the probabilities for the e^- to be LH or RH are equal. LH e^-, more strongly coupled to the Z, will make more of them (corresponding to A_{LR}). Then a μ^- from Z decay going forward, i.e. in the same direction as the incident e^- (red μ^- arrow in the figure) is LH, while it would be RH if going backward (blue arrow). The former configuration is preferred. Overall, this induces for the μ^- a *forward–backward asymmetry*, A_{FB}^{mu}, at the level of A_{LR} squared, with slightly more μ^- forward. Since opposite to the μ^- there is always a μ^+, this also implies a *charge asymmetry*.

Using tau lepton pairs instead of muons, one can measure the tau polarization from its decay products, providing another evaluation of the weak mixing angle.

Accurate measurements of such asymmetries for fermion pairs from Z decay were one major test of the SM. Performing this measurement with quarks, appearing as jets, was more involved and the full power of the detectors was needed to identify and measure the charge of the parent quark [Highlights 7.7 and 7.8]. This allowed to measure, e.g. the A_{FB} asymmetry for beauty-antibeauty, A_{FB}^b.

The accuracies reached were such that the effect of loops beyond the lowest order SM predictions could be measured [BOX 5.1]. The mixing angle should be the same at the same energy, irrespective of very different final states and systematic uncertainties. The agreement turned out to be very good, with however a 3σ disagreement between the SLD A_{FB}^b and LEP A_{FB}^b. The accuracy obtained for the weak angle, i.e. one measure of the quality of SM testing, improved by a factor 35 during LEP era, providing sensitivity to virtual particles in loops with up to 13σ significance.

For top, and to a lesser extent, for the BEH boson, good agreement between indirect (LEP) and direct mass measurements (Tevatron, LHC) was a stringent proof of the validity of the SM. Figure 7.4 in the LEP introduction shows the increasing accuracy of indirect (green) and direct (colour) top mass results over time.

LEP could not produce longitudinally polarized beams, but it could polarize the beams transversely. At a precise frequency, a periodic e-m excitation of the beam would *resonantly depolarize* it. This effect was used to determine accurately the *beam energy and thus the Z mass.*

"The search for the top could be compared to that of a bush hunter, his ear to the ground" (G. Altarelli). Once the steps of the elephant "top" were recognized, LEP turned to the much more discrete pace of the tiger — the BEH boson. Assuming the validity of the SM, the boson was felt to be light: ≤ 200 GeV at the end of LEP.

LEP made many outstanding contributions to QCD [Box 4.2], offering in particular a variety of measurements of the strong coupling α_s and a clear proof of its energy dependence. The accurate LEP1 measurements of the couplings of the three forces provided evidence that they might unify at very high energy in a SUSY version of the SM. This suggested a Grand Unification of the forces near 10^{16} GeV, a situation prevailing in our Universe about 10^{-36} s after the Big Bang. It also suggested that SUSY may be more than just an elegant idea.

To resolve anomalies concerning the tau, LEP, in an international venture with other experiments (e.g. BES in China), showed it to be a normal lepton, a recurrence of the first two, albeit much heavier. Its hadronic decays could then be used to test QCD, and as input for estimating muon $g - 2$ [Highlight 2.4].

With 17 million hadronic Z^0's recorded, LEP1 was also a tau, charm and beauty factory, and made major contributions to heavy quark physics. The clean topology, the strong boost imparted to the B particle and the fact that all beauty species appear in Z decays were major assets. With the micro-vertex detectors, the performance of their identification exceeded expectations. It contributed to a first hint of CP violation in the world of beauty. On the fascinating particle-antiparticle oscillations much progress occurred. These topics, probably a key to our existence, have been pursued at B factories and now at LHC.

LEP2

LEP2 measured directly the W mass as 80.412 ± 0.042 GeV, while LEP1 (together with SLC) had nailed it indirectly as 80.373 ± 0.033 GeV, another beautiful success of the SM. The measurement of the W pair production offered a spectacular demonstration of the weak bosons self-interaction.

For SUSY, LEP2 set a lower mass limit for some SUSY partners close to the maximum E_b and excluded a LSP (Lightest SUSY Particle, dark matter candidate) lighter than ~45 GeV. The SUSY torch is now passed to the LHC.

Concerning the direct search for the BEH boson [27] it was shown that, with a good beauty tag, LEP2 could discover it up to $M_h = 2\ E_b - M_Z$, a mass limit that LEP2 indeed reached at each energy. Within the framework of SUSY the existence of a light SM-like BEH boson is predicted. In 1994 the upper limit of its mass was estimated to be 130 GeV in the simplest versions of SUSY. A beam energy of 110 GeV was thus required to verify the existence of such a boson.

Supersymmetry (SUSY) **Box 7.2**

Supersymmetry (SUSY) is an attempt to restore the striking asymmetry in the Standard Model [Box 6.4] between constituent fermions and field bosons, by associating with each of the known particles a partner of the opposite category. If SUSY is real, it cannot be a perfect "symmetry" (it must be "broken"), since such partners do not exist having the same mass. SUSY would cure some of the shortcomings and paradoxes of the SM. Mathematically the theory is well founded. In brief, it says that any "step" made in space-time (translation, rotation, boost) can be decomposed into two, from space-time to an entity called *superspace* and back to space-time. Its realization requires the partners quoted above, but says nothing about their mass and the mechanism of breaking the symmetry.

A priori doubling the number of elementary particles does not seem to be a progress. However unbroken SUSY has in fact one parameter less than the SM. Long ago the prediction and discovery of antiparticles, leading to such a doubling, was a decisive step in our understanding of particle physics. But SUSY breaking reintroduces many parameters, which renders this theory difficult to ever falsify.

Searches for SUSY at colliders assume that some partners may be light enough to be accessible. Most assume that the Lightest SUSY Partner (LSP), being unable to decay to a lighter one, is stable, invisible and leads to missing energy-momentum in the final state. Being stable the LSP may also be a cosmic remnant of the early Universe, constituting its *Dark Matter,* and can be searched for as such [1].

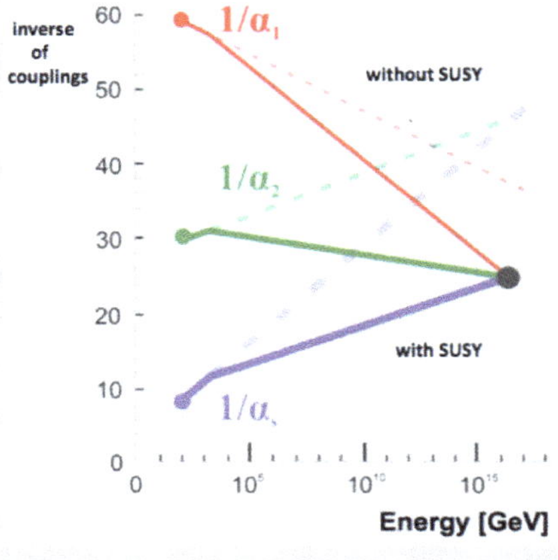

Light SUSY would solve the puzzle of the SM, the relatively light mass of the BEH boson, which radiative corrections [Box 5.1] should push to a higher value. Fermions and bosons populating the loops in a nearly symmetric way would much reduce these corrections.

A further virtue of SUSY and its only strict requirement (and possibility of falsification), is to predict (in its minimal version) a light SM-like boson, < 130 GeV. The 125 GeV boson (see LHC) is thus an encouragement but not a proof. Another hint is that in SUSY the coupling constants of the three forces meet exactly at very high energy as shown in the figure.

The search for SUSY partners and LSP has so far been unsuccessful. SUSY partners of the gluon and quarks of the first two families, if they exist, must be heavier than ≈ 1.5 TeV (unless the LSP is heavier than ≈ 0.8 TeV, in which case no limit is set yet). The limits on partners of the 3rd family and of BEH and EW gauge bosons are lower. The search for SUSY partners is a priority at high luminosity and high energy LHC.

[1] H. E. Haber and G. L. Kane, Is Nature Supersymmetric?, *Scientific American*, June 1986.

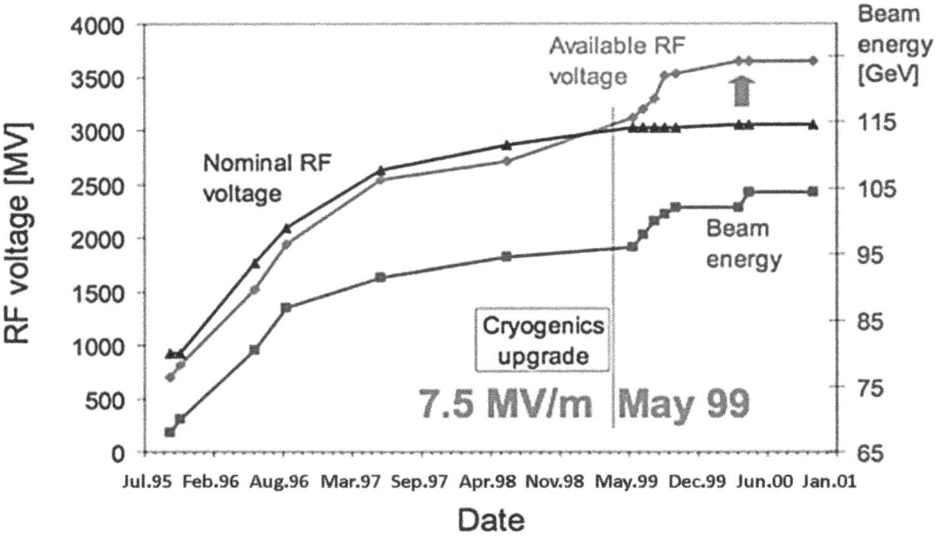

Fig. 7.5. Increase in energy of the LEP beams.

In 2000, following the remarkable improvement of the RF, LEP2 delivered a useful beam energy of 103 GeV (Fig. 7.5). An indication of a weak signal at M_h = 115 GeV was claimed, most insistently by ALEPH. Twelve years later the LHC showed that this signal was spurious and discovered the boson at 125 GeV/c^2.

SUSY or not, LEP2 missed the boson by 10 GeV. Replacing all less performing RF cavities with high-gradient Nb-coated superconducting ones, as proposed in 1996 [28], would have provided the required increment of energy. The production of cavities was stopped in 1996 to free resources for LHC.

The LEP legacy to its off-spring LHC was thus that the SM BEH boson should appear in a mass range between the lower limit of the direct search (114 GeV/c^2) and the indirect upper limits from electroweak measurements, about 200 GeV/c^2, reducing to 150 GeV/c^2 in 2007 when including the Tevatron results.

In retrospect, it is instructive to compare the expectations of LEP with the results. All measurements exceeded, often by far, the expectations [27]. The achieved error on the Z mass was 2.1 MeV instead of 20 to 50 MeV expected. For the W mass is was 33 MeV instead of 100 MeV. The number of neutrino families was determined with an error of 0.008 compared to 0.3 first anticipated. All asymmetries [Box 7.1] were also measured about 3 times more precisely than foreseen. In addition to the gain from combining the results of the four experiments, this was due to a number of important facts:

- The outstanding performance of the machine, and the excellent collaboration of its team with the experimentalists;
- The relative ease of performing experiments at LEP. Due to the low level of background the interesting events were easy to select and quite clean (Fig. 7.6). Accurate knowledge of E_b permitted to use kinematic constraints;
- The diameter of the vacuum chamber in the experimental intersections could be reduced thanks to clever shielding against synchrotron radiation, so that micro-vertex detectors could be placed close to the collision point to improve b-tagging;
- Redundancy of experimental procedures. Efficiencies could be measured instead of being evaluated from Monte Carlo runs, reducing systematic errors;
- Development of new techniques, some emerging or being improved during LEP lifetime, e.g. the luminometers and micro-vertex detectors;
- Outstanding work of theorists, in close collaboration with the experimentalists, which were key to testing so beautifully the fundamental features of the SM.

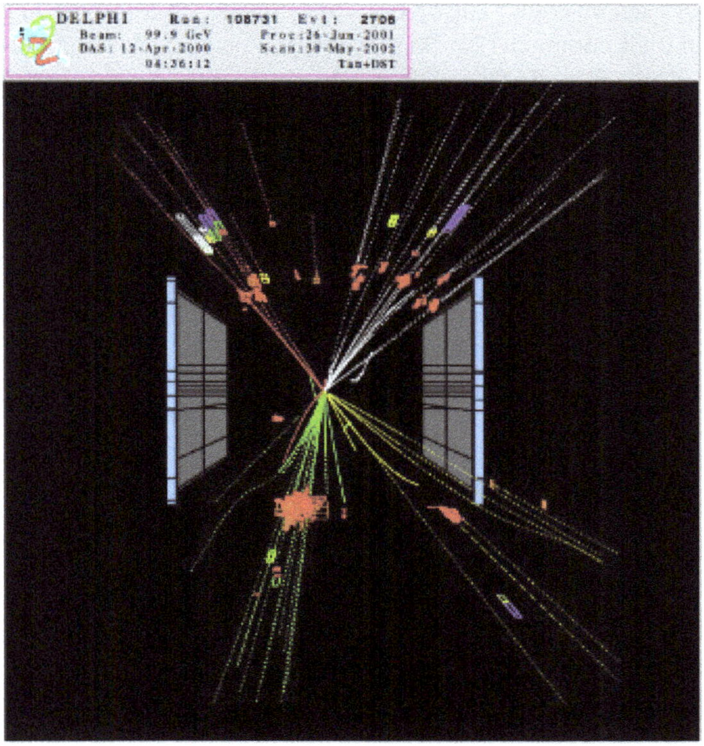

Fig. 7.6. A typical event as recorded at LEP.

7.2 Concrete Stuffing for the LEP Magnets

Jean-Pierre Gourber

With LEP, CERN was confronted for the first time with the problem of producing a large number of magnets with a very low bending field. At the maximum design energy of 125 GeV, the field of the main dipole magnets was only 0.135 T, to be compared with the 1.5 T to 1.8 T of classical magnets for protons machines, but it had nevertheless to be very precise and uniform from unit to unit. CERN engineers faced up to the challenge of satisfying this unusual requirement by developing innovative solutions that would also result in substantial savings in cost.

The regular lattice periods in arcs were 79 m long with bending magnets made of six equal C-shaped dipole yokes (cores) between focusing quadrupoles. With the shorter cells at the arc ends, there were in total 3304 cores. These were excited by passing current through long bars of extruded hollow aluminium, insulated with glass-epoxy, and connected in series giving a single powering circuit (see Fig. 7.7). This solution was cheaper than traditional coils and minimized the space required for joints between magnets. The 5.75 m long cores were assembled in pairs at the surface with their common 12 m long excitation bars and lowered into the tunnel via a large elliptical shaft. The excitation bar interconnection was made *in situ* by TIG welding with special care to ensure continuity of the water cooling channel.

The cores were made as usual by stacking low carbon steel laminations. The transverse dimension of the gap was imposed by the dimensions of the vacuum chamber, the required 10^{-4} field homogeneity, and the mechanical stability of the flux return (backleg) during punching. Because of the low field, the laminations

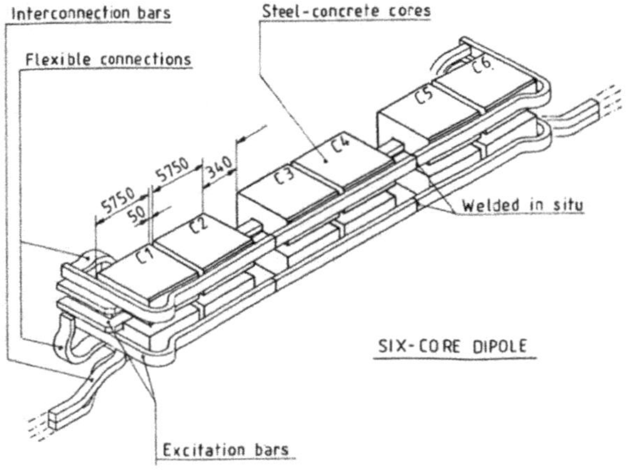

Fig. 7.7. Assembly of dipole magnet yokes at LEP.

could be spaced so that the field level in the flux return would reach the optimum value of 1.5 T at the maximum field, corresponding to 125 GeV. The spaces had to be filled with material cheaper than steel and the manufacturing process had to be economical. Somewhat surprisingly, concrete was proposed as the "stuffing" of choice of the voids between the iron laminations [29, 30]!

The 1.5 mm thick laminations were precision punched from decarburized steel sheet having good magnetic properties. The punching die also pressed suitable indentations which, by alternating the laminations, ensured the required optimum spacing of 4 mm in the stack (Fig. 7.8). The stack of laminations with its two 15 mm thick end plates was then placed in a mould and filled with a low shrinkage mortar. Four longitudinal rods anchored to the end plates were tensioned to provide compressive pre-stress of 0.5 MPa after de-moulding. This ensured that all points remained in compression during manipulation of the cores. The cores behaved like pre-stressed concrete beams; they were only half the weight of full steel cores and being very rigid they could be supported on three feet, making for easy alignment. The magnetic properties of these magnets were nevertheless very close to those of full steel magnets. The local variation of the field in the beam aperture due to the lamination spacing were less than 10^{-7}, and hence negligible.

The punching of the laminations by a single firm and the parallel fabrication of the cores by two civil engineering firms went smoothly and to the schedule of three years. On delivery to CERN, the *magnetic* geometry of the gap of each core was measured using a specially developed carriage. Complementing these survey measurements one core out of ten was equipped with a provisional set of excitation

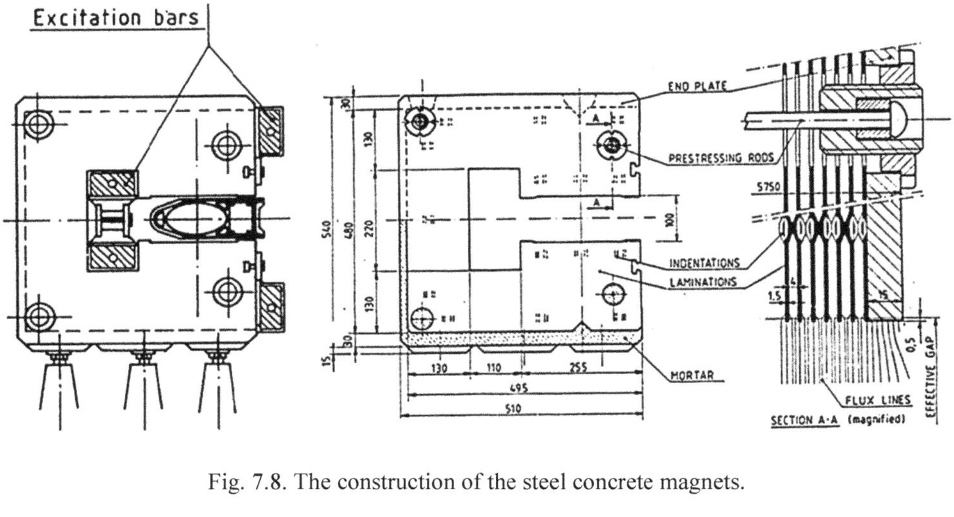

Fig. 7.8. The construction of the steel concrete magnets.

bars and its field measured. Results met specifications. However, when these measurements were repeated some months later there was a bad surprise: despite the use of a mortar carefully chosen for its low shrinkage, some residual shrinkage was occurring long after the cores were dry. The longitudinal effect of shrinkage had been anticipated in the design of the core, the supports and the excitation bars. But the shrinkage also had an unforeseen effect: it put the laminations into transverse compression. Such stress decreases the magnetic permeability of the steel, affecting the field in the gap [31]. Such a drift was unacceptable for an accurate energy calibration of LEP. The solution was to submit every core, one year after manufacture, to five cycles of a slight opening/closing of the gap, to provoke micro-fissures in the mortar and relieve the compression in the back leg. In addition, each core was equipped with a flux loop embedded in the lower pole. These loops, which were connected in series in the ring, were used to monitor the field in situ so that the influence of temperature variations, and eventual further ageing of the mortar, could be taken into account when determining beam energy.

7.3 Pumping LEP: Sticky Tape for Molecules

Cristoforo Benvenuti

Storage rings need a low residual gas pressure to minimize beam–gas interactions and keep the particles circulating for hours. For an electron storage ring, such as LEP, the problem is aggravated by the synchrotron radiation produced by the beams, which hits the vacuum chamber walls and desorbs a large amount of gas.

The small conductance of the vacuum chamber limits the flow of the gas molecules to the pumps, so reducing their pumping efficiency. At LEP, pumps with one metre spacing would have been needed to circumvent this problem. An elegant alternative solution adopted in similar previous projects consists in replacing the lumped pumps by a linear sputter-ion pump inserted in the dipole bending magnets. In this case pumping is ensured by ionizing the gas by a discharge ignited between two electrodes, and burying the ions in the cathode. The discharge is maintained at low pressure thanks to a high voltage and a high magnetic field applied to the pump. However, the field of the bending magnets at the LEP injection energy was too low to ignite the gas discharge in the pumps. Another solution had to be found. Finally, a getter strip [Box 7.3] inserted all along 23 km of the LEP chambers was adopted, as shown in Fig. 7.9 [32].

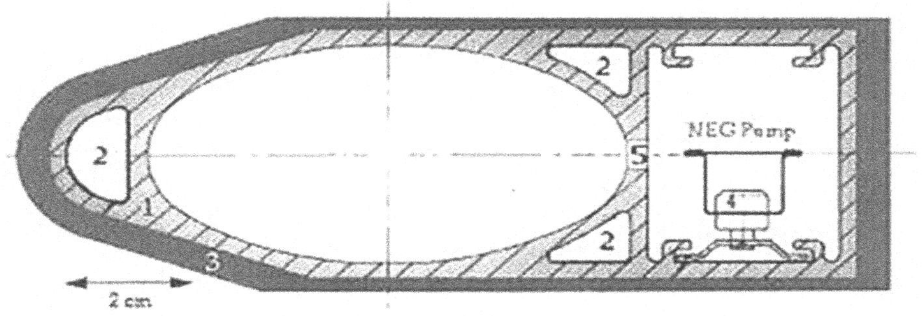

Fig. 7.9. Cross-section of the LEP dipole chamber with the getter pump. 1: extruded Al profile, 2: cooling channels, 3: lead shielding for synchrotron radiation, 4: ceramic insulators, 5: pumping slots.

Getters **Box 7.3**

Most metals and metal alloys display chemical reactivity with some gases, at least with oxygen (which is responsible for their surface oxidation). Materials able to trap the majority of gases in the form of stable chemical compounds are called getters and are widely used in vacuum technology. However, rare gases and methane do not react with getters, so additional pumping of a different nature must be added.

Getter applications were pioneered by D. G. Fitzgerald (England, 1883) and A. Malugnani (Italy, 1894), both for use in incandescent lamps. Since then getters have been inserted in all vacuum sealed devices to continuously pump the gas produced by components of the device.

A clean getter surface traps gas molecules, but in doing so its reactivity decreases and finally vanishes at saturation. To restore pumping two different strategies may be adopted, which define two different getter families. Evaporable Getters (EG), for which pumping is restored by coating the saturated layer by a fresh getter film, and Non-Evaporable Getters (NEG), for which the surface is cleaned by heating. Heating provides the energy needed to diffuse the trapped gas molecules from the surface into the getter bulk. The two important practical characteristics of NEGs are the surface area, which defines how much gas may be pumped before saturation, and the activation temperature (Ta) required for cleaning the NEG surface.

EG s have been used for Ultra High Vacuum (UHV) applications since the 1950s in the form of Titanium sublimation pumps, but these could not be adapted to provide linear pumping. On the other hand, NEGs were used only at relatively high pressures ($10^{-4}/10^{-6}$ Pa), at which the fast surface saturation imposes continuous NEG heating, not applicable in the case of LEP because it would upset the circulating beams. LEP required pressures lower than that, and the NEG behaviour at low pressures without continuous heating remained to be demonstrated at the time (it has since been confirmed).

Following the approval of the LEP project, vigorous development work was undertaken to explore the possibility of Non-Evaporable Getter (NEG) pumping. At that time only one NEG type was commercially available, in the form of a metal strip coated on both faces with 0.1 mm of getter powder. The getter material was a Zr-Al alloy with a daunting activation temperature of 750°C. The strip width chosen for performance evaluation was 30 mm.

Very quickly after the beginning of the experimental investigation it became clear that pressures as low as 10^{-10} Pa could be obtained by this NEG kept at room temperature. Since in this case gas diffusion does not take place, it was vital to ascertain how much gas could be pumped before reaching saturation.

The answer to this question is given in Fig. 7.10 [33]. The results were reassuring. About 10% of the initial performance was still available after pumping, per metre of NEG strip, 10 Pa litre of H_2 and CO, the most important components of the accelerator residual pressure. In practical units, 10 Pa litre of gas corresponds to about 0.1 cc of gas at atmospheric pressure, a huge amount by UHV standards. This result is a consequence of the active NEG surface area being about 100 times larger than its geometrical area, due to the high NEG porosity.

Although these results were comforting, the wide performance difference for different gases was puzzling and this poor understanding was not acceptable for a multi-billion project. For this reason, a large effort was invested in understanding

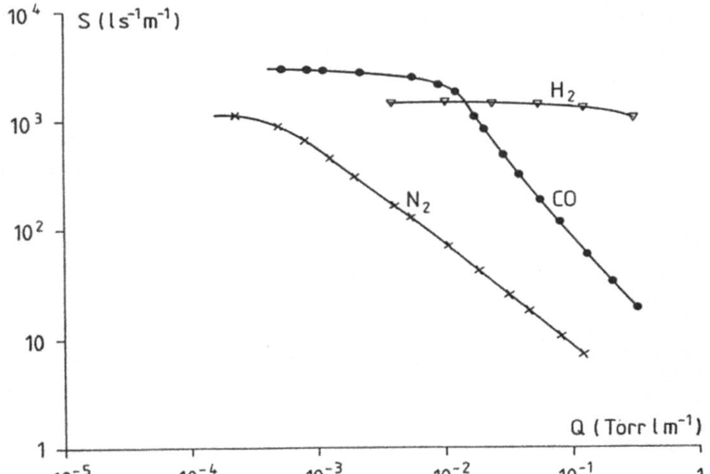

Fig. 7.10. Variation of the pumping speed S as a function of the pumped quantity of different gases. The measured sample is the strip (1 m long, 30 mm wide) adopted for the pumping of LEP. The getter is a Zr-Al alloy. (1 Torr = 133 Pascal)

the NEG pumping behaviour for individual gases and gas mixtures [34]. It was found that H_2 diffuses into the NEG bulk even at room temperature, while for heavier gases, which stick to the surface, pumping at room temperature depends on the NEG porosity and on the number of getter atoms needed to trap a gas molecule. The large difference of the pumping curves shown for CO and N_2 in Fig. 7.10 is a consequence of the fact that a CO molecule occupies one NEG atom, while six adjacent, free atoms are required to adsorb a molecule of N_2, for which the surface saturation is therefore much faster.

The strategy resulting from these studies was to rely on NEG pumping at room temperature and to apply regeneration heating whenever the remaining pumping was too low for the desired performance of the accelerator. Based on this strategy, a few LEP chambers were equipped with different prototypes of NEG pumps and finally the model shown in Fig. 7.9 was adopted. In this pump the NEG strip is electrically isolated from the chamber with ceramic pins inserted into a metal frame. At the two ends of each chamber the NEG strip is connected to electric feedthroughs which allow the NEG to be heated up to the temperature of 750°C by applying an electric current of about 100 A.

The last point to be clarified was the NEG pump behaviour in real accelerator conditions. How often would the NEG have to be heated to restore its pumping? Would the required heating frequency upset LEP operation?

To answer these questions a chamber equipped with a linear NEG pump was installed in PETRA, an electron storage ring similar to LEP already in operation at DESY (Hamburg). The test was conclusive: the behaviour of the NEG pump was fully satisfactory, leading to its adoption for the LEP project.

About 2000 chambers, most of which were 12 m long, were equipped with a NEG pump as shown in Fig. 7.9. The vacuum pressure after a 24 h baking cycle at 150°C was specified to be less than 2×10^{-9} Pa. Acceptance testing was done in a dedicated laboratory (Fig. 7.11) where many units were tested simultaneously.

At LEP the average pressure in the absence of beams was below 10^{-9} Pa. Pumping of rare gases and methane, at the level of 1% of the NEG pumping, was supplied by locally installed sputter-ion pumps. Fortunately, degassing induced by synchrotron radiation at an electron storage ring decreases with operation due to surface cleaning. For this reason, the frequency of the NEG heating cycles quickly became so low that they only had to be applied when the accelerator was stopped for maintenance.

The NEG pump worked perfectly during the entire ten years of LEP operation.

Fig. 7.11. The test laboratory of the LEP chambers. About 2000 aluminium chambers were tested in batches of 12. Only those providing a pressure lower than 2×10^{-9} Pa after bake-out were accepted.

7.4 Superconducting Skin Boosts Accelerator Cavity Performance

Cristoforo Benvenuti

If the walls of radiofrequency (RF) accelerator cavities are superconducting (SC), the resistive RF losses on the cavity walls are orders of magnitude less than those of normal conducting cavities. However, some power is still absorbed and it increases proportionally to the square of the accelerating field [Box 7.4]. When this field is increased, a value is reached at which the internal surfaces of the cavity exceed the transition temperature T_c (9.3 K for niobium), and the cavity becomes normal conducting (it "quenches").

Among the various factors which contribute to the temperature of the cavity surface, the thermal conductivity of the cavity wall material plays a major role. Niobium (Nb), the usual SC metal of choice for these cavities, is not a good thermal conductor, and it becomes even worse in the SC state, because the paired electrons do not transport heat. Copper would be a much better choice. Although copper is not a superconductor, it can be coated with a superconducting film [20]. This film does not need to be thick: About 10^{-3} mm would suffice to shield the underlying copper from the RF power. Copper would also provide the additional bonus of a much lower cost.

RF behaviour of superconductors **Box 7.4**

In the presence of radio frequency (RF) fields the electrical impedance of a superconductor (SC) is not zero, as it is for DC applications (e.g. magnets). The reason is that not all the conduction electrons are coupled in resistance-free Cooper pairs, even at temperatures lower than the SC transition temperature T_c. While in DC applications the current carried by the Cooper pairs [Box 8.1] short-circuits that carried by unpaired electrons, in RF conditions the unpaired electrons behave as electrons in a normal conductor and experience resistive losses. The population of normal conducting electrons decreases exponentially with the ratio of the temperature of operation to T_c, so materials with a high T_c represent a better choice. The negative role of the normal electrons is further reduced by choosing metals with low resistivity at low temperature.

While high T_c and low resistivity in the normal state are desired characteristics of a superconductor for RF accelerating cavities, essential practical features are ductility and weldability, required for manufacturing cavities from sheet metal. Niobium (Nb), with T_c = 9.3 K, reasonable electrical conductivity and mechanical behaviour similar to that of copper is the most widely used material for this application.

The "figure of merit" Q of an accelerating cavity is proportional to the ratio of the RF power stored in the cavity to the power dissipated resistively on the cavity walls in one RF cycle. At liquid helium temperature Nb offers a Q value about 5 orders of magnitude greater than that of copper. This difference more than compensates for the thermodynamic cost of helium liquefaction and the complexity of cryogenics.

Niobium is a good choice in spite of its very modest critical magnetic field (about 0.2 T) compared to that of the Nb-Ti alloy used for SC magnets (about 14 T). This is because the field of 0.2 T is only reached for accelerating gradients of over 40 MV/m, greater than that needed for circular e^+e^- colliders and linear colliders such as the ILC.

The real limitation of Nb lies in its modest thermal conductivity which defines the temperature gradient through the cavity wall and may result in a cavity "quenching" whenever T_c is exceeded locally. This was the main justification for developing a Nb-coated copper cavity. An added bonus of this approach is that it makes possible the manufacture of cavities using SC materials not suitable for forming from bulk.

For these reasons a vigorous development program was undertaken in parallel to that of the traditional bulk Nb approach in view of exploring the possible use of Nb film technology for the cavities to be used for the LEP upgrade from the initial energy of 50 GeV to about 100 GeV. When this work was started in 1980 little information on thin Nb films was available, and no one had succeeded in obtaining T_c = 9.3 K. A deeper analysis showed that all these films had been produced in sputtering systems with poor vacuum (10^{-4} to 10^{-5} Pa). Niobium is very reactive and during coating its purity is spoiled by trapping residual gas molecules. By improving the process vacuum and adopting standard UHV procedures the nominal T_c was immediately obtained on small samples. However, this was only the first difficulty. Since the cavity is an almost closed vessel of ellipsoidal geometry, obtaining a coating of uniform thickness is far from trivial. Furthermore, the copper surface cleanliness is crucial to ensure good film adhesion. Finally, the coating process requires clean room conditions to avoid surface defects produced by trapping particles floating in the ambient air.

Test cavities were initially coated by bias diode sputtering, which was later replaced by cylindrical magnetron sputtering [35]. The test cavity was a single cell of 500 MHz frequency shown in Fig. 7.12. The cathode was a stainless steel cylinder covered with a Nb liner, from which atoms were sputtered off during the coating by positive ion bombardment. These ions were produced by triggering a discharge in argon or krypton gas at about 10^{-2} Pa, thanks to a negative bias of 400 V on the cathode. The cylindrical cathode contained a solenoid producing an axial field of about 1.4 T at 20 mm from the its surface. This field imparted a circular motion to the discharge electrons, increasing their ionisation efficiency and the sputtering rate. During operation the solenoid was cooled using Freon.

A nice feature of this cathode structure is that by adjusting the cathode diameter and the length of the magnet it is possible to obtain a very uniform coating thickness, because at the cavity equator the higher perpendicular Nb emission compensates for a lower deposition rate due to the larger distance from the cathode.

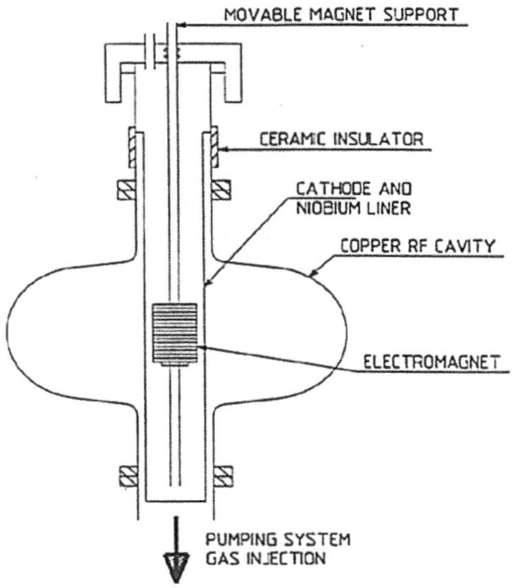

Fig. 7.12. The sputtering configuration.

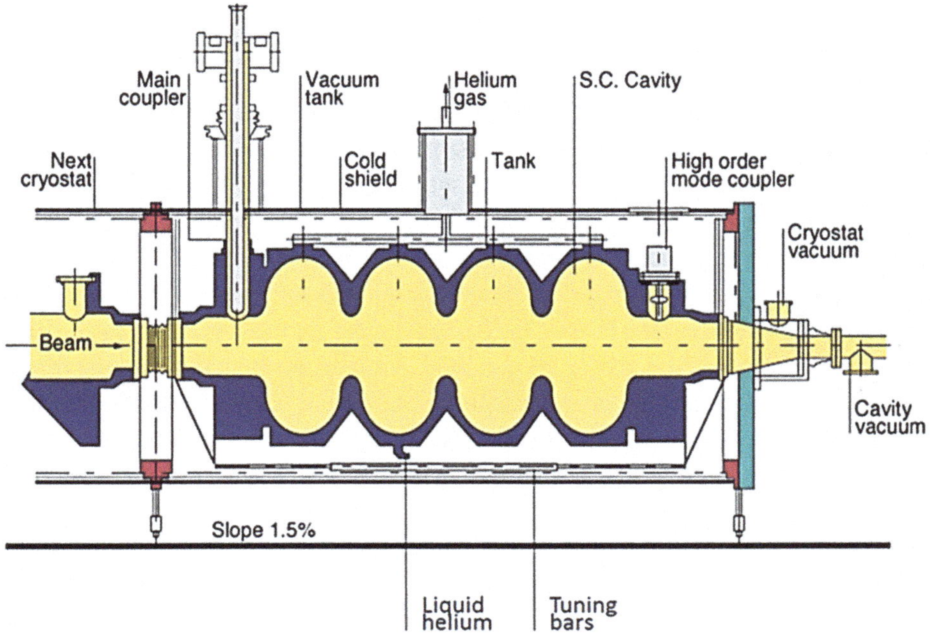

Fig. 7.13. The LEP accelerating RF cavity in the cryostat. The cell diameter is 752 mm and the mid-cell spacing is 0.43 m ($\lambda_{RF}/2$).

After the initial development on single cell test cavities, the coating process was applied to LEP 4-cell 352 MHz cavities (Fig. 7.13) quickly exceeding the 5 MV/m field initially specified for bulk Nb cavities, so as to allow the specified field to be raised to 6 MV/m [36]. A total of 288 cavities were produced by three European manufacturers to whom the CERN know-how had been transferred. Finally, these cavities reached very reliably an average accelerating field of 7.5 MV/m, and the LEP energy was gradually increased up to the maximum achieved value of 104.5 GeV [37].

The coating approach could be extended to other materials of superior superconducting properties. Promising results were obtained with Nb_3Sn coating, but it was too late to envisage its application to LEP. These studies have since also been taken over by other laboratories [38].

7.5 Measuring the (Accelerator) World

Michel Mayoud

When commonly speaking of "very accurate measurement" of objects, most people think of metrological control in terms of dimensions and/or position on a bench, and much less of geodesy for describing the earth in terms of measuring angles to within one arc-second, and distances to a few parts in a million. The technology derived from geodesy was also used for micro-geodetic networks as seen in engineering surveys for the construction and deformation control of large objects such as dams. This is why CERN called for geodesists to take care of the sub-millimetre metrology of the PS machine. As accelerators grew and the relative size of the magnet aperture decreased (to contain cost), they required ever tighter positioning tolerances: 0.6 mm for the PS, 0.2 mm for the ISR, 0.15 mm for the SPS, 0.12 mm for LEP; down to 0.01 mm at intersection points, microns for certain LHC components and nanometres for future linear colliders. As in other domains, CERN undertook successive programmes of development, innovating in techniques and instruments for the metrology of accelerators and experiments.

In this vein, prior to tackling the requirements of geodesy at LEP, the team had built up experience and developed a succession of specific devices and procedures for previous accelerators. This included measurement of length using calibrated invar wires; the first application of wire offset measurements, where the offset to a stretched nylon wire was observed with a microscope on a micrometric carriage, providing a simple way to align objects with an accuracy of 0.1 mm; the application of electronics and sensors, and electronic clinometers (developed in-house) [39]. The geodetic reference network for the SPS was measured by trilateration (using calibrated electro-optical devices for the measurement of lengths), with data processed according to a local spherical model. The SPS was the first machine to be housed in a deep tunnel, and a major challenge was to guide the boring machine within the centimetre tolerances imposed by the project. CERN used state-of-the-art commercial gyro-theodolites, gained experience, and acquired widely recognised expertise in underground geodesy.

Nevertheless, given the size and the location of LEP a new chapter in geodesy technology had to be opened. The geometrical tolerances were tighter: The project called for a relative accuracy of 0.1 mm for the alignment of components along the 27 km circumference of the machine and the best possible absolute accuracy (i.e. within a few mm) with respect to the theoretical geometry. While for the SPS the spherical approximation was sufficient to express the effects of the curvature of the earth, for LEP, which lies partly under the Jura mountains, it was necessary to take into account the variation in gravity, in particular the impact on the vertical direction, due to the distribution of mass. This meant that a new geodetic and geophysical CERN coordinate system had to be defined, into which all accelerator components had to be inscribed. Data were derived from astro-geodetic measurements using accurate zenith cameras (supplied by ETH Zurich), and a precise digital mass model was compiled for integration into the official Swiss model. The data allowed to define the plane on which the 27 km machine was located in measurable space, and was essential for correcting angular and gyroscopic measurements of the underground network used for accurately guiding the tunnelling machines along the 3.3 km sectors between control points. By using the best commercial gyro-theodolite combined with CERN instruments and techniques developed specifically for the task, the azimuthal guiding errors were contained to a radial offset of 10 mm (rms). This geodesy and metrology was also highly dependent on a unique commercial instrument that uses two polarised laser beams (He-Ne and He-Cd) to eliminate first order effects of fluctuations of temperature and pressure of the air, and was designed for measuring long distances with an accuracy of 1 mm at 10 km, the maximum range being about 15 km. By using this tool, the LEP external geodetic network was at the time the most accurate ever obtained for such a size, and was for a while a testing ground for geodesists worldwide [40]. LEP metrology was thus based on well-defined control points — essential for the optimal control of the underground metrological network [41].

A major outcome of a CERN collaboration with IAUB (l'Institut Astronomique de l'Université de Berne) was that millimetre accuracies can be obtained from GPS measurements, when the data is processed with related orbital, atmospheric and ionospheric data from satellite information, combined with local parameters [42].

As concerns instrumentation for the alignment at installation of about 5000 machine components, all CERN instruments were upgraded to allow a full computer-aided alignment process. The metrology of experiments was also significantly improved by designing new complex 3D networks, measured by all available means — with no constraint in setting oblique reference lines or planes in the design. Hydrostatic levels and clinometers were also installed for monitoring height, slope and tilt changes of huge equipment such as the 12 000 t CMS magnet.

It was thanks to the accuracy of the alignment of the LEP collider and a profound understanding of the impact of local geology and global phenomena that effects such as that of the tides described in the next highlight could be clearly evaluated, and that the experiments could record data of far greater precision than anticipated.

7.6 Precise Energy Measurement: Heed the Moon

Albert Hofmann

In LEP the circumference, C, was determined by the orbit passing through the centre of all the quadrupoles, but the effective orbit length of the ultra-relativistic circulating beam was determined by the frequency of the RF system, which fixed the revolution frequency. If C changed by ΔC say, with the frequency kept constant the beam would move transversely in the magnets. This shift would move the beams off centre in the quadrupoles entailing a change in energy $\Delta E/E = \alpha_c^{-1}\, \Delta C/C$ where α_c is a beam parameter known as the momentum compaction factor. This latter factor was between 1.5×10^{-4} and 4×10^{-4} for LEP, so the energy calibration of the machine was extremely sensitive to geological motion caused for example by tides and variations in the water table which change C [43].

Tides are caused by the variation of the gravitational attraction of celestial bodies, principally the sun and the moon. The tidal forces produce local stretching in the horizontal plane and resulted in a peak-to-peak variation in the circumference of the machine of up to 1 mm. Following these calculations an experiment was performed during a period of full moon, the result of which is shown in Fig. 7.14. Thereafter the information generally available on the tides was put into a model that was used to correct for these effects in the determination of the true energy of the accelerator.

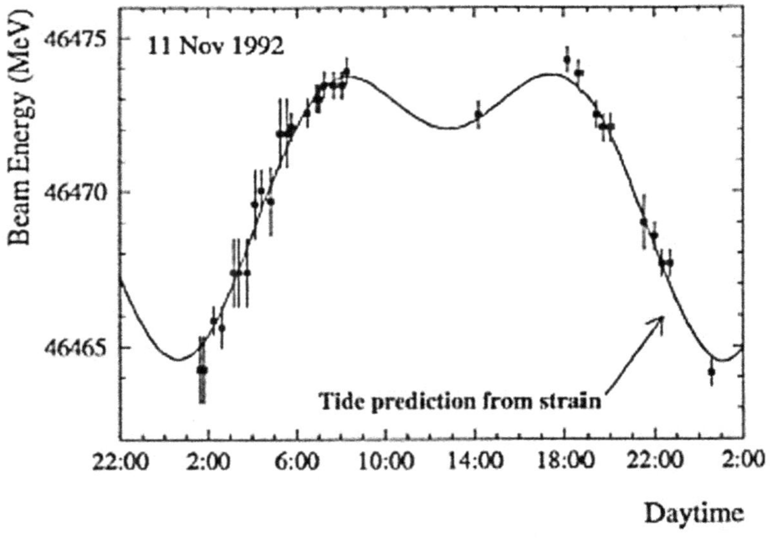

Fig. 7.14. Variation in the energy of LEP beams during a day of full moon. One can see the excellent agreement with the change in energy predicted by the calculation of horizontal strain due to the tides.

In addition to the daily and monthly tidal movements, LEP was also affected by long-term changes in circumference. The movement was observed by monitoring the radial movement of the beam relative to the quadrupoles using the beam position monitors during the annual period of operation of the accelerator for physics. Fluctuations seen during a typical period from May to November are shown in Fig. 7.15, displaying changes of up to 2 mm, a profile found to be very similar from year to year. The diameter tended to be larger during the summer months, and some of the changes were clearly correlated with rainfall. This monitoring of seasonal fluctuations was quite important for the experiments, as the resulting variation in energy was significant, being up to $\Delta E/E = 5 \times 10^{-4}$. Although this did not require the development of new equipment, it illustrates the skill of the operations team in understanding the machine performance in the utmost detail and led to a factor 10 improvement in the determination of the Z mass.

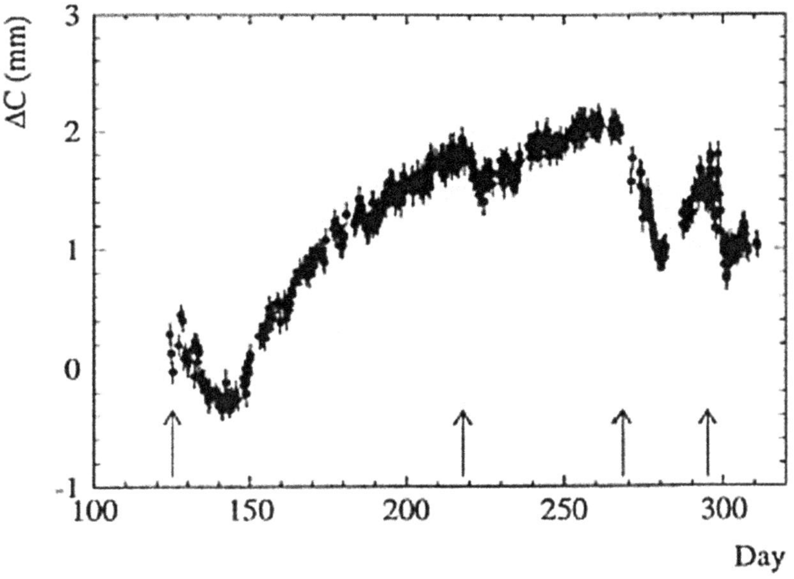

Fig. 7.15. Change of LEP circumference in 1996. A total drift of 2 mm was observed. It gradually increased during the summer, with rapid movements following heavy rain (indicated with arrows).

7.7 The LEP Silicon Vertex Detectors: Right on Target

Hans Dijkstra

After their successful use in fixed target experiments for charmed physics starting in the early 1980s [Highlight 5.9], innovative silicon strip detectors were developed and successfully operated in experiments at colliders like LEP. One major motivation was the study of particles containing heavy flavour quarks, charm or beauty. These particles are short-lived with lifetime τ of the order of 10^{-12} s (1 ps). To tag such particles the transverse accuracy of extrapolation of the tracks of the particle decay to the interaction point (IP) must be better than their characteristic offset $c\tau = 300$ μm. This requires excellent spatial resolution of the detectors, a minimal distance to the IP and minimal thickness of material traversed by the particles. Material affects the direction of the particles, whence deteriorating the extrapolation accuracy. A collider imposes further constraints. The innermost radial position of the detectors is fixed by the radius of the beam pipe, the outermost one by the size of the first tracking chamber. Installation and removal of the complete Si detector system must be possible independent of the other detectors. Layers of silicon strip detectors were arranged in concentric cylinders around the beam pipe. The radius r of the detector wafer, given by the holding frame, provided the r coordinate. The detectors had readout (RO) strips on both sides, parallel (for Φ coordinate) and perpendicular (for z coordinate) to the beam, respectively, allowing space measurements. The RO electronics was normally mounted along two edges of the detector wafer, adding extra material and distributing the heat load throughout the volume of the detector. Solving these problems was a major challenge at LEP, as described here.

A major direction of research addressed the two-sided RO of the Si-detectors. This is made difficult by an e^- accumulation layer at the Si-SiO_2 interface on the ohmic or n-side of the detector, which lowers the inter-strip resistance, hence sharing the charge over many RO strips. The strips must thus be insulated from one another. To disrupt the e^- layer, ALEPH and DELPHI used blocking p+ electrodes between the n strips. DELPHI used AC-coupled readout strips with a width larger than the n-strip implants, and at a negative potential compared to these implants for insulation.

The ALEPH Silicon Vertex Detector [44] was the first detector operating in a collider to use the two-sided RO. The achieved spatial resolution of the complete device (intrinsic plus alignment) was 12 μm in the r–Φ view and 12 μm in the longitudinal z view. Two detectors each were mounted onto one electrical building block, the "module", and two modules were mounted together lengthwise to form the basic mechanical building block of the system, the "face". The inner layer consisted of 9 mechanically independent faces, the outer layer of 15 faces, making a total of 2 × 24 electrically and mechanically identical and independent modules. Each module contained two identical Si wafers, the hybrids with RO electronics and connections. Most components were custom-made.

The design of the detectors was the result of an intense R&D program carried out over several years. It introduced a novel biasing scheme which made it possible to deplete the entire detector volume using only two contacts, one for the p+ side and one for the n+ side. Due to the capacitive charge sharing between RO strips, only every fourth p+ strip on the junction side and every second n+ strip on the ohmic side were connected to the readout electronics.

With only a modest degradation in terms of material and an acceptable ambiguity level, the RO traces were routed on separate thin substrates. The Z-strips were wire bonded to diagonal RO strips at the edge of the detector, the signals being carried to the electronics in a zig-zag geometry, using additional Z-strips to link the diagonal ones (Fig. 7.16).

The RO chip was a custom-designed with 64 parallel channels arranged in a RO pitch of 100 μm (CAMEX64, CMOS Amplifier with MultiplEXing 64 channels). It was developed in CMOS technology in collaboration between the Max Planck Institut für Physik, Munich and the Fraunhofer Institut, Duisburg. The readout electronics for the Φ and Z-strips of each module consisted of two different ceramic printed circuits equipped with their readout chips and surface-mount components.

DELPHI [45] operated with a silicon strip microvertex detector from the start of LEP in 1989, initially consisting of two concentric layers of single-sided (SS) detectors providing high precision Φ measurements. In 1991 a third layer was added after the installation of a smaller diameter beam pipe. The detector achieved an averaged single hit precision of about 8 μm and a detection efficiency of 98%.

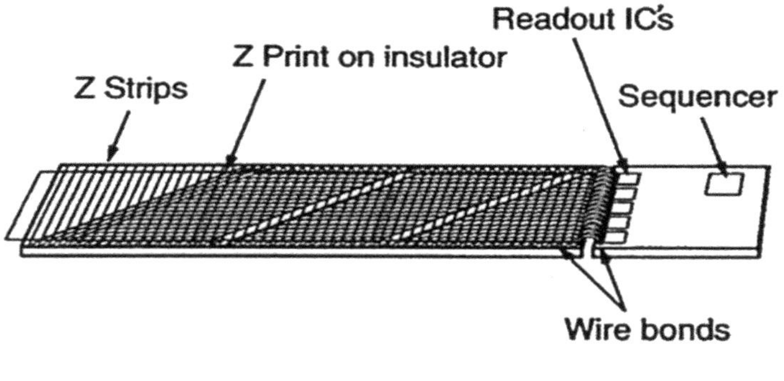

Fig. 7.16. The Z-strips are wire bonded to diagonal RO strips at the edge of the detector. (Courtesy C. Damerell)

The 1994 upgrade of the detector provided 3-coordinate measurements without degrading performance. It used double-sided (DS) detectors with orthogonally oriented RO strips on opposite faces of the detector wafer. In order to maintain the same material thickness, DELPHI developed DS detectors with a second metal layer on the ohmic surface making it possible to read-out the signal from both sides at the same end of the detector. In 1996 a further upgrade increased the solid angle coverage to improve the LEP2 BEH boson searches.

A process was developed using two metal layers, about 1 μm thick, separated by an insulating layer. The first layer coupled capacitively to the n+ side. Two different approaches to insulate the second metal layer were used in the industrial production: a low temperature deposition of oxide over the whole surface including the metal and an application of a film of Polyimide with an electric permittivity of 0.3 pF/cm. In both approaches tiny holes were opened in the insulating layer, such that the second metal layer, deposited on top could make the desired connections to the first metal layer and the orthogonal readout lines. The top lines ran parallel to the p+ diodes on the other side of the silicon. In this way all the signals could be read out at a common edge of the detector.

Constructing DS double metal layer detectors with integrated coupling capacitors and polysilicon resistors required 12 to 15 different masks compared with the 5 masks needed for a SS detector. Figure 7.17 shows a perspective view of the structure. Details are discussed in [45].

One more trick was used by DELPHI. Traditionally, the corresponding RO lines of the two detectors forming a half-module were "daisy-chained" together. However, since the RO lines of both sides are at the same potential one can join the n-side of one to the p-side of the other. This so-called flipped module design equalized the noise on the two sides and resolved ambiguities measuring the polarity of the deposited charge.

DELPHI also introduced the first forward pixel detectors at an e^+e^- collider [46].

These few technical details give a taste of the many challenges that were faced in order to optimize the use of this type of detector. Other aspects included cooling, radiation hardness, monitoring of radiation dose, shielding against electrical noise, data acquisition, offline reconstruction and alignment.

The other two LEP experiments, L3 and OPAL, also had excellent micro-vertex detectors [47, 48], while the rival SLD experiment, at the Stanford Linear Collider, used a novel CCD detector concept [49], benefiting from a smaller beam pipe radius. These vertex detectors were essential for all LEP heavy flavour physics.

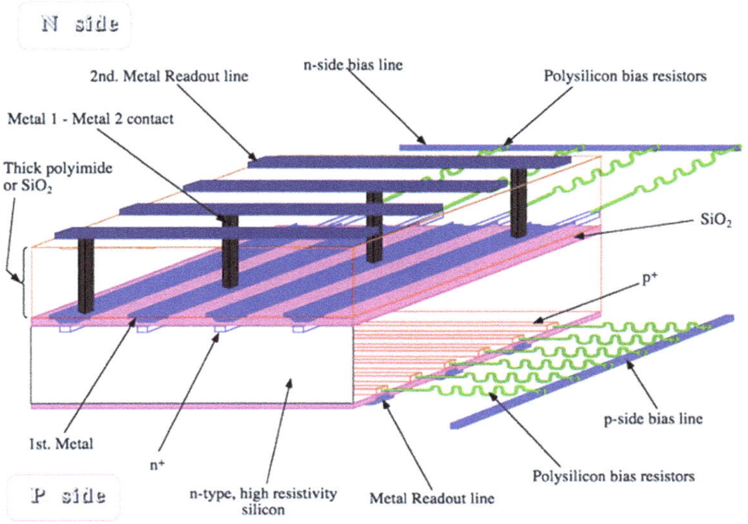

Fig. 7.17. Conceptual view of the structure of a DS double metal layer detector.

7.8 DELPHI RICH: The Luminous Footprint of Particles

Paul Baillon

Muons and electrons can be identified by their specific interactions with matter. Charged hadrons, however, are frequently identified through their characteristic mass, inferred with a combined measurement of their velocity and momentum. Besides deriving the velocity from time of flight or an energy loss measurement, a powerful method is to measure the Cherenkov radiation [Box 3.2]. When a charged particle traverses a transparent dielectric material (e.g. a liquid or gas) at velocity v greater than c/n (n being the refractive index of the material), it excites the material to emit "Cherenkov" radiation with a characteristic angle θ, where $\cos\theta = c/vn$. Measurement of θ gives the velocity v. Some detectors used this angular pattern to identify hadrons in beams (see CEDAR in SPS chapter). For spectrometers this idea became reality with the insight that gaseous detectors, such as MWPCs [Highlight 4.8] could be made to measure individual photons of the Cherenkov light [50]. The trick is to add certain organic vapours to the chamber gas. These vapours, a favourite being tetrakis-dimethyl-amino ethylene (TMAE), are ionized by Cherenkov light in the ultraviolet spectrum, and the ionization electron is subsequently detected in the gaseous detector. This remarkable invention led to a proposal [51] to build a large 4π quasi-spherical Cherenkov detector around an interaction point (IP), providing a very precise velocity measurement. Conceptually, see Fig. 7.18 (left), the Cherenkov angle would be measured by focusing the light cone into a ring and projecting it onto the MWPC. A rather crude momentum measurement would suffice for an adequate detection of the particles. This concept was as elegant as it was revolutionary at the time. It led however to a more realistic, yet still very ambitious incarnation at the LEP experiment DELPHI, the Ring Imaging Cherenkov, RICH [52]. The isotropic sphere was replaced by a more conventional geometry, a cylinder (Barrel) and two end caps (EC), which could be integrated in an otherwise classical spectrometer. It represented a major advance in experimentation, allowing hadron identification in most directions around the IP.

RICH at DELPHI

The measurement of the velocity is most sensitive close to the threshold of light production, where the angle varies rapidly with momentum. To cover a range of momenta, one should use several detectors with different indices. As a compromise, the Delphi RICH (Fig. 7.18, right) had two radiators, one liquid and one gaseous. Between them, a single Drift Tube (DT), a TPC with quartz windows, detected both the photo-electrons (PE) from Cherenkov light coming directly from the liquid, and that focused back from the gas by parabolic mirrors. An electric field in the DT caused the PEs to drift along its length of 155 cm to MWPCs. Wire and cathode pads gave two coordinates. The third came from the drift time between beam collision and PE detection, the drift speed being known.

In the following the main technological challenges are discussed with emphasis on the Barrel RICH, developed and built at CERN. For the Forward RICH see [53].

The Drift Tube was the heart of the RICH. Designed along the ideas of [50], it used the property of TMAE to emit a free electron when hit by an ultraviolet photon (> 6 eV). Its vapour pressure (about 100 Pa at 40°C) is sufficient to absorb most of the photons within 2 cm. The range of effective wavelengths imposed the use of quartz windows and adequately transparent fluids. To achieve the necessary drift conditions, a very homogeneous electrical field of 530 volt/cm was generated by applying a voltage of up to 80 kV on conductive chromium strips (0.1 mm) surrounding the quartz windows at a pitch of 6 mm. Chromium was evaporated in a large vacuum bell from a bloc heated by an electron beam and reached the quartz windows through an appropriate and precisely positioned mask. A conductive paint ensured connection between quartz faces. In the drift gas (75% methane, 25% ethane), the drift velocity was 6.67 cm/μsec. A challenge was the purity of the mixture — it had to exclude any electronegative gas capturing electrons, such

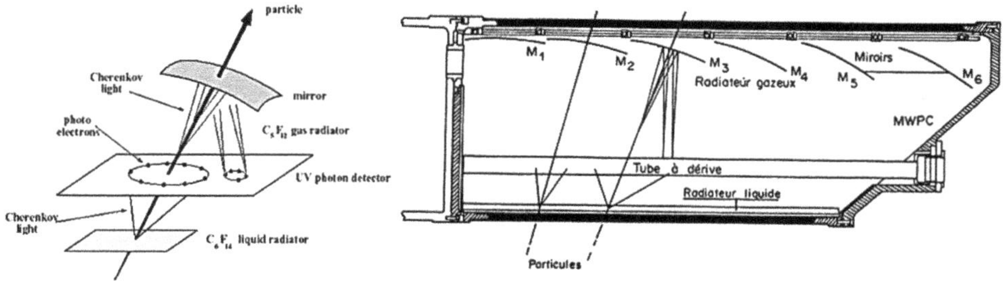

Fig. 7.18. Sketch of RICH. Left: the principle. Right: barrel RICH of DELPHI.

as oxygen, which also destroys TMAE. The liquid radiator, 1 cm of C_6F_{14} (index 1.28) was contained in a box, closed on one side by a quartz window, 12 cm away from the first DT, and on which copper wires with a 6 mm pitch were positioned to improve the homogeneity of the field in the DT. The Cherenkov light propagated in the liquid over about 1cm, coaxial with the particle direction, and with a maximum opening angle of $\approx$ 670 mrad. This cone refracts when entering the gas space and intercepts the DT along a figure depending on the incidence angle.

Measuring the resulting PEs (typically 12) gave the Cherenkov angle in the liquid with a precision of $\sigma \approx 5$ mrad. Due to the chemical activity of freon, with time, the pipes of the filling system lost some tightness and liquid was leaking into the gas. A distillation plant was built to separate the fluids, such that the RICH could operate successfully up to the end of LEP.

After the DT, particles crossed a gas radiator of $\approx$ 40 cm of freon gas C_5F_{12} (index 1.0018 at 1 bar), emitting Cherenkov light at a small angle (maximum opening $\approx$ 62 mrad). The light was reflected back by parabolic mirrors and focused onto the DTs, resulting in an almost circular ring pattern of about 2.4 cm radius and with typically 8 PEs. The required accuracy on the Cherenkov angle was about 5 mrad rms per PE, to which various sources of error contributed roughly equally. Therefore, mirrors with an optical quality in the mrad range were adequate. Optimally, the focal point of the mirrors should be on the DT, with close to normal incidence, which minimizes various errors. Off-axis parabolic mirrors were thus needed. Best achievable reflectivity in the UV was mandatory as was a very smooth reflecting surface to avoid diffusion, with a local deviation from smoothness of rms < 1.6 nm. Ordinary glass pane obtained by the industrial floating technique has an rms of 0.2 nm. Thus the RICH mirrors were made of 6 mm thick float glass, slumped on a parabolic mould machined on a precision lathe. The glass plate, cut to the correct circular shape, was placed in a large oven. The top of the mould was heated at 610°C and the air below it was evacuated, so that the glass would accurately adopt its shape. Once so shaped, the glass plate was cut to the desired dimensions and coated under high vacuum with pure aluminium (80 nm) protected by a layer of MgF_2 (30 nm). The average reflectivity achieved in the UV was 80%. This technique, pioneered by the inventors of the RICH [50] and developed at CERN [54, 55], was transferred to industry, which produced 300 mirrors for the Barrel RICH.

Given their quality, the mould was later used to make 18 mirrors for an astroparticle experiment in the Pyrenees. They focused the Cherenkov light produced by very energetic γ rays showering in the upper atmosphere. These γ rays are emitted by a neutron star, a remnant from a star collapse observed 1000 years ago, located in the CRAB nebula [56].

The success of the RICH (Barrel and EC) during a decade of DELPHI data taking was to allow to identify hadrons over a wide momentum range. Figure 7.19 illustrates the complementarity between the RICH identification (left) and that obtained from the ionisation measurement in the DELPHI TPC at lower momenta (right), and see Box 5.2. The DELPHI RICH initiated the development of similar instruments, e.g. those at LHC [Highlight 8.10].

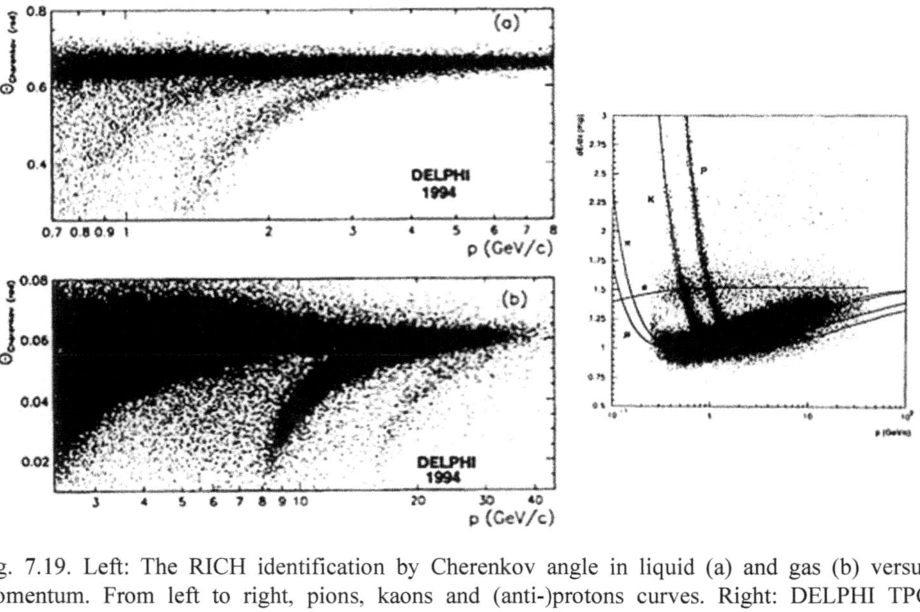

Fig. 7.19. Left: The RICH identification by Cherenkov angle in liquid (a) and gas (b) versus momentum. From left to right, pions, kaons and (anti-)protons curves. Right: DELPHI TPC identification via measurement of *dE*/*dx*.

7.9 BGO for the L3 Experiment: Betting on Precision

Paul Lecoq

At the time the LEP programme was launched the top quark had yet to be discovered, and its mass was expected to be within the energy range of the collider. The electromagnetic calorimeter of the L3 experiment was therefore designed to achieve the best possible energy resolution down to 100 MeV in order to explore the detailed spectroscopy of the toponium system, the putative bound state of the top–antitop quarks. A similar situation had been encountered a few years earlier, after the discovery of the J/ψ, the charm–anticharm bound system, by B. Richter and S. Ting (also spokesman of L3): the high precision study of charmonium states took advantage of the unprecedented performance of the thallium-doped sodium-iodide (NaI(Tl)) Crystal Ball at PEP. This success convinced the community that a homogeneous crystal-based electromagnetic (e-m) calorimeter would be the best way to achieve good energy resolution in the 100 MeV to 10 GeV energy range. Indeed, one hallmark of the L3 concept was the state-of-the-art performance of the calorimeter in terms of energy and angular resolution for photons and electrons over this wide energy range.

The L3 collaboration "put its money" on a recently discovered crystal, bismuth germanate (BGO). This material had been known as a scintillator for less than ten years [57]. In 1983 measurements in a test beam at CERN had demonstrated that BGO, with PIN silicon photodiode readout, had excellent properties as a homogeneous e-m calorimeter [58]. The reason for the choice of BGO was motivated by the much higher stopping power for photons and electrons compared to NaI(Tl). This meant that a compact, highly granular and high performance calorimeter could be built with minimum impact on the dimensions and cost of the external detectors (hadron calorimeter and muon spectrometer). Moreover, besides having good scintillation properties, BGO is not hygroscopic, unlike NaI(Tl).

The calorimeter was installed in a 0.5 T solenoidal magnet of almost 10 metres in diameter, and surrounding a high precision vertex detector [Highlight 7.7]. It consisted of nearly 12 000 BGO crystals with readout provided by photodiodes insensitive to magnetic field. It was composed of a cylindrical barrel with 7680 crystals and two endcaps of 1536 crystals each in a pointing geometry with a small offset of 10 mrad to avoid photon leakage in the gaps between crystals. Each 24 cm long crystal had the form of a truncated pyramid, about 2×2 cm^2 at the inner face and 3×3 cm^2 at the exit.

Transforming the BGO crystal from laboratory curiosity to mass-producible technology meeting the stringent L3 experimental specifications required a new style of R&D. A large-scale multidisciplinary effort had to be organized, a novelty for detector R&D in high energy physics. This involved a number of institutes of the L3 collaboration, but also experts in crystallography, solid state physics and luminescence. A decade later this same collaboration would become the seed of the Crystal Clear collaboration, which made major contributions to the development and construction of the Lead Tungstate based calorimeter of the CMS experiment at the LHC [Highlight 8.8]. As a result of this effort (and thanks to the contacts of the L3 spokesman), technology for the mass production of BGO was developed at the Shanghai Institute of Ceramics (SIC), where 12 000 BGO crystals of excellent and reproducible quality were produced over a period of three years with monthly batches of 130 to 400 crystals.

The 10 t of germanium oxide needed for this production were provided as an in-kind contribution to the experiment from the strategic reserve of the former Soviet Union. The bismuth oxide was provided by China. The very pure bismuth and germanium oxides (impurity level $< 10^{-6}$) were carefully mixed in the correct stoichiometric proportions. The resulting polycrystalline powder of $Bi_4Ge_3O_{12}$ (BGO) was then poured into a platinum foil, folded and welded in the shape of the crystal (trunk of pyramid) and only slightly larger than the final crystal to minimize the losses at the mechanical processing phase. These crucibles were introduced vertically in an oven for growing the crystal, with a small seed crystal in contact with the powder. After heating-up the crucible at a temperature slightly higher than the BGO melting point (1040°C) a temperature gradient was slowly displaced along the crucible from the seed to the end. The long term stability of the electricity supply was crucial, and the interest of the Shanghai authorities in the project certainly helped in ensuring it was stable.

An accurate ($\approx$ 50 μm), fast, cheap and safe method for cutting and polishing the crystals was developed at CERN and transferred to SIC. Diamond disks were used to saw the ingots to the required dimensions. The surface finish was good and reproducible enough to proceed directly to simultaneous mechanical polishing of nine crystals on a spinning table. Automatic mechanical and optical characterization benches were also developed at CERN and installed in Shanghai for systematic evaluation of all crystals prior to shipment. The yield of the crystal growth procedure was over 99%.

Because of the 0.5 T magnetic field the BGO crystals could not be read out by regular photomultipliers. Instead, two 1.5 cm^2 PIN photodiodes with high (70%) quantum efficiency were glued to the rear face of each crystal. The absence of gain and stringent requirement of good resolution at energies down to 100 MeV imposed severe constraints on noise in the readout electronics. A charge sensitive amplifier was mounted directly on the back of the crystals, using a low noise FET in cascade mode, with an output rise time (300 ns) matching the decay time of BGO scintillation.

Material in front of and between crystals was minimized by using carbon fibre honeycomb material, an audacious choice at that time (see Fig. 7.20). Support was provided by thin (200 μm) carbon fibre-epoxy walls between crystals and a cylindrical inner tube coupled at each end of the barrel to a conical funnel, which carried the 10 t weight of the full barrel calorimeter at four bearing pads. The light yield of BGO depends on its temperature (−1.55% per °C), so the mechanical structure also featured a complex cooling and thermal regulation system to remove heat produced by the front end electronics close to the crystals, and stabilize the temperature to within 0.5°C. This unique calorimeter, the success of a very large world-wide collaboration, achieved the original design goals: low energy 100 MeV photons could be measured with 5 MeV precision, high energy photons in the region of 10 GeV to better than 1% [24].

The mass production of BGO became a mature technology so that the excellent properties of this crystal could be exploited in other domains, such as nuclear and space experiments, as well as in medical imaging devices, where BGO became the crystal of choice for the second generation of PET scanners for about two decades.

BGO is now progressively replaced by LSO (lutetium orthosilicate), a crystal discovered in the 1990s, having a higher light yield and a shorter decay time.

Fig. 7.20. The BGO calorimeter barrel during assembly at CERN.

7.10 The Magnetic Cavern of L3

Alain Hervé

The L3 detector was designed to measure with unprecedented accuracy and resolution electrons, photons and muons produced in the particle collisions. These particles passed through a sequence of detector systems where they were tracked and identified according to their characteristic signatures: the vertex-chamber, the electromagnetic calorimeter, the hadronic calorimeter and the muon chambers. This mission called for a highly specific architecture of the detector, in particular the experimental hall, the unconventional magnet, the muon system and the support of the system of inner detectors surrounding the beam pipe at the collision point [24] (see Fig. 7.21).

The original idea was to integrate the magnet yoke into the rock of the cavern, whence the name "magnetic cavern" which would stay with L3. Finally, it was decided for ease of maintenance of the muon chambers and to access equipment with a travelling crane to increase the size of the cavern. The 21.4 m diameter, 26.5 m long experimental hall is oriented longitudinally along the LEP beam line that enters the hall with a slope of 1.39%: the complete detector is inclined to follow this slope. A 23 m diameter and 52 m deep access shaft connects to one end of the experimental hall, for installing experimental equipment. After the installation of the major large components the shaft was used for the installation of experimental services (water, cables, gas, ventilation, etc.), and as technical space for installing power converters, cryogenics, monitoring equipment and offices.

Observing muons with state-of-the-art momentum resolution was one of the L3 priorities: an important ingredient was the largest magnet ever built for a particle physics experiment. The octagonal shaped magnet has a yoke consisting of 6400 t of low carbon steel and an 1100 t aluminum coil. Given the size of the coil, it was built from individual octagonal aluminum sections, each one being built at CERN. These sections were subsequently integrated into the yoke. A current of thirty thousand amperes in the aluminum coil generated a uniform field of 0.5 T. The magnetic circuit of the octagonal yoke had to be closed with massive iron endplates. These endplates, each weighing 700 t, were split in the middle and could be opened like doors to give access to the muon detectors for maintenance. The magnet is lined with an active thermal screen providing a thermally controlled environment to the particle detectors. The yoke rests on a concrete cradle integrated in the floor of the detector hall.

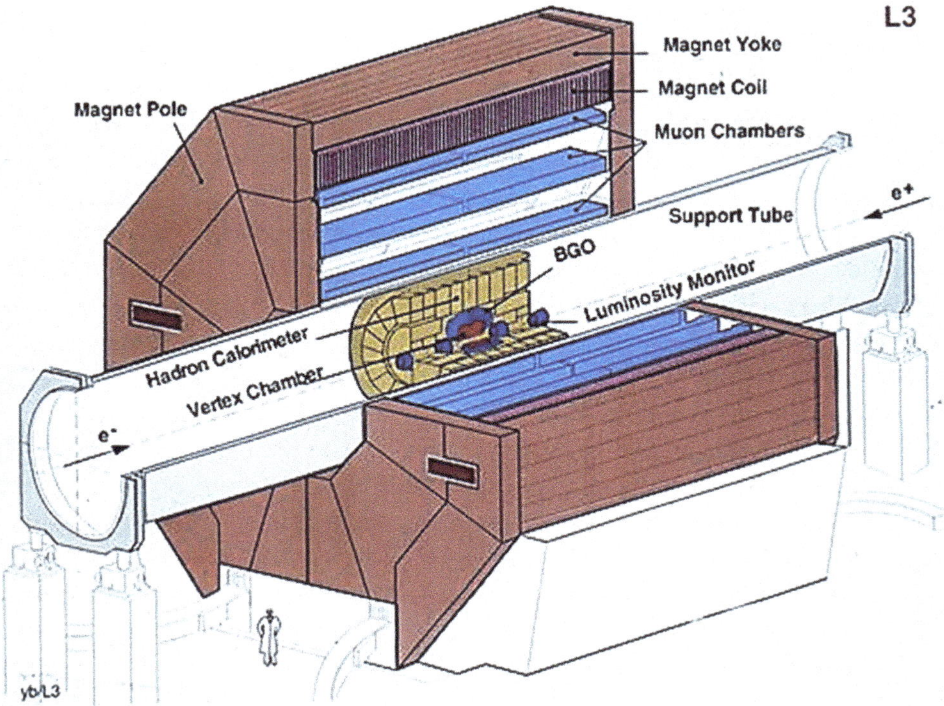

Fig. 7.21. Overview of the L3 detector [24].

Inside the magnet precision Muon Chambers were arranged in three concentric layers, to measure the trajectories of muons in the magnetic field over a volume of 1000 m^3. To achieve the design momentum resolution these chambers were built with mechanical tolerances of 30 μm over an area of 3 m × 6 m. Sixteen octants of muon chambers were supported on two support tubes. Several precision straightness monitors, using LEDs and quadrant photodiodes were specially developed to ensure accurate mutual alignment of the chambers within an octant. At the time this muon system was the most advanced of its kind: its success encouraged the LHC collaboration ATLAS to adopt the concept and take it to the next level of dimension and performance [Highlight 8.12].

The 300 t support tube, 30 m long and 4.45 m in diameter was another rather daring innovation that contributed to the performance of the muon spectrometer. It passed through the magnet coaxial with the beam line. It rested on continuously adjustable servo-controlled jacks placed on concrete pillars. The part of the support tube inside the magnet was of non-magnetic stainless steel, the remaining portion in carbon steel. In this way the support tube rendered all detector systems (600 t in total) mechanically independent of the magnet.

The hadron calorimeter measured the energy of elementary particles, jets of particles, and contributed to muon identification [Box 3.2]. It consisted of 300 t of depleted uranium plates sandwiched between some 8000 proportional chambers.

The electromagnetic calorimeter was another “world première”, designed to measure the energies and positions of electrons and photons from 100 MeV to 100 GeV with high resolution [Highlight 7.9].

Charged particles were tracked in the time expansion chamber which had to be of limited radius to fit inside the BGO calorimeter. It featured an interesting variant of the MWPC [Highlight 4.8] measuring tracks with 40 μm accuracy in a volume of one cubic metre. At the beginning of 1991, the radius of the LEP beam pipe was reduced from 8.0 cm to 5.5 cm. L3 took advantage of this extra space by installing a Silicon Microstrip Detector (SMD) to upgrade its central tracking capability.

The L3 experiment was an important precursor to the LHC experiments. The precision crystal electromagnetic calorimeter was a precursor of the CMS electromagnetic calorimeter [Highlight 8.8]. The alignment of muon chambers using optical systems was further developed for the ATLAS muon spectrometer.

L3 was the only LEP experiment installed stationary at the interaction point (as is the now case for the LHC experiments). The other LEP experiments were assembled adjacent to the accelerator, to be pulled in and out on rails so that LEP could continue to be operated if an experiment required extensive maintenance. New procedures for installation and maintenance had to be developed to take into account the interaction with the machine. To gain time in the installation sequence, heavy parts of the detector, such as the support tube and the hadron calorimeter barrel (weighing each around 300 t) were completed in the surface hall prior to installation in the cavern using rented heavy lifting equipment. Following this experience, the same technique was used extensively for the CMS experiment.

The L3 experimental area and large magnet has been reused to house the ALICE LHC experiment. This was a magnificent legacy, allowing L3 infrastructure to be re-used, in true CERN fashion, for ALICE, a state-of-the-art experiment dedicated to the study of heavy ion collisions at the highest possible energy.

References

1. B. Richter, Very high energy electron-positron colliding beams for the study of the weak Interactions, *Nucl. Instr. & Meth.* **136**, 47 (1976), and SLAC-PUB-1738.
2. C. Pellegrini, C. Rubbia and A. Ruggiero, Comments on the desirability and feasibility study for e^+e^- colliding rings with $L = 10^{32}\ s^{-1}s^{-2}$ qnd E_{cm} =120 GeV, CERN-NP Int. Rep. 75-11 (1975). https://cds.cern.ch/record/1161844?ln=en.
3. LEP Summer Study, Les Houches, France and CERN, Sept. 1978, CERN-79-01-V-2, (1979). https://cds.cern.ch/record/133077?ln=en.

4. A. Zichichi (Ed.), ECFA-LEP Working Group report ("The white book") ECFA 79/79 (1980), reproduced in https://cds.cern.ch/record/816523?ln=en.
5. M. Bourquin (Ed.), General meeting on LEP, Villars, ECFA-91-054 (1981). https://cds.cern.ch/record/98884?ln=en.
   T. Ekelof and G. Flugge (Eds.), International conference on experimentation at LEP, Uppsala, *Physica Scripta* **23** 4 1 & 2 (1981).
6. J. Adams, CERN Annual Report 1980, p. 25 (1981). https://cds.cern.ch/record/1475744?ln=en.
7. J. Ellis and R. Peccei (eds.), *LEP Physics Jamboree - LEP Exp. Committee*, CERN-1986-002 (CERN, Geneva, 1986). http://dx.doi.org/10.5170/CERN-1986-002.
8. K. Johnsen *et al.* (eds.), *Design Concept for a 100 GeV e^+-e^- Storage Ring (LEP)*, CERN-1977-014 (CERN, Geneva, 1977). http://dx.doi.org/10.5170/CERN-1977-014;
   LEP Study Group, *Design Study of a 15 – 100 GeV e^+-e^- Colliding Beam Machine (LEP)*, "Blue Book", ISR-LEP 78/17 (CERN, Geneva, 1978). http://cds.cern.ch/record/101331;
   LEP Study Group, *Design Study of a 22 – 130 GeV e^+-e^- Colliding Beam Machine (LEP)*, "Pink Book", ISR-LEP 79/33 (CERN, Geneva, 1979). http://cds.cern.ch/record/101333.
9. CERN, LEP phase 1 and cost breakdown, CERN/SPC/0472, (1981), https://cds.cern.ch/record/61975?ln=en.
10. CERN, Etude d'impact du projet LEP sur l'environnement, (1982). https://cds.cern.ch/record/99730?ln=en.
11. LEP design report ("The green book"), Vol. II, The LEP main ring, CERN-LEP/84-01 (1984). http://cds.cern.ch/record/102083?ln=en.
12. H. Laporte, La construction du LEP, *Revue Travaux (Science et Industrie)*, Paris (1988).
13. G. Bachy *et al.*, The LEP collider: construction, project status and outlook, *Particle Accelerators,* **26**, pp. 19-32 (1990).
14. H. Schopper, *LEP, Lord of the Collider Rings at CERN* (Springer, Berlin, 2009).
15. E. Picasso, LEP construction, (2009). https//cds.cern.ch/record/1228538?In=Fr.
    E. Picasso, A few memories from the days at LEP, *Eur.Phys.J.* **H36** pp. 551-562 (2012) DOI: 10.1140/epjh/e2011-20050-0.
16. K. Hubner, Designing and building LEP, *Physics Reports* **403-404**, 177-188 (2004);
    S. Myers (ed.), *The LEP Collider: from Design to Approval and Commissioning*, CERN-1991-008 (CERN, Geneva, 1991). http://dx.doi.org/10.5170/CERN-1991-008.
17. C. Wyss (Ed.) LEP design report, Vol. III, LEP2, CERN-AC-96-01-LEP-2 (1996). http://cds.cern.ch/record/314187?ln=en.
18. LEP design report, Vol. I, The LEP injector chain, CERN-LEP-TH-83-29 (1983). http://cds.cern.ch/record/98881?ln=en.
19. T. Taylor, Technological aspects of the LEP low-beta insertions, *IEEE* **NS-32** (1985).
20. P. Bernard, Superconducting RF cavities, *Lectures in honour of E. Picasso, Pisa* (1999), CERN-OPEN-2010-001. https://cds.cern.ch/record/1231370?ln=de.
21. ALEPH, *Nucl. Instr. & Meth. A*, **294**, 121-178 (1990), *Nucl. Instr. & Meth. A*, **303**, 393 (1991), CERN-EP-90-2
22. DELPHI, *Nucl. Instr. & Meth. A*, **303**, 233-276 (1991), CERN-PPE-90-128, CERN- EF-90-5.
23. OPAL, *Nucl. Instr. & Meth. A*, **305**, 275-319 (1991), CERN-PPE-90-114.
24. B. Adeva *et al.*, The construction of the L3 experiment, *Nucl. Instr. & Meth.* A **289**, pp. 35–102 (1990).
25. ADLO Collaborations, Lower bound for the Standard Model Higgs boson mass from combining the results of the four LEP experiments, CERN-EP/98-046 (1998).

26. ADLO Collaborations *et al.*, Precision ElectroWeak measurements on the Z resonance, *Physics Reports*, **427**, 257-454 (2006).
27. D. Treille, LEP/SLC: What did we expect? What did we achieve? A very quick historical review, *Nucl. Phys. Proc. Suppl.* **109 B** 1-16 (2002).
28. G. Altarelli (ed.), *Workshop on Physics at LEP2*, CERN-1996-001-V-1 (CERN, Geneva, 1996). http://dx.doi.org/10.5170/CERN-1996-001-V-1.
29. J.P. Gourber and C. Wyss, Prototypes of LEP bending magnets with steel-concrete cores, *IEEE Trans. Nucl. Sci.* **NS-28**, 3, 2867 (1981).
30. L. Resegotti, The LEP magnet system, *(MT-8), J. Phys. Colloques*, **45**, C1-233-C1-239 (1984).
31. J. Billan, J.P. Gourber and K.N. Henrichsen, Geometry and remanent field measurement of the LEP dipole magnets, (MT-8), *J. Phys. Colloques*, **45**, C1-953-C1-956 (1984).
32. LEP-Vacuum Group, LEP vacuum system: present status, *Proc. 11th Int. Vacuum Congress,* Köln, Germany (1989).
33. C. Benvenuti and F. Francia, Room temperature pumping characteristics of a Zr-Al Non-Evaporable Getter (NEG), *J. Vac. Sci. Technol.* **A6**, 2528 (1988).
34. C. Benvenuti and F. Francia, Room temperature pumping characteristics for gas mixtures of a Zr-Al nonevaporable getter, *J. Vac. Sci. Technol.* **A6**, 3864 (1990).
35. C. Benvenuti, N. Circelli and M. Hauer, Niobium films for superconducting accelerating cavities, *Appl. Phys. Lett*, **45** (5) p. 583 (1984).
36. C. Benvenuti *et al.*, Preparation of niobium coated superconducting RF cavities for the Large Electron Positron Collider, *Proc. III workshop on RF Supercond.*, **ANL-PHY-88**, 445 (1988).
37. A. Butterworth *et al.*, The LEP superconducting RF system, *Nucl. Instr. & Meth. A*, **587**, 151–177 (2008), and E. Jensen, RF cavity design, CAS 2013, Trondheim, https://cds.cern.ch/record/1982429/files/arXiv:1601.05230.pdf
38. The LNN-INFN international workshop on thin films, LNL Laboratories, Padova (2006).
39. S. Turner, (ed.), *Lecture Notes in Earth Sciences n°12,* pp. 207-394 (Springer, Berlin, 1986).
40. G. Beutler and W. Gurtner, Test of GPS on the CERN-LEP Control Network, *Schriftenreihe der Universität der Bundeswehr München, Heft 20-2*, pp. 337–385, (1985).
41. J. Gervaise and M. Mayoud, in *Advances of Accelerator Physics and Technologies,* (World Scientific, Singapore, 1993), pp. 546–584.
42. M. Mayoud, in *Frontiers of Accelerator Technology,* (World Scientific, 1996), pp. 717–742.
43. L. Arnaudon *et al.*, Effects of terrestrial tides on the LEP beam energy, *Nucl. Instr. & Meth.* A**,** **357**, 249, (1995).
44. B. Mours *et al.,* The Design, construction and performance of the ALEPH silicon vertex detector, *Nucl. Instr. & Meth.* A **379**, pp.101–115 (1996).
45. V. Chabaud *et al.,* The DELPHI silicon strip microvertex detector with double sided readout, *Nucl. Instr. & Meth.* A **368**, pp. 314–332 (1996).
46. K.H. Becks *et al.*, The DELPHI pixels, *Nucl. Instr. & Meth.* A **386**, pp. 11–17 (1997).
47. M. Acciarri *et al.* (for the L3 SMD Collaboration), The L3 silicon microvertex detector, *Nucl. Instr. & Meth.* A **360**, 103–109 (1995).
48. S. Anderson *et al.* (for the OPAL Collaboration), The extended OPAL silicon strip microvertex detector, *Nucl. Instr. & Meth.* A **403**, pp. 326–350 (1998).
49. K. Abe *et al.*, Design and performance of the SLD vertex detector, a 307 Mpixel tracking system *Nucl. Instr. & Meth.* A 400, pp. 287–343 (1997).
50. J. Seguinot and T. Ypsilantis, Photoionization and Cherenkov ring imaging, *Nucl. Instr. & Meth.* **142**, p. 377 (1977).

51. T. Ekelof, J. Seguinot, J. Tocqueville and T. Ypsilantis, The Cherenkov Ring-Imaging Detector: Recent Progress analysis and Future Development, *Physica Scripta* **23**, 718 (1981).
52. W. Adam *et al.*, The Ring imaging Cherenkov detector of DELPHI, *Nucl. Instr. & Meth.* A **343**, pp. 68–73 (1994).
53. W. Adam *et al.*, Performance of the forward RICH detector system at DELPHI, *Nuclear Science* **4**, No. 3, 583 (1993).
54. P. Baillon *et al.*, An improved method for manufacturing accurate and cheap glass parabolic mirrors, *Nucl. Instr. & Meth.* A **276**, 492 (1989).
55. P. Baillon *et al.*, Production of 300 paraboloidal mirrors with high reflectivity for use in the barrel RICH counter in DELPHI at LEP, *Nucl. Instr. & Meth.* A **277**, 338 (1989).
56. P. Baillon *et al.*, Gamma ray spectrum of the Crab nebula in the multi TeV region, *Astroparticle Physics* **1**, p. 341 (1993).
57. M.J. Weber and R.R. Monchamp, Luminescence of $Bi_4\,Ge_3\,O_{12}$: spectral and decay properties, *J. Appl. Phys.* **44**, p 5495 (1973).
58. E. Lorenz, Status of BGO development and perspectives of BGO calorimeters in high energy physics, *Nucl. Instr. & Meth.* A **225**, p. 500 (1984).

Chapter 8

The Large Hadron Collider (LHC): The Energy Frontier

Giorgio Brianti and Peter Jenni

8.1 Introduction

The accelerator

The CERN Large Hadron Collider (LHC) complex, which includes the 27 km long superconducting collider itself and several huge particle detectors, is the largest and most sophisticated research instrument ever built. We describe here the main features of this marvel of technology [1] and the highlights of the developments which paved its way, from conception in the early 1980s to the first years of exploitation from 2009 to 2013. This period was marked by the discovery of the Higgs boson, announced in July 2012.

In the second half of the 1970s the European Committee for Future Accelerators (ECFA) chose the Large Electron Positron Collider (LEP) as CERN's flagship machine for a thorough experimental study of the Standard Model. The planned beam energy was about 100 GeV, the tunnel circumference about 27 km. Already at that time the sentiment prevailed that such a construction represented a major and perhaps unique undertaking, also in view of a possible subsequent use of the tunnel for a proton machine. While the construction of the LEP tunnel was underway, in the USA the design and construction of the Superconducting Super Collider (SSC) for proton–proton collisions with a beam energy of 20 TeV was proceeding at high speed. The question arose as to what CERN should do for a proton/hadron machine. CERN had not only a tunnel (in construction) but also an existing well-proven chain of injectors (linear accelerator, booster, PS and SPS), which, after some modifications, could be used for injecting protons and ions into a collider in the LEP tunnel [2]. In this context, we started the feasibility study of a proton-proton collider in the LEP tunnel at the end of 1982. Though modest, it prompted the interest of the user community and led to the Lausanne ECFA-CERN workshop in 1984 devoted to the Large Hadron

Collider in the LEP tunnel [3]. Three years later the Workshop on physics at future accelerators in La Thuile, Italy [4] and the report to the CERN Council of the long-range planning committee [5] chaired by Carlo Rubbia, 1984 Nobel Prize laureate and strong supporter of hadron colliders, put the LHC squarely on the map as the next CERN project after LEP.

The project received the first approval by the CERN Council in December 1994 and final approval for the present configuration in 1996 [6]. Lyndon Evans was appointed project leader and led the construction to completion. Commissioning started in 2008, but was soon interrupted by a major technical incident. After the repair, the first collisions were observed in November 2009, and a successful three-year run started in 2010. Globally, it has represented a considerable advance in the technology of large superconducting accelerators.

In a proton–proton collider the two counter-rotating beams travel in separate channels in the arcs around the ring and consist of many bunches of particles separated by a short distance; bunches are brought into collision at the centre of the experimental facilities with a small crossing angle. For the LHC the maximum number of bunches is 2808, the separation 7.5 m and the crossing angle 285 μrad. The layout of the collider is shown in Fig. 8.1. There are eight arcs, where the beams are bent (the bending radius is 2804 m), and eight straight-sections used for utilities or collision insertions; four insertions are equipped with detectors.

The two most important parameters of a proton-proton collider are the energy of the collisions and the luminosity. The energy is related to the discovery potential of new particles of higher mass and the luminosity to the production of a relevant number of the desired events in a reasonable running time. The luminosity is a measure of the quality of the beams and is proportional to the rate of collisions [Box 6.1].

The same basic collider structure can be used as a proton–proton collider with centre-of-mass energy 14 TeV and luminosity of 10^{34} cm^{-2} s^{-1} or as a lead ion-lead ion collider with a centre-of-mass energy of 1148 TeV (2.76 TeV per nucleon) and luminosity of 10^{27} cm^{-2} s^{-1}. Protons can also be made to collide with Pb ions.

The ***beam energy*** is proportional to the product of B (the magnetic field of the main dipole magnets) and ρ (the bending radius of the arcs). As ρ was determined by the dimensions of the pre-existing LEP tunnel, it was mandatory to aim for the highest possible field: the magnets had to be superconducting [Boxes 4.3 and 8.1]. Prior to the LHC, three large accelerators were based on such magnets: the Tevatron (FNAL), HERA (DESY) and RHIC (BNL). All of these made use of well-proven Nb-Ti cable cooled with helium to a temperature slightly above 4.2 K (He I). In these machines the magnetic field was ≤ 5 T [6]. The question was how to obtain a higher field with commercially available material and technology.

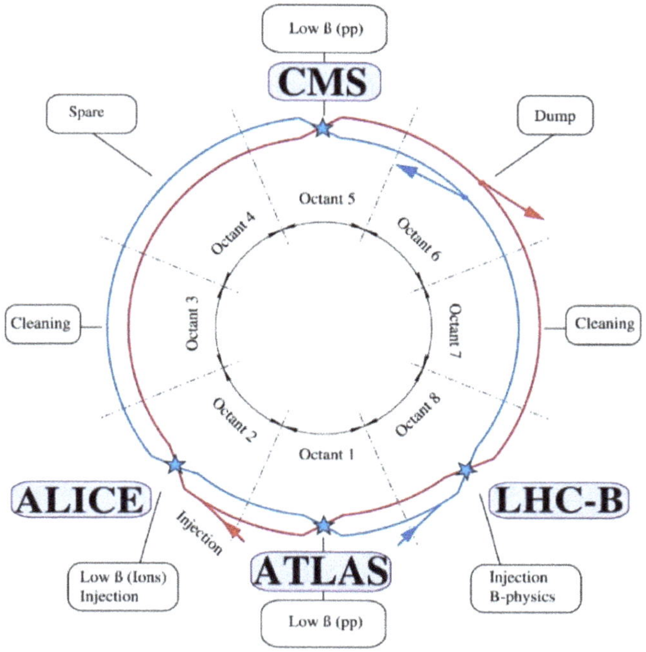

Fig. 8.1. Layout of LHC Collider.

As explained below, the choice was to use Nb-Ti superconductors cooled to a lower temperature, namely 1.9 K provided by superfluid helium (He II). At the required current density the gain due to the lower temperature is about 1.5 T, or an additional 1.3 TeV in energy — over and above the 6.8 T being proposed for the SSC at the time (itself based on intervening improvements of Nb-Ti superconducting material). Before the LHC, the technology of Nb-Ti cooled to 1.9 K by superfluid helium had already been used for a large magnet system (though of a different and more compact geometry), the Tokamak TORE SUPRA, at CEA-Cadarache (France). It should be stressed that the solution adopted for the LHC reconciled two requirements, the quest for the highest possible magnetic field and a design such that the magnets could be constructed in existing industry. With a dipole field of 8.3 T this design provides a nominal beam energy of 7 TeV. Of the almost ten thousand magnets required for the collider, the 1232 main dipoles (length 16.5 m, weight 34 t, max. field 8.3 T) represented a particularly challenging effort of design and construction [Highlight 8.2].

Operating Nb-Ti conductors at 1.9 K offers another important advantage at the expense of a more complex cryogenic system. Due to the particular properties of superfluid helium such as high heat capacity, low viscosity, and exceptionally good thermal conductivity, the long string of magnets can be cooled via a static bath of helium by heat exchange with a very small flow of liquid. This avoids depending on pumps — components of limited reliability and efficiency — for producing a large continuous flow of cold helium. The total cold mass operating at 1.9 K is 37500 t. This temperature is lower than the background temperature of the Universe [7] [Highlight 8.3].

The considerable kinetic energy of each beam (362 MJ) and the large electromagnetic energy stored in the magnets (11 GJ for the complete system) require a sophisticated protection system to prevent damage in case of beam guiding problems or resistive transition (magnet "quench"). Each of the two beams has sufficient energy to melt 500 kg of copper, and the electromagnetic energy of the magnets, if not properly discharged, can provoke substantial damage as in the case of the 2008 incident.

The nominal excitation current of the strings of dipoles and quadrupoles in the arcs is 12 kA. To optimize system cost the individually powered quadrupoles of the matching sections were designed to operate at a more modest 6 kA. Nevertheless, the total current to be transferred from room temperature down to the helium environment of the superconducting magnets was still about 3 MA. To reduce cryogenic load it was therefore decided to use High Temperature Superconducting (HTS) current leads. Since over a thousand such leads of three current ratings were required, a design suitable for mass production was developed at CERN. It has been the first large scale application of HTS [Highlight 8.4].

Another important aspect is the beam vacuum, for which requirements are imposed by beam lifetime and background in the experiments. The interactions between the protons and rest gas are driven by two processes, i.e. single proton-nuclear collisions in which a proton is lost, and multiple small-angle Coulomb scattering, which provokes an increase of the size of the beam and a corresponding decrease of the luminosity. To ensure a beam lifetime of about a day, the residual pressure in the vacuum chamber should not exceed 10^{-8}–10^{-9} Pa. The vacuum chamber in the LHC consists basically of two types: (i) chambers made from stainless steel at 1.9 K in the arcs; (ii) chambers made from copper operating at ambient temperature in the total of 6 km of straight sections.

The vacuum system has to address a dynamic effect created by electrons, the electron cloud phenomenon. These electrons, produced by synchrotron radiation or by the ionization of the rest gas, can be accelerated to the walls of the vacuum chamber, producing secondary electrons, which in turn can be accelerated by the subsequent bunches, impinging on the chamber walls and releasing further electrons with the risk of a run-away situation. This can give rise to beam blow-up and instability, and deposit heat in the cryogenic system, so it has to be controlled. This is done by lowering the secondary emission yield of the chamber walls. Different strategies are applied in the cold and warm sections.

In the cold sections the vacuum chamber is at the same temperature as the superconducting dipoles enclosing the chamber ("cold bore design"). For the pumping, one takes advantage of the fact that the vapour pressures of all residual gases are lower than 10^{-10} Pa at 1.9 K, a feature already explored in the ISR. Such pressures are sufficiently low for machine operation. The electron cloud is fought by so-called beam scrubbing, i.e. using gradually increasing beam currents until the walls have reached the desired degree of cleanliness. However, the circulating beam would deposit up to 1.7 W/m on the vacuum chamber (due to synchrotron radiation, image current, residual electron cloud effects, etc.) which should not be absorbed at 1.9 K. Therefore, the magnet cold bore is shielded with an inserted beam screen kept at 5 to 20 K by separate circulation of helium. The beam screen has holes (about 4% of the screen surface) to allow the gas molecules to reach the surrounding cold bore.

For the warm sectors, the successful performance of Non Evaporable Getter (NEG) pumping for LEP was a strong motivation to envisage NEG pumping also for these LHC sectors. As the LHC chambers are too small to host the LEP getter pump, a way was found to coat the chamber walls with a film of NEG to provide pumping and control of the electron cloud effect [Highlight 8.5].

A view of a section of the installed LHC machine is shown in Fig. 8.2.

As stated previously, besides the energy of the circulating particles, the luminosity L is an important parameter of the LHC, but it is the luminosity integrated over a given period, say one year, that is crucial for data-taking [Box 6.1]. The luminosity varies with time, since the beams degrade during the collision run. The integrated luminosity is simply the integral $\int L(t)dt$. These parameters are the key figures of merit of a collider.

The LHC was commissioned in 2010 with 3.5 TeV protons; operating at 4 TeV it has reached a luminosity of 0.8×10^{34} cm^{-2} s^{-1} and provided up to 23 fb^{-1} per experiment in 2012. Since mid-2015 the LHC is operating with a beam energy of 6.5 TeV and has exceeded the nominal peak luminosity of 1×10^{34} cm^{-2} s^{-1}. The luminosity will be gradually doubled in the coming years promising an integrated luminosity of 300 fb^{-1} before the accelerator is shut-down for a substantial upgrade turning it into the High Luminosity LHC (HL-LHC) by around 2025 (Chapter 12). The expected peak luminosities at 7 TeV with the nominal 25 ns between bunches

Fig. 8.2. Partial view of the LHC, showing a string of dipoles and an interconnection.

leads to 30 to 60 inelastic interactions per bunch crossing, which determines approximately the number of events the detector has to be able to resolve. For very sensitive experiments this number is decreased by reducing the luminosity in their interaction region. A small vertical separation is introduced which changes with time to "level" the luminosity, hence optimizing the integrated luminosity.

The number of circulating proton bunches is very large (2808) with a bunch spacing of 7.5 m, corresponding to 25 ns. This means that the beams have to be brought into collision at a crossing angle of 285 μrad. This angle, while introducing a moderate reduction of the luminosity and a negligible loss in centre-of-mass energy, eliminates the unwanted bunch collisions near the collision point.

Lead-lead or lead-proton collisions have also been part of the programme since 2010. This mode of operation is regularly scheduled, though only for a fraction of the time foreseen for proton-proton operation. In 2015, the design luminosity of 1×10^{27} cm^{-2} s^{-1} and an integrated luminosity of 30 pb^{-1} was reached with 5.0 TeV centre-of-mass energy; these numbers refer to lead nucleon-pair collisions.

The joy of witnessing the first proton collisions following the repair of the accelerator, and the anticipation after so many years of work on both the collider and detectors, is recorded on the faces of scientists in the CERN Control Centre (CCC) in March 2010 [Fig. 8.3].

Fig. 8.3. The CERN Control Centre (CCC) witnesses the first 7 TeV collisions in March 2010.

The physics programme and the detectors

Enthusiasm for the extraordinary physics potential of a LHC arose in the community at the Lausanne ECFA-CERN workshop in 1984 [3]. Studying the Brout–Englert–Higgs (BEH) mechanism [Box 8.2] and searching for the associated so-called "Higgs boson" was clearly a priority for the future of particle physics, but this was expected to be very challenging. The boson mass is not predicted by the Standard Model and could have been as high as 1 TeV. This required exploring a broad range of mass — well-suited for a high energy p–p collider where the energy spectrum of the constituents of the protons (quarks and gluons) allow all possible Higgs masses to be searched for simultaneously. As detectable cross-sections were predicted to be very low, a very high luminosity collider was mandatory.

The Higgs search was by far not the only reason for the great interest generated by the LHC. Already in the 1980s there was clear motivation from speculative theories predicting physics Beyond the Standard Model (BSM), the most popular being Supersymmetry (SUSY) with its characteristic missing transverse energy signatures (E_T^{miss}) due to the escaping lightest neutral SUSY particle (LSP) [Box 7.2]. Other hypothetical models predicted new heavy resonances, or leptoquarks (particles with both lepton and baryon numbers), or even substructure to quarks,

A theory of superconductivity **Box 8.1**

In 1957 John Bardeen, Leon Cooper and Robert Schrieffer (Nobel prize 1972) published the first microscopic theory of superconductivity (BCS Theory [1]) based on quantum mechanics treatment of the behaviour of conduction electrons in a metal in relation with lattice vibrations (phonons). Originally, superconductivity (SC) was explained by macroscopic descriptions such as those of London [2] and Ginzburg and Landau [3]. In the BCS theory, below a critical temperature T_c pairs of electrons (of opposite spin) form so-called "Cooper pairs". Pairing is mediated by phonons in the ion lattice, to assume an energy state which is lower than that of the sum of the single electrons, and therefore favoured. As the pairs have zero net spin, other such pairs, unlike single electrons, can accumulate in the same state to form a condensate. Below T_c the energy of vibrations of the metal lattice is insufficient to change the state of the condensate which can then carry the current freely through lattice without expending energy — in contrast to normal unpaired electrons that experience resistance via encounters with the phonons in the lattice. The process of formation and dissolution of Cooper pairs is continuous, the net result being resistance-free transport of current (DC) through the metal (up to the critical magnetic field $\mu_0 H_c$). For RF SC see Box 7.4.

Superconductors come in two types, *Type I* (most pure metals), which are perfectly diamagnetic, and exclude magnetic field up to the critical field H_c, and *Type II* (alloys, compounds, ceramics and some metals), which feature two critical fields H_{c1} and H_{c2}; up to H_{c1} they behave like Type I materials, but between H_{c1} and H_{c2} magnetic flux can penetrate locally in the form of *vortices*. While $\mu_0 H_{c1}$ is typically a fraction of a tesla, $\mu_0 H_{c2}$ ranges from about 15 T for Nb-Ti to about 24 T for Nb_3Sn, and up to 100 T for some ceramic materials. Only Type II material is useful for large-scale applications.

Whereas for Type I superconductivity the material should be pure, Type II materials must feature discontinuities (e.g. grain boundaries, dislocations and point defects) to encourage the *pinning* of magnetic vortices. In order to increase the critical current I_c, the manufacturing process therefore includes steps to decrease grain size (e.g. by cold work). Such pinning centres may also be introduced by doping with additives.

The BCS theory was first developed to explain Type I superconductivity. With a few tweaks the model could be applied to the Type II materials of the time, up to $T_c \approx 24$ K, beyond which temperature superconductivity was reckoned to be impossible. So there was great surprise when high temperature superconductors were discovered in 1986. BCS theory is both elegant and attractive, but despite many efforts it has not yet been possible to adapt it convincingly to explain the behaviour of these new materials, and there may be a place for a different theory that covers the phenomenon in a more satisfactory way [4]. However, the BCS theory of superconductivity, derived from the reasoning of solid state physics, also inspired particle physics theorists to postulate the BEH field and its associated boson [Box 8.2], and therein also resides major worth.

[1] J. Bardeen, L. N. Cooper and J. R. Schrieffer, *Phys. Rev.* **108**, 1175-1204 (1957).
[2] F. London and H. London, *Proc. Royal Soc. A: Math., Phys. & Eng, Sc.* **149**, 856 (1935).
[3] V. L. Ginzburg and L. D. Landau, *Zh. Eksperim. I Teor.* **20,** 1064 (1950).
[4] J. E. Hirsch, *Physica Scripta* **80** 3, 1-11 (2009).

The Brout–Englert–Higgs (BEH) mechanism **Box 8.2**

At the heart of the SM [Box 6.4] is the freedom to perform at any point of space-time an arbitrary rotation in internal spaces, called a "gauge" transformation (GT), leaving the physics invariant. Such an operation implies replacing one particle by another, e.g. e^- by its neutrino, or a strange quark by a charm quark. Intuitively, the invariance can be expected to work better if one swaps similar particles, e.g. of zero mass. Indeed, the first version of the SM worked only for massless particles, not describing our world: this simple form of Gauge Invariance (GI) does not provide a good description.

Can one reconcile the concept of GI with a quantum field theory describing Nature?

QED with its massless photon as force carrier shows the property of GI and provided a very helpful hint as did the theory of superconductivity, which is based on electro-magnetism. Inside a superconductor (SC) there exists a field due to the Cooper pairs [Box 8.1], made of two electrons interacting through the lattice. This field locks the phase of an electron over a macroscopic distance and thus effectively prevents the possibility to perform an arbitrary local GT. The BEH mechanism is similar, positing the existence of a new field permeating all of our universe and acting in the weak isospin internal space to impose there an orientation, preventing arbitrary rotations in that space. The analogy between the BEH mechanism and SC can be pushed quite far [1]. In the BCS theory of superconductivity the photon becomes massive by interacting with the field of Cooper pairs: this is the Meissner effect, i.e. the inability of a magnetic field to penetrate the SC beyond the penetration depth d, as if its quantum, the photon, had become massive in proportion to $1/d$. Similarly, the W and Z bosons acquire mass from the BEH field, whose non-zero value ("vacuum expectation value" VEV) plays the role of the Cooper pair field, described in a similar mathematical framework. The BEH field gives also the mass to the fermions, which depends on the strength of their interaction with the BEH field. However, we lack an understanding of these couplings, and the fundamental origin of the mass values remains a mystery.

It is now believed that some 10^{-11} seconds after the Big Bang the BEH field underwent a phase transition (Fig. 1.4), which changed its VEV from zero to a finite value, responsible for giving mass to all elementary particles from this moment onwards. Implementing the BEH mechanism into the early SM version was the decisive step in turning the SM into a description of our world.

The Higgs boson is the particle manifestation of an excitation of this BEH field. Its mass is linked to its property of self-interaction (SI). Not knowing the strength of SI, the SM only provides general guidelines, but cannot predict its mass (see however Box 7.2). It took the energy and luminosity of the LHC collisions to produce and observe this boson with a mass of ~130 protons at CERN in 2012 (see 8.1). Following this discovery F. Englert and P. Higgs were awarded the physics Nobel Prize in 2013 [2].

The importance of the discovery of the Higgs field goes beyond the SM. Unlike the gravitational or e-m field, it is a novel "scalar" field, not characterized by a direction in our space. Its detailed study may contribute to our understanding beyond the SM.

[1] http://www.slac.stanford.edu/pubs/beamline/26/1/26-1-dixon.pdf (1996).
[2] http://www.nobelprize.org/nobel_prizes/physics/laureates/2013/.

not to mention many other exotic ideas. It was also realized early on that the LHC would produce huge numbers of heavy flavour particles, opening a new frontier in precision flavour physics. Unprecedented exploratory steps could be made by studying the quark-gluon plasma at high density and temperatures when colliding heavy ions, such as fully ionized lead nuclei. Motivated by these physics prospects, CERN opted ultimately for an experimental programme with two very large general-purpose detectors (ATLAS and CMS), two large detectors optimized respectively for flavour physics (LHCb) and for heavy ion collisions (ALICE), complemented later by three much smaller specialized experiments (TOTEM, total cross-section and forward physics; LHCf, measuring forward neutral particle production; MoEDAL, monopole search).

The detection of the Higgs boson played a particularly important role in the design of the *general-purpose experiments* ATLAS and CMS. In the region of low mass ($114 < m_H < 150$ GeV), the two channels considered most suited for unambiguous discovery were the decay to two photons and the decay to two Z bosons, each decaying in turn into e^+e^- or $\mu^+ \mu^-$, where one or both of the Z bosons could be virtual. As the natural width of the low-mass Higgs boson is less than 10 MeV, the width of any observed peak would be entirely dominated by the instrumental mass resolution. This meant that in designing the general-purpose detectors, considerable importance was placed on the magnetic field strength, on the precision tracking systems and on high-resolution electromagnetic (e-m) calorimeters. The high-mass region, as well as the signatures from supersymmetry, required excellent resolution for jets and missing transverse energy (E_T^{miss}), implying also almost complete 4π hadronic calorimetry coverage.

The choice of the magnetic field configuration determined the overall design for these experiments. It was also well understood that optimizing the chance of making discoveries at the new "magic" energy scale of the LHC - and in the harsh conditions generated by about a billion protons interactions every second, and several tens of events every bunch crossing — would require the invention of new technologies while at the same time pushing existing ones to their limits [8]. In fact, a prevalent saying was "We think we know how to build a high energy, high luminosity hadron collider — but we don't have the technology to build a detector for it". In reality of course both turned out to be difficult and demanded technological breakthroughs. That the general-purpose experiments have worked so well since the start-up of the LHC is a testimony to the enthusiasm and competence of many thousands of engineers and physicists, and to the difficult yet correct technological choices and decisions made during the construction of these experiments. Indeed, the very same elements mentioned above were crucial in the recent discovery of a Higgs boson.

Experimentation at the LHC **Box 8.3**

Protons have radius $R = 0.88$ fermi (1 fermi = 10^{-13} cm) and interact strongly with each other when $\lesssim 2R$ apart, corresponding to an area of 100×10^{-27} cm^2, or a cross-section of 100 millibarn (1 barn (b) = 10^{-24} cm^2). Total pp cross section is $\sim$ 70 mb. This fact, and the design luminosity of 10^{34} cm^{-2}s^{-1}, impose severe constraints on the detectors:

– Every 25 ns bunches of protons collide, producing about 1000 particles. These must be measured individually, so the detectors must be finely subdivided;

– 31×10^6 bunch crossings/s (2808 bunches x 11.2 kHz) must be examined. Within a few μs, the electronics select about 10^5 events based on their potential interest for physics (first level trigger). In a second computer-based selection about 200 events are kept. With $\sim$ 1 MB/event, a volume of 100 GB/s is analysed and 200 MB/s archived;

– Per year of data taking, typically 10^7 s, 2×10^9 events, corresponding to 2 petabytes (PB), are recorded and analysed. Such analysis takes about 100 s per event for a PC, implying 20000 equivalent PC units operating continuously. For the 4 experiments, data (re)processing and Monte Carlo data generates 30 PB/year;

– High radiation levels are a strong concern: up to 2×10^7 tracks per cm^2/s traverse the detectors. Each deposits ~2 MeV/ g cm^2 [Box 5.2]. This leads to an irradiation of about 100 kGy/year. Detectors and electronics are designed to work for ≥ 10 years.

Trackers must have low mass to avoid perturbing particles — another major constraint.

At the LHC, the focus is on energetic collisions of the point-like (size < 10^{-4} fermi) proton constituents, the partons, gluons and quarks, for which the associated wavelength is about 10^{-3} fermi at the 100 GeV energy scale. So cross sections 10^{-6} times smaller than the total cross section must be considered (see figure, to be compared with Fig. 12.12 for an e^+e^- collider). The process rate also depends on the properties of the partons within the proton and how they interact (strongly or weakly). To produce heavy or energetic particles, the colliding partons must carry a large fraction x of the proton momentum — less probable to occur for larger x. Producing the Higgs boson [Box 8.2] proceeds through a loop of heavy virtual particles [Box 5.1] and is very infrequent. To beat background, clean rare decay modes are used. To observe one Higgs event in the 4-lepton channel means registering $\sim 10^{13}$ pp interactions.

To address such challenges, in 1987 CERN launched a world-wide R&D programme and set up an external, independent LHC Committee with the mandate to scrutinize, monitor and evaluate progress (Chapters 8 and 11).

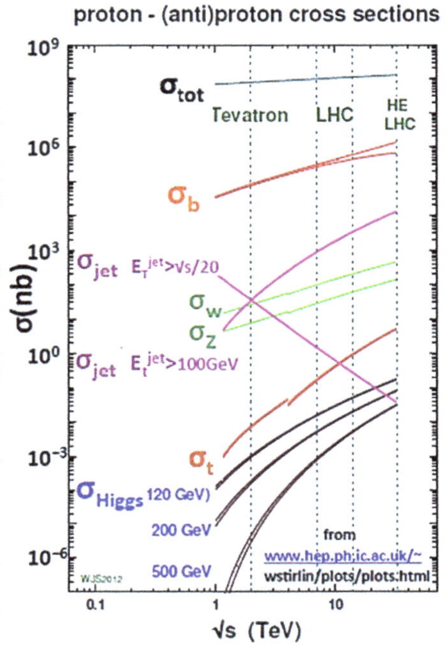

Very different challenges were faced by the two *special-purpose experiments* LHCb and ALICE, which are reflected in their totally different designs. They are operated at lower luminosity, avoiding multiple collisions per bunch crossing, and employing a range of detectors optimized for particle identification.

These four sophisticated instruments were developed in a worldwide coordinated R&D effort, culminating in large scale prototype measurement campaigns in particle beams at CERN and other accelerator laboratories, over the 1990s and early 2000s. Construction of the various components took place over about 10 years, starting in the second half of the 1990s, in universities, national laboratories, and industry. Typically, after initial local testing, components were delivered to CERN; following reception tests they would be assembled and installed in the experimental caverns where they would be finally commissioned. Figure 8.4 shows the four detectors during their final installation phase.

Below the four detectors are briefly introduced, followed by selected technological highlights. These may address an aspect of a specific experiment or one of concern to all of them, such as electronics and signal processing. The selection of highlights is not exhaustive: they are meant to give a flavour of the many challenges that the experiments had to tackle in realizing the LHC research agenda. The four collaborations have published detailed and comprehensive technical descriptions of their detectors as finally built and operated during the first years of LHC [9–12].

ATLAS [9]

The design of the ATLAS detector is based on a large superconducting air-cored toroidal magnet system for the muon spectrometer complemented with a smaller superconducting solenoid around the inner tracking cylinder. The momentum of the muons are precisely measured as they travel unperturbed by material for more than 5 m in the toroid. About 1200 large muon drift tube chambers of various shapes, with a total area of 5000 m^2, measure the impact positions with an accuracy of better than 0.1 mm. Another set of about 4200 timing chambers are used to provide a fast muon trigger (resistive plate chambers in the barrel, thin gap chambers in the end-caps).

In the field-free region between solenoid and toroids there is a highly granular e-m calorimeter using a novel lead–liquid argon (LAr) sampling technique [Highlight 8.8], complemented by full-coverage hadronic sampling calorimeters. For the latter a plastic scintillator–iron sampling technique, also of novel geometry, is used in the barrel part of the experiment, and LAr calorimeters cover the end-cap region near the beam axis where particle flux and exposure to radiation is highest. The calorimeters have 200 000 e-m cells and 10 000 hadronic cells.

Reconstruction of all charged particles, including those of displaced secondary vertices, is done in the inner tracking detector, which combines highly granular pixel (50 μm × 400 μm, 80 million channels) and microstrip (13 cm × 80 μm, 6 million channels) silicon semiconductor sensors close to the beam axis [Highlights 8.6 and 8.7], and a 'straw tube' gaseous detector (4 mm diameter, 350 000 channels) that provides about 35 signal hits per track. The latter also uses information from transition radiation for the identification of electrons.

Figure 8.4(a) shows one end of the cylindrical barrel detector after about 4 years of *in situ* installation work in the underground cavern. The ends of four of the barrel toroid coils are visible, illustrating the eightfold symmetry of the structure. The relatively lightweight overall structure of the detector adds up to 7000 t, covering a volume of 22 m in diameter and 46 m in length.

CMS [10]

The CMS detector features prominently a state-of-the-art superconducting 4 tesla solenoid. This single magnet [Highlight 8.12] serves both the high precision inner tracking chambers as well as the external muon detector, which is instrumented with large gaseous drift chambers in the barrel and cathode strip chambers in the end-caps complemented by resistive plate trigger chambers, embedded in the return yoke. With this configuration a compact overall design could be achieved, with the diameter limited to 15 m. The magnet yoke makes up most of the total detector weight of 12 500 t.

Muon detection, and their most accurate measurement, was a priority criterion for the CMS design, in addition to precision measurements for photons and electrons, achieved with a new type of radiation-hard e-m crystal calorimeter, the largest ever built. The challenging, but very successful development and manufacture of the 75 848 lead tungstate scintillating crystals in the final set-up took more than a decade [Highlight 8.8].

For charged particle tracking the collaboration opted for precise position measurements for each trajectory (13 layers with a position resolution of ~ 15 μm per measurement) leading to a large number of cells distributed inside a cylindrical volume 5.8 m long and 2.5 m in diameter: 66 million 100 × 150 μm^2 silicon pixels and 9.3 million silicon microstrips from about 10 cm × 80 μm to 20 cm × 180 μm. With a silicon detector area of 198 m^2 the CMS tracker is the largest ever built [Highlights 8.6 and 8.7].

The hadron calorimeter, comprising ~3000 small solid angle projective towers covering almost the full solid angle, is built from alternate plates of ~5 cm brass absorber and ~4 mm thick scintillator plates that sample the shower energy. The scintillation light is measured with photo-detectors (hybrid photodiodes) that can operate in the strong magnetic field, as the calorimeters are inside the solenoid.

The iron yoke of the CMS detector is sectioned into five barrel wheels and three endcap disks at each end. This enabled the detector to be assembled and tested in a large surface hall while the underground cavern was being prepared. The sections, weighing between 350 t and 2000 t were then lowered into the cavern (Fig. 8.4(b)) between October 2006 and January 2008, using a gantry crane system equipped with strand jacks: a pioneering use of this technology to simplify the underground assembly of large experiments.

LHCb [11]

The LHCb detector concept takes advantage of the dominant rate of beauty production close to the beam directions, for practical reasons on just one side of the interaction region. The layout of LHCb therefore resembles a fixed target spectrometer, however with very unique features.

A silicon strip vertex locator (VELO) detector can be positioned during stable beams very close to the interaction region and beam line in order to measure accurately primary and secondary vertices, selecting events with b-quarks. Particle identification to identify unambiguously the various final states is achieved by two ring imaging Cherenkov detectors [Highlight 8.10], whereas the momentum measurements are based on a large-aperture warm dipole magnet generating an integrated field of 4 T m for trajectories going through all tracking stations (silicon strips and straw tube drift chambers) of the spectrometer. Calorimetry uses sampling scintillator lead (e-m) and coarser scintillator iron technology. The muon detector behind the calorimeters, an absorber with iron plates providing a total of 20 interaction lengths, sampled by four chamber layers, completes the LHCb spectrometer. Figure 8.4(c) shows a picture of the fully installed LHCb detector in LHC Point 8.

ALICE [12]

The ALICE detector has to cope with events having extremely high multiplicity, characteristic of heavy ion collisions, with charged particle measurements down to the lowest technically possible momenta. Novel particle identification has been developed for many of the specific heavy ion physics signatures.

The ALICE Collaboration has reused the huge former L3 warm solenoid magnet providing a field of 0.5 tesla over a large central volume [Highlight 7.10]. Inside the magnet with its 10 000 ton iron yoke is located, as central tracking detector, the world's largest Time Projection Chamber (TPC) of 5.6 m in diameter and 5.4 m in length, which provides precision tracking as well as particle identification by *dE/dx* ionization. The innermost region, inside the TPC, facing the most extreme particle density region around the collision point, is instrumented with an inner tracking system consisting of silicon pixel, silicon drift and double-sided silicon strip detectors. Several detector systems dedicated to particle identification cover the volume outside of the TPC: transition radiation and Cherenkov radiation detectors, and a state of the art time-of-flight system [Highlight 8.9]. High resolution e-m calorimetry for photon measurements is implemented with lead tungstate scintillating crystals, similar to those of CMS.

A muon spectrometer starting with a massive 4 m long sophisticated hadron absorber cone, and featuring a large classical dipole magnet, covers on one side the solid angle from 2 to 9 degrees with respect to the beam direction. A front view of the ALICE detector nearing installation completion is shown in Fig. 8.4(d).

Fig. 8.4. The four LHC detectors during the installation phase. Upper row left (a) ATLAS, right (b) CMS, lower row left (c) LHCb, and right (d) ALICE.

Table 8.1. The expected and observed quality of measurements in ATLAS and CMS, expressed in standard deviations of the excess (signal) from the background-only hypothesis, for $m_H = 125$ GeV, for various decay modes.

Experiment	ATLAS		CMS	
Decay mode	Expected (σ)	Observed (σ)	Expected (σ)	Observed (σ)
γγ	4.6	5.2	5.3	5.6
ZZ	6.2	8.1	6.3	6.5
WW	5.8	6.1	5.4	4.7
bb	2.6	1.4	2.6	2.0
ττ	3.4	4.5	3.9	3.8

Complementarity of technologies

As pointed out above, rather different technological choices have been made by the LHC experiments, and yet all performed at or even beyond specifications. Particularly striking are the different technologies implemented on major detector systems by ATLAS and CMS, thereby offering significant complementarity for the instrumental approaches to the searches for the Higgs boson and for BSM physics. This aspect was an important consideration in CERN's approval procedure for the LHC experiments, with the aim of addressing the difficult LHC research program with different and largely independent experimental strategies.

Table 8.1 illustrates that indeed the two experiments have achieved very similar results for the Higgs boson discovery and observation in different final states, and this despite the large conceptual differences in the detector designs. Shown in the table [13] are the observed and expected significances of the signals expressed in the number of standard deviations (σ).

8.2 Superconducting Magnets: Powerful, Precise, Plentiful

Romeo Perin

The LHC dipole magnet (Fig. 8.5) is one of the most complex objects designed and fabricated on a large scale (1232 installed) in the history of particle accelerators. Major challenges for the designers were the choice of materials, assembled at room temperature, shrinking by 2 to 5 mm/m during cool-down to the working temperature of 1.9 K (−271°C), and the precision required for the magnetic field, B, in the presence of large electromagnetic forces (proportional to B^2) tending to open and flatten the coils. Sophisticated methods and equipment also had to be developed for handling the large magnetic stored energy in case of a transition from the superconducting to the normal conducting state (*quench*).

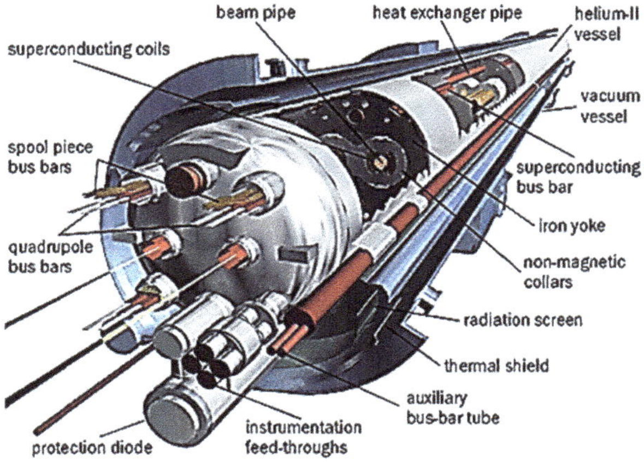

Fig. 8.5. Cut-away view of the 15 m long superconducting LHC dipole magnet.

The dipole must provide an accurate field from 0.5 to 8.3 tesla (T). For comparison the magnetic field in the core of a transformer is about 1 T. Previous large-scale production of superconducting magnets for accelerators was for devices operating at up to about 5 T, so R&D was clearly necessary. Initially, with laboratory resources focused on the construction of LEP, this was done with limited means and contributions from institutes and industry, but could continue more vigorously following project approval in 1994 [14]. The presence at CERN of experienced staff in superconducting magnets and cryogenics was vital for developing the technology. It was necessary to extend the state-of-the-art for practically every feature of the magnet, requiring exhaustive calculations and an extensive mock-up and model programme to arrive at cost-effective solutions. Importantly, thanks to good cooperation with scientists in the USA, CERN could benefit from the R&D on the conductors and 6 T magnets that had been done in the framework of the Superconducting Super Collider (SSC) project.

Already early in the programme a twin-aperture magnet (Fig. 8.6) was favoured, due both to the limited transverse space in the LEP tunnel and to the efficiency of the design [15]. In this design concept, first proposed by John Blewett of BNL in 1971, the magnetic flux in one aperture is partly returned via the other. For the LHC magnet this resulted *inter alia* in a cost reduction of about 25%.

After a series of tests, the important decision on the superconductor was taken. Well-proven niobium-titanium alloy (Nb-Ti) was chosen rather than niobium-tin (Nb_3Sn). Due to the brittleness of Nb_3Sn, coils have to be wound before the heat treatment required for forming the superconductor — a risky procedure, and costly for mass production. So despite being less performant at high fields, Nb-Ti, was

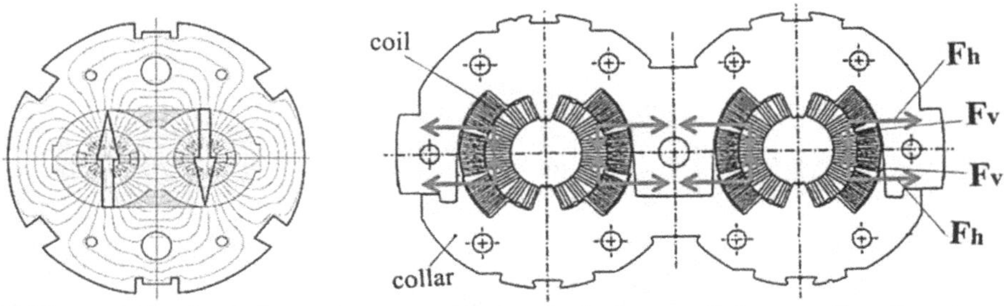

Fig. 8.6. Field and forces in the twin aperture magnet. The direction of the field is opposite in the two apertures — required for bending the trajectories of counter-rotating beams of particles of like sign. The horizontal components of the forces on the adjacent coils of the two apertures cancel; at the design field of 9 T, 2 F_h = 400 t/m and F_v = 100 t/m. Pre-stress on the coils and structure stiffness must ensure minimising movements due to these forces.

adopted — cooled to 1.9 K in superfluid helium (He II) to improve its current-carrying capacity. He II is also a very good conductor of heat (~ 1000 times better than copper), enabling very efficient cooling of the coils.

The stability of the beam in the collider depends crucially on the field quality of the dipole and quadrupole magnets, determined by the position of the conductors around the aperture rather than the accuracy of the surface of the poles as for normal iron-dominated magnets. The position of the conductor must be computed, the magnets manufactured to these specifications and maintained during operation in the presence of the strong electromagnetic forces [16].

The cross-section of the dipole magnet [17] is shown in Fig. 8.7. The coils are wound around a bore of diameter 56 mm and the force and flux containment structure consist of laminated stainless steel collars, a laminated iron yoke split vertically and a stainless steel shrinking cylinder. The latter also functions as the helium vessel, and is slightly curved in the horizontal plane to align the magnet apertures with the beam trajectory. The iron yoke serves to stiffen the structure and to shield the stray field; it also enhances the useful field by about 10 %. Static superfluid helium at atmospheric pressure permeates the magnet, cooled by vaporization of superfluid helium in the heat exchanger tube at the top of the iron yoke [Highlight 8.3]. Heat in-leak is reduced to a minimum by enclosing the cold mass in thermal shields covered with superinsulation under vacuum. The cold mass diameter is 570 mm and the complete magnet in its vacuum vessel has diameter 914 mm, and length 16.5 m. It weighs 34 t.

The 40 turns of cable are grouped in blocks separated by copper wedges to approximate a pure dipole field [6, 16]. For the design field of 9 T the current

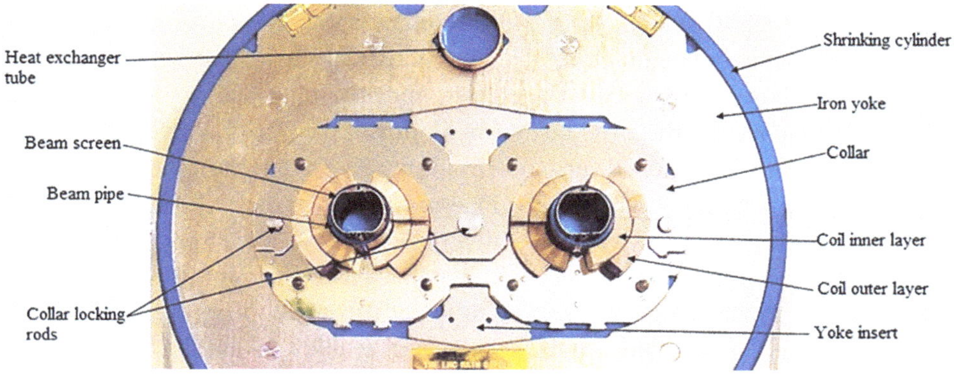

Fig. 8.7. Model showing part of the dipole magnet cross-section.

required is 12840 A. This highlights the value of superconductivity: if the cable were to be normal conducting copper it would require a cross-section a thousand times larger than the 25 mm^2 of the superconducting cable and power consumption would be at least 50 times greater (with superconducting magnet systems it is the refrigeration plants that consume most of the power — independently of field level).

If a quench occurs the current circulating in the stabilizing copper of the cables generates a turn-to-turn voltage of 20 V and voltage to ground of up to 2 kV. Following extensive studies and tests, polyimide tape was chosen for the insulation. Each coil layer was cured in a mould under pressure at 190°C. Tolerances on conductor position and coil dimensions were 0.05 mm.

The integrated bending strength of each of the 1232 magnets had to be equal to within one part in 1000. This called for careful positioning of components, with checks at every stage. After collaring, field quality was verified by (room-temperature) magnetic measurement, and if necessary corrected by shimming.

When powered, the coils must not experience tensile stress and must therefore be pre-stressed to resist the electromagnetic forces [16]. Pre-stress is provided by compressing the coils with accurately stamped collars. The collars are locked by inserting calibrated rods (Fig. 8.7). To avoid over-stressing the insulation, compression at the top of the coil is chosen to be zero at design field; a maximum pressure of 90 MPa (about 9 kg/mm^2) occurs on the median plane.

At 9 T the twin-aperture dipole stores 8×10^6 joules (8 MJ) of magnetic energy. The energy stored in a string of 154 magnets is 1200 MJ — enough to lift the Eiffel Tower by about 15 m! Quenches happen, and could destroy a magnet and endanger the whole string if not controlled. Quenches are detected by monitoring the appearance of resistive voltages; on detection the quench is spread in the coil by firing heaters that drive it above the critical temperature. Each magnet is equipped with a large semiconductor diode which conducts at the voltage induced by the resistance of the quenching coil; the current in the string by-passes the quenching magnet. Concurrently, the LHC beams are dumped, the power converter is turned off and a resistor is switched into the circuit to extract the energy of the magnet string which is completely discharged in about 100 seconds [18].

The magnets were produced in industry to build-to-print specifications and with sustained follow-up by CERN. To ensure uniformity of the magnets assembled by three manufacturers and to benefit from economy of scale, CERN procured directly most of the critical components (cables, insulation, spacers, steel laminations, collars, cold bore tubes, shrinking cylinder halves, diodes, etc.) and supplied them to the three companies [19].

Before installation in the LHC tunnel, the magnets were cooled down to 1.9 K and tested [20]. Field quality was measured by means of a specially developed equipment [21] consisting of 16 m long sets of rotating ceramic shafts carrying the measuring coils. Two of these shafts are inserted in room temperature enclosures installed in the magnet apertures and driven by a twin rotating unit.

Superconducting wires and cables

The conductors for the inner and the outer coil layers are flat cables made of composite strands consisting of thousands of Nb-Ti filaments embedded in pure copper (Fig. 8.8). For example, the strand for the cable of the inner layer of the coil has a diameter of 1.065 mm and contains 8900 Nb-Ti filaments. The copper serves to stabilize the superconductor, separate the filaments and to conduct the current briefly in the event of a quench. Fine filaments are important to limit the effect of currents that can be induced in the superconductor and will therefore persist. Such currents are proportional to the diameter of the filaments and are a source of field errors. Filament sizes of 6 μm to 7 μm were chosen for the strands as being the smallest that could be guaranteed without compromising the yield of the manufacturing process, and hence the cost.

Wire is made by stacking Nb-Ti bars and copper rods in a copper can, co-extrusion and drawing. The industrial process starts with 900 mm long, 200 mm diameter bars of Nb-Ti alloy wrapped by a thin sheet of pure niobium; each bar is turned into 30 km long filaments in a sequence of wire drawing and heat treatments. To limit eddy currents in the cable, yet provide sufficient contact for current sharing, strands are thinly coated with an alloy having lower conductivity than copper. Cabling had to be strictly controlled, to guarantee dimensional tolerances and smooth edges (to avoid damaging the insulation). A machine was developed for on-line optical control.

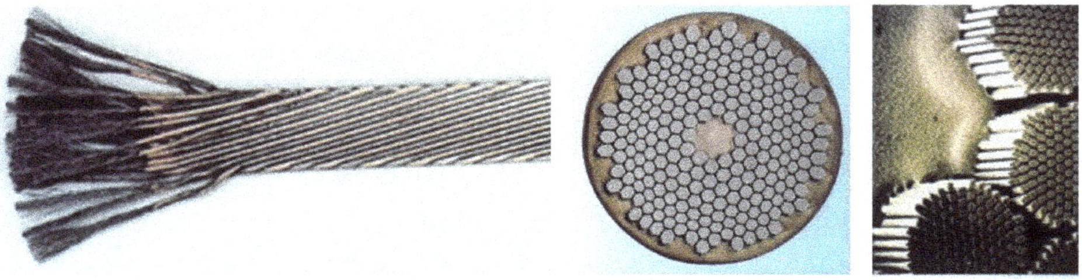

Fig. 8.8. A flat LHC cable, a strand and groups of Nb-Ti filaments in a strand [1].

The critical current of the 28-strand inner cable at 10 T is $\geq$ 13750 A.

The total quantity of cable procured amounted to about 7000 km (1200 t). The length of strand produced was about 270 000 km. To maintain the precision in the transverse dimensions of the strands and cables to some μm for such a production was an extraordinary industrial achievement [22].

In order to detect deviations from the specified process that could affect the strand and cable quality, strict Statistical Process Control (SPC) was applied at all steps of the production. To ensure uniformity, CERN handled the procurement of the 470 t of Nb-Ti alloy and 26 t of Nb sheet for the six suppliers of strand.

8.3 LHC Cryogenics: Quantum Fluids at Work

Philippe Lebrun

The LHC cryogenic system [23] had to face several unprecedented challenges, both qualitative and quantitative. Here we present the technological solutions developed to meet these challenges reliably, economically and efficiently.

The first challenge is the low temperature of operation of the magnets — 1.9 K — driven by the need to improve the current-carrying capacity of the Nb-Ti superconductor at high field. As this temperature is below that of the "lambda transition" between the two liquid phases of helium (Fig. 8.9), this implies operation in helium II, which exhibits particular transport properties associated with its superfluid state: very low viscosity, high heat capacity and excellent heat conductivity. These properties have been studied in detail since the discovery of superfluid helium by Piotr Kapitsa, John Allen and Don Misener in 1937, and can be profitably applied to the cooling of superconducting magnets [24]. Moreover, provided the superfluid helium can permeate the magnet coils via percolation channels designed into the electrical insulation, the stability of the superconductor

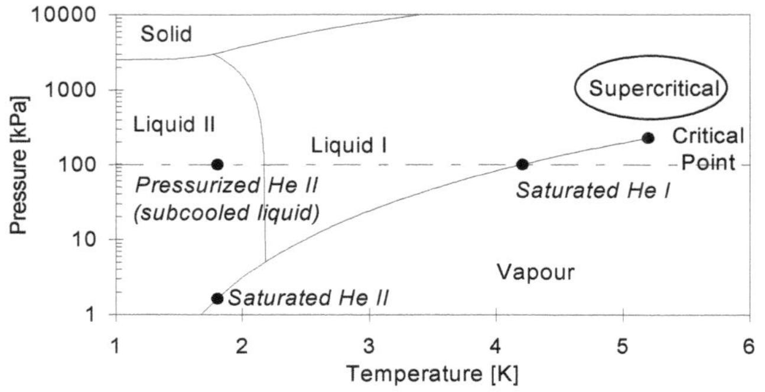

Fig. 8.9. Phase diagram of helium: the magnets operate in baths of pressurized He II, cooled by saturated He II flowing inside the heat exchanger tube.

against thermal disturbances is greatly enhanced, an essential benefit as its specific heat is an order of magnitude smaller at 1.9 K than in normal helium at 4.5 K.

It can be seen in the phase diagram (Fig. 8.9) that the saturation pressure of superfluid helium is below 5 kPa (50 mbar), which raises two concerns: (i) cryostats containing superfluid helium at saturation operate far below atmospheric pressure, with risk of air in-leaks, contamination of the helium and clogging by solid air; (ii) the dielectric strength of gaseous helium is greatly reduced at such pressure, increasing the risk of electrical breakdown. To circumvent these shortcomings, the LHC superconducting magnets operate in baths of "pressurized" He II, i.e. above saturation pressure, in fact in sub-cooled liquid at about atmospheric pressure [25]. The very large (but finite) heat conductivity of helium II ensures that each bath is quasi-isothermal; it cannot however be used to transport heat over long distances in the accelerator tunnel. The LHC magnet cooling scheme is thus composed of a heat exchanger tube [26] threading its way along the magnet chain, in which a small two-phase flow of saturated helium II gradually vaporizes as it absorbs heat from the pressurized helium II baths (Fig. 8.10).

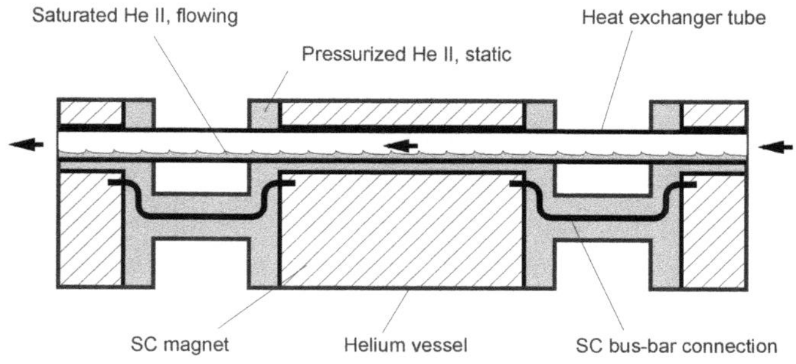

Fig. 8.10. Principle of the LHC superfluid helium cooling scheme [1].

This scheme combines the benefit of magnets operating in static, isothermal superfluid helium close to atmospheric pressure, with that of fixed temperature imposed by the saturated helium II inside the tube. Moreover, the small, stratified two-phase flow inside the tube can be driven solely by gravity (in the sloping LHC tunnel) and/or by flowing vapour, thus avoiding the use of circulation pumps. Finally, in case of resistive transition of a magnet, the rise in temperature of the helium bath leads to dry-out inside the heat exchanger tube, thus limiting thermal propagation to neighbouring magnets.

The second challenge of the LHC cryogenic system, a novelty for hadron accelerators, stems from the dynamic heat loads induced by the circulation of the particle beams, due to synchrotron radiation, dissipation of image currents in the resistive wall of the beam pipes, loss of particles scattered by the residual gas as well as by the collisions in the interaction points (only affecting the sections of the accelerator close to the experiments), and deposition of energy by electrons resonantly accelerated in the electric field of the circulating bunches, the so-called "electron cloud". In nominal operation, the beam-induced heat loads can be up to 1.7 W/m, i.e. an order of magnitude higher than the allowed heat in-leak at 1.9 K. To reduce their thermodynamic impact, it is therefore necessary to absorb most of these losses on beam screens [27] fitted in the magnet apertures and cooled at higher temperature, from 5 K to 20 K by forced circulation of supercritical helium. The beam screen tubes, made of non-magnetic austenitic stainless steel to resist eddy-current forces at resistive transitions of the magnets, are internally coated with 75 μm of high-conductivity copper to reduce the wall impedance seen by the beam and slotted to act as cryopump baffles, shielding the 1.9 K pumping surface of the cold bore from radiation and particles lost from the beams (Fig. 8.11).

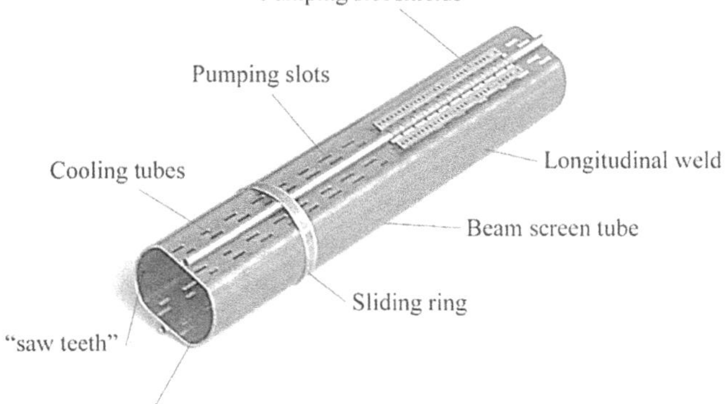

Fig. 8.11. The LHC beam screen.

The large size of the collider, impacting heat loads and refrigeration duties, and its underground implantation, limiting access to eight points around its perimeter, with sector lengths of 3.3 km in-between access points and depths exceeding 100 m below ground, constitute a third challenge. The LHC ring is cooled by eight large-capacity cryogenic plants, located at five points around the perimeter, each normally cooling a 3.3 km sector, but able to serve two adjacent sectors to provide some redundancy at partial load. Each plant (Fig. 8.12) provides a mix of liquefaction and refrigeration duties at 50–75 K and 4.5–20 K, amounting to an equivalent entropic capacity of 18 kW at 4.5 K with excellent efficiency reaching 28% of the Carnot cycle [28]. The cryogenic plants are complemented at their cold end by eight 1.8 K stages producing each 2.4 kW of refrigeration power by expansion to the 1.6 kPa (16 mbar) saturation pressure of helium. Low-pressure helium vapour is compressed by a train of cold hydrodynamic compressors up to a fraction of atmospheric pressure, followed by room-temperature screw compressors operating at sub-atmospheric suction pressure. This arrangement combines good efficiency, limited capital expenditure and compliance to the strongly variable demand resulting from the dynamic heat loads [29].

The issue of cryogen supply and inventory management [30] completes this brief overview. The system contains 135 tons of helium, of which about 60% are in the magnets when the machine is in operation, the rest being shared between the distribution pipework, the cryogenic plants and the minimum reserve in the buffer storage vessels. Helium is procured from the market and delivered to CERN in standard liquid transport containers. Upon warm-up of the machine, the helium must be stored and its purity preserved. Long-term storage is done at room temperature in 250 m^3 gas vessels at 2 GPa (20 bar), which can only accept about half of the inventory. Part of the helium can also be stored, for limited amounts of time, in 120 000 litre vacuum-insulated liquid tanks at atmospheric pressure. The rest is re-injected in the market via "virtual storage" contracts with the gas vendors. This strategy allows receiving the amounts needed for operation in due time, while limiting the capital expenditure. Liquid nitrogen is used for precooling the 37 500 t mass of the magnets, a task incommensurable with the installed capacity of the cryogenic plants. For this purpose, 10 000 t of liquid nitrogen supplied by 50 m^3 semi-trailers are vaporized at rates of up to 40 t per hour, to produce 4.8 MW of precooling power down to about 100 K, thus permitting to bring the whole machine from room to cryogenic temperature in about three weeks.

Fig. 8.12. Cryogenic plant providing 18 kW @ 4.5 K. Left: compressor station. Right: cold box.

8.4 Current Leads: High Temperature Superconductors to the Fore

Amalia Ballarino

Powering the superconducting magnet circuits requires the transfer of about 3 MA of current from the power converters (at room temperature) to the liquid helium environment. Good electrical conductors are also good conductors of heat, so the devices (current leads) used to transfer the current must be designed to minimize the heat in-leak yet safely carry the current. Conventional current leads achieve this by using the helium vapour that is boiled off at the foot of the lead to cool a combined conductor/heat exchanger. Such "self-cooled" leads transfer about 1 W of heat to the liquid helium bath per kA of current (compared with about 48 W/kA for a lead without vapour cooling). The interest of including a High Temperature Superconducting (HTS) section in the leads is that resistive heating in that section is absent and thermal conduction can be low, so heat in-leak is reduced [31].

Already in 1992, just six years after the discovery of High Temperature Superconductivity [32], conceptual studies were underway at CERN on using HTS in the current leads for the LHC, current leads being an evident application [Box 8.4]. Targeted tests revealed that it should be possible to make reliable HTS leads [33], and in 1999 the decision was taken to use such leads at the LHC. The aim was to obtain a saving with respect to conventional leads of a factor 10 on the required liquefaction power. Over a thousand such current leads would be needed, 64 rated at 13 kA for the main dipoles and quadrupoles, 474 rated at 6 kA for the quadrupoles of the matching sections and 520 rated at 600 A for correctors.

High temperature superconductors (HTS) **Box 8.4**

Up to 1986, practical low temperature superconductors (LTS), which first appeared in the 1960s, had been studied in depth [Boxes 4.3 and 8.1], and had been adopted for use in detector and accelerator magnets — and crucially a commodity for the magnetic resonance imaging (MRI) industry. But these materials only work at very low temperature; most magnets are cooled using liquid helium (4.2 K), which is expensive to buy and to keep cold. With this background the unforeseen discovery of ceramic materials (cuprates) that are superconducting at higher temperatures, even beyond that of abundant and cheaply available liquid nitrogen (boiling at 77 K, and a good electrical insulator), caused great excitement in the applied science community. But for accelerator scientists, the superior *high field characteristics* of these so-called "high temperature superconductors" (HTS) at *low temperatures* are far more important, as they offer the promise of magnets operating at 20 T and even higher.

The crystal structure of an HTS compound is highly anisotropic, with a stack of layers of oxides, interspersed with atoms of copper oxide The superconducting current passes preferentially in the planes between the layers rather than perpendicular to it and the intensity depends on the number of layers. HTS behaviour is not explained by BCS theory. Many types have been found, but only two developed for practical use:

BSCCO, which has two main phases, Bi-2212 and Bi-2223 (the numbers refer to atoms of Bi, Sr, Ca, Cu respectively), have critical temperatures of 85 K and 110 K. These were developed rather quickly, and are referred to as first-generation HTS. The multi-filamentary conductors are made via powder-in-tube (PIT) technology; the matrix is silver to avoid reaction with the precursor powders and let oxygen permeate during the heat treatment process. Bi-2212 can be made in the form of wires that can be cabled (before heat treatment at about 800°C, under pressure). Bi-2223 is produced in the form of tape, about 4 mm wide. The silver matrix provides good electrical stabilization. A variant using a matrix of silver-gold alloy to reduce thermal conductivity is used for HTS current leads (to limit the heat conduction into the cold environment). The cost of silver limits the likelihood of widespread large scale application.

REBCO, which has one main phase, RE-123, where RE refers to rare earth (often replaced by Y for Yttrium) and the numbers to atoms of RE, Ba, and Cu respectively, has a critical temperature (with Y) of 93 K. This second generation conductor is produced in the form of 12 mm wide tape that can be split into the "standard" 4 mm wide type. The superconducting layer is only 1 to 3 μm thick, deposited on a strong metallic substrate; a layer of copper provides electrical stabilization. At 4.5 K and 20 T current densities of 400 A/mm^2 are currently achieved. Flat cables can be assembled from shaped ribbons cut from wide tape (Roebel cables [Highlight 12.3]). This material is presently very expensive to produce, but is favoured by industry due to a potential for cost reduction.

Magnesium di-boride (MgB_2, $T_c = 39$ K) is not a true HTS, but is easier to produce in the form of multi-filamentary wire, can be cabled, and is very affordable. Although brittle and more sensitive to magnetic fields than Nb-Ti [Box 4.3], it works efficiently at 20–25 K making it well-suited for power transmission [Highlight 12.3].

The LHC HTS current lead programme was the first large-scale, mission-oriented, application of HTS materials [Box 8.4]. The work started with in-depth studies of various types of superconductor, including bulk BSCCO and YBCO materials [33], before converging on the choice of multi-filamentary BSCCO 2223 embedded in silver-gold alloy (Ag5.3wt%Au) tape. The tape was being actively developed for other projects such as power transmission, for which the BSCCO is embedded in pure silver. Silver is necessary for texturing the superconductor, but pure silver is an excellent conductor of heat. Its thermal conductivity is reduced by doping with gold. Specific methods were developed for optimizing leads combining an HTS section and a gas-cooled section. Taking advantage of the availability in the LHC of helium gas at 20 K, it was decided to operate the HTS section in the range 4.2 K to 50 K, and the gas-cooled section between 50 K and ambient temperature [31]. Prototypes were assembled and tested at CERN, where all the technological processes were developed and qualified. The design was based on the application of standard workshop practice and assembly tolerances, to enable normal engineering firms to tender. A 13 kA lead is shown in Fig. 8.13.

The procurement of the series relied on build-to-print CERN drawings and detailed assembly procedures. The 64 HTS current leads rated at 13 kA were manufactured in industry, while the large series of 6 kA and 600 A leads were made at the Budker Institute of Nuclear Physics (BINP), Novosibirsk. Close technical follow-up, with transfer of know-how acquired during the development activity, assured the quality of the production of the current leads, which was completed in a period of about two years. The required 31 km of BSCCO 2223 tape was specified and procured by CERN, where it was also soldered into stacks. These were measured in cryogenic conditions via an external contract prior to delivery to the lead manufacturers. Contracts were also placed for cold acceptance testing of all leads before integration in the LHC.

Since installation the current leads have undergone several thermal cycles and thousands of electrical cycles, as well as suffering some untoward events (e.g. stopping the flow of helium) generating the resistive transition of the HTS elements [34]. The leads were designed to withstand such events: the HTS material is shunted with sufficient normal material to carry the current during a fast ramp down (time constant 105 s in a dipole circuit) and quench protection thresholds were specifically defined to cope with the requirements of the HTS material.

In view of the success of this project, other users have moved to designing HTS leads for manufacture following standard workshop practice. For example, major aspects of the LHC design have been adopted by ITER for their current leads [35].

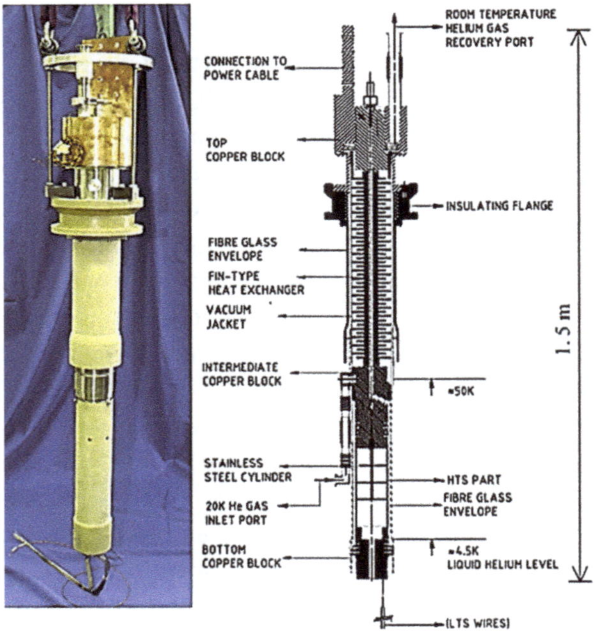

HTS section of a 13 kA lead. The stacks of BSCCO 2223 are soldered to a stainless steel support tube with brazed copper terminals.

Detail showing the joint between HTS stacks and Nb-Ti wires (leading to the cold busbar system) at the cold end of the 13 kA lead.

Fig. 8.13. 13 kA HTS current lead for the LHC

8.5 A Pumping Vacuum Chamber: Ultimate Simplicity

Cristoforo Benvenuti

The successful performance of Non Evaporable Getter (NEG) pumping for LEP (Chapter 7) was a strong motivation to envisage also applying a similar solution to the LHC warm sectors. But the LHC warm chambers were too small to host the LEP getter pump. The obvious solution would be to coat the chamber walls with a thin NEG film [36], all the more because such a solution could also be adopted for the chambers in the experimental areas.

It was clear that a thin film NEG coating, if this could be done, would be more effective than the LEP NEG pump for the following reasons: (i) the NEG surface would be much larger, resulting in a lower pressure, and (ii) the coating would inhibit the degassing of the underlying chamber, the major gas source in the accelerator. In addition, the NEG coating pump would be intrinsically safe (no moving parts, no risk of electric short circuits) and simpler (no power supplies, no electric feedthroughs required). To cope with the machine gas load a coating thickness of 1–2 μm would be sufficient using typical getters. However, heating the film for NEG activation implies heating the chamber, therefore activation should be feasible at a temperature that also the Aluminium chambers in the experimental areas can withstand ($\leq 200°C$). This temperature was too low for the activation of the available getters, so a suitable NEG alloy had to be found.

The coating process envisaged (cylindrical magnetron sputtering) brought another problem. The coating is obtained by ion bombardment of a cathode. These ions are produced by triggering a discharge in argon or krypton gas (at a pressure of about 10^{-1} Pa), obtained by applying a negative bias voltage of 400 V to the cathode, a rod a few mm in diameter, placed at the centre of the chamber (Fig. 8.14). Producing such cathodes of the different alloys to be explored has a cost and takes time. Furthermore, some of the alloys of the usual getter elements, such as Ti or Zr, are not ductile and the production of the cathode is problematic. To overcome this difficulty, cathodes were made by twisting together wires of pure metals, readily available from industry, as shown in the figure. The atoms extracted by sputtering from the different wires recombine on the chamber walls to provide the desired alloy. In this way the elemental composition of the alloy may be varied and, in addition, for a given choice of metals the relative composition may be changed by changing the diameter of the wires. A smaller wire results in a lower content of the corresponding metal in the final alloy. In this way about 20 different alloys were tested until finally a Ti-Zr-V alloy was found to offer an activation temperature of 180°C, low enough for LHC. For this final choice wires of equal diameter were used, resulting in the same atomic content of these metals [37]. As this technology was not available in industry, the ~ 6 km of room temperature LHC chambers were coated at CERN. Some coated chambers were tested under different conditions showing that extremely low pressures could be achieved (below 10^{-10} Pa). Furthermore, the degassing induced by synchrotron radiation was much lower than for uncoated chambers. Finally, the secondary emission of electrons was reduced by the coating to such a level that the electron cloud was no longer a problem.

Fig. 8.14. Sputtering configuration with a cathode consisting of twisted wires.

Thanks to the NEG coating the vacuum performance has been very satisfactory since the beginning of the LHC operation. This innovation is protected by a CERN patent which has been licensed to several companies and institutes and has since been adopted in accelerators worldwide.

8.6 Vertex Detectors at LHC: In Search of Beauty

Leonardo Rossi

Ordinary matter is made of stable elementary charged particles: up and down quarks and electrons. Heavier replicas of these elementary particles are produced in high energy particle collisions, but their fate is to decay to their stable partners after a lifetime ranging from 5×10^{-25} to 2×10^{-6} s. Two of the heavy replicas, the beauty quark and the tau lepton, are especially interesting, because, belonging together with the top quark to the third family, they may open a window on physics beyond the Standard Model [Box 6.4]. At LHC relativistic beauty quarks and tau leptons are produced abundantly and, having a lifetime of about one picosecond (10^{-12} s), they decay after travelling a few millimetres.

Extrapolating with high precision to the collision points of protons at the LHC to identify and measure decay vertices is therefore of paramount importance. This requires spatial accuracies of ~ 0.01 mm, which are routinely obtained with highly segmented silicon detectors [Highlight 7.7]. The task becomes difficult because of the environment of the LHC proton–proton collisions. At the nominal collision rate of 10^9 interactions per second about 1000 charged particles traverse the vertex detector each time the proton bunches collide at the centre of the experiment. This happens at a frequency of 40 MHz and gives both a large instantaneous hit rate and a large accumulated irradiation dose, which must be considered for long-term operation.

Vertex detectors at LHC must be fast (signal collected in 25 ns), highly segmented (able to cope with > 1000 particles), precise (about 0.01 mm spatial accuracy), radiation resistant (up to 500 kGy) and as thin as practical to minimize loss of information for the downstream detectors. Both ATLAS and CMS chose hybrid silicon pixel detectors [38] as their vertex detector. This technique, nurtured by the RD19 collaboration at CERN in the '90s, had been first employed, on a much smaller scale and with less demanding specifications, in the experiments WA97 and DELPHI. It consists of a sensor directly coupled to its mating electronics read-out chip. The sensor is a matrix of silicon diodes implanted on high resistivity wafers; the read-out chip is a corresponding matrix of front-end electronics with amplifiers and discriminators followed by a coding circuit. The mating between the two parts is obtained through high-density "bump-bonding", an operation consisting in the deposition on both matrices of tiny metallic droplets, which are then fused to establish electrical contact between each diode and its front-end electronics channel. The silicon diodes are reverse polarized with a field of ~ 3×10^3 V/cm, allowing complete charge collection within 10 ns, to provide a current pulse of ~ 0.5 μA to the front-end amplifier. This arrangement provides true 3-D hit information and robust pattern recognition capability even in a dense track environment. The very small sensitive volume (~5×10^{-6} cm^3) of each pixel minimizes the dark current and provides resistance to large radiation doses.

Figure 8.15(a) shows an individual pixel bump-bonded to its mating front-end channel, while Fig. 8.15(b) shows the exploded view of a module, the basic building block of a pixel detector. ATLAS and CMS were forced to push the hybrid pixel technique far beyond previously established characteristics. Many challenges had to be overcome; the most important being related to instantaneous and integral track density, and to the need to minimize the amount of material. Once installed as the innermost part of each experiment access is difficult, so some degree of redundancy is necessary to minimize the risks of losing decay vertex information.

About 5000 pixel/cm^2 are required to deal with the space resolution and the doses expected. This required pushing the bump-bonding technology far beyond standard specifications used for the consumer electronics. ATLAS chose 50 μm × 400 μm pixels and digital read-out while CMS decided on 100 μm × 150 μm pixels

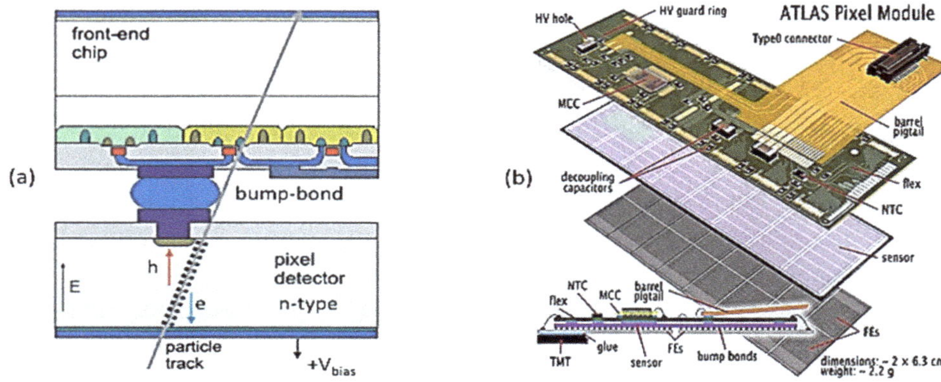

Fig. 8.15. (a) Sketch of one pixel mated to its electronic channel through bump-bonding; (b) exploded view of an ATLAS pixel module where one sensor is bump-bonded to 16 front-end chips. The signals from the 47532 individual pixels are coded in the front-end chips and routed to the read-out electronics via a flexible circuit.

with analogue read-out for charge interpolation, the sensor thickness being ~ 0.3 mm in both experiments. The optimization of charge interpolation pushed CMS to spread the ionization charge between pixels, while ATLAS chose to concentrate most of the charge on one pixel. The different charge spread is obtained through proper orientation of the modules in the magnetic field. The ATLAS design privileges robustness while that of CMS privileges accuracy.

The amplifiers must have a fast rise time, to pass the discriminator threshold (typically set at ~ 15% of the average signal) within the bunch-crossing period of 25 ns. But the time to return to baseline can be 100 times longer as the probability for a pixel to be hit in a bunch crossing is very low ($< 10^{-4}$). If the return-to-baseline signal slope is constant (as in the case of ATLAS) the signal length can be used to measure the pulse height, to give the specific energy loss in silicon. The front-end electronics has to meet two main specifications: fit all the required functionalities within $< 2 \times 10^{-4}$ cm^2 and guarantee that the circuit behaviour remains constant (or tuneable) up to doses of 500 kGy. Fortunately, the two requirements could be satisfied by adopting deep-submicron CMOS electronics, made available by IBM through a framework contract with CERN. With gate length of 0.25 μm, not only the designers can arrange ~1000 transistors per pixel cell, but the oxide layers become so thin that electrons can tunnel out. The dominant effect of the accumulated dose on the electronics (threshold shifts due to accumulation of charges in the gate and field oxide of CMOS transistors) can therefore be avoided.

The ATLAS and CMS vertex detectors operate in a solenoidal magnetic field of 2 T and 4 T, respectively. They provide three space point measurements over the first ~ 12 cm of the particle trajectories, with each coordinate measured to an accuracy of ~ 0.01 mm in the track bending plane. Each measurement layer represents ~ 2.5% of a radiation length (X_0), mostly because the high-speed front-end electronics is power-hungry (~ 0.5 W/cm^2), and the detector must operate at below –5°C to minimize the radiation damage to the sensors. These requirements determine most of the material in a layer, i.e. conductors to bring the power in and cooling system to take it out, the active part of a layer (sensor and electronics) being < 20% of the overall X_0. ATLAS has 1744 identical modules covering ~ 1.7 m^2 with ~ 8×10^7 channels, CMS has 1776 modules of seven different sizes covering ~ 1 m^2 with ~ 6.5×10^7 channels. Both pixel systems meet the demanding LHC specifications with very similar spatial resolution and efficiency, greatly contributing to tagging of events involving beauty quarks and tau leptons.

The ALICE experiment is also equipped with a hybrid pixel vertex detector specifically designed for complex topologies emerging from lead-lead collisions. The environment is quite different: the interaction frequency is only ~ 4 kHz, but for each collision there may be up to 50 tracks/cm^2 (mostly low-momentum). The radiation damage is therefore less of an issue, but detector granularity and the amount of material per layer are primary concerns. The pixel size (50×425 μm^2) could not be shrunk further, as it is limited both by the bump density and the electronics gate length. ALICE made then a special effort to minimize the amount of material per layer, which was reduced to < 1% X_0, using very thin mechanical supports and distributing signals with flexible circuits made of aluminium tracks on Kapton supports, specially realized at CERN. The ALICE vertex detector consists of a barrel having two layers, at ~ 4 and ~ 8 cm radius, housed in a common carbon fibre support; 120 modules cover ~ 0.2 m^2 with ~ 10^7 channels.

In LHCb the vertex detector plays a key role, since the main physics objective of the experiment is the study of beauty hadrons. Within the acceptance region of the experiment, this study is possible with a luminosity of ~ 10^{32} cm^{-2}s^{-1}. LHCb has therefore to sustain a rate that is about halfway between ATLAS/CMS and ALICE. As it is not confronted with the high multiplicity of the Pb-Pb interactions this opens the possibility to use established silicon microstrip detector technology — but requires a special arrangement to optimize vertex resolution. The proton–proton interactions in LHC take place within a cylinder of ~ 0.1 mm radius and ~ 15 cm length. The vertex detector must cover this region and must be as close as possible to the beams to minimize errors due to track extrapolation. The solution adopted has been named VELO (Vertex Locator) and is made of 42 half-moon-shaped silicon microstrip detectors with pitch varying from ~ 40 μm (at the inner radius of 8 mm) to ~ 100 μm (at the outer radius of 42 mm). Each detector is a doublet of 0.3 mm thick sensors, one measuring the radial coordinate and the other the azimuthal coordinate of the tracks. VELO extends for ~ 1 m and a small cut-out in the centre of sensors allows the main LHC beam to pass through unimpeded. The total number of channels is ~ 1.7×10^5, all with analogue read-out to optimize spatial resolution through charge interpolation.

The main challenge of the VELO project was to come as close as possible to the interaction region and to minimize the amount of material from the interaction to the first measurement. The VELO system is placed inside the LHC vacuum pipe, but, to protect the integrity of the primary LHC vacuum system, the sensors are separated from the beam volume by a 0.3 mm thick aluminium shield. The silicon detectors must be protected from damage caused by the LHC beam (especially during injection and tuning). The VELO system is therefore remotely retractable by 35 mm and is placed close to the beam only when safe operation is assured.

Vertex resolutions of ~ 0.01 mm in the transverse plane and < 0.1 mm in the beam direction are routinely obtained — performance that contributes to LHCb as a being successful competitor to experiments at e^+e^- beauty factories.

8.7 Large Silicon Trackers: Fast, Precise, Efficient

Luigi Rolandi

Precise tracking is indispensable for any collider experiment. Lepton signatures, based on the combined information of the central tracker and the e-m calorimeter, or the muon system, are privileged signals for potential discoveries allowing to separate new phenomena from the huge background. At the LHC, Z and W bosons are identified via their leptonic decays and are used both to calibrate the detectors and to search for new physics phenomena. About 60% of the energy of jets is due to charged particles: by combining measurement of their momenta with the calorimetric information the jet energy can be determined with good resolution.

The capability to reconstruct detached vertices, combining the information of the central tracker and the vertex detector, allows the identification of hadrons with a b quark. This is an important tool for the precise study of the top quark and for discovery physics in all cases when the new particles have a preferential decay to heavy quarks, such as Higgs bosons or supersymmetric particles [Box 7.2].

The transverse momentum p_t of a charged particle curling in a magnetic field B is derived from the sagitta s of the projection of length L on the plane transverse to the magnetic field

$$s = 750\ \mu\text{m}\left(\frac{B}{1\,\text{T}}\right)\left(\frac{L}{1\,\text{m}}\right)^2\left(\frac{50\ \text{GeV}}{p_t}\right).$$

The transverse momentum resolution, i.e. the momentum measurement precision, is proportional to $1/(BL^2)$. The trajectory is measured at N points with spatial resolution σ, and the momentum resolution scales as $\sigma/\sqrt{N}$. With a magnetic field of 4 T and some 15 points having a resolution of $\sigma = 40$ μm each the momentum resolution is 1% at $p_t = 50$ GeV. With this resolution the width of the peak of a 125 GeV mass Higgs boson decaying to four muons is about 1% and the signal to background ratio in the interval 110–130 GeV is near 2.

The high luminosity and the bunch crossing frequency of the LHC pose severe constraints on the response time of the tracking devices [Box 8.3]. Typically, some 800 particles are produced together with the rare particles of physics interest in the many collisions at each bunch crossing and traverse the detector simultaneously. The pattern recognition algorithm must be able to reconstruct all these tracks belonging to the same collision event. Tracks from different collisions in a bunch crossing can be identified because they originate from a different primary collision vertex. Since the collision area of LHC has a Gaussian distribution along the beam direction with a σ of about 6 cm, the vertices of the different inelastic collisions are separated by about 1 cm on average. Figure 8.16 shows a 4-muon event selected by the ATLAS analysis searching for the Higgs boson.

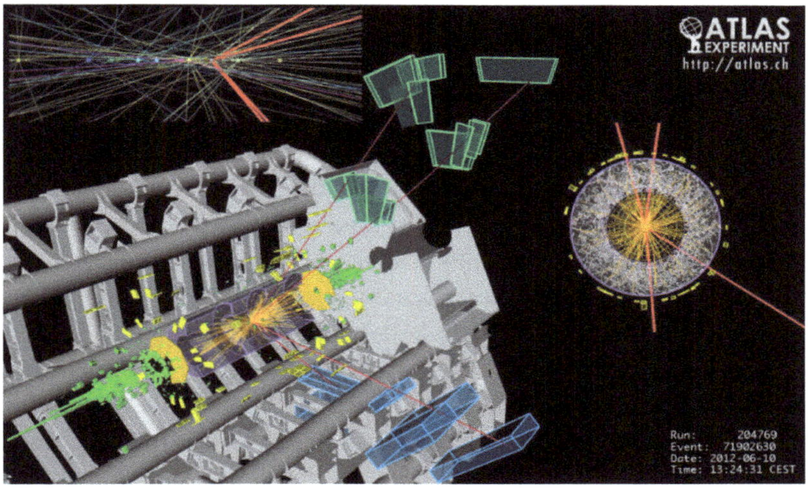

Fig. 8.16. Event display of a 4μ Higgs boson decay candidate reconstructed by the ATLAS detector. The muon tracks are shown in red and cross the muon chambers located in the external part of the detector. The insert on the right is a projection perpendicular to the beams showing all charged tracks reconstructed in the same bunch crossing (yellow lines). The insert on the top left shows a region of about 10 cm length where the beams cross with more than 10 recostructed vertices and their associated tracks.

The complexity of pattern recognition increases as a function of "occupancy", the average number of hits per event in one elementary detection element. High efficiency pattern recognition requires occupancy below 1% and correspondingly small detection elements. A 10 mm^2 surface covered by a 10 cm long and 100 μm wide silicon detector strip placed at 20 cm from the beam line is hit by charged particles only once every 100 bunch crossings at nominal LHC luminosity.

ATLAS and CMS use 300 micron thick silicon detectors for the innermost part of their tracking systems. In these detectors charge collection time is comparable to bunch crossing period and using fast read-out electronics the pile-up is limited to one bunch crossing. This calls for very large surfaces of the silicon detector. The ATLAS SCT [9] has 15 912 silicon sensors mounted in 4088 modules for a total silicon surface of 63 m^2 and 6 million electronic channels. CMS also uses silicon for the outermost part of detector: the silicon strip tracker [10] has 24 244 sensors mounted in 15 148 modules for a total silicon surface of 198 m^2 and 9.3 million electronic channels. The concept of these large silicon trackers was revolutionary; their development and construction represented an enormous engineering and logistic advance, considering that the largest silicon trackers existing at that time had a surface of some 6 m^2 [Highlight 5.9].

The silicon sensors for CMS were manufactured for the first time on 6-inch wafers in a standard planar process. This led to significant cost reduction per unit area compared to standard 4-inch wafers, making possible the construction of large (~ 85 cm^2) sensors for the outer layers, and greatly reducing their total number.

The precise (~ 10 μm) assembly of the sensors and the readout electronics into modules is challenging, and was traditionally performed by trained technicians. This was not possible for the CMS tracker because of the huge number of modules, so the experiment commissioned an automated mounting system based on a precision X–Y coordinate machine with large two-dimensional coverage, video pattern recognition capability and innovative glue-dispensing and pick-and-place tools. These robotic machines could assemble three modules precisely in less than an hour. Several were transferred to collaborating institutes where over 15000 modules were assembled in one year. The total of 25 million electrical connections between strips of two silicon sensors mounted on the same module and between sensors and readout electronics were made using fast (1 Hz) bonding machines.

Particles produced by collisions and secondary interactions in the detectors irradiate the detectors. All elements of the detector, including read-out electronics and cables must therefore be appropriately radiation resistant. The radiation dose (especially high close to the interaction point) degrades the detector response during the 10-year design lifetime. Radiation damage is mitigated by housing the detector in a dry nitrogen gas environment and cooling the sensors to –10°C.

The central trackers of ATLAS and CMS are similar. About 2 m in diameter and 6 meter in length, they enclose the central part of the detector and cover the polar angular range from 10° to 170°. Both operate in a solenoidal magnetic field, 4 T for CMS and 2 T ATLAS. In both trackers the innermost detector layers are built with silicon pixels, with a pixel size of about 0.02 mm^2 to provide space points with ~ 10 μm resolution and occupancies smaller than 0.06%. The intermediate layers use silicon strip detectors with high strip density; they provide 6–8 measurements with about 25 μm resolution of the coordinate perpendicular to the strip and have occupancies below 2%. Some detector layers are mounted with strips rotated by a small angle to provide spatial resolution along the strips. The technology used for the outer layers is different: CMS uses 500 μm thick silicon strip detectors with coarse pitch providing eight measurement points with resolution of < 50 μm while ATLAS uses a Transition Radiation Tracker with 4 mm diameter gas straw tubes providing 35 points with about 130 μm resolution.

The huge number of front-end electronics channels located on the detectors consumes about 30 kW of electrical power that requires cooling and results in considerable additional material. The total amount of material varies from about 40% of a radiation length to near 2 radiation lengths at places where the services (cables, pipes) are concentrated. The large amount of material spread along the trajectory of electrons affects the measurement of their energy in the calorimeter. Hadrons are also affected: depending on the angle, some 5%–20% are not reconstructed due to interaction within the tracker volume. The tracker material limits the momentum resolution at low momentum due to multiple scattering.

The intrinsic resolution of the tracking detectors is usually better than that of the detector assembly. The position of the detectors varies with time due to changing environmental conditions (temperature, magnetic field strength, etc. that mainly occur during LHC stops). Alignment procedures are used to measure and monitor their position over time to recover the intrinsic resolution. The corrections from the nominal to the real positions of all detector elements is computed using a large set of reconstructed trajectories of particles from pp collisions and cosmic rays [39]. The statistical error of this method is small, provided a sufficiently large set of tracks is used. A subset of silicon detectors is constantly monitored with optical alignment systems. As stability is imperative, the silicon detectors are mounted on carbon fibre supports with thermal expansion coefficients smaller than 10^{-6}. Depending on the number of independent detectors to be aligned, the simultaneous fit of a large number of parameters can be computationally challenging. The most complex case is the CMS silicon tracker with about 2×10^5 parameters.

The ATLAS and CMS trackers have been operational since the beginning of LHC data taking and perform at design specifications [40, 41]. Less than 3% of the channels are defective. The track quality agrees with simulation to within 2%. Medians of the track-hit residual distributions after alignment indicate that the individual modules are aligned to substantially better than 10 μm. The muon track reconstruction efficiency is at the 99% level. The muon momentum resolution measured with di-muon events agrees with simulation to better than 5%, and the muon momentum scale is known to better than 0.1% [9, 10].

The success of the development and operation of these silicon trackers is the basis for the next generation at the LHC, capable of handling the increase in LHC luminosity with collision rates in excess of 5×10^9 collisions per second. This will require Si trackers with increased radiation resistance, readout electronics with ten times the data transmission capacity and novel approaches for electrical powering.

8.8 Two Approaches to High Resolution Electromagnetic Calorimetry

Tejinder Virdee

The motivation for, and the importance of, high-resolution electromagnetic calorimeters at the LHC arose from the desire to cover the full range of allowed masses for a putative standard model (SM) Higgs boson. In the early 1990s, it was believed that a SM Higgs below a mass of about 100 GeV should be detectable at LEPII, where eventually a limit of 114.4 GeV was obtained.

In the (low) mass interval 100 GeV–130 GeV the dominant decay mode of the SM Higgs boson is into a $b\bar{b}$ pair. It is particularly difficult to cover this region at the LHC, because this decay mode is masked by copious production of $b\bar{b}$ pairs in pp collisions. In the studies performed in the early 1990s the two-photon decay was considered to be the final state most likely to give an early significant signal of a low mass SM Higgs boson at the LHC, provided the energies of these photons could be measured with instruments of an unsurpassed quality. This mode became the benchmark to evaluate candidate devices for electromagnetic calorimetry. Furthermore, in the range 130 GeV–500 GeV, the SM Higgs should be detectable via its decay into two Z bosons that subsequently decay, leading to a final state with either 4e, or 2e2μ, or 4 μ, where electrons, as well as muons, play a central role.

In July 2012 the clearest signals for the discovery of the Higgs boson at $m_H \simeq 125$ GeV at the LHC appeared in the same two channels mentioned above.

The design of ATLAS and CMS electromagnetic calorimeters

In the early 1990s, there was no candidate technology for electromagnetic calorimeters that could meet at the same time the diverse requirements such as:

- Very good energy resolution ($\sigma/E \leq 1\%$ at 100 GeV), and accurate enough determination of the angle between the two photons so as not to degrade much the instrumental di-photon mass resolution and hence clearly observe a small and narrow signal on a large background.
- Fine lateral and longitudinal segmentation.
- Fast response: at design operation bunches of protons cross every 25 ns and some 20 pairs of protons would be colliding in every bunch crossing, in addition to the interaction of interest. Thus, the products of the collision of interest may be confused with those from other collisions in the same bunch crossing. This confusion can be reduced by using highly granular calorimeters with good time resolution ($\ll$25 ns), giving low occupancy (fraction of detector elements that contain useful information), at the price of a large numbers of detector channels.

- Radiation resistance to ensure stable response throughout the lifetime of the experiment. At the smallest angles (~ 6°), the radiation levels reach a dose of 100 kGy during a ten year operation of the detector. This level of radiation severely limited the choice of possible detection media.

When a high energy photon or electron traverses a dense (high-Z) medium, they give up their energy by creating a shower of secondary photons and electrons and positrons [Box 6.3]. The parameters that determine the thickness needed and the lateral extent of the electromagnetic showers are the radiation length (X_0) and Molière radius (R_M), respectively. Typically, $25X_0$ are needed to contain high-energy showers and R_M should be as small as possible. The electrons and positrons either excite the atoms (whereupon scintillation light is emitted through de-excitation) or ionize the atoms (creating electron–ion pairs which separate rapidly in an externally applied high electric field). The amount of scintillation light or electron–ion pairs generated is proportional to the energy of the original particles.

Several R&D programmes were launched, particularly on scintillating inorganic crystals (in view of improving their radiation resistance) and on noble liquids, working as ionization detectors (in view of improving their speed and granularity).

The ATLAS and CMS Collaborations chose two very different instruments, the sampling "accordion" liquid argon calorimeter of ATLAS [9] and the homogeneous $PbWO_4$ crystal calorimeter of CMS [10] yet meeting overall the same physics goal. Technically they differ in many ways as shown in Table 8.2.

Table 8.2. Some parameters of interest for the e-m calorimeters of the ATLAS and CMS experiments.

Parameter	**ATLAS**	**CMS**
Intrinsic energy resolution		
Stochastic term ($a/\sqrt{E}$, E in GeV)	10%	3%
Constant term (c)	0.2%	0.5%
e-m cluster energy equivalent of noise GeV (b/E, E in GeV)	0.25	0.20
Effective radiation length (cm)	2.3	0.9
Effective Molière radius (cm)	4.5	2.2
Segmentation		
Lateral (major sampling) $\Delta\eta\times\Delta\phi$	0.025×0.025	0.017×0.017
Number of Longitudinal Samplings	4	1
Speed of response	Max. drift time = 400 ns	Time constant = 20 ns
Electronics peaking time	50 ns	45 ns
Material in front of ECAL ($\lvert\eta\rvert$ = 0 (1.7)) X_0 (excluding solenoid in the case of ATLAS)	0.35 (1.35)	0.35 (1.5)
Sensitivity to radiation	Relatively insensitive	Sensitive (see text)

It is clear that a compromise had to be struck by both ATLAS and CMS in terms of the technical characteristics — some which lead to a "better" performance in one experiment or the other. Nevertheless, both experiments have achieved a similar overall physics performance as evident from the estimated values of the *expected* $H \to \gamma\gamma$ signal significance from the Run 1 data of 4.6σ and 5.1σ for ATLAS and CMS, respectively.

A further challenge was the design of front-end electronics that had to have low energy equivalent of electronics noise, low-power consumption, be radiation resistant, and capable of covering a dynamic range (~16 bits) from a few tens of MeV to several TeV (energy of the largest anticipated signals, for example from a putative Z' decaying in an electron-positron pair).

As opposed to momentum measurements in a magnetic field, the relative resolution of calorimeters improves with energy, typically as its square root. The energy resolution is usually parameterized as: $\sigma/E = a/\sqrt{E} + b/E + c$, where a is the "stochastic" term, b the "noise" term and c the "constant" term. The challenge is to simultaneously keep all three terms small. For homogenous electromagnetic calorimeters it is possible to achieve a value of a few percent whereas in sampling electromagnetic calorimeters $a \sim 10\%$ is normally achievable. However, at high energies, a major limitation arises from the size of the constant term that is related to uniformity and time-stability of response, and importantly the accuracy of the inter-cell calibration. In practice, for large systems it is challenging to reach overall accuracies significantly below 1% at high energies (~ 100 GeV or higher), even for the most performant electromagnetic calorimeters.

ATLAS accordion liquid argon sampling electromagnetic calorimeter

Liquid argon calorimetry offers several attractive performance features, including a response that is intrinsically linear, stable over time and tolerant to high levels of radiation. In liquid argon the secondary charged particles in showers produce electron–ion pairs. The signal develops on the electrodes as soon as the freed charges start drifting in the liquid, i.e. instantaneously. The charges drift for a time of ~ 400 ns in a 2 mm liquid argon gap at a field of ~ 2 kV/cm. A shorter effective response time can be achieved by electronic pulse shaping [Highlight 4.10].

The challenge of a liquid argon calorimeter for use at the LHC lay in the implementation of fine segmentation and fast speed of response. Conventionally, in sampling calorimeters, the absorber and active planes had been oriented perpendicular to the incident particles. But in such a geometry, it is difficult to realise fine lateral segmentation, with small size cells and towers that should point towards the interaction region. Longitudinal subdivision should be implemented without introducing insensitive regions and a large number of penetrating interconnections which degrade the performance. To overcome these shortcomings a novel absorber-electrode configuration, known as the "accordion" (Fig. 8.17(a) and (b)), was introduced, in which the particles traverse the chambers at angles of around 45°. This accordion geometry enables implementation of high granularity and seamless coverage. The electrodes can easily be grouped into towers at the front or at the rear of the calorimeters, such that adjacent modules can be tightly packed, without introducing cracks between modules. The absorber is made of lead plates, clad with thin stainless steel sheets for structural rigidity as shown in Fig. 8.17(a) and (b), sandwiching copper-clad Kapton electrodes positioned centrally by precision-cut honeycomb. Details of the sampling structure can also be seen in Fig. 8.17(b). Longitudinal subdivision and pointing towards the interaction region is achieved by the way the electrodes are etched (see Fig. 8.17(b)).

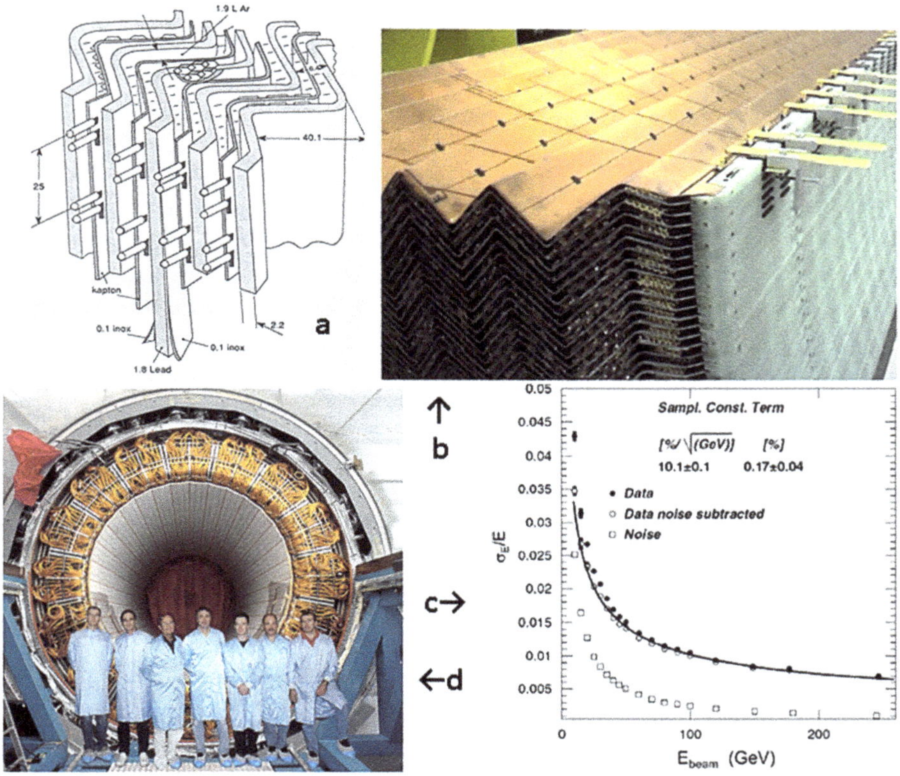

Fig. 8.17. Clockwise from top left: (a) The conceptual drawing of the ATLAS "accordion" liquid argon calorimeter, (b) partial view of a calorimeter module, (c) the energy resolution measured using a prototype for electrons in test beam, (d) the completed barrel part of the liquid argon calorimeter before lowering into the underground experiment cavern.

Considerable engineering challenges were overcome as the several innovative features resulted in major production issues. The most difficult was the fabrication of large electrodes, ~ 2 m in length containing nearly 1000 resistive pads each. This problem was solved by careful monitoring of the production on-site by experts from the collaboration. Approximately 20 000 m^2 of honeycomb spacers have been used to maintain the flexible electrodes in the centre of the gap between absorbers. To avoid major problems with the high-voltage behaviour of assembled modules, a rigorous and careful cleaning procedure for all parts, especially the honeycomb, had to be implemented. Radiation-tolerant electronics had to be produced for all components in the cavern, including the front-end for nearly 200 000 channels, housed near the cold-to-warm signal feed-throughs.

The performance of the liquid argon calorimeter in test beams is illustrated in Fig. 8.17(c). The completed barrel part of the liquid argon calorimeter before lowering into the ATLAS underground experiment cavern is illustrated in Fig. 8.17(d). The result related to the physics benchmark reaction — that of the di-photon mass spectrum, clearly showing the presence of the Higgs boson, is illustrated in [13].

CMS lead tungstate scintillating crystal calorimeter

Lead tungstate scintillating crystals offer several advantageous performance features. Compact size — a length of 23 cm is sufficient to absorb the energy of high energy electrons and photons, good tolerance to high levels of radiation, and fast response — 90% of the scintillation light is emitted within 20 ns. But it also has a number of drawbacks. The light yield is low and strongly temperature dependent (–2%/°C). The low yield brings the need for photo-devices that can provide gain in high magnetic field — ruling out photomultipliers for the barrel region. *Silicon avalanche photodiodes* (APDs) were eventually chosen because they can provide sufficient gain in magnetic fields. The amount of light collected is affected by loss of transparency due to colour centres created by radiation.

In 1994, tests on a small number of lead tungstate crystals coupled to APDs yielded very encouraging results, convincing the CMS collaboration to adopt this technique. Some examples of the technical challenge faced by CMS are given here. Between 1993 and 1998 much research and development work was carried out to improve the transparency and the radiation hardness of the crystals. The relative amounts of lead oxide and tungsten oxide in the crystals had to be optimized; an affordable means of ensuring adequate purity of the raw chemicals had to be set up; and any remaining defects in the crystals structure had to be compensated by identifying and introducing specific dopants. Although the intrinsic scintillation mechanism is not affected by radiation, tests in particle beams revealed a loss of transparency dependent on the absorbed dose rate. It was also found that the crystals show recovery of transparency when the irradiation ceased, revealing a dynamic behaviour of the yield of collected light. CMS is able to maintain the good performance by deploying a powerful laser-light monitoring system to precisely track, and correct for, changes in the transparency of each and every crystal. The mass manufacture of crystals took place between 1998 and 2008. The crystals were manufactured largely in Russia, but with some in China. In Russia a high-volume crystal growing (Fig. 8.18(a) and (b)) and processing capacity had to be put in place in a factory previously in the military-industrial sector. The last of ~ 75 000 crystals were delivered in March 2008, and the full electromagnetic calorimeter was installed by mid-August 2008.

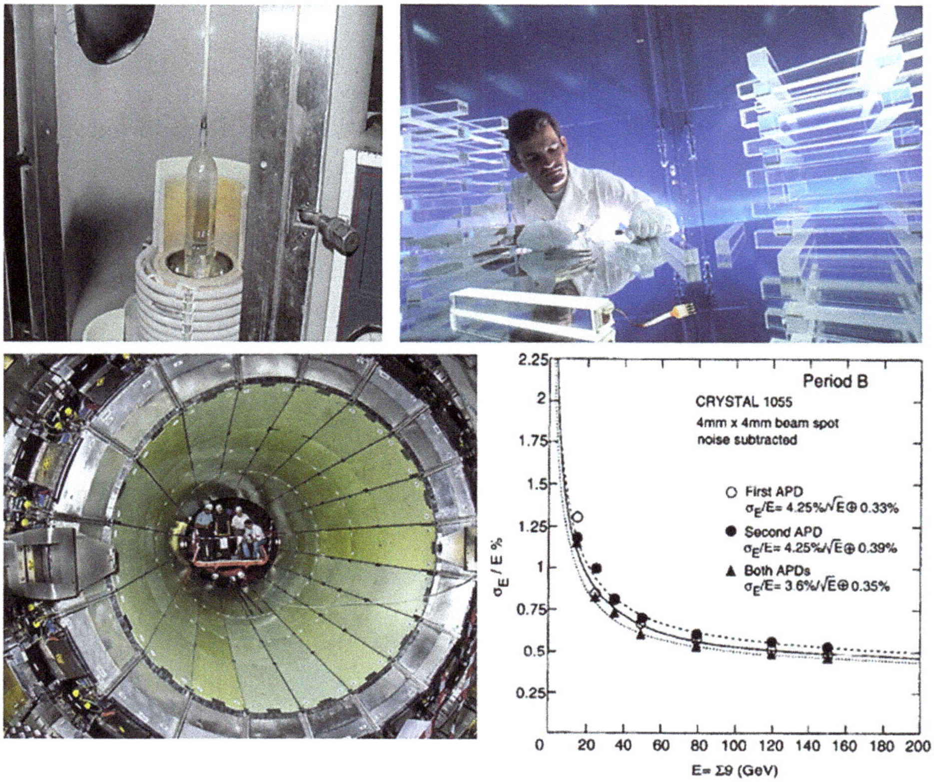

Fig. 8.18. Clockwise from top left: (a) An ingot of lead tungstate crystals just after the 3-day growing cycle, (b) a set of cut and polished crystals, and one in the foreground with two APDs mounted, (c) the energy resolution measured using a prototype for electrons in test beam (d) the completed barrel part of the calorimeter just after installation in the CMS underground experiment cavern.

Engineering challenges included the stabilization of the temperature of about 100 t of crystals and 130 000 APDs to 0.1°C (achieved by water cooling) and stabilizing the bias voltage of the APDs to some tens of mV. This was necessary because the APD gain has large temperature (−2.4%/°C), and bias voltage dependence (3%/V). The goal was to keep these contributions to the energy resolution at the ‰ level. The crystals were supported by specially developed, lightweight carbon-fibre or glass-fibre structures and integrated with high performance electronics. The performance of the lead tungstate calorimeter in test beams is illustrated in Fig. 8.18(c). Figure 8.18(d) shows the installed barrel part of the crystals calorimeter in the CMS underground cavern. The result related to the physics benchmark reaction (that of the di-photon mass spectrum) clearly showing the presence of the Higgs boson, is illustrated in [13].

Concluding remarks

It is remarkable, if not surprising, that the ATLAS and CMS Collaborations have chosen two very different techniques for their electromagnetic calorimeters, yet addressing and meeting overall the same physics goals. This did not happen by design or by a committee decision. It happened, because even in these 3000-plus strong collaborations the physicists and engineers use their experience, familiarity, conviction and "taste" in selecting an experimental approach and shaping the decisions. In each of the two collaborations members were familiar with the respective technology and were able to convince their collaborations that the technology could be honed to achieve the required performance. "Required" means achieving a measurement accuracy which approaches the intrinsic performance limited only by the fundamental aspects. There was little room for even the slightest degradation, lest the discovery of the Higgs boson be jeopardized. That both collaborations reported, on the same day of July 4, 2012, the discovery of the Higgs boson with the same experimental significance is a tribute to the foresight, intuition, perseverance and dedication of many hundreds of students, technicians, engineers and physicists worldwide.

8.9 Multigap Resistive Plate Chamber: Chronometry of Particles

Crispin Williams

The ALICE experiment at the CERN LHC is designed to study nuclear matter under extreme conditions of temperature and density produced in collisions of heavy ions. Knowledge of the identity of the particles emerging from such a collision is crucial for this research. That this is not a trivial task can be appreciated by observing Fig. 8.19, which displays a typical heavy ion collision. Among the thousands of particles, which are, for example, the protons, pions or kaons?

Most of these particles are produced with momenta less than a few GeV/c. For these particles a frequently used technique is the "Time-of-Flight" (ToF) velocity measurement. The curvature of the tracks in a magnetic field determines the particle momentum (i.e. "relativistic" mass × velocity). If the velocity is also measured, its mass can be determined; hence its identity is inferred. All that is needed is to surround the central tracker with a detector capable of precisely measuring the arrival time of the particles. This, together with the time the collision occurred and the distance to the point of collision gives the time-of-flight (and its velocity) and thus, with the momentum measurement, its mass. The detector must be highly segmented, so that each of the thousands of particle tracks has its own individual time stamp. This detector must also work in a strong magnetic field.

In the late 1990s ALICE started to develop its experimental concept. ToF-identification was an essential part, but none of the existing techniques were remotely close to meeting the novel and stringent specifications: a new detector was needed. The breakthrough was the "The Multigap Resistive Plate Chamber" [42].

The principle of this detector is as simple as it is elegant. It is a derivative of the time-honoured gaseous parallel plate detector. A single gap Resistive Plate Chamber (RPC) is shown in cross section in Fig. 8.20. A strong electric field is applied across this gap. A charged particle traversing the gap ionizes the gas molecules producing positive ions and electrons. Typically, a few primary ionisation collisions occur in the 2 mm gap. The electrons are accelerated in the strong electric field, collide with and further ionize the gas molecules leading to

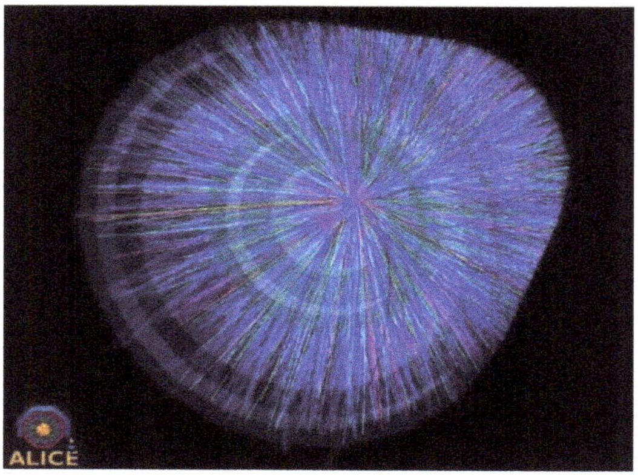

Fig. 8.19. A lead-lead collision observed in the ALICE experiment. Each of the lines represents a particle track. (Source: ALICE).

an avalanche [Highlight 4.8]. The movement of these charges induces signals at the pick-up electrodes. Registering the induced signal provides a measurement of the position of the particle and the arrival time. The time resolution is quite respectable being typically 2×10^{-9} sec (2 ns). But this was not good enough for ALICE where a time resolution of < 0.1 ns (100 ps) was required.

Progress came through a detailed understanding of the processes influencing the time resolution. Only avalanches that are initiated close to the cathode can grow to a sufficient size able to produce a detectable signal. Fluctuations in its initial growth give rise to time jitter (i.e. uncertainty) of the observed signal. Decreasing the width of the gap decreases the time jitter; but this also decreases the efficiency of the detector. The *Multigap Resistive Plate Chamber (MRPC)* is a solution invented and developed by the LAA project at CERN. The gas is divided into several small gas gaps (five in this example) with a single pair of readout electrodes, as shown in Fig. 8.21. The voltage is applied to the outer surfaces of the two external plates. The internal plates are allowed to float electrically, which considerably eases construction. The MRPC is made by stacking sheets of resistive material (window glass in the case of the ALICE TOF barrel). The gap size is controlled by running a fishing line (!) across the surface. This is the short version of a long story of heated debates between experts, on whether such a device would work and how exactly it should be constructed. Details are revealed in [42].

A crucial aspect of precise timing is the readout technique for a highly segmented large-area detector. Suppression of electronic noise is mandatory in order not to spoil the timing resolution. This is best achieved with a 'differential' readout, where both pickup electrodes are separated from the electrical ground of the system, and connected to a differential amplifier. An ultra-fast differential

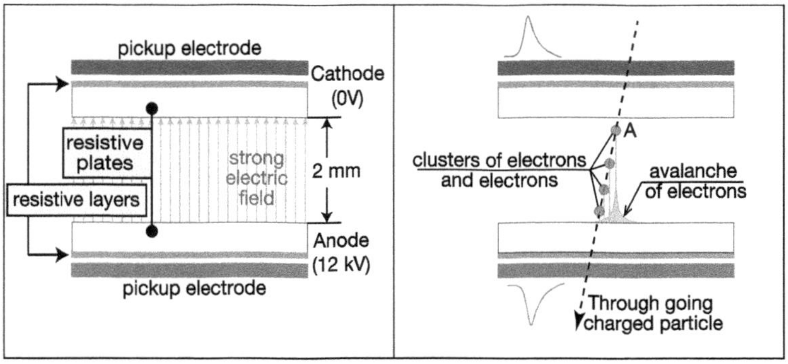

Fig. 8.20. Cross section of single 2 mm gap RPC. A traversing charged particle creates clusters of electrons and ions in the gas. The electrons avalanche across the gap due to the high electric field.

amplifier-discriminator, known as the NINO, was developed in 250 nm CMOS technology and has become the standard for precise timing. Together with these developments, MRPCs have been shown to be excellent high performance TOF detectors meeting the requirement of a systems time resolution below 100 ps [42].

For ALICE a TOF cylindrical detector of radius 3.7 m with a length of 7 m was built surrounding the Time Projection Chamber (TPC). The MRPCs have ten 250 micron gaps; the detector planes are segmented into 2.5×3.5 cm^2 pads, resulting in a total of 160,000 signal channels. All MRPCs are operated at the same applied voltage and the same discriminator threshold. Figure 8.22 shows a typical plot of $\beta = v/c$ (measured by the TOF) versus the momenta measured by the TPC. At low momenta, the dE/dx measurements from the TPC provide additional identification power [Box 5.2]. A similar detector based on these MRPCs has been built for the STAR heavy ion experiment at RHIC. A TOF barrel built from MRPCs will be installed for the BES III experiment at IHEP, Beijing. The CBM experiment at GSI/FAIR is planning to build a 120 m^2 wall of MRPCs for TOF measurements [43]. There are also proposals to build a neutron detector with precise timing with MRPCs embedded between iron plates. MRPCs are also being considered as detectors of 511 keV gamma photons for Positron-Emission-Tomography (PET), where its excellent time resolution would improve the imaging quality.

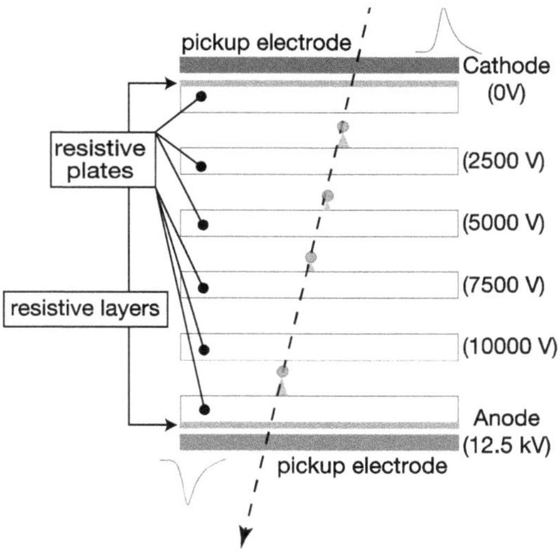

Fig. 8.21. Schematic cross-section of a Multigap Resistive Plate Chamber (MRPC). Voltage is only applied to the outer surface of the two external plates. All internal plates are electrically floating.

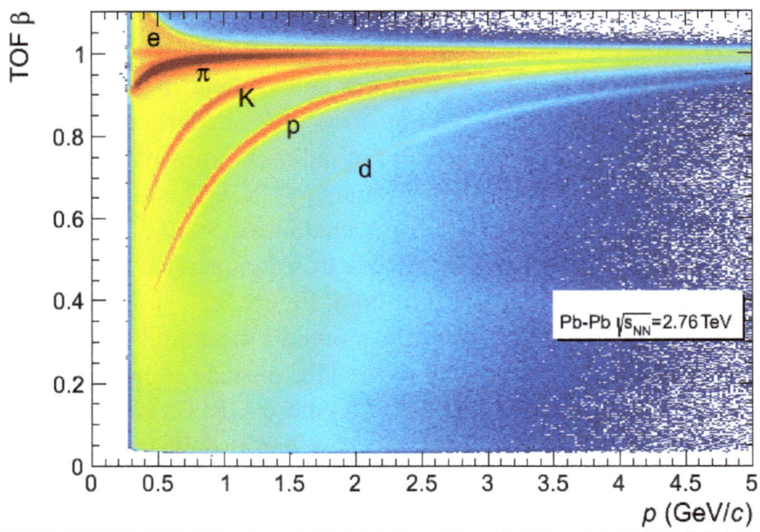

Fig. 8.22. Distribution of particles with given velocity $\beta = v/c$ (measured by the TOF barrel) versus their momenta. (Source: ALICE).

8.10 The LHCb RICH: Lord of the Cherenkov Rings

Carmelo D'Ambrosio and Olav Ullaland

The LHCb experiment is dedicated to the study of CP violation [Box 3.4] and the rare decays of heavy flavours [Box 6.4]. A key to these studies is the identification of the many particles produced in the complex decay chains of particles containing the b-quark. The Ring Imaging Cherenkov (RICH) detector system provides charged particle identification over a wide momentum range, from 2 to 100 GeV/c.

A major requirement for charged hadron identification in a flavour-physics experiment is the reduction of combinatorial background. Many of the interesting decay modes of b- and c-flavoured hadrons involve hadronic multibody final states. At hadron colliders like the LHC, the most abundantly produced charged particle is the pion. The heavy flavour decays of interest typically contain a number of kaons, pions and protons. It is therefore important when reconstructing the invariant mass of the decaying particle to select the charged hadrons of interest in order to reduce the combinatorial background. Knowledge of the particle identity also helps to distinguish between final states of otherwise identical topology. An example is the two-body hadronic decays, $B \to h^+h^-$, where h indicates a charged hadron. In this case there are many contributions, as illustrated in Fig. 8.23, including $B^0 \to \pi^+\pi^-$, $B_s^0 \to K^+K^-$ and other decay modes of the B^0, B_s^0 and Λ_b [44].

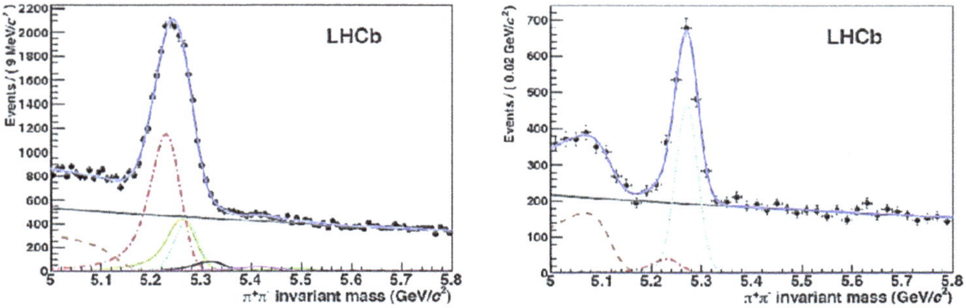

Fig. 8.23. Invariant mass distribution for B $\rightarrow h^+h^-$ decays in the LHCb data before the use of the RICH information (left), and after applying RICH particle identification (right). The signal under study is the decay $B_0 \rightarrow \pi^+\pi^-$, represented by the turquoise dotted line. The contributions from different b-hadron decay modes ($B_0 \rightarrow K\pi$ red dashed-dotted line, $B_0 \rightarrow$ 3-body orange dashed-dashed line, $B_s \rightarrow KK$ yellow line, $B_s \rightarrow K\pi$ brown line, $\Lambda_b \rightarrow pK$ purple line, $\Lambda_b \rightarrow p\pi$ green line), are eliminated by positive identification of pions, kaons and protons and only the signal and two background contributions remain visible in the plot on the right. The grey solid line is the combinatorial background [44].

A measurement of the Cherenkov photon emission angle (Θ_C) of a charged particle, together with the knowledge of the refractive index of the Cherenkov medium ($n(\lambda)$), determines its velocity (v), where $v/c = 1/(n \cos\Theta_C)$. A Cherenkov detector is not a stand-alone detector. An accurate determination of the mass is obtained by coupling this information with the knowledge of the momentum and the trajectory of the particle [Boxes 3.2 and 6.3]. As the Cherenkov light is produced along a cone at a polar angle Θ_C relative to the charged particle trajectory, a spherical or parabolic mirror is used to project these photons in a well-defined, ring-like pattern onto a photon detector plane. This is the basic principle behind a Ring Imaging Cherenkov detector [Highlight 7.8].

The LHCb RICH Particle Identifier comprises two separate detectors, RICH 1 and RICH 2, (Fig. 8.24). The first, RICH 1, is upstream of the magnet and covers the momentum range from ~ 2 to 60 GeV/c. RICH 2 is placed just in front of the calorimeters and covers momenta up to 100 GeV/c. The main task of the two detectors is to efficiently identify charged hadrons, as shown in Fig. 8.23.

Some first generation RICH detectors had problems with ease and reliability of operation. Common to these devices was detection of photons in the far-ultraviolet (VUV) range, with wavelength between 160 mm and 200 nm. The refractive index increases rapidly in this region and uncertainty in determination

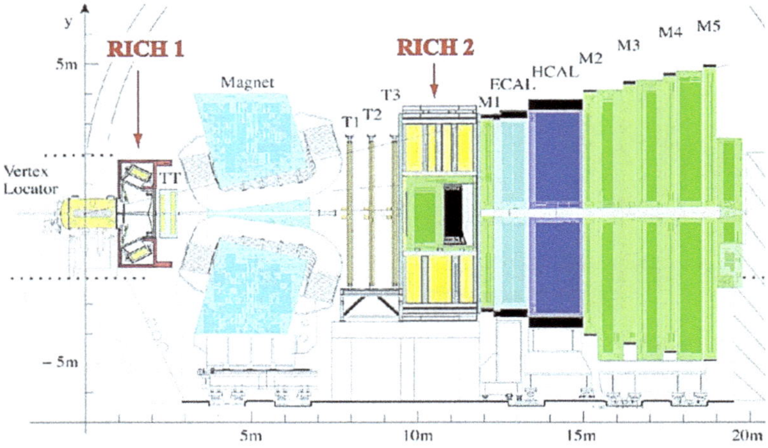

Fig. 8.24. Side view of the LHCb experiment.

of the Cherenkov angle is dominated by the chromaticity of the radiator gas.[a] Operating in the VUV region also meant that impurities, such as O_2, H_2O or CO_2, at the ppm level in the gas would absorb the photons. A very high purity gas was therefore required.

The reason for the detection of photons in the VUV region was simply that no single detector for visible photons with a spatial resolution better than a centimetre had yet been developed. Experiments had to rely on specially adapted gas chamber detectors [Highlight 4.8], which required a photon energy in the VUV region to ionize the gas and generate the signal. This was acceptable, as the number of Cherenkov photons grows quadratically with the photon energy so that sufficient numbers could be detected. However, calculations showed that detecting photons in the visible range (VIS) would decrease the chromatic effects on the overall error and thus pave the way to much more precise detectors. Specifically, the error on the Cherenkov angle is given by $\sigma_\theta/\sqrt{N}$ where σ_θ is the single photon Cherenkov angle error and N the number of detected photons. When working in the VUV, experiments tried to maximize N. In the visible range, giving up photons while simultaneously decreasing σ_θ results in a much improved Cherenkov ring resolution. Such a change of perspective was not easy, as it required a complete re-appraisal of the available tools, instruments, hardware and software, involving all RICH system components, from particle-sensitive hardware to electronics and data acquisition and processing. For the hardware the following steps were taken:

- Choice of gases having intrinsically low chromaticity with good sensitivity to particle momenta between 2 GeV/c and 100 GeV/c. The gases chosen were

[a] RICH detectors can also use liquid or solid radiators. Here only gas radiators are considered.

C_4F_{10} in RICH 1 and CF_4 in RICH 2. These gases are relatively easy to handle and, by working in the visible range, acceptable gas impurities are in the percent range rather than the few parts per million (or billion) for VUV.

- Development of efficient single-photon detectors working in the VIS with high spatial resolution (pixel sizes from 0.25 to 2.5 mm), called pixel-HPD (HPD for Hybrid Photon Detector). Their development had been started in the 1990s and was pursued at CERN, triggered by the need for precise photon detectors for scintillating fibres. The basic idea was to insert in a vacuum envelope a silicon pixel detector and electronic chip to directly detect and digitize photo-electron signals produced by the Cherenkov photons. This is not an easy task and it took more than 10 years to realize the first mature example of this completely new class of photon detector. They feature high spatial resolution, an impressive signal to noise ratio and can be assembled in arrays. There are 484 pixel-HPDs in the RICHs covering a surface of ~3.6 m^2.
- Development of thin, light, precise and highly reflective mirrors, together with their support and alignment structures, to focus the photons on the pixel HPDs. Mirrors had to be thin to minimize material traversed by the particles. A mirror array contains hundreds of components totalling a reflective surface of ~19 m^2. They are aligned to a precision of 0.05 mrad to form an almost perfect sphere and minimize the error on σ_θ of the single photon Cherenkov angle.

RICH 1 features carbon fibre composite mirrors and RICH 2 has 6 mm thin glass mirror arrays. They are held in place either by carbon fibre (RICH 1) or by aluminium honeycomb (RICH 2) structures with Polycarbonate micrometric adjustments. With these technologies the mechanical structures for these large RICH detectors represent only ~2.5% and ~10% of a radiation length, respectively. To give an idea of its size, the RICH 2 volume is larger than 100 m^3. After about ten years of development, all the mirrors show excellent alignment stability and tilting errors lower than 0.1 mrad after online corrections. The result of this effort is best illustrated with an example: Fig. 8.25 shows the "world record" distribution of σ_θ, obtained with RICH 2 in 2012 [45].

The ability to detect sharp Cherenkov rings in the LHCb RICH system allows to detect and identify several rings simultaneously. This process is referred to as pattern recognition. A new "global search algorithm" based on statistics and probability was developed. Instead of searching for rings, the probability for each detected photon to belong to a possible ring is evaluated. By constructing and minimizing a likelihood function, the mass and hence the identity of the particles is assigned. The RICH system can handle thousands of photons produced by hundreds of particles up to a million times per second and deliver information about the particles crossing the detector and their underlying physics processes.

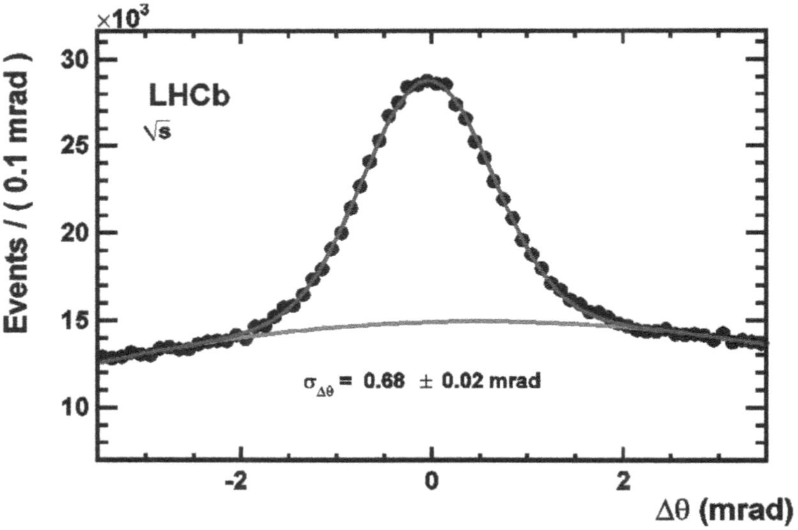

Fig. 8.25. Single-photon Cherenkov angle precision of 0.68 mrad for LHCb RICH 2 [45].

For 2020, LHCb is planning to further improve the RICH detectors aiming for operation at 10 times the present collision rate of $3 \times 10^7\ s^{-1}$ at LHCb and 40 times the readout rate. Beyond that, one can anticipate a new "*quantum leap*" producing the RICH detectors of the future, when new components, ideas and a good dose of human ingenuity will all synergize, thoughtfully and skilfully in the conducive laboratory environment of CERN.

8.11 Signal Processing: Taming the LHC Data Avalanche

Philippe Farthouat

When the concept of the LHC program began to take shape, physicists and engineers were confronted with a seemingly unconquerable world of experimentation: proton bunches colliding 40 million times a second, entailing a billion collisions, 10^{11} particles, and data volume approaching 100 TB (10^{14} Bytes) per second [Box 8.3]. It would take a worldwide-coordinated R&D programme to possibly find solutions to the design and construction of the detectors matching such requirements. Particularly worrisome was the question whether the signal and data processing systems for the hundred million detector channels could be developed and built. It took courage, sweat, stamina, ingenuity, collaboration with industry and a fair dose of luck to write a happy end to this story, as told below.

Each of the several large sub-detectors of an LHC experiment has its custom-designed signal processing system, taking care of the necessary analogue signal conditioning (amplification, filtering, etc.), analogue-to-digital conversion, digital signal processing (data compression, digital filtering, etc.), and interface to the data acquisition system of the experiment. The LHC characteristics, the trigger and data acquisition functional models, the harsh environment and large number of channels are factors that drove the definition and design of the readout electronics and led to some spectacular technical developments [46].

Each experiment contains a large number of readout channels: up to 100 million for the inner trackers, several hundred thousand for the calorimeters and about one million for the muon spectrometers. These astronomical numbers could not be handled without integrating a large fraction of the readout electronics directly into the detector (front-end electronics, FEE), and extensive multiplexing before transmission to the off-detector electronics (back-end electronics, BEE) located in service rooms at distances of up to 100 m away from the detectors. The high density of signal channels demanded the FEE to be implemented with Application Specific Integrated Circuits (ASICs). Located on the detector, they are subject to extreme environmental conditions, in particular radiation and high magnetic fields.

In nominal conditions, the LHC delivers bunches of protons colliding in the centre of the experiments every 25 ns and ideally, one would record all the activity in the detectors every 25 ns. However, with an event data size of a 1 MB or more, a readout and storage bandwidth of several tens of terabytes per second would be required. In addition, most of the events are of little interest for physics studies. It took — a famous example — about 10^{15} collisions before the Higgs boson discovery was announced, based on a few hundred particles identified as Higgs boson decays. Clearly, registering all the information is not the way to proceed.

In the strategy adopted only the events of potential physics interest are selected out of the billion collisions per second. This selection is probably the most critical part of the data processing system. The experiment teams developed a multi-level selection ("trigger") approach. The level-1 trigger uses hardware processors, implemented off-detector in state-of-the-art Field Programmable Gate arrays (FPGAs) and exploits only a subset of information, e.g. from the calorimeters and the muon spectrometers in the case of ATLAS and CMS. Selection criteria are energetic muons, electrons, photons or events with a high multiplicity of particles, signalled by a large energy deposit in the calorimeter, and provide the basis for a decision to select or reject an event. It delivers a YES (Level-1 accept signal [L1A]) or NO decision every 25 ns within a fixed decision time interval ("latency") in the range 2 μs–3.2 μs and at an average rate of up to 100 kHz for ATLAS and CMS, and a latency of 4 μs and 1 MHz trigger rate for LHCb.

During this latency, about 100 new bunch crossings occur and the signals corresponding to these bunch crossings must be stored in so-called level-1 buffers. The correct data retrieval from the level-1 buffers relies on the fact that the L1A has a fixed latency time of N bunch crossings with respect to the interaction time. When a L1A signal is received by the front-end electronics, the data of the event giving rise to this L1A were therefore stored in the level-1 buffer location corresponding to N bunch crossings earlier. These data are then selectively retrieved and transmitted to the back-end electronics. However, the complete information of an event is spread over several thousand locations and must subsequently be correctly assembled. For this purpose, the entire readout system is synchronized with the 40 MHz bunch crossing (BC) clock of the LHC machine.

Following this transmission to the off-detector electronics the subsequent selection proceeds in the high-level trigger. This is software-based and implemented in arrays ("farms") of processors (PCs) accessing the data through industrial-standard networks. In the high-level trigger either a subset of data of potential interest as flagged by level-1 trigger information or complete data of an event are analysed in more detail (Event Filter), ultimately reducing the flow of events to a few hundred Hz. These events are permanently stored for further physics analysis. The high-level trigger needs up to 1 s computation time per event. Farms of ~ 2000 PCs with more than 30 000 cores working in parallel perform the processing tasks. As an example, the data flow and selection for ATLAS is shown in Fig. 8.26.

There is a large radiation background when the LHC operates and electronics is strongly affected by this radiation. As an example, threshold voltages in CMOS ASICs are sensitive to the total ionizing dose, the current gains of bipolar transistors are sensitive to displacement damages caused by non-ionizing energy loss, and energetic particles may produce single-event upsets in the digital electronics. Very nasty destructive events such as latch-up or gate burnout may also occur. The radiation levels in the centre of the experiments reach up to 1 MGy and several 10^{14} 1-MeV-neutron-equivalent/cm^2 for a 10 years of operation.

The front-end ASICs were designed either in radiation-hard processes or in deep sub-micron CMOS technologies with special design rules, such as enclosed layout transistors (ELT), a technique invented years ago, but revived and extensively used for the LHC experiments. All sensitive digital elements were implemented in triplicated logic with procedures such that an upset in one circuit element would not disturb the circuit functionality. Commercial Off-The-Shelf Components (COTS) could be used for detectors receiving lower radiation after thorough radiation qualification.

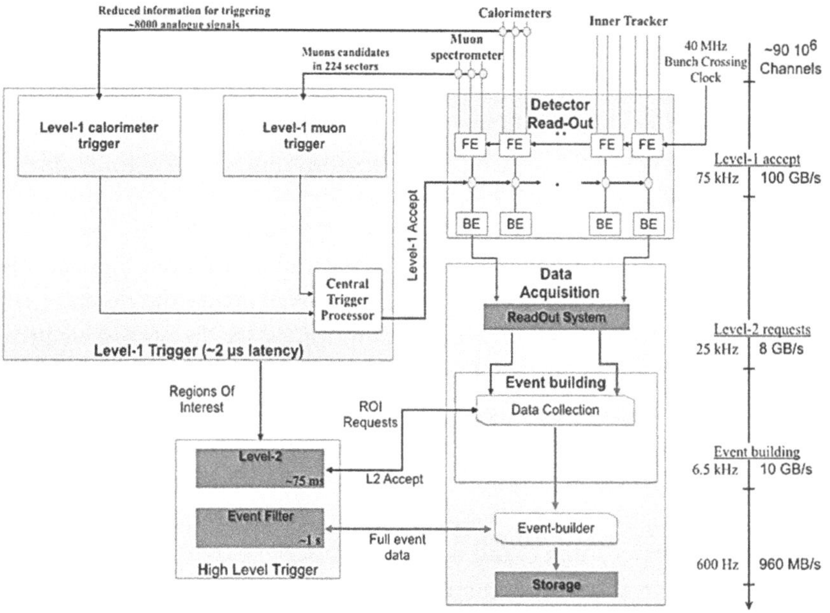

Fig. 8.26. Example of readout, trigger and data acquisition architecture (ATLAS). The front-end electronics (FEE) records the detector signals at the bunch crossing frequency (40 MHz). The information is retained during the Level-1 trigger latency. The data corresponding to a Level-1 Accept are sent to the back-end electronics (BEE) through dedicated front-end links and then, after data formatting and processing, to the data acquisition system. A selected subset of data is sent to the level-2 trigger which makes a decision in about 75 ms. The data fragments of the events to be kept are assembled in the event builder. The complete events are then processed in the Event Filter (making a decision in about 1 s) and the fraction of accepted events are stored for later physics analysis. The required data bandwidths at each level as well as the rates involved are shown.

Developing this radiation-tolerant electronics remained for many years the biggest hurdle towards functioning signal processing systems. Ultimately, solutions were adopted in collaboration with semiconductor foundries, exploiting in new ways the features that were introduced for standard commercial products.

Besides the complex analogue processing, which is different for the various detectors of the experiment and which processes input signals as low as a few thousand electrons, the front-end electronics contains the level-1 buffer. Usually it also handles the analogue to digital conversion (effective resolution up to 10 bits and dynamic range up to 64 000) and the data compression, reducing the data volume to be transmitted and hence the number of readout links.

In most cases digital optical links were selected because of their high bandwidth and very low material mass (optical fibres are by far lighter than copper links). It required the radiation hardness qualification of optical devices (edge emitting lasers and PIN diodes), the development of low-mass packages and for the CMS tracker the development of an analogue optical link. A high-speed radiation-hard serializer able to feed an optical link at 1.6 gigabit per second has also been developed and extensively used. The installed density of optical links is unique: 20 000 are installed in ATLAS and more than 100 000 in CMS.

The power dissipated by the front-end electronics required special attention because it occurs in confined places; bringing power inside the detector requires thick copper cables. The total power consumption of the front-end electronics is ~ 350 kW for each ATLAS and CMS. Operated at low voltages, tens of kA are needed. In order to reduce the size of the cables penetrating the detector volume, DC-DC converters located at places with low magnetic field in the experimental caverns, followed by linear voltage regulators in the detector volume are used. This again required the development, often in collaboration with industry, of radiation tolerant devices that could be installed in the experimental caverns.

The development and deployment of this signal processing system with its large number of radiation-hard ASICs (1.5 million ASICs have been installed in ATLAS and CMS) represents a unique achievement, even more so, considering the radiation levels, to which the readout electronics of the inner detectors are exposed.

This success, unique in so many ways, is a tribute to a world-wide collaboration of dedicated physics, engineering institutes and industry, with CERN assuming a substantial load and the coordination of these activities.

8.12 Giant Magnets for Giant Detectors

Alain Hervé, Peter Jenni and Thomas Taylor

All four major experiments, ALICE, ATLAS, CMS, and LHCb, feature large spectrometer magnets [47]. These devices are the costliest single components of the detectors, and many of the other components rely on their structure for support. In contrast with LEP, where 3 of the 4 detectors were assembled in a "garage position", the LHC experiments were installed in their final position on the beam. When the detectors are fully assembled the magnets are difficult to access; it follows that any significant repair would require a long shutdown, affecting the entire machine. This gave rise to two important constraints: (i) all the magnets had to be delivered and installed on time, and (ii) the magnet systems had to be designed for utmost reliability. Although the magnets were to be designed, manufactured, tested and installed under the technical and financial responsibility of the collaborations, they also had to satisfy constraints imposed by CERN: powering, cooling, cryogenics and controls would be supplied and maintained by the laboratory. For reasons of economy this infrastructure had to be standardized. Common safety standards also had to be applied. CERN therefore allocated experienced technical staff to the collaborations to help with the magnet projects, to ensure compatibility with the interfaces and to provide the essential leadership for ensuring success (Chapter 11).

Following the experience with LEP, CERN was already associated with studies for the LHC detector magnets before the collaborations gelled and the magnet designs were frozen. It was realized that the major detectors would require larger superconducting magnets than had been made before, and that they would have to run at higher current, probably around 20 kA. The architecture of the conductors for magnets for experiments — cabled Nb-Ti stabilized with pure aluminium — was well established in previous work (at CEA, Saclay, CERN, INFN, KEK and RAL), but only for currents of up to about 5 kA. The superconducting cable could be similar to that developed for the LHC dipole magnets [Highlight 8.2], but co-extrusion of the cable with the aluminium stabilizer would call for gaining access to very large extrusion presses, and extensive testing to verify that the superconductor was not degraded by overheating during the process. It was also essential to ensure good bonding between the cable and the high purity aluminium, and to be able to control the quality of the bond on production lengths. Collaborations were set up with ETH Zürich and CEA Saclay to develop and validate such conductor for use in windings for a practical magnet [48].

For the winding being envisaged for the powerful CMS magnet, designed for operation at 4 T, the high-purity Al used to stabilize the superconductor was not sufficiently strong to withstand the stress on the conductor when the magnet was excited to full field (4 T). One solution proposed was to surround the pure Al/Nb-Ti insert described above with aluminium alloy, to be co-extruded simultaneously [49]. This was found to be impractical as it required heating to > 500°C causing massive degradation to the superconductor. The alternative was to solder or weld Al alloy flanges to the pre-extruded insert — a process that became the subject of another development in collaboration with ETHZ. Working together with industry it was found that by using electron-beam welding the process was both feasible and practical and resulted in a superconductor degradation of less than 5%.

The strong conductor issue was approached in a different way for the ATLAS central solenoid (CS). A doped pure aluminium [50], sufficiently strong but with low resistivity was developed for this magnet (an in-kind contribution of Japan).

The conductors for the CMS and ATLAS magnets are shown in Fig. 8.27.

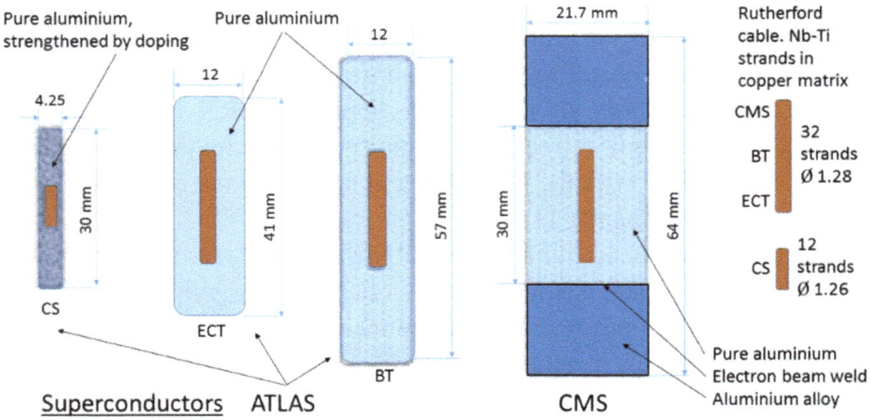

Fig. 8.27. Comparison of the superconductors used for ATLAS and CMS.

CMS magnet

The CMS magnet (Fig. 8.28) consists of a 6 m diameter, 12.5 m long solenoid coil surrounded by a steel yoke [51]. The coil is special in that unlike previous magnets of this type that featured single layer coils, it has four layers. For ease of manufacture (and transport) it is divided into five sub-units; the coils are wound on the inside of the support cylinder [52]. The 12 000 t yoke is made of plates between which detectors are installed to track muons emanating from the interaction. The steel, which serves as a general supporting structure for the detector, carries the return flux and is magnetized, so the tracks are curved and momentum can be determined. The coils are cooled by liquid helium in parallel tubes between manifolds top and bottom, forming a thermo-siphon configuration, following that of the ALEPH magnet at LEP. The magnet was designed for 4 T, up to which field it was tested, but it is run at 3.8 T to prolong its lifetime. Besides being presently the single magnet having the largest stored energy (2.6 GJ) in the world, it also features a record ratio of stored energy to cold mass (about 12 kJ/kg) for this type of magnet. This is tolerable due the inductive heating of the external mandrel if a quench should occur, rapidly spreading the quench to the whole winding. In normal circumstances half the energy is extracted in case of a quench.

The magnet was assembled and tested in the surface building above the cavern at LHC point 5, and lowered into position in the cavern in eleven pieces of mass between 600 and 2000 t. This procedure was based on the experience gained with lowering the support tube for the L3 experiment at LEP [Highlight 7.10]. The magnet itself has been working flawlessly since its installation in 2008, but some auxiliary equipment has had to be repaired.

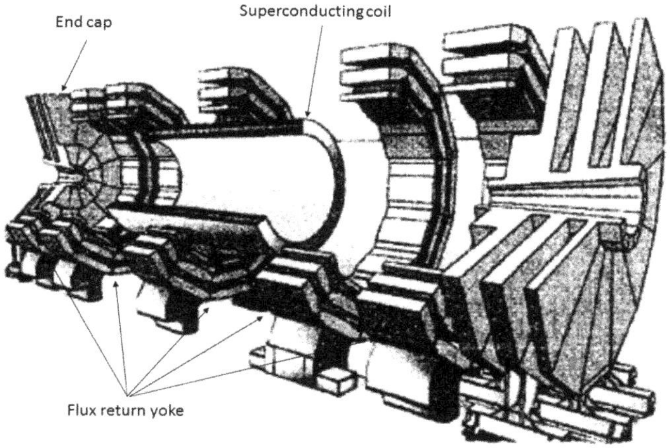

Fig. 8.28. Schematic exploded view of the CMS solenoid magnet.

ATLAS magnet system

ATLAS features a unique configuration of four large superconducting magnets [53]. Most imposing is the 26 m long air-cored barrel toroid (BT), presently the largest magnet in the world, consisting of eight rectangular superconducting coils, each enclosed in its own cryostat. The unusual architecture of this experiment makes it possible to decouple precise muon momentum measurements, made beyond the calorimeters, from momentum measurements in the inner tracking system. The toroidal configuration (Fig. 8.29), completed with two endcap toroids (ECT) [54], provides increasing field integral with decreasing forward angle, corresponding to nearly constant sensitivity in the measurement of transverse momentum over the angular momentum of particles from a proton collider. The low-mass superconducting air-cored solution provides high muon resolution thanks to minimal multiple scattering. A thin, 2 T superconducting solenoid (CS) [55], 5.8 m long, diameter 2.4 m, surrounds the inner tracking detector. Its insulating vacuum enclosure also serves the LAr electromagnetic calorimeter [Highlight 8.8] — a novel approach that minimizes material thickness. For the muons the two independent momentum measurements are combined to get the best resolution. The configuration and size of the system made this project challenging with regard to cost containment and installation. Major CERN contributions were the modification of the original design to simplify (to make more affordable) the BT coil casing of the barrel toroid, and changes to enable powering the three toroidal magnets in series, rather than separately. The stored energy of the BT is 1.1 GJ, and that of the ECTs is 0.5 GJ.

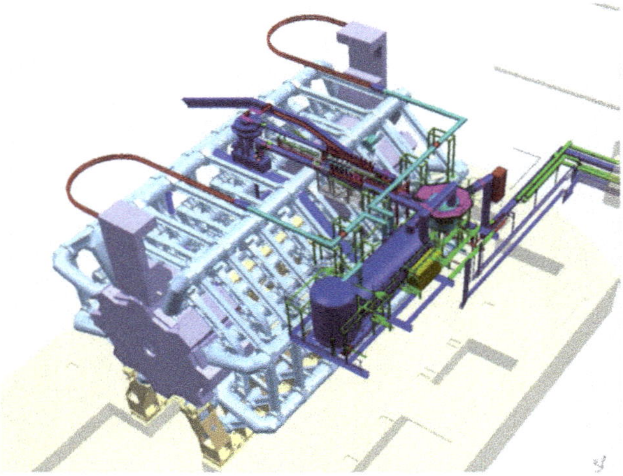

Fig. 8.29. Schematic view of the toroidal magnet system including services connections. The eight BT coils are visible with the two castellated ECTs inserted on both ends.

The three toroids are cooled by pumping supercritical helium through pipes attached to the coil casing, requiring a powerful pump. The dedicated refrigeration plant is located in a side-cavern.

The complete magnet system was commissioned in the cavern in 2008, and has worked reliably during all data taking periods. It is the largest and most complex superconducting magnet system ever built for a particle physics experiment.

LHCb magnet

Like ALICE, the LHCb detector also features a forward dipole. The first design of the magnet was based on the use of superconducting coils. However, at that stage the design was conceptual and the LHCC magnet advisory group (MAG) judged that it would be difficult to flesh out the design and build and test the magnet in the allotted time. Moreover, a quick study showed that sufficient performance could be obtained with resistive coils, avoiding the necessity of local cryogenics, by making a judicious adjustment to the steel yoke [56]. The idea was to maximize the integrated bending field by having sloping poles (Fig. 8.30), in a similar way to a previous magnet that had been designed at CERN for use at the ISR. The LHCC strongly urged the collaboration to pass to the resistive version, which they did. The magnet was designed at CERN and the coils were specified and purchased via competitive tender, with a strong technical follow-up by CERN. The total mass of the magnet is 1450 t; it consumes 4.2 MW when run at an excitation current of 5.85 kA to deliver an integrated field of 3.6 T m.

Fig. 8.30. The LHCb magnet (left) and the Alice dipole magnet (right) assembled, ready for testing.

ALICE solenoid and dipole magnet

This experiment uses the large solenoid designed and constructed at CERN for the LEP L3 experiment [57], complemented with a new large dipole magnet installed downstream for analysing particle (in particular di-muon) trajectories emerging at small angles [58]. The solenoid was overhauled and fitted with an improved water cooling system. The dipole magnet was designed by JINR (Dubna) in collaboration with CERN; steel for the yoke was an in-kind contribution from Russia. The coils were finish-designed at CERN and purchased via competitive tender. The delivery schedule was held thanks to strong technical follow-up and support by CERN (Fig. 8.30). The total mass of the magnet is 800 t; it consumes 5 MW at an excitation current of 6 kA, to deliver a bending strength of 3 T m.

References

1. L. Evans (ed.), *The Large Hadron Collider: a Marvel of Technology* (EPFL Press, 2009).
2. G. Brianti, The Large Hadron Collider project: historical account, *Phys. Rep.* **403–404** (2004).
3. M. Jacob (ed.), *ECFA-CERN Workshop on a Large Hadron Collider in the LEP Tunnel, Lausanne and Geneva*, CERN 84-10-V-1, ECFA-84-085-V-1 (CERN, Geneva, 1984). http://www.iaea.org/inis/collection/NCLCollectionStore/_Public/16/034/16018264.pdf; M. Jacob (ed.), *ECFA-CERN Workshop on a Large Hadron Collider in the LEP Tunnel, Lausanne and Geneva*, CERN 84-10-V-2, ECFA-84-085-V-2 (CERN, Geneva, 1984). http://www.iaea.org/inis/collection/NCLCollectionStore/_Public/16/034/16034963.pdf.
4. J.H. Mulvey (ed.), *Workshop on Physics at Future Accelerators, La Thuile, Italy*, CERN-1987-007 (CERN, Geneva, 1987). https://cds.cern.ch/record/179654?ln=en.
5. Report of the long-range planning committee, CERN/1658. August 1987, see CERN/1670 (1987).

6. *LHC Design Report. Vol. 1: The Main Ring*, CERN-2004-003 (CERN, Geneva, 2004). http://dx.doi.org/10.5170/CERN-2004-003-V-1.
7. P. Lebrun, Superfluid helium as technical coolant, *LHC Project Report 125,* (1997).
8. J.H. Mulvey (ed.), *The Feasibility of Experiments at High Luminosity at the Large Hadron Collider: report of the High-Luminosity Study Group to the CERN Long-Range Planning Committee*, CERN-1988-002 (CERN, Geneva, 1988). http://dx.doi.org/10.5170/CERN-1988-002.
9. ATLAS Collaboration, The ATLAS Experiment at the LHC, *JINST 3,* S08003 (2008).
10. CMS Collaboration, The CMS Experiment at the LHC, *JINST 3*, S08004 (2008).
11. LHCb Collaboration, The LHCb Detector at the LHC, *JINST 3*, S08005 (2008).
12. ALICE Collaboration, The ALICE Experiment at the LHC, *JINST 3*, S08002 (2008).
13. P. Jenni and T. S. Virdee, The discovery of the Higgs Boson at the LHC, in *60 Years of CERN Experiments*, ed. H. Schopper and L. Di Lella, (World Scientific, 2015) ISBN 978-9814644143.
14. R. Perin, Status of LHC programme and magnet development, *IEEE Trans. Appl. Supercond.* **5**, 189–195 (1995).
15. D. Leroy, R. Perin, G. de Rijk and W. Thomi, Design of a high-field twin-aperture superconducting dipole model, *IEEE Trans. Magnetics*, **24**, 1373–76 (1987).
16. R. Perin, pp. 71-92 in *Proc. CAS-CERN Accelerator School: Superconductivity in Particle Accelerators, Hamburg, Germany, May 1995*, S. Turner (ed.), CERN-1996-003 (CERN, Geneva, 1996). http://dx.doi.org/10.5170/CERN-1996-03.71.
17. P. Fessia *et al.*, Selection of the cross-section design for the LHC main dipole, *IEEE Trans. Appl. Supercond.* **10** (2000).
18. A. Siemko, *Safeguarding superconducting magnets*, pp. 35–37, CERN Courier, September 2013.
19. F. Savary *et al.*, Description of the main features of the series production of the LHC main dipole magnets, *IEEE Trans. Appl. Supercond.* **18**, No. 2, (2008).
20. V. Chohan, Operation for LHC cryomagnet tests: concerns, challenges & successful collaboration, *2007 Asian Particle Accelerator Conference (APAC)*, RRCAT, Indore, India, pp. 84–88 (2008).
21. J. Billan *et al.,* Twin Rotating Coils for Cold Magnetic Measurements of 15 m Long LHC Dipoles, *IEEE Trans. Appl. Supercond.* **10,** 1422–26 (2000).
22. D. Leroy, Review of the R&D and supply of the LHC superconducting cables, *IEEE Trans. Appl. Supercond.* **15**, 1454–60 (2006).
23. P. Lebrun, Cryogenics for the Large Hadron Collider, *IEEE Trans. Appl. Supercond.* **10**, 1500–06 (2000).
24. G. Claudet and R. Aymar, Tore Supra and helium II cooling of large high-field magnets, *Adv. Cryo. Eng.* **35A**, 55–57 (1990).
25. P. Lebrun and L. Tavian, pp. 453-476 in *Proc. CAS-CERN Accelerator School: Superconductivity for Accelerators, Erice, Italy, April 2013*, R. Bailey (ed.), CERN-2014-005 (CERN, Geneva, 2014). http://dx.doi.org/10.5170/CERN-2014-005.453.
26. P. Lebrun, L. Serio, L. Tavian and R. van Weelderen, Cooling strings of superconducting devices below 2 K: the helium II bayonet heat exchanger, *Adv. Cryo. Eng.* **43A**, 419–42 (1998).
27. V. Baglin, P. Lebrun, L. Tavian and R. van Weelderen, Cryogenic beam screens for high-energy particle accelerators, *Proc. ICEC24 – ICMC 2012*, Fukuoka (Cryogenics and Superconductivity Society of Japan, 2013) pp. 629–634.

28. S. Claudet, P. Gayet and U. Wagner, Specification of four new large 4.5 K refrigerators for the LHC, *Adv. Cryo. Eng.* **45B**, 1269–1276 (2000).
29. S. Claudet *et al.*, Specification of eight 2400 W at 1.8 K refrigeration units for the LHC, Proc. ICEC 18 Mumbai (Narosa Publishing, 2000) pp. 215–221.
30. D. Delikaris *et al.*, Cryogenic operation methodology and cryogen management at CERN over the past 15 years, *Proc. ICEC 22 – ICMC 2008* (KIASC, Seoul 2009) pp. 33–38.
31. A. Ballarino, Large capacity current leads, *Physica C* **468**, 2143–2148 (2008).
32. J.G. Bednorz and K.A. Müller, Possible high Tc superconductivity in the Ba–La–Cu–O system, *Zeitschrift für Physik B Condensed Matter,* **64** (2), 189–193 (June 1986).
33. A. Ballarino, HTS current leads for the LHC magnet powering system, *Physica C* **372–376**, 1413–1418 (2002).
34. A. Ballarino, Operation of 1074 HTS current leads at the LHC: overview of three years of powering the accelerator, *IEEE Trans. Appl. Supercond.* **23** (3), 4801904 (2013).
35. P. Bauer *et al.*, R&D towards HTS current leads for ITER, *IEEE Trans. Appl. Supercond.* **19** (3), 1500–1503 (2009).
36. C. Benvenuti, Non-Evaporable getters: from pumping strips to thin film coatings, *Proc. EPAC 98* (1998) pp. 200–204.
37. C. Benvenuti, P. Chiggiato, F. Cicoira and Y. L'Aminot, Non-evaporable getter films for UHV applications, *J. Vac. Sci. Technol. A* **16** (1), 148 (1998).
38. L. Rossi, P. Fischer, T. Rohe and N. Wermes, *Pixel Detectors, From Fundamentals to Applications* (Springer, 2006).
39. CMS Collaboration, Alignment of the CMS tracker with LHC and cosmic ray data, *JINST* **9**, P06009 arXiv:1403.2286 (2004).
40. ATLAS Collaboration, Performance of the ATLAS inner detector track and vertex reconstruction in the high pile-up LHC environment, *ATLAS-conf-2012-042*, (2012). http://cdsweb.cern/record/1435196
41. CMS Collaboration, Description and performance of track and primary-vertex reconstruction with the CMS tracker, *JINST* **9**, P10009 (2014).
42. E. Cerrón Zeballos *et al.*, A new type of resistive plate chamber: the multigap RPC, *Nucl. Instr. & Meth. A* **374**, 132 (1996).
43. M.C.S. Williams, Particle identification using time of flight (2012), *J. Phys. G: Nucl. Part. Phys.* **39** 123001 (2012).
44. R. Aaij *et al.*, (LHCb collaboration), Measurement of b-hadron branching fractions for two-body decays into charmless charged hadrons, *JHEP* **10**, 037 (2012).
45. M. Adinolfi *et al.*, The LHCb RICH Collaboration, *Eur Phys. J. C* **73**, 2431 (2013).
46. D. Green, *At the Leading Edge: The ATLAS and CMS LHC Experiments* (World Scientific, 2010).
47. T.M. Taylor, The magnets for the LHC experiments, *IEEE Trans. Appl. Supercond.* **10** 1, 342–346 (2000).
48. I.L. Horvarth *et al.*, The CMS conductor, *IEEE Trans. Appl. Supercond.* **10** 1, 395–398 (2000).
49. J.C. Lottin *et al.*, Conceptual design of the CMS 4 T solenoid, *Adv. in Cryo. Eng.* **41** 819–825 (1996).
50. K. Wada *et al.*, Development of high-strength and high RRR aluminium–stabilized superconductor for the ATLAS solenoid, *IEEE Trans. Appl. Supercond.* **10** 1, 373–376 (2000).
51. A. Herve, The CMS detector solenoid, *IEEE Trans. Appl. Supercond.* **10** 1, 389–394 (2000).

52. P. Fabbricatore *et al.*, The construction of CMS superconducting coil, *IEEE Trans Appl. Supercond.* **14** 2, 552–555 (2004).
53. H.H.J. ten Kate, The superconducting magnet system for the ATLAS detector at CERN, *IEEE Trans. Appl. Supercond.* **10** 1, 347–352 (2000).
54. D.E. Baynham *et al.*, ATLAS end cap toroid magnets cryostat design, manufacture and installation at CERN, *IEEE Trans. Appl. Supercond.* **14** 2, 522–525 (2004).
55. A. Yamamoto *et al.*, The ATLAS central solenoid, *Nucl. Instr. & Meth. A* **584** 53–74 (2008).
56. M. Losasso *et al.*, Test and field map of LHCb dipole magnet, *IEEE Trans. Appl. Supercond.* **16** 2, 1700–1703 (2006).
57. F. Wittgenstein *et al.*, Construction of the L3 magnet, *Proc. 11th Int. Conf. on Magnet Tech. (MT-11)*, 130–135 (1989).
58. D. Swoboda *et al.*, Status of the ALICE magnet system, *IEEE Trans. Appl. Supercond.* **12** 1, 432–437 (2002).

Chapter 9

Data Handling and Communication

Frédéric Hemmer and Pier Giorgio Innocenti

9.1 Introduction

Accelerators and experiments in high energy physics cannot be conceived today without the extensive use of computers. For accelerators, computers are used for the simulation of the performance, and together with data communication for the operation, control and performance monitoring. For the experiments they are essential for simulating the expected performance, selecting the events of interest and analysing the data. Data volumes are generated in all these steps and are very large, requiring careful attention to data storage and data retrieval, as well as to efficient programming. Digital communication is needed for cooperative working between processors and between the physicists.

This chapter traces the evolution of computer usage at CERN, with an emphasis on the impact on experimentation, from SC to LHC [1].

In 1958 the computer age started (Fig. 9.1) at CERN with the operation of the Ferranti Mercury, a machine based on vacuum tubes [3]. It was used for calculations by engineers and physicists and for processing the data produced by the scanning and measurement of pictures (recorded on photographic film) obtained by bubble chambers and optical spark chambers.

Processing of bubble chamber pictures

The arrival of the IBM 709 in 1961 marked a progress in the processing of bubble chamber and spark chamber pictures. While still using vacuum tubes, it had for the time modern peripherals (card punches, tabulators and magnetic tapes). Importantly, it used FORTRAN as a programming language, in which the first general purpose programs for track reconstruction and kinematics were written. Two years later, the IBM 7090 arrived, based on transistor electronics and with good input/output facilities.

Digitizers for bubble chamber and spark chamber pictures were linked on-line to the computer to speed up the input process, a substantial improvement over transfer by paper tape; processing of data from film represented the main usage at that time.

Remote access to the central computers became available in the late 1960s [Highlight 9.3].

The era of central mainframe computers lasted for more than thirty years. The CERN Computer Centre operated a number of IBM machines, the CDC 6600 and 7600 and a Cray X-MP supercomputer. Given the demanding applications and the knowledge of its staff, CERN was a highly appreciated partner for the computer manufactures, who were relying on CERN experience to improve their products.

The FORTRAN "culture" linked to bubble chamber data processing produced a library of programs used worldwide by the high energy physics community; this approach was extended to processing of data produced by electronic experiments. Advanced code management methods were developed [Highlight 9.6].

Fig. 9.1. Wim Klein, the first CERN computer, was operational in the Theory Division before the Ferranti Mercury [2].

Electronic experiments

In 1963 the appearance of the spark chambers with electronic readout prompted radical change in the use of computers. Before that time, electronics experiments, as opposed to experiments using photographic film, manually recorded event rates and logical combinations of signals, such as time coincidences, produced by detectors with coarse space resolution. The novelty of the digitized spark chamber was that, after a trigger, it produced prompt hits (electronic pulses) of particles in the detectors, with good space resolution; the hits were suitable for immediate digitization and recording by a computer.

The data acquisition for the Missing Mass Spectrometer [4], tested at the SC in 1963 with acoustic spark chambers, displayed all the features of a modern particle physics experiment: a selective fast trigger, fine space resolution, a digitizer for each data channel, a link [5] to a computer which recorded the event on tape, performed full reconstruction and displayed the results. The computer was the Ferranti Mercury: It had a relatively slow input capability, but its high processing power permitted the complete reconstruction and analysis of one out of ten events recorded on tape. Immediate display of the results from the sample offered a novel valuable monitoring of experiment performance.

The Ferranti Mercury ended its useful life with that experiment. Minicomputers using solid state electronics became popular in data acquisition in the sixties, but could not perform a complete analysis of a data sample, due to their limited processing power and to the lack of a FORTRAN compiler: FORTRAN had become the *de facto* standard language for data analysis.

In the early 1970s, the deployment of multiwire proportional chambers [Highlight 4.8] in many experiments brought a large increase in data volume: These detectors have typically many readout channels to achieve good space resolution and have high rate capability. Processing a large number of complex events, such as those produced on fixed target at the SPS or by colliding beams at the ISR, required innovative software methods. This trend towards more complex events, each consisting of a large data set, has continued, with a notable increase due to the very fine grain fast silicon detectors, culminating at the LHC.

With the increasing complexity of the detectors and data selection criteria, computers became part of the trigger strategy. Triggers, earlier obtained from timing coincidences and Boolean logic, now used simple hardware processors and computers performing calculations as required for instance by pattern recognition. The minicomputers available at the time had marginal computing power for this application, and required lengthy re-programming and adaption of the algorithms used in the full off-line reconstruction.

A solution came from the use of special processors (370/168E) developed at the Stanford Linear Accelerator Centre (SLAC) [6]. Running on custom designed hardware, they emulated a reduced set of the instructions of the IBM 370 mainframe, which were needed to execute the reconstruction programs used in the program libraries: the available reconstruction programs could then be used with only minor modification. One such processor was used as third level trigger to select a high transverse momentum particle in an experiment with the Split Field Magnet detector at the ISR [7] in the late 1970s. Clusters of 168E and of 3081E (emulating the IBM 370/3081) were extensively used by the UA1 trigger [8] and off-line reconstruction, and by other experiments.

The introduction of emulators was the first successful example of a full precision arithmetic algorithm applied to real time selection of events. These developments, favoured by the availability of affordable large computing power, started a slow evolution of data taking, from recording the raw hits in the detectors after a coarse selection, to putting to permanent storage only the events selected by a complete pattern recognition carried out in real time. In this mode of operation, the reliability of trigger and reconstruction is of paramount importance: event topology, detector acceptance and response as well as particle reconstruction efficiency and precision must be understood in detail to avoid irretrievable losses and biases.

This was a new situation. Up to the 1970s, an experiment was designed using analytic calculations for the performance evaluation. Any shortcomings of the initial design would be "repaired" by appropriate analysis of the stored raw data: The prevailing mood was "you will sort it out later with software!" This new situation created a need, already at the design stage, for a complete simulation of the entire physics process of interest and of the underlying background: event generation, tracking of particles in the detector, interaction with the detector materials, signal generation and processing, digitization, many levels of trigger, pattern recognition, momentum or energy of all particles and finally comparison with the initial set of events [Highlight 9.5]. Besides being essential for detector design, simulation became indispensable for data analysis: simulated events, produced by reliable "event generators" and tracked with full detail through the detectors are compared with real data.

In the 1990s, projects were carried out in collaboration with industry to explore how computer architectures differing from the sequential von Neumann machine could exploit some form of fine-grain parallelism, but results were not convincing. The structure of the data, closely linked to the topology of the detector, consisting of independent elements, favoured a coarse grain parallelism, easily implemented by arrays of (sequential) commodity processors such as personal computers. In the event selection, the first level trigger identifies a subset of data (ROI, Region Of Interest) where a detailed calculation should be performed and assigns this task to a single processor. Another ROI from a different event is assigned to another processor and the calculations proceed in parallel. When the result of an ROI points to a potentially interesting event, the corresponding full data set is assigned to a processor for further calculations. At the end of a successful selection the event is sent to permanent storage. The selectivity of the filter must ensure that the volume of stored data is manageable. This architecture has proved to be efficient and economical, and is used by LHC experiments [Highlight 8.11].

Off-line processing of data has followed a similar trend, moving from mainframes to event parallelism on clusters of workstations and PCs with efficient access to data storage.

In an experiment the bulk of the data comes from two sources:

- digitized signals from the detectors;
- data generated by simulation.

The raw data is first *reconstructed* to obtain a physics view of the event including elements such as particle tracks, computation of physical quantities, identification of particles, and classification of the event according to its physics characteristics. The output from this stage, the *event summary data* (ESD), forms the basis for further stages of analysis, including a comparison with corresponding ESDs generated by simulation.

The analysis is a distributed process involving many groups within the collaboration studying different physics processes, or the behaviour of the detector itself, all requiring access to the ESD. These groups will in general extract events of interest, perform additional processing, and create new datasets that are shared among the members of the group or within a small team.

There are several characteristics of the data processing which influence the design of the computing services:

- The volume of the data is large; currently, each year the LHC experiments generate about 30 Petabytes (30×10^{15} bytes) of new data, raw, simulated and processed;
- The bulk of the data is shared between many groups, trawling simultaneously through large subsets of events;
- Only the raw (detector) data is static. Everything else is reprocessed as understanding of the physics, the detector, and the analysis programs evolves, and as the focus of physics interest changes;
- The larger data sets (raw, event summary, simulated) are read-only: updates are applied to *new versions* of the data set and the events are independent of

each other. Thus there is ample scope for parallel processing and replicas are relatively easy to manage.

The understanding and analysis of the data requires deep knowledge of both the physics and the detector. The active participation of specialists from many different areas is needed to adapt the reconstruction and simulation algorithms as new phenomena are studied. The large data sets are in no way self-explanatory: they are inseparable from the collaboration. As a result, data security is concerned more with protecting against accidental corruption or loss, rather than with encryption or restricting access.

The evolution of commodity processors in recent years, with many execution cores, an extended instruction set allowing operations on vectors and special purpose co-processors, requires re-thinking of the simple event parallelism. In the real time selection process, FPGAs (Field Programmable Gate Arrays) and video co-processors are used to prepare the data for the processor/core which performs pattern recognition, with large gains in event processing speed; some of the results of pre-processing replace raw data to permanent storage. In simulation many tracks can be handled simultaneously by using the vector capability of the processors.

Data analysis techniques have evolved with the complexity of detectors and the increased size of the data samples. In the bubble chamber time typically one had to extract a relatively narrow signal peaking on top of a uniform background, from a histogram representing a one-dimensional distribution. The signal was localized by "cuts" and its significance could be assessed by simple statistical methods.

In the LHC experiments the (rare) signals of interest are masked by a large and complex background arising from different processes. The search must be performed in a multi-dimensional parameter space, by choosing a set of efficient variables to enhance the signal-to-background, typically done by multivariate analysis (MVA). Different tools, such as neural networks or boosted decision trees, are used. Large sets of simulated signal and background events are needed to train the tools: Machine Learning (ML) becomes a step in the discovery process [9].

Significant progress has been made in recent years regarding the understanding of, and implementation of services and solutions for, long-term data preservation for future re-use [10]. Funding agencies and indeed the general public are now beginning to understand the need for preservation and sharing of data, which represent the "digital memory" of the experiments, as a part of the scientific legacy. This "memory" contains the raw data recorded by the experiments and also a significant amount of metadata, software programs and "knowledge". The latter is tightly coupled to the collaboration and its members. The timescale for data preservation is determined by the availability of data from new experiments exploring the same physics: as the lifetime of products in information technology is short, hardware and software obsolescence may have to be faced.

Data re-use is undoubtedly valuable for its discovery potential. However, it has an intrinsic limitation: the "raw data" kept for permanent storage are the result of the trigger process (taking place before the initial data recording) designed for specific physics reactions. The bulk of the data produced by the collisions are irreversibly lost and may carry with them undiscovered information.

Over the years, experiments have become larger and more complex, and the size of the collaborations reflects the size and ambitions of the experiments. Collaborating groups are located on all continents. This mode of operation is made possible by fast long distance networking [Highlight 9.4] and by software tools that ensure data coherence. It goes without saying that the Web [Highlight 9.7] and other means of digital communications have been essential to the success of these distributed collaborations.

9.2 Computing Clusters and Data Storage: The New Factory and Warehouse

Les Robertson

Infrastructure for innovation

Since the earliest days of CERN the availability of computational and data storage capacity has been one of the most important factors in enabling the extraction of physics from the data collected by the experiments. It has evolved from the automated analysis of bubble chamber photographs in the 1960s to the worldwide grid used today to distribute the mass of data from the Large Hadron Collider for processing in 170 sites in 42 countries where high energy physicists have access to computing resources. Funding has always been limited, driving a continuous search for new technologies to provide more cost-effective solutions — in data storage, networking and processing power.

In the 1960s and early 1970s the clear choice for computational capacity was the super-computer, designed for fast floating point calculations, though not particularly good for handling large volumes of data. CERN's innovative role in that era included working with the manufacturer to optimize mathematical libraries to exploit best the detailed architecture of the machine. The beginnings of the hierarchical mass storage management architecture were developed, the evolution of which is still continuing today. In the 1970s CERN already had a managed two-level data buffer ("cache") for its vast magnetic tape library.

During these early years the rather visionary idea emerged that physicists needed more than just access to a large computer to run their programs. They needed to acquire, store and analyse data from experiments and simulations, and do this from all parts of the CERN site. Experimental data files needed to be transported and processed on central batch systems, with control of the job flow and delivery of the results. With the first accelerators, data was transported on magnetic tapes to the computer centre "bicycle online" and physicists carried large trays of cards and reams of paper to and from their offices. But in 1968 the first of a series of CERN-developed site networks was implemented culminating ten years later in the high speed general purpose packet switched CERNET [Highlight 9.3].

By the mid-70s supercomputer architecture had evolved towards support for vectorizable codes. Sterling efforts to exploit this hardware for High Energy Physics (HEP) algorithms were unsuccessful and commercial mainframes with their simpler architecture became the cost-effective solution, ushering in a period of heterogeneity as the competitive acquisition criteria led over the years to the installation of systems from different manufacturers with their proprietary architectures, operating systems and local network media and protocols. The challenge was to interconnect these systems to the site network and, later, to the growing number of incompatible wide area networks that were appearing in support of science in different countries and regions.

By the end of the 1980s and the start-up of the LEP collider, CERN had acquired in-depth experience in storage management, networking and distributed processing. This was a key factor in three major developments that proved essential for HEP data processing over the following decades: cluster computing, extending this to the wide area as a data grid, and of course the Worldwide Web.

Hierarchical mass storage service for the CDC 7600 supercomputer

In March 1972 CERN installed serial number 3 of the fastest supercomputer of its day, the Control Data Corporation 7600. There were three large disks connected to the system providing about 2 Gigabytes (GB) of online storage. Another 600 Megabytes (MB) of disk storage was available on the "front-end" computers mainly used for holding data which was required to be permanently online such as executable files, but the physics data were stored on magnetic tape which had to be copied to the 7600 disks before being accessed. By the end of 1974 there were about 60,000 magnetic tapes in the tape vault, each with a nominal capacity of 20 to 40 MB [11]. In order to manage the movement of data between tape and the 7600 a two-tier storage system was developed in 1973. The user referred to data by the tape number and the system arranged for the tape to be recovered from the vault, mounted on a tape drive on one of the front-end computers and copied into a cache on the 7600 disks. The cache was managed using criteria such as size of file, age, and time since last access. A third tier was introduced a few years later as a tape reel cache, enabling frequently used tapes to be available for immediate mounting. The system was called FIND [12] and at the time it was not described as a *Hierarchical Storage Management* (HSM) system, a term that was coined only later with the arrival of robotic tape handling devices, but it already had all of the characteristics of an HSM.

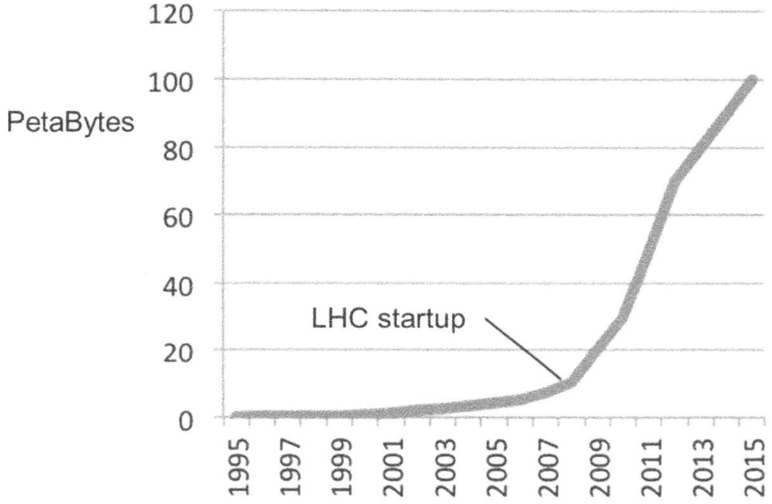

Fig. 9.2. Growth of mass storage usage at CERN.

The basic concept developed in 1973 has remained at the heart of storage management at CERN until the present day. There have been several generations of software integrating new technologies like robotic tape handlers and databases as the volume of data exploded (Fig. 9.2), and evolving to the cluster model with a fully distributed cache [13].

BETEL — Extending cluster, processors and data, across the wide area

In 1993 CERN led a project which implemented the first international demonstration of a computer cluster extended to the wide area using network technology with comparable performance to the local area networks of the time. The project, BETEL (Broadband Exchange over Trans-European Links) [14], used 34 Mbit/s ATM[a] (Asynchronous Transfer Mode) links provided by France Telecom and Telecom Switzerland, to interconnect the 100 Mbit/s FDDI (Fiber Distributed Data Interface) networks at CERN, EPFL (Lausanne, Switzerland), IN2P3 (Lyon, France) and EUROCOM (Nice, France).

[a]ATM was a standard used by telephone companies during the 1990s to provide high throughput data traffic combined with low latency characteristics suitable for voice and video. By the end of the decade it was being displaced.

Using this infrastructure the CERN physics cluster was extended to a similar cluster in Lyon, including access to the magnetic tape services at both sites. After enhancing the RFIO (Remote File Input Output) protocol [15] used for data access to take account of the long network latency, the performance goals were achieved enabling seamless access to all disk and tape resources from both sites. The project also demonstrated interactive physics analysis using CERN's Physics Analysis Workstation (PAW) system and a tele-teaching application used by users at EPFL and EUROCOM.

SHIFT — An early implementation of cluster computing

The computational needs of the experiments planned for the LEP collider from its start in 1989 far exceeded the capacity that could be provided within CERN's computing budget using the mainframe technology that was the backbone of the computer centre. In HEP data processing the major part of computational resources are used to process very large numbers of independent events. Events can therefore be processed in parallel, and using systems with different performance characteristics: the requirement is *high throughput* rather than *high performance*. A small team was given the job of looking for a distributed solution that could exploit low cost components integrated with the central computers and with the storage infrastructure essential for efficient data analysis.

There was good experience with microprocessors for online use of the experiments, but these had limited floating-point performance. Specialized processors had also been developed, implementing an instruction set sufficient for certain physics codes but without general I/O and network capability [16]. More promising were the "personal workstations" being used at CERN in a limited role for interactive graphics applications, and based on the new single-chip reduced instruction set (RISC[b]) processors. Personal workstations with the graphics functionality removed gave an order of magnitude improvement in price/performance over the mainframes. The only problem was how to integrate them into the physics analysis service.

The project SHIFT (*Scalable Heterogeneous Integrated Facility Testbed*) [17] developed a straightforward architecture providing scalability and accepting heterogeneity that would enable the service to expand smoothly as the underlying data storage and computational resources grew. It was seen as essential to construct the system with off-the-shelf hardware components, in order to be able to exploit new technologies and cost opportunities as soon as they arose. The key idea was to define separate services for mass storage, data cache management (disk) and

[b]Compared with traditional processors of that period RISC processors provided higher performance by using a simplified instruction set that enabled execution to be completed in fewer microprocessor cycles per instruction.

computation loosely integrated across a high performance network backbone. Each service would be implemented as a set of independent servers to ensure scalability. The software was designed to be portable across systems providing a Unix-like system interface and a TCP/IP[c] (Transmission Control Protocol/Internet Protocol) network service. The fundamental components that had to be developed were: a distributed cache management system integrated with the site mass storage service; a remote file access system [15], later, re-routing in the event of a link failure; a job scheduler to manage the workload across the cluster, initially based on a system developed by NASA.

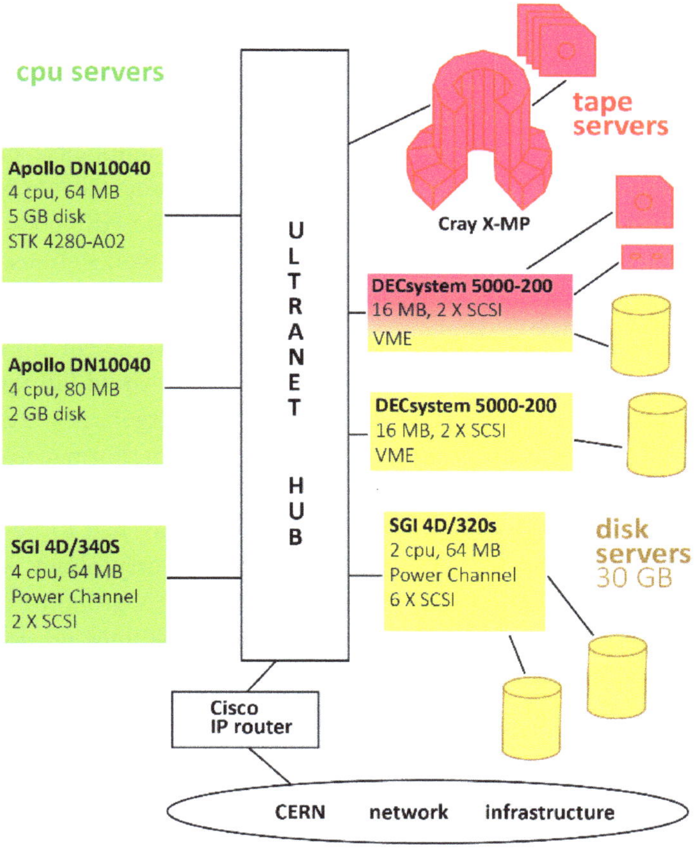

Fig. 9.3. Pilot SHIFT configuration at CERN in 1991.

---

[c]TCP/IP is the most widely used communications protocol. It was invented and developed in the 1970s at DARPA (Defense Advanced Research Projects Agency), USA.

The initial implementation of the architecture using Hewlett-Packard Apollo DN10.000 computers as disk and computer servers, could already rival the largest mainframe in terms of aggregate processing capacity, but lacked efficient access to mass storage and was only used for simulation work. The cluster was soon enhanced with the addition of systems from Silicon Graphics Inc., Digital Equipment Corporation, SUN Microsystems and a Cray X-MP as the mass storage server, all interconnected by a high-speed network from UltraNet. Figure 9.3 shows the configuration at the beginning of 1991. The unusual choice of a supercomputer as a mass storage server was made because, in common with the other systems in the cluster, it used a version of Unix as operating system, had a very good implementation of TCP/IP network protocols, and could be connected to the UltraNet backbone. From this early start, the cluster grew rapidly, and within a few years supplanted the mainframes as the workhorse of physics simulation and analysis. The architecture, in large part because of its simplicity, proved its flexibility, absorbing successive generations of disk and processor technologies. PCs were introduced in 1997 as soon as their performance on HEP codes became acceptable and they in turn soon displaced the RISC-based workstations.

The Worldwide LHC Computing GRID — A Cluster of Clusters

When planning the facilities for analysing the data from LHC began in 1998, once again it was clear that the budget for computing at CERN would fall far short of supplying the needs. The CERN cluster was by that time largely powered by PCs which seemed to offer a very promising future for raw performance and cost effectiveness — no better technology was on the horizon. However, many other HEP sites had also installed powerful clusters. A neat way had to be found of interconnecting them without the physicist having to worry about where the available resources were and where the data was located. A group of physicists and computing experts from a broad selection of HEP institutes got together to study possible solutions, producing a recommendation in 2000 for a multi-tier hierarchy of inter-connected regional centres, each with its own cluster and storage [18]. The tiers were defined by the type and volume of data that would be stored there: the top level, Tier-0 (CERN, and Budapest from 2013), would store a copy of all of the raw and simulated data and of the processed data required by all of the physicists in an experiment; Tier-1 sites would be large operations with mass storage services, and each Tier-1 would hold a copy of a fraction of the data held at CERN (ensuring at least two copies of all important data); lower tier sites would store only temporary copies of the main datasets, obtaining copies of datasets as required from a higher tier. The end-user would not have to know where the data was located or where there were free resources. The work would be automatically split into suitable processing units and directed to an appropriate regional centre.

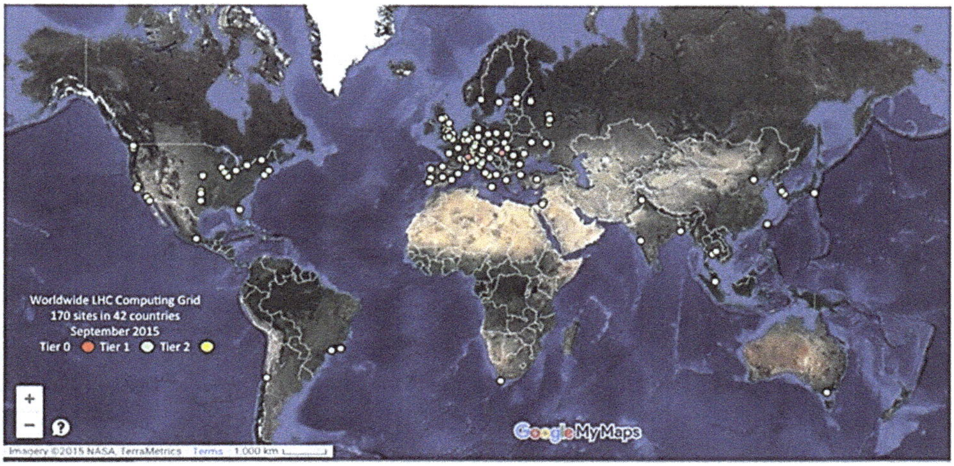

Fig. 9.4. A grid of 170 interconnected computing clusters in 42 different countries provide the processing capacity for analysing LHC data (Sept. 2015). Budapest and CERN are Tier 0 sites.

While the responsibilities for operating services and managing data were defined as a *hierarchy* which would have implications for planning inter-site network bandwidth, it was important that the lower-level software should view the system as fully inter-connected, providing complete flexibility to job schedulers, data management services and other higher-level components: the logical architecture would be a *grid*. The *middleware* (software) required to operate the grid would be complex and rather than developing a special LHC solution it was decided to start with the *Computational Grid* technology being developed in the GLOBUS project in the USA [19]. The Italian Nuclear Physics Institute, INFN, already had experience with the Globus toolkit. The grid concept looked to be of general interest for scientific applications and CERN took the lead in launching an international multi-science project, the *European Data Grid* (EDG) [20], which received funding from the European Commission. EDG was a proof of concept project which developed an initial implementation of the middleware needed to complement the Globus toolkit and deployed this on a demonstration grid. A similar project, the Grid Physics Network (GriPhyN) was begun in the USA.

The grid concept and experience with early prototypes was promising and in September 2001 CERN set up the *LHC Computing Grid* (LCG) Project to move ahead with planning and deploying a grid on the scale required for LHC era. LCG collaborated closely with EDG, GriPhyN and successor grid projects, but with the clear goal of focusing on the functionality, delivery schedules, reliability metrics, interoperability and other factors needed to ensure an operational grid with the performance and timescales imposed by the LHC experiments. The first LCG deployment was in September 2003 with 14 sites, followed by rapid growth in the number of sites and the installed resources. The differing goals of the R&D grid projects and LCG with its emphasis on service stability would be the cause of considerable conflict in the ensuing years, particularly concerning the functionality and timescales of the middleware. With the deadline of LHC start-up approaching, a compromise was worked out and the Worldwide LHC Computing Grid was in round-the-clock operation at 135 sites when data began to flow in 2008 [21].

A major concern in 2000 was the network bandwidth that would be available (and affordable) between the centres and much effort was invested in working with national and international bodies that provided network capacity for science to ensure that their solutions took account of the relatively high requirements of LHC. This work paid off and by the time that LHC began operation CERN was well connected to a large, mainly fibre, international infrastructure [Highlight 9.4]. Indeed, today, fifteen years after the first proposal for the distributed model, the available bandwidth enables a much more flexible model for data distribution: instead of scheduling work to where the data is located it is now possible in many cases to move data on demand to where there are free computational resources.

9.3 Local Area Networks: Organizing Interconnection

Ben Segal

The ever closer and increasingly important interconnection of particle experimentation and accelerators with Information Technology (IT) is the main theme of this chapter. In this highlight we illustrate one essential facet: the evolution of network concepts.

Innovating in the Early Years — 1960s and 1970s

CERN rarely develops computer networking technology *per se*, but is often an early adopter of the latest technology, operating it at its technical limits. This was different, however, during the 1960s and 1970s, when CERN had no choice but to innovate due to the limited offer of suitable commercial products. During these years several networking systems were developed in-house, presented here in chronological order, and reflecting the rapid evolution of IT and its growing importance for high energy physics. A comprehensive survey is given in [22].

In the early 1960s CERN's first computer, the Ferranti Mercury, was connected by a one-kilometre *data link* to electronic experiments in the PS experimental hall, making it the first computer at CERN to analyse on-line experimental data in "real-time". The complete link interfacing was designed and built in-house.

CERN's first distributed computer system (1968–1978) was FOCUS, which linked several experiments to the Control Data Corporation mainframes in the Computer Centre [23]. The link used 2 Mbit/s twisted pair serial lines via a smaller CDC machine serving as a concentrator and front-end. This allowed physicists at their remote experiment sites to manage their files, to execute batch jobs on the mainframes, and to retrieve their results.

The next step was the OMNET network, CERN's first packet switched network (1971 to the early 80s) [24]. It eventually interconnected around 30 user minicomputers as well as the central IBM and CDC mainframes, via dedicated switching nodes. A small purpose-built CERN operating system, SMO, in a high-level language, PL-11, had to be developed for these nodes. The network protocol software, and data link hardware and interfaces were developed and built at CERN.

From 1971, twelve Remote Input Output Stations (RIOS) allowed remote batch job submission to the central CDC mainframes from remote CERN sites. Users no longer needed to transport and physically input cards and output printouts to and from the Computer Centre — a great gain in efficiency. Each minicomputer-based RIOS emulated a standard CDC remote batch terminal, but was much cheaper and more performant. Through one particularly important innovation, the human operator, who had been needed to initiate and supervise a computing job, was emulated (and replaced) by RIOS software. This strategy was successful and further expanded [25].

A technological prescient development was CERNET, a packet switched network, designed from 1975 and deployed from 1977 to 1988 [26]. CERNET offered a "Line Protocol" between switched interfaces, capable of up to 8.5 Mbit/s over short parallel connections, or serial line speeds of around 2.5 Mbit/s over twisted-pair cabling of up to 2 km (with repeaters). An "End-to-End Protocol" connected host to host applications. Once fully deployed, CERNET interconnected the central IBM and Control Data mainframes and over 100 minicomputers. The main CERNET application was the "File Access Protocol", but a "Virtual Terminal Protocol" (remote logon) accessed the main IBM system.

Not to be outdone by the experiments the accelerator builders used IT technology in innovative ways too. For example, *Accelerator Control Networks* were designed and built to control the SPS. This was based on a star topology, with a message handling computer at the centre of the star [Highlight 5.2].

Profiting from the IT Revolution: 1980s and 1990s

Starting in the 1980s, the computing and networking world was shaken by several revolutions. CERN seized this opportunity, following these developments and collaborating closely with industry in a bid to anticipate the optimal choices for hardware and software, as outlined below.

Choice of TCP/IP and Internet

Throughout the 1970s and 1980s there was a bitter contest between "US" (TCP/IP Internet) protocols and "European" (ISO-OSI) network protocols, as well as between diverse commercial network standards, Despite being in Europe and despite much pressure, CERN decided to test TCP/IP. Starting in 1984, the installation of TCP/IP on most of CERN's computing platforms was as innovative as it was audacious. By the end of the 1980s TCP/IP and the Internet had clearly become dominant and CERN could benefit fully from its courageous choice. CERN joined the global Internet in 1989 [27]. The later Web and Grid developments at CERN all depended vitally on this work [Highlights 9.2 and 9.7].

Choice of Ethernet

The first widely used Local Area Network (LAN) was the 12 Mbit/s token ring network supporting the Apollo Domain and its powerful workstations, starting in 1982. It used thin coaxial TV cable, requiring dedicated installations. Ethernet (10 Mbit/s) arrived in 1983, using different types of coaxial cables, followed by a token passing ring at 4 Mbit/s for LEP controls in 1986 [28] and much later by FDDI at 100 Mbit/s on optical fibre cables. Fortunately, Ethernet was soon recognized as the future dominant standard.

CERN's central Ethernet services

A centrally-managed Ethernet service was first set up in the mid-1980s on coaxial cables at 10 Mbit/s. Early difficulties due to the multiple network protocols present at that time had to be overcome, but the rapid evolution of Ethernet technology allowed CERN to homogenize its network infrastructure. The media (and prices!) improved dramatically and network speeds passed from 10 Mbit/s to 100 Mbit/s, and finally to Gigabit Ethernet. Today, Ethernet networking is simply a matter of choosing and installing commercial products. LHC controls are based on Gigabit Ethernet, running mostly via optical fibre [29], and distributing machine synchronization signals with sub-nanosecond accuracy [30].

Choice of Unix/Linux

Unix first appeared at CERN in 1981 but was hardly noticed in spite of its important role in connecting CERN, and later all of Switzerland, to the worldwide email and news network USENET. It was widely mistrusted compared to proprietary operating systems. Unix really came out of the closet when chosen for the Cray XMP installation in 1987. CERN and Cray worked together to develop the new UNICOS system, the first to support a full-fledged mainframe, and later to be the model for the SHIFT "distributed mainframe". Unix was chosen as the operating system for LEP controls in 1988.

While Unix had proved itself at CERN by the late 1980s, it was still for the most part a commercially supported product. The open source version, Linux, supported by the online programmer community, was mistrusted, raising fears that "we will have to fix its bugs ourselves". When finally accepted, during the final migration of SHIFT to PC/Linux platforms in the late 1990s, it rapidly became dominant.

Choice of the "SHIFT" architecture

It was a game- and life-changing choice to use commodity CPU, storage and network hardware to build a "distributed mainframe" in a scalable, flexible and economic way, starting in 1990. CERN's "Scalable Heterogeneous Integrated FaciliTy", SHIFT, was the precursor of today's cluster and cloud computing systems which completely replaced mainframes.

The networking issues for SHIFT

SHIFT's CPU, Disk and Tape server nodes had to be linked by a "Network Backplane" with, for the time, extreme throughput requirements (10 Mbit/s for a small system). Early on, it also had to deal with the high CPU power consumed by network I/O at the nodes. This was solved by the UltraNet [31] system. Later the backplane became a hybrid system containing elements of Ethernet (10 Mbit/s), FDDI (100 Mbit/s), UltraNet (600 Mbit/s) and HiPPI (800 Mbit/s) [32], until switched Gigabit Ethernet became sufficient for the task. CERN contributed to the further development, which led to the faster 6400 Mbit/s Gigabyte System Network (GSN) [33], a precursor of InfiniBand [34] now used for the data acquisition of the LHC experiments.

9.4 High-Speed Worldwide Networking: Accelerating Protocols

François Fluckiger

Public research networking created the Internet. Rarely in history did the academic community have such a dramatic and profound impact on the economy, and on society. As part of this community, the role of CERN was essential. CERN is well-known for its invention of the Web, but less so for its contribution to Internet infrastructure, central to the development of the worldwide Internet, and for its leading role in the development of high bit-rate networking.

Paving the way to the Internet and Web

In the early 1980s, CERN's external connections amounted to two minuscule leased lines operated at 9.6 kilobits per second (kbit/s), one to the Rutherford Appleton Laboratory (RAL), the other to CEA, Saclay. Ten years later, CERN had become the centre of a large star-shaped network, by far the largest internet hub in Europe. Today the LHC Optical Private Network (LHCOPN), a network connecting CERN to the LHC T1 sites operates multiple 100 Gigabit per second (Gbit/s) links. How did CERN get there?

Facing technological challenges has frequently been at the root of developing leading-edge technologies for CERN engineers. In turn, mastering a technology provides the confidence necessary to undertake very ambitious projects. Such was the case for networking, where CERN developed as early as the 1970s outstanding expertise on network protocols.

Facing the need for an open (i.e. based on non-proprietary technology), high bit-rate communication system to interconnect on-site a variety of computer brands — mainly in order to transfer files — CERN launched CERNET [26], the most ambitious networking project of its time. Communication between computers requires the use of common rules. These rules are called communication protocols. They are building blocks stacked one on top of the other in a layered structure, and are in practice implemented as modules of hardware, software or a mix of the two. They broadly divide into two types. The first type ensures the transport of raw information data between two network points, independent of its significance or use. The second type of protocol offers the end user a tangible service, e.g. email. CERNET implemented a genuine and complete suite of layered protocols, ranging from the physical and electrical plug to the CERN File Transfer Protocol running on half a dozen different brands of computers. CERNET was put into operation in 1977. This is a unique example of a complete multi-brand networking stack developed by a single organization. It was also the fastest packet-switching network of its time [Highlight 9.3].

Probably more important for the future was the type of technology adopted by CERN for one of the layers, called the packet layer. All modern networks use the packet principle where the data stream is chopped into individual pieces called packets. Packet protocols are divided into two fundamentally opposite classes: the connection-oriented class where the two end-computers must establish a connection before an information packet can be sent — the equivalent of a telephone call, and the connectionless class, where packets are independent of each other, as letters posted in a letter-box, and can be sent at any time. Connection-oriented networks check first via the call set-up mechanism that a fixed end-to-end route is available. The main advantage lies in the fact that commercial network operators know the duration and the amount of traffic exchanged over each connection and they can charge the user on a per-call basis. The drawbacks are the extra delay to setup the call before any useful data can be sent and the additional complexity. Connectionless networks are not aware that a series of packets may belong to a flow. As letters, individual packets carry the full destination address. The drawback is that no prior check is made that a non-congested route is available when new packets are admitted. The chief advantages are the no-delay admittance of packets, higher resilience as routes are not pre-assigned, and more generally, simplicity.

CERN opted in the early 1970s for the latter, the connectionless approach. A prescient decision, since connectionless is precisely the philosophy of the Internet Protocol, the famous IP. By creating, 15 years before the explosion of IP, a deep expertise in IP-like technology, CERN paved the way for its future role in the Internet infrastructure and also for its invention of the Web.

The emergence of HEPNet

In the early 1980s, only RAL and Saclay were connected to CERN, via slow analogue leased lines funded by these two organizations. They were mainly used for remote login to CERN mainframes. In 1983, a proposal was made by CERN to connect more sites and offer additional services (File Transfer, Job submission and email in addition to remote login). This initiative was coined HEPnet. As alternative to leased lines the use of the emerging public packet switched networks was proposed by the national PTTs, monopolies at that time. The service offered by the PTTs was based on X.25 technology, a connection-oriented protocol, which is the exact opposite of CERNET and IP. CERN connected to the Swiss PTT X.25 network, allowing computers to make calls — as over a telephone network — to and from any other HEP site connected to their domestic X.25 network. To facilitate file transfers, CERN developed a gateway between CERNET and X.25, which was released in 1983. This development was one of the very few operational gateways between connectionless and connection-oriented networks.

The main drawback of X.25 networks was their per-call charging principle modelled by the PTTs on the telephone, making difficult any planning of the telecommunication costs. The X.25 PTT service was progressively abandoned and new leased lines — initially analogue, then digital at 64 kbit/s — connected additional HEP sites to CERN. In parallel, in 1982 IBM proposed to several research computer centres in Europe, to establish and fund for four years a network based on the BITNET (Because It's Time NETwork) model in the USA. The initiative was called EARN (European Academic and Research Network) [35]. Additional analogue links were installed and CERN became in 1983 the largest hub of EARN. At that time, all of these links operated in a proprietary way.

CERN: Precursor in Internet Technology

In 1985, a TCP/IP coordinator was appointed at CERN and the LEP control system adopted this technology. CERN was one of the first sites in the world to receive for evaluation early versions of IP routers from the future market leader. It was at the end of 1988 that CERN started to operate leased lines with the TCP/IP protocols. By then, more leased lines had been set up, extending to Boston and Tel-Aviv. In summer 1988, a CERN team went to Beijing to organize the very first academic connection between China and Western Europe. But more than the number of connections, it was the tremendous increase of the bandwidth that changed dramatically the resulting services. In 1988, most analogue links had been replaced by digital circuits, at multiples of 64 kbit/s. Furthermore, by funding the first European 2 Mbit/s academic link in Europe, between CNAF Bologna and CERN, INFN propelled external networking into the Megabit per second era. No ground links of that bit-rate were available, but experimental satellite channels were. From 1981 to 1983, the STELLA project evaluated protocol aspects of 2 Mbit/s satellite channels between CERN, Pisa and RAL.

CERN becomes the largest Network Hub in Europe

In 1990, the very first Internet transatlantic megabit link (1.5 Mbit/s) was set up between CERN and Cornell University. A full TCP/IP connection to the US National Science Foundation Network (NSFNet) was established, the link being the result of EASYNet, a new IBM initiative in Europe, three years after the end of their EARN funding.

In 1991, the star network around CERN became the largest internet hub in Europe: 80% of the Internet capacity in Europe for international traffic was installed at CERN (Fig. 9.5). It is no surprise that the performance of Tim Berners-Lee's first Web server impressed the world: it was at the heart of the European Internet, just a few hops from most destinations. Why did this occur? There was an absolute need for high bit-rate access to CERN from HEP centres so that physicists could work remotely almost as if they were on-site at CERN. In addition, networks grew like crystals: as one hub started "crystallizing", everyone wanted to connect to it being inspired by the novel opportunities offered, Therefore, non-HEP scientific centres started to set up links to CERN, often serving as relays to other domestic HEP sites. Finally, a networking hub, being a central point of failure, requires extremely reliable operation. Inherited from the accelerator culture, CERN had a tradition of excellence in 24 hours/7 days operation that was recognized worldwide.

CERN initiates IP address allocation in Europe

Another key contribution of CERN to the Internet infrastructure was the attribution of IP addresses. Any device connected to the Internet needs a unique identifier: the famous IP address. Before 1989, all addresses were allocated by the Internet Assigned Numbers Authority (IANA) in the USA. It was CERN, which convened at its site in December 1988 the first meeting to discuss the creation of a structure in Europe to allocate the European addresses. Six months later, the first meeting of RIPE (Réseaux IP Européens) took place in Amsterdam. RIPE is still, 25 years later, the authority that allocates Internet resources and services in Europe, in the Middle East and parts of Central Asia [36].

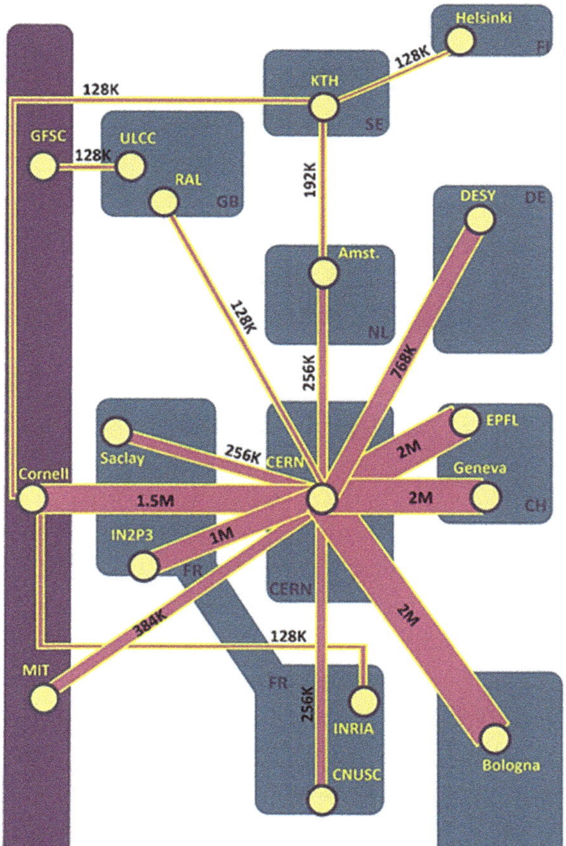

Fig. 9.5. The star network around CERN in 1991. Only high throughput data links are shown: for all links see http://cds.cern.ch/record/2054391.

Having reached its zenith in 1991, the CERN share in the Internet infrastructure slowly decreased, for two reasons. First, though prepared to remain a major Internet Hub, CERN considered that it would be healthier for the Internet if other hubs emerged. Second, in many countries, National Research and Education Networks (NRENs) had emerged, to which most HEP centres were connected [37]. These NRENs were, and still are, interconnected via the pan-European academic backbones operated by the DANTE organization, initiated by the European Commission [38]. The connection of CERN to these successive backbones moved from 34 Mbit/s in 1997 to 10 Gbit/s in 2005. In that period, CERN played a similar role to that of the European NRENs. However, there was still a need to push the technological limits and in 2003, in the context of the EU-sponsored CERN DataTAG project, the world record for data transfer was broken by CERN, together with the California Institute of Technology (Caltech), the Los Alamos National Laboratory (LANL) and SLAC, with the successful transfer of 1 terabyte of data in less than an hour over the 10,037 km between CERN and California.

The Gigabit era

To meet the requirements of the Worldwide LHC Computing Grid (WLCG) [Highlight 9.2], CERN and its HEP partners returned to the approach of HEP-dedicated links between the CERN and the major HEP centres (Fig. 9.6).

These dedicated ultra-high bit-rate links are provided by GÉANT [37], a pan-European research and education network supported by the EU, and by other NRENs: they form the LHCOPN (LHC Optical Private Network). The minimum bit-rate is 10 Gbit/s and several circuits operate at higher speed including a dual 2 × 100 Gbit/s transatlantic link to the Energy Science Network (ESNet) in the US.

Through the deployment of the Internet infrastructure technology, CERN, together with the rest of the academic community, has also had an indirect impact. Academic organizations — be they universities or research centres — are sometimes compared to industrial companies in unfavourable terms regarding aspects such as efficiency, organization and rigour. This is perhaps viewed as a price to pay for creativity — a perception inherited from the old idea that inventiveness and organization are antagonistic qualities. The academic community was sometimes viewed as being only able to invent new technologies, but not to contribute to their development. The development of the academic and research networks has demonstrated that the academic community is capable of deploying globally extremely complex systems. They are capable of delivering professionally operated systems, which can revolutionize the economy in a few years and permeate all levels of society.

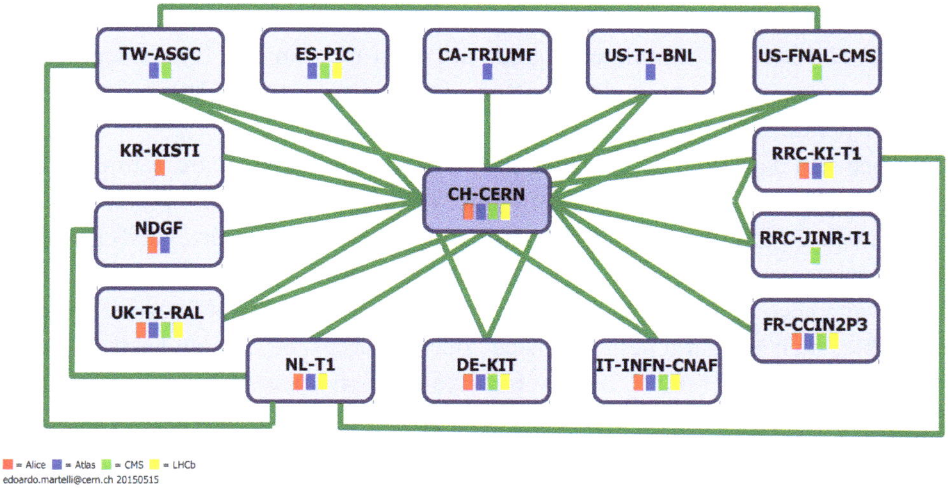

Fig. 9.6. LHCOPN topology. Details at http://cds.cern.ch/record/1997396?ln=en.

9.5 Detector Simulation: Events Before the Event

René Brun

Simulation[d] of a particle physics experiment relies on a chain of models which follow the evolution of the particles produced in the interaction up to the signals generated in the detectors. These can be:

- Models of the physical process representing the primary goal of the experiment and of all the known processes that could be considered as background;
- A model of the detector (geometry, materials, magnetic and electric fields) and of the path and interactions of each particle through the detector elements;
- Models of the signals generated by the particles in each detector type, followed by the electronic processing to produce digital information;
- A model of the selection procedures (trigger) which retain the wanted topology and reject background.

All models in the various steps are complex and contain many variables: simulation of an experiment is possible only by using Monte Carlo methods.

In the end, the simulation chain produces "data" in the same format as the real experiment. Models of the physical processes are needed to test the viability of the initial idea. Geometry and acceptance go in step with the overall design of the

[d]A simulation uses a mathematical description, or model, of a real system in the form of a computer program (Encyclopaedia Britannica).

apparatus. Signal generation is useful to understand in detail the behaviour of individual detectors. Simulated signals are used to train and tune selection and pattern recognition procedures. Simulated raw "data" help testing the reconstruction software.

Once real data become available and are processed, results are compared with the output of the simulation chain. In order to extract efficiently a potential signal from the background, one must identify the most efficient (uncorrelated) variables. Multivariate analysis (MVA) treats many statistical variables simultaneously, taking into account the effects of all variables on the response of interest. The tools available, such as neural nets, "boosted" decision trees, etc., go under the name of Machine Learning (ML). These tools make extensive use of simulated "data" to train their decision algorithms. TMVA [39] (Toolkit for Multivariate Data Analysis with ROOT) is used by the LHC experiments for this task.

As the understanding of the experiment improves, more refined simulations may be desirable, requiring running the whole cycle several times.

The full simulation cycle is very demanding of computer resources. Different methods of "fast" simulation have been devised to replace the full simulation of sub-systems by introducing simplified assumptions on detector geometry or by skipping some Monte Carlo calculations and using the results from formulae or from a stored catalogue. Computing time can thus be reduced by a large factor.

Fast simulation is not a replacement of full detector simulation. It must be tuned/validated with full simulation results (while full simulation is itself tuned/validated with data containing well-known reference signals). Fast simulation is used primarily for quick and approximate estimates of signal and background rates. It is also useful for an initial survey of complex, multi-dimensional parameter spaces which require a very large (simulated) data set.

Simulation is essential in designing experiments and interpreting the results, not only in high energy physics (HEP), but also in many other branches of science and technology. Simulating the known phenomena with good precision is a necessary prerequisite to the discovery of new effects, by comparing simulation with data.

Event generators have been developed for simulating interactions of different primary particles and a wide range of energies. An interaction process is described mathematically by a "matrix element" which must be integrated over the multidimensional internal parameters phase space to provide the total and differential cross-sections. Each point of this phase space is associated with a probability. According to this probability events are randomly generated.

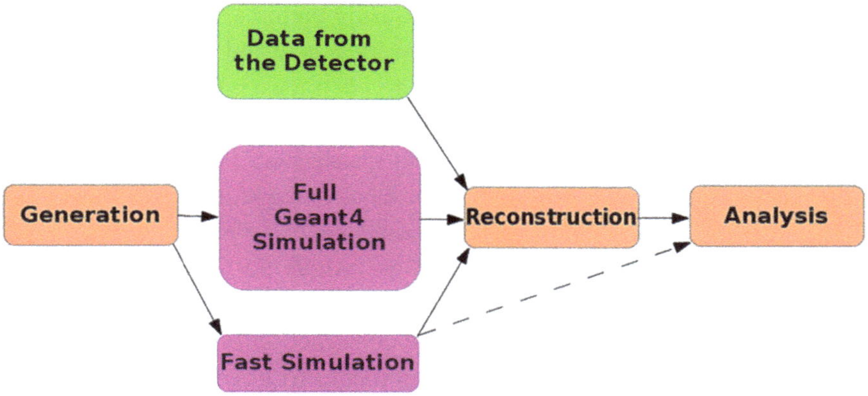

Fig. 9.7. Simulation in the LHC experiments.

As an example, consider the steps involved in the generation of events in proton-proton interactions as they happen at the LHC. The initial state consists of colliding protons, but the most interesting interactions are the hard collisions between constituent partons [Boxes 6.4 and 8.3]: the simulation must first take into account the distribution of partons inside a proton, as described by its structure functions. Then the specific "hard interaction" between partons has to be selected. As the most interesting processes (e.g. Higgs physics) are very rare compared to the dominant QCD background, the latter must be correctly simulated, which implies considering many relevant loop diagrams [Boxes 4.2 and 5.1]. Other partons belonging to the initial protons may interact "softly", and must be also simulated as they contribute to the complexity of the event. Following the initial hard interaction, the bare partons do not exist in nature (they are confined inside the hadrons) and must be "dressed", to form the known hadrons or mesons. This goes in steps: parton shower (describing their radiation), hadronization (grouping partons into particles) and decay of unstable particles. The task of comparing simulated events with data is further complicated by the fact that at the LHC several proton-proton interactions take place at each beam crossing.

Event generators must consider the physics complexity at the appropriate level together with the most efficient simulation methods. The goal is to generate enough simulated events to match the number of real events with affordable computing resources. An event generation package widely used for high energy collisions of electrons, positrons, protons and antiprotons is PYTHIA [40]; other packages have been developed for specific processes or initial states (e.g. heavy ion collisions).

Up to 1975 the major software development for HEP was associated with bubble chamber experiments. Event kinematics was reconstructed from the information obtained by scanning and measuring pictures. With early electronics-based experiments, the reconstruction software was tested and run only after data taking. Events consisted of straight or curved tracks hitting detector "planes" (digitized spark chambers or multiwire proportional chambers) and showers in calorimeter "blocks" (sodium iodide or lead-glass). EGS [41] from SLAC was the state-of-the-art program for simulating an electromagnetic shower in the simple geometry of these calorimeters. EGS users had to implement essentially two functions: to tell the system "where I am" or "where do I go". With the growing complexity of the detectors one soon realized that full simulation (tracking + calorimetry) was essential for testing the reconstruction programs and computing the geometrical acceptance. In 1975 a combined effort between the Omega, NA3 and NA4 experiments at the SPS led to the Geant1 and later Geant2 simulation tools (Generation of Events ANd Tracks). These provided a simple environment to propagate particles produced by an event generator, specific to each experiment, through the various detector modules, taking into account magnetic fields and multiple scattering. In 1981, a major upgrade to Geant3 [42] provided a tool to describe geometry with an automatic transport mechanism. Geant3 was interfaced to EGS3, then EGS4 and several hadronic shower packages. Geant3 included major developments and became the main simulation tool for the LEP detectors, for experiments at the Tevatron and RHIC and for the design of LHC detectors.

In 1994, the Geant4 [43] system was implemented in C++, based on the same principles as Geant3 as concerns detector description and transport. Today it is the cornerstone for all detector simulations in the HEP world. During phase1 of the LHC, about 55% of the total CPU time on the GRID machines was dedicated to detector simulation. It has become essential to improve the performance of the simulation tools, and a new project GeantV was launched in 2012 with as objective to gain possibly a factor 10 in speed. Both Geant3 and Geant4 had and still have a strong impact outside the HEP world, contributing to the design and analysis of many devices in astrophysics, medicine, hadron therapy, materials science.

Another development should be mentioned in the field of simulation: FLUKA [44]. Started as a program to calculate radiation shielding at the PS in the early 1960s, it was refined for the same purpose to cope with the higher energies of the SPS. It went through a major upgrade which made it a general purpose tool for calculations of particle transport and interactions with matter, covering an extended range of applications from proton and electron accelerator shielding to target design, calorimetry, activation, dosimetry, detector design, cosmic rays, neutrino physics and radiotherapy.

9.6 Data Analysis and Programming Environment: Distilling Information

René Brun

Data storage formats and analysis tools

Data from bubble chambers and optical spark chambers were recorded on photographic film. Points on selected tracks were measured on the film and stored on magnetic tape as "raw data" in relatively simple formats specific to each experiment. Raw data from the early experiments using digitized spark chambers and simple calorimeters were also stored in experiment-specific formats. Then the reconstruction program was processing these events and writing a "Data Summary Tape" (DST) with compact data types for the reconstructed tracks (direction, momentum) or calorimeter information. The next step was data analysis: The task was to assign or calculate particle masses and other quantities of interest, store them and display selected information in histograms. During the 1960s, the operation was optimized for the bubble chamber experiments by producing a suite of programs THRESH, GRIND, and SLICE, implementing the three stages of geometry, kinematics and assignment of mass to each particle: they could be used by all experiments with little customizing [45]. During the early 1980s, it was realized that this process could be simplified by storing the quantities of interest in "n-tuples" (a simple row of n variables put into tables). The Physics Analysis Workstation PAW [46] system developed in 1983 was based on row-wise n-tuples. A user could interactively process the n-tuples data with simple queries. Histograms were automatically viewed in many possible graphics formats on the user workstation. The system became rapidly very popular for data analysis and n-tuples grew larger. PAW n-tuples became the main analysis format for most experiments at CERN and other labs between 1988 and 1998.

In 1995, the ROOT project was started based on the same philosophy as PAW, but implemented in C++ and able to support more complex data types [47]. By 2000, ROOT had a built-in system capable of accepting most C++ constructs found in the LHC experiments. ROOT had been selected in 1998 for the Tevatron experiments at Fermilab, followed in 1999 by the RHIC experiments at BNL. All LHC experiments, as well as most HEP or related laboratories in the world adopted the ROOT system. ROOT was progressively interfaced with many other tools (graphics, user interfaces, statistics, mathematical libraries, etc.), becoming de-facto the new CERN Program Library. About 20,000 people download the ROOT binary or source files every month and about 9,000 people are registered to the ROOT Forum. Besides being used by more and more people in the scientific domain, ROOT is also used in the car and oil industry, and in finance.

Programming Languages

For about four decades, FORTRAN was the main programming language. Simple to learn and well adapted to small or medium size single process and sequential applications, it was the language of CERNLIB, the CERN Program Library [48], built from contributions of users developing simple algorithms, linear algebra or general utilities with up to a few hundred lines of code. CERNLIB was very popular in the scientific community: during the 1970s, it included task-oriented packages with a few thousand lines of code, e.g. GD3 for basic graphics, HBOOK [49] for histogramming, MINUIT [50] for fitting and minimization. At the same time data structure management systems were developed. These contained both numerical information and logical information about the object they describe, and included HYDRA [45] for bubble chamber experiments and ZBOOK [51] for electronic experiments. The ZEBRA system (ZBOOK+HYDRA) appeared in 1982, enabling the creation and management of a complex network of dynamic data structures, and saving them to files in machine independent format.

With the advent of FORTRAN90, it would have been possible to upgrade ZEBRA, taking advantage of the new features (similar to C structures), but this did not happen: FORTRAN was less and less competitive with the new software programming systems that provided user interfaces, graphics, network communications and interfaces with the operating system and hardware devices. With the increasing number of personal workstations and network connections, more and more applications required calls to the operating system, file and network system. These interfaces were typically implemented in the C language. For example, in the PAW system a substantial fraction of the user interface and graphics was in C.

The years 1990–1994 witnessed many discussions about the future language for HEP. By the end of 1994, it was agreed that the programming language for HEP was going to be C++. Two Research and Development projects were launched: RD44 for the development of Geant4 and RD45 for the evaluation of Object-Oriented data bases. Starting in 1995 the ROOT system also opted for C++.

Programming Environment

The size of programs has grown with time at a tremendous rate. In the early 1970s, large programs were around 10,000 lines of code. CERNLIB was some 30,000 lines of FORTRAN and assembler in 1970. The HBOOK or HYDRA packages contained around 15,000 lines in 1975. CERNLIB reached a maximum of about 150,000 lines in 1985, including the major packages. ZEBRA stabilized with about 30,000 lines in 1985. Geant3 started with ~10,000 lines in 1981 and reached ~200,000 lines in 1993; PAW reached a maximum in 1994 of ~250,000 lines. During the 1970s, an experiment-specific code (essentially the reconstruction program) was less than 10,000 lines, reaching about 100,000 lines at the start of LEP in 1989. Today, Geant4 contains ~1.5 million lines, ROOT ~3 million lines, with specific software on top of these packages weighing in at ~5 million lines per experiment — not counting the analysis software by thousands of physicists. The memory used by these programs was around a few tens of kilobytes in the 1970s. The full simulation of the OPAL detector planned for LEP in 1981, including electromagnetic and hadronic showers, could run on central computers having only 1 MB of memory. From LEP to LHC experiments have grown in size and complexity, and the signal to background ratio has become much more demanding: today one sees simulation systems requiring 3000 times more memory! (Fig. 9.8)

Managing the source code of HEP software has always been a concern, even when systems had only a few thousand lines of code, because the software development has always been a joint effort in experiments. The PATCHY system [52] was developed at the end of the 1960s for bubble chamber software. A PAM file (PATCHY Master file) was common and read-only for all developers. Each developer prepared his/her own file with changes or additions to the master file. A PAM file was updated periodically by user consensus. PATCHY remained the preferred solution in the 1980s until an interactive version of PATCHY, called CMZ, appeared in 1989, making it easier to develop software on distributed systems by using workstation with nice text editors. PATCHY/CMZ were replaced

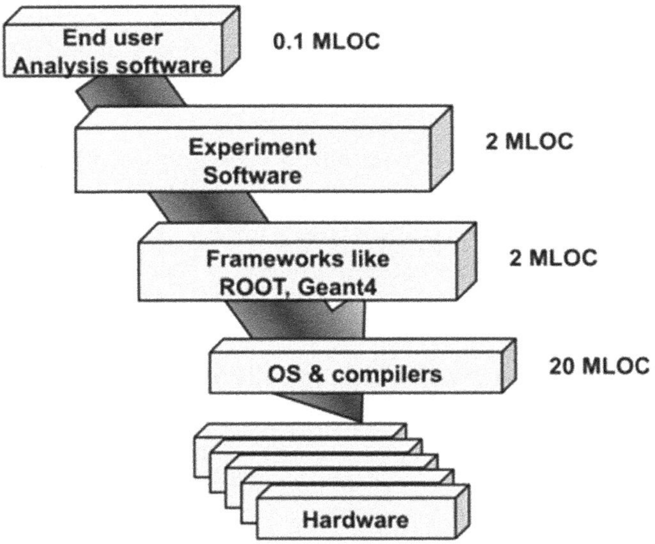

Fig. 9.8. The Software hierarchy (MLOC ↔ Million Lines Of Code).

by the open source CVS (Concurrent Versioning System) appearing at the end of the 1990s. With CVS, it became possible to edit/manage a central code repository shared by hundred/thousand developers, showing differences between successive versions and making easier code releases. Around 2005 an evolution of CVS, called SVN (Sub Version System) was adopted for all developments in LHC computing (and everywhere else). The GIT system replaced SVN around 2012, offering improved capability to manage successive versions of code developed and used by thousands of people. This trend will likely continue, following the continuous requirements to manage many millions of lines of code.

Error reporting systems have also evolved in the wake of the growth of the software size. From pure oral reports in coffee meetings or paper mail, huge progress followed the installation of the local area CERNET, followed by BITNET in the early 1980s, which brought worldwide e-mail exchange. The first web-based systems appeared around 1995 allowing the visualization of the source code and its documentation. The first common error reporting and managing system SAVANNAH appeared in 2000, followed by the current JIRA system in 2012. Several projects also host a discussion forum where users can report questions and problems, so that more experienced users can help (e.g. the ROOT forum).

Outlook

Computing at CERN has followed and continues to follow the general trends in the field. The computing languages C++, Python, Java are now the main languages in science, industry, communication and information. This approach facilitates communication among these various fields, but is also a important asset for the careers of young scientists who seek employment outside High Energy Physics (HEP).

General tools like ROOT will continue to evolve following the general trends not specific to HEP like user interfaces, graphics and networking. A considerable effort is devoted to integrate machine learning libraries and techniques to facilitate the ever more complex data analysis phases. The Open Data philosophy facilitating the use of experiments data for teaching will also require simple and common user interfaces, professional documentation and data descriptions.

The detector simulation with the Geant family will have to evolve to cope with the high luminosity machines and increasing energies. Factors of 10 to 100 in processing speed will have to be gained, taking advantage of parallel architectures and by integrating fast simulation algorithms into the full and therefore necessarily slow simulation system. Machine learning techniques will enter this field to optimize automatically the simulation and analysis processes.

9.7 World Wide Web: Global Networking

François Fluckiger

The World Wide Web was born at CERN. As is the case with many births, the coming of the Web was a combination of chance and necessity.

In June 1980, a team of half a dozen "rent-a-programmers" were hired by CERN from a British electronics company to work for a few months on the PS Control System. Luckily, in addition to writing software, they had been asked to thoroughly document their work. This was a complex task as most modules were inter-dependant. One of the young engineers proposed to use a special technique to reflect the dependencies and references between the multiple parts of his own documentation. The small system he wrote was called Enquire. When he left CERN at the end of the contract, the floppy disk with the code of the Enquire system ended up in a drawer and no-one ever reused it [53]. The technique that had been prototyped for the Enquire system was *hypertext* and the name of the young engineer was Tim Berners-Lee. The hypertext concept was not invented by him, and Enquire was not its first implementation. The principle had been laid down in 1945 and the term was coined in 1963 in the context of the Xanadu project [54], followed by the first demonstration of a hypertext system in 1968. Yet, in 1980 two things had happened: the hypertext idea had entered CERN, and a young creative engineer had discovered CERN's pioneering tradition.

It was also by chance that the band of engineers, including Tim, was exposed to novel technology invented by the SPS Controls Group and exported to the PS Controls Group in the late 1970s: remote execution. A program was able to send a small part of itself to another computer and request its remote execution. An extension of this facility came naturally: the possibility for a programmer to include a special call to trigger a remote execution. The PS Control System was one of the first in the world to use Remote Procedure Calls (RPCs) in an operational environment. The RPC expertise created at CERN in the late 70s proved to be essential to the development of Web technology ten years later. Thus, Tim Berners-Lee left CERN in December 1980 with two complementary assets: practical application of hypertext expertise, and exposure to CERN RPC expertise.

CERN's unique expertise in Document Format

Tim returned to the UK, where he furthered his knowledge of RPC and more generally of networking. He applied for a fellowship at CERN in 1984 and joined the On-Line Computing Group. Involved in the LEP Data Acquisition System, he proposed using the RPC technology on which he had become an expert. But in parallel, and without any formal mandate, he started to reflect on how to solve the problem of the huge mass of diverse digital information CERN and the HEP community at large had to deal with. Naturally, he thought of a hypertext system, but unlike Enquire, not a centralized system that, he had realized, would never scale, but a fully distributed system. And the technique to support the decentralized approach was RPC. Hypertext plus RPC: he started to merge the two base technologies he had been working on over the past eight years.

To implement his idea of a document access system, Tim Berners-Lee also needed a standard format for describing text documents, which would be able to efficiently link to other documents. The chance again was that at CERN, world-class expertise existed on document standards. In particular, CERN experts had been major contributors to the Standard Generalized Markup Language (SGML) released by the International Standards Organization (ISO). The Web inventor benefited from the collaboration of the CERN SGML expert to design the Web document format standard, the famous Hypertext Markup Language (HTML).

CERN's unique expertise in Internet Technology

With HTML, one component of the Web technology was designed. But two others were missing. First a set of communications rules was needed between two computers to exchange documents: how to request a document, how to transport a document if available. Such a set of rules is called a protocol. Tim had to invent a Hypertext Transport Protocol — what he called HTTP. Second, a mechanism was needed to unambiguously and uniquely identify any document or resource; but how to form such a universal and unique identifier, when society is not able to uniquely name human beings on planet Earth? Tim had to invent a Universal Resource Identifier (URI) including a particular type of it to locate the resource, the Universal Resource Locator (URL).

The next chance was that the Web met the Internet. By the mid-1980s CERN started to evaluate and use the emerging TCP/IP protocol, and had gained expertise in Internet technology. By 1988, it became clear that IP (the Internet Protocol) was the protocol of choice for future packet networks. This is just what Tim was looking for. Before IP, the other wide-area technologies (such as IBM's SNA, X.25 or ATM, the Asynchronous Transfer Mode) were based on the connection-oriented principle: before a first packet could be sent, a telephone-like call had to be engaged between any two distant computers. This is not practical for an RPC-based system like the Web that needs very fast interaction. In contrast, IP allows a packet, which may carry a request for a new Web page, to be sent without prior call. Tim Berners-Lee was then able to build his HTTP protocol on the light and efficient Internet Protocol. The chance was also that the Internet is provided with mechanisms unique in the history of telecommunication, one of these being that any system or object connected to the Internet can have a unique name, independent of its real IP address. And in addition, an automatic system (Domain

Name System, DNS) allows the transformation of these names into addresses. To identify documents or resources contained in IP systems, Tim Berners-Lee modelled his URIs/URLs on Internet names.

Keeping it free and protected with open source

After his first proposal in 1989 (to which he never received a formal reply), Tim started the development that led to the opening of the first server in December 1990. It was again chance that at that time CERN was at the heart of a big star-shaped hub, having the greatest capacity for Internet transfer in Europe. The response time of the CERN Web server was impressive, and collaboration with other servers was boosted by excellent connectivity. Growth was astonishing: the number of Web servers grew from 26 at the end of 1992 to 200 in October 1993.

Then chance intervened yet again. In 1993, Tim Berners-Lee considered it necessary to make Web software more widely available. On 30 April 1993, CERN relinquished all intellectual property rights to the software and put it in the Public Domain. At that time, concepts like open software and public domain were in their infancy. In summer 1994, Tim Berners-Lee left CERN to join MIT and create the World Wide Web Consortium (W3C), keeping a European branch, initially at CERN and then at INRIA. After Tim's departure, the CERN technical team was taken over by new management [55]. The team was preparing the release of version 3 of the CERN server software, but by that time, the principles of open source licencing had become better known at CERN, and the risk of appropriation by third parties was more clearly understood. Indeed, any company could have taken the Web software and (after a minor change) prevented others from using it freely. As luck had it, over those 18 months, this did not happen. In November 1994, version 3 was released under a formal open source licence: CERN retained the intellectual property but gave to anyone the right to use, modify and distribute it freely. To protect freedom and avoid appropriation, all subsequent releases have been distributed by MIT under a similar open source licence.[e]

Satisfying an unforeseen need

As stated above, the coming of the Web was a combination of chance and necessity. The Web fulfilled a genuine need — that of the community of particle physicists, made up of more than ten thousand scientists, spread throughout the whole world. The Web was a necessity: the HEP community needed it, but did not know it needed it [55]. This is the wont of a disruptive invention: no-one requests it, but once there, no-one can do without it. Indeed, Tim Berners-Lee's 1989 proposal triggered little interest. One has to admit that re-reading the text, it is hard

---

[e]An open-source license is a type of license for computer software that allows the source code, blueprint or design to be used, modified and/or shared under defined terms and conditions.

to figure out that it would revolutionize the world. Until 1994, the project received only limited support from CERN [53] but it was allowed to continue, despite the lack of interest from potential users. In 1992 Tim presented the project to the HEPnet Requirements Committee, a gathering of CERN users tasked to formulate the needs in networking. No-one really understood the benefits. Only when they started to use it did the physicists understand that it was an unconscious necessity!

What the Web brought to users is the abolition of login tyranny. Before the Web, it was already possible to access remote information (files, programmes, text, etc.) but on two conditions: first to know and remember on which remote computer the information was stored, and second to be registered on the home computer and to login (or otherwise provide your credentials) on the target computer (with the exception of the special "Anonymous File Transfer" function). The conjunction of clickable hyperlinks and login-less servers removed these two barriers.

Mastering more than one domain

Beyond chance and necessity, there are other reasons for the success of the World Wide Web. One is the "think-global" spirit. In Tim Berners-Lee's design, the planetary dimension was present right from the beginning (Fig. 9.9). Even though today's universality was beyond the wildest dreams, he designed the Web for a global community of physicists, with a permanent concern as to its simplicity and scalability. Also to his merit, Tim Berners-Lee mastered several cutting-edge technical skills, in particular a deep understanding of both networking and document structure techniques. Progress in science and technology is often made by those who excel in more than one domain: in our increasingly complex world, compartmentalization of individuals into specialities can be a brake on creativity. The dual invention by the same person of the Hypertext Markup Language (HTML) and of the Hypertext Transport Protocol (HTTP), two designs that are a perfect match, is a fitting illustration that will leave its mark in history.

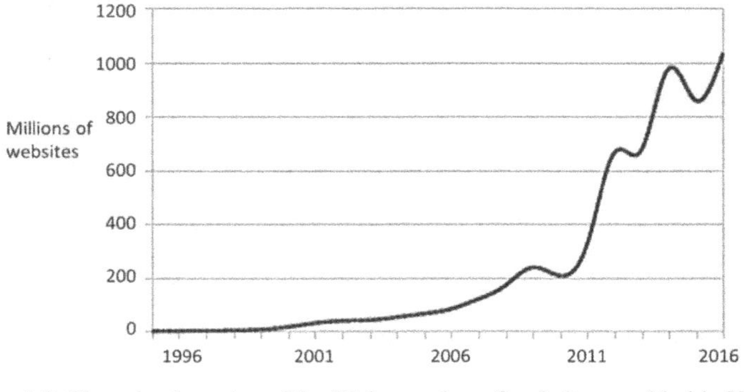

Fig. 9.9. The relentless rise of the Web: number of websites worldwide [56].

References

1. R. Brun, F. Carminati, G. Galli Carminati, *From the Web to the Grid and Beyond – High Energy Physics Driven Computing Paradigms* (Springer, 2012).
2. CERN Courier, 7, Issue 9, p. 173, September 1967; https://cds.cern.ch/record/1728889/.
3. P. Zanella, 30 years of computing at CERN, *12th CERN School of Computing* (1989) pp.1-29; http://cds.cern.ch/record/205360?ln=en.
4. H. Blieden *et al.*, in *Informal Meeting on Film-less Spark-chamber Techniques and Associated Computer Use, CERN, Geneva, Switzerland, March 1964*, CERN-1964-030 (CERN, Geneva, 1964) pp. 49-56. http://dx.doi.org/10.5170/CERN-1964-030.49.
5. F. Marciano *et al.*, pp. 245-249 in *Informal Meeting on Film-less Spark-chamber Techniques and Associated Computer Use, CERN, Geneva, Switzerland, March 1964*, CERN-1964-030 (CERN, Geneva, 1964). http://dx. doi.org/10.5170/CERN-1964-030.245.
6. P. F. Kunz, *Nucl. Instr. & Meth.* **135**, 435 (1976).
7. C. Bertuzzi *et al.*, pp. 329-335 in *Topical Conference on the Application of Microprocessors to High-Energy Physics Experiments, CERN, Geneva, Switzerland, May 1981*, CERN-1981-007. http://dx.doi.org/10.5170/CERN-1981-007.329.
8. S. Cittolin and the UA1 Collaboration, *Nucl. Instr. & Meth. A* **263**, 174-179 (1988).
9. D. Castelvecchi, Artificial intelligence called in to tackle LHC data deluge, *Nature* **528**, 18-19 (2015); http://www.nature.com/news/artificial-intelligence-called-in-to-tackle-lhc-data-deluge-1.18922.
10. J. Shiers, Data preservation is a journey, *CERN Courier*, May 2016, and references therein, http://cerncourier.com/cws/article/cern/65032.
11. T. Bloch, The CERN Computer Centre, *report CERN/DD/74/35* (1974); http://cds.cern.ch/record/1050339/files/dd-74-35.pdf.
12. O. Martin and A. Tengvald, Performance measurements on the CDC 7600 at CERN, *report CERN/DD/75/10* (1975); https://cds.cern.ch/record/307944/files/CM-P00059674.pdf.
13. G. Lo Presti *et al.*, CASTOR: A Distributed Storage Resource Facility for High Performance Data Processing at CERN, *Proc. 24th IEEE Conf. on Mass Storage Systems and Technologies* IEEE Computer Society (2007) pp. 275-280.
14. B. Segal *et al.*, SHIFT-BETEL: A (very) distributed mainframe, *Proc. Conf. Computing in High Energy Physics '94*, San Francisco, California (1994) pp. 188-191.
15. J.-P. Baud *et al.*, Mainframe services from Gigabit- networked workstations, *Proc. Summer 1992 USENIX Conference*, June 1992, San Antonio, Texas, U.S.A.; http://usenix.org/legacy/publications/library/proceedings/sa92/baud.
16. A. Fucci *et al.*, pp. 341-354 in *Topical Conf. on the Application of Microprocessors to High-energy Physics Experiments, CERN, Geneva, Switzerland, May 1981*, CERN-1981-007 (CERN, Geneva 1981). http://dx.doi.org/10.5170/CERN-1981-007.341.
17. J.-P. Baud *et al.*, SHIFT, the scalable heterogeneous integrated facility for HEP computing, *Proc. International Conference on Computing in High Energy Physics, Tsukuba, Japan*, Universal Academy Press, Tokyo (1991) pp. 571-582.
18. M. Aderholz *et al.*, Models of networked analysis at regional centres for LHC experiments (MONARC*), report CERN/LCB 2000-001*(2000), http://cds.cern.ch/record/510694/files/Phase2Report.pdf.
19. I. Foster and C. Kesselman. *The GRID: Blueprint for a New Computing Infrastructure,* (Morgan Kaufmann Publishers, San Francisco, 1998).

20. F. Gagliardi *et al.*, The EU DataGrid project: building and operating a large scale grid infrastructure, in *Engineering the Grid: Status and Perspective, eds.* B. Di Martino, J. Dongarra, A. Hoisie, L.Y. Yang, and H. Zima (American Scientific Publishers, 2006); http://www.gridpp.ac.uk/papers/Perf2002Paper.pdf.
21. L. Robertson, *Computing Services for LHC – from Clusters to Grids*, Chapter 3 of ref. [1].
22. H. Slettenhaar, Data communications at the CERN Computer Centre, *report DD/74/26* (1974); http://cds.cern.ch/record/418287/files/dd-74-26.pdf.
23. J.M. Gerard *et al.*, The use of the FOCUS system to monitor an experiment at the CERN Intersecting Storage Rings, *report DD/73/13* (1973); http://cds.cern.ch/record/1050741/files/dd-73-13.pdf.
24. C. Parkman and G. Lee (eds.), *OMNET — High Speed Data Communications for PDP-11 Computers*, CERN-1979-011 (CERN, Geneva, 1979). http://dx.doi.org/10.5170/CERN-1979-011.
25. T. Bruins *et al.* (eds.), *SUPERMUX — A Multi-Host Front End Concentrator System for Asynchronous Consoles*, CERN-1974-019 (CERN, Geneva, 1974). http://dx.doi.org/10.5170/CERN-1974-019.
26. J.M. Gerard (ed.) *CERNET — a High-Speed Packet-Switching Network*, CERN-1981-012 (CERN, Geneva, 1981). http://dx.doi.org/10.5170/CERN-1981-012.
27. B.M. Segal, A Short History of Internet Protocols at CERN (1995); http://cnlart.web.cern.ch/cnlart/2001/001/tcpip/
28. P. G. Innocenti, The LEP control system: architecture, features and performance, *Proc. 14th Int. Conf. on High-Energy Accelerators*, Tsukuba, Japan, (1989) pp. 885-892; http://cds.cern.ch/record/210879/files/p183.pdf.
29. R. J. Laukner, R, Rausch, SPS and LEP controls: status and evolution towards the LHC era, *Proc. Int. Conf. on Accelerator and Large Experimental Physics Control Systems,* Chicago, U.S.A. (1995); http://ccdb5fs.kek.jp/cgi-bin/img/allpdf?199603225.
30. J. Serrano *et al.*, The white rabbit project, *Proc. INIB2013*, Oxford, UK (2013); http://accelconf.web.cern.ch/AccelConf/IBIC2013/papers/thbl2.pdf.
31. R. Beach, UltraNet: An Architecture for Gigabit Networking, *Proc. 15th Conf. on Local Area Networks,* IEEE (1990) pp. 232–248.
32. Ben Segal *et al.*, HiPPI in the CERN Computer Centre, CERN Report CN/96/13; http://cds.cern.ch/record/311057/files/cn-96-013.pdf?version=1.
33. CERN Press Release, http://press.cern/press-releases/2000/09/cern-first-demonstrate-10-gigabitsec-network-four-major-computer.
34. P. Grun, Introduction to Infiniband for end users, *Infiniband Trade Association*, April 2010; https://cw.infinibandta.org/document/dl/7268.
35. O. Martin, *The hidden Prehistory of European Research Networking*, E-Book (Trafford Publishing, 2012); http://oliviermartin.authorsxpress.com/author/oliviermartin/.
36. F .Fluckiger, CERN celebrates another key contribution to the Internet, *CERN Computer Newsletter,* June issue (2009); http://cerncourier.com/cws/article/cnl/39610.
37. H. Davies and B. Bressan, *A History of International Research Networking* (Wiley, 2010); http://eu.wiley.com/WileyCDA/WileyTitle/productCd-352732710X,subjectCd-CM12.html.
38. F. Fluckiger, The European researchers' network *La Recherche,* **328** (2000) (*translation*); https://cern.ch/Fluckiger/Articles/F.Fluckiger-The_European_Researchers_Network.pdf.
39. https://root.cern.ch/tmva.

40. T. Sjöstrand *et al.*, An introduction to PYTHIA 8.2, *Comput. Phys.Commun.* **191**, 159 (2015); http://home.thep.lu.se/~torbjorn/pdfdoc/pythia8200.pdf.
41. R.L. Ford and W. R. Nelson, The EGS Code System, *report SLAC-210*, UC-32 (1978); http://slac.stanford.edu/pubs/slacreports/reports03/slac-r-210.pdf.
42. R.Brun *et al.*, Geant3 user's guide (1987); http://cds.cern.ch/record/1119728?ln=en.
43. S. Agostinelli *et al.*, Geant4, a simulation toolkit, *Nucl. Instr. & Meth.* **506** (2003).
    J. Allison *et al.*, Geant4 developments and applications, *Trans. Nucl. Science* **53** (2006).
44. A. Ferrari *et al.* (eds.), *FLUKA: a Multi-Particle Transport Code*, CERN 2005-010 (CERN, Geneva, 2005); http://dx.doi.org/10.5170/CERN-2005-010.
45. R.K. Bock and J. Zoll, Bubble chamber experiments: central computers in the analysis, *CERN Courier* **12**, issue 3 (1972); http://cds.cern.ch/record/1729464/files/vol12-issue3.pdf.
46. R. Brun *et al.*, PAW - A general purpose portable software tool for data analysis and presentation, *Computer Physics Communications* **57**, 432-437 (1989).
47. I. Antcheva *et al.*, ROOT - A C++ framework for petabyte data storage, statistical analysis and visualization, *Computer Physics Communications* **180**, 2499-2512 (2009).
48. CERN Program Library, http://cernlib.web.cern.ch/cernlib.
49. R. Brun and D. Lienart, HBOOK user guide, version 4, *CERN program library* **Y250** (1988).
50. F. James, MINUIT. Function minimization and error analysis, Reference manual, *CERN Program Library, long write-up* **D506** (1994);
    https://root.cern.ch/d35c7d8c.web.cern.files/minuit.pdf.
51. R. Brun *et al.*, ZBOOK – User guide and reference manual, *CERN program library* **Q210** (1984).
52. J. Zoll, PATCHY reference manual: version 5.04 (1995);
    http://cds.cern.ch/record/446554?ln=en.
53. B.E. Carpenter, *Network Geeks, How they Built the Internet* (Springer, 2013);
    http://www.springer.com/us/book/9781447150244.
54. J. Gillies, R. Cailliau, *How the Web Was Born, The Story of the World Wide Web* (Oxford University Press, 2000).
55. D. Raggett, My involvement with the early days of the Web, *W3C 10th Anniversary Conference*, Boston (2004);
    http://www.w3.org/People/Raggett/the-early-days-of-the-Web.htm.
56. http://news.netcraft.com/archives/2016/05/26/may-2016-web-server-survey.html.

Chapter 10

Knowledge and Technology: Sharing With Society

Cristoforo Benvenuti, Christine Sutton and Horst Wenninger

10.1 A Core Mission of CERN

Education and basic research are large costs to society, which governments, in general, accept as a necessary long-term investment. Still, it is fair to ask how this investment can be used to the best overall benefit for society.

Many excellent reports and talks [1–5] describe how basic research, motivated by "curiosity" — the desire to extend our understanding of nature — contributes as much to innovation as do the applied sciences, which focus on solving specific problems. Both stimulate industry and advance technology.

Consider the core technologies of high energy particle physics. Accelerator applications range from killing bacteria and treating cancer, to welding and cutting, hardening materials, reducing harmful gas emissions from thermal power plants, and much more. Particle detectors are not only ubiquitous in medical imaging; they are also used, for example, by volcanologists and archaeologists to image sites of potential interest. Computer simulation toolkits developed for particle physics have become essential in other fields, including medicine, aerospace, and nano-science.

Complex instruments and cutting-edge technologies are required for the construction and operation of accelerators and intricate detectors, as well as for managing the wealth of data they produce. When existing technologies are inadequate to meet requirements, developments are triggered that often give rise to technological innovations. Most famously, the World Wide Web, which was invented at CERN in the late 1980s to allow automatic sharing of information for scientists, has revolutionized global communication with enormous social and economic impact. Three decades later, data-handling and analysis needs for research at the LHC continue to push forward the evolution of worldwide distributed grid computing, bringing benefits to bio-computing and health sciences, astrophysics, environmental studies, and others.

One of the missions of CERN is to disseminate the results of its research and technological development, to trigger and prepare possible applications outside particle physics and to be a forum for the exchange and scrutiny of ideas for this purpose. Since its origins, CERN has shared knowledge with students, visitors, collaborating institutes and industrial partners [6]. Today, the Organization and the collaborating universities and institutions are together enhancing their efforts in knowledge-transfer and education, aiming to trigger innovation and improve the understanding of basic research as a motor for new and relevant technologies.[a]

Knowledge-transfer mechanisms

The commitment to formalizing the transfer of knowledge acquired at CERN has grown steadily over the decades, with an increasing availability of services and support for inventors and external partners, and for multidisciplinary ventures [7]. During this time, a number of knowledge-transfer mechanisms have been implemented in order to profit from the worldwide network of competencies in universities and high-tech industries in the Organization's Member and Associate Member States. The following gives an overview of these mechanisms together with illustrative examples.

Knowledge transfer through people

CERN's 12000 user-scientists are the most important knowledge-transfer mechanism, together with the educational programmes for students, fellows, and technical trainees. The Laboratory is a hub for knowledge sharing, with visiting researchers, teachers and industrial partners taking expertise back to institutes, universities and industry. Via training, education, and large-scale collaboration, CERN propagates this culture of knowledge sharing, which is considered one of its major contributions to society.

Training and education form one of CERN's core missions. A prominent example is the Summer Student Programme for physics and engineering students (Fig. 10.1), who join research teams for several weeks and attend lectures in particle physics, accelerator technologies and computing. The technical student programme allows engineering and applied physics students to spend up to 14 months at CERN. The Laboratory also provides the research base for large numbers of doctoral and diploma students from institutes around the world, who work on topics in accelerator physics, applied physics, and experimental and theoretical particle physics. Each year, close to a thousand doctoral degrees are awarded on the basis of work carried out at CERN experiments and accelerators.

---

[a]The editorial team thanks Giovanni Anelli for his contributions to this chapter.

Fig. 10.1. Each year CERN welcomes some 300 undergraduate students to its long-established Summer Student Programme, where they work with research teams and attend lectures in particle physics, accelerators and computing.

The skills thus acquired are highly attractive for work in other fields. One study found that more than 40% of students went on to work in the private sector [8].

CERN has a long-standing tradition in organizing conferences, workshops and high-level schools in the Member States and beyond. The CERN School of Physics, which originated in the early 1960s, went on to become the European School of High Energy Physics, and inspired the similar CERN–Latin-American Schools and the Asia Europe–Pacific Schools. The CERN School of Computing and the CERN Accelerator School followed in the 1970s and 1980s, respectively. CERN initiated the European Particle Accelerator Conference series, which has merged with other major events to become the International Particle Accelerator Conference. The Laboratory also brings its research and technology to the public and into classrooms, through guided tours, exhibitions and programmes for several hundred high school teachers each year, including courses, frequently in their national language.

For the effective transfer of expertise to different fields, it is essential not only to identify potentially interesting technologies, but also to assess their relevance to society. This can be achieved through multidisciplinary networks and collaborations, where scientists from different fields and industry discuss ideas to establish a common roadmap. This has proved particularly true in healthcare. A prime example is the initiative taken to set up the inaugural meeting for a light-ion-therapy network in Europe, ENLIGHT [9], at CERN in 2002, which implemented the concept of multidisciplinary collaboration, and led to growing interest, for example, in hadron therapy.

A more recent development is IdeaSquare [10], a facility located in a dedicated building that brings together researchers, engineers, industry and students to stimulate new ideas inspired by CERN's on-going detector and accelerator R&D. IdeaSquare also hosts innovation-related courses and events, such as the Challenge Based Innovation (CBI) course. Benefiting from the technical knowledge of the researchers and engineers working at CERN, this dedicated MSc-level programme is targeted at multidisciplinary student teams.

Knowledge transfer through purchasing and outsourcing contracts

Procurement contracts with industry amount to nearly 60% of CERN's annual budget. These can range from off-the-shelf products (e.g. office furniture) to R&D contracts requiring cutting-edge technologies for the accelerators. According to surveys [11], a majority of companies that have supplied products or services to CERN have indicated that their contract with the Organization led to an increase in their expertise, often resulting in new business opportunities. The benefit to industry in terms of "economic utility", defined as increased turnover plus cost savings, was estimated to be about three times the value of the CERN contracts. Traditionally, this has been the major route to technology transfer.

The CERN Member States expect a "fair return" for their financial contribution to the Organization through industrial contracts. To assist CERN in these efforts, links to industry are maintained via "industrial liaison officers" from each Member State who participate in the knowledge-transfer activities.

Knowledge transfer through joint ventures and collaboration agreements

CERN has a well-established tradition of collaboration with research institutes and companies, with the primary objective of generating technological results for use both by the partners and at the Laboratory, but also with an eye to the potential for commercial exploitation [12]. Through R&D ventures and agreements, CERN has often acted as a catalyst and an active promoter of international standards, notably in electronics and information technology.

Since its foundation, CERN has played an important role in the exchange of ideas between scientists who are studying a topic or planning, constructing, and performing experiments and analysing the results. Over the years, about a thousand experiments have taken place at the Laboratory, with both CERN and the users' institutions benefiting from the mutual experience [13]. In addition, the Organization recognizes experiments in fields allied to particle physics, such as astroparticle physics, that are being performed elsewhere, and are not necessarily accelerator-based. They obtain a modest level of support, including the use of test beams, provided this does not affect the in-house programme. As of 2016, there are 26 such "recognized experiments", including, for example, AMS, IceCube, MICE, and LIGO. CERN also benefits from the associated contacts.

The Organization also has a tradition of providing staff and facilities for development work on behalf of other institutes. In an early example, the organization provided tangible help in establishing the European Southern Observatory (ESO). With its experience in building large-scale apparatus and in dealing with industry, CERN was able to bring useful expertise to the project and for several years, the ESO Telescope Project Division was based at CERN.

In other instances, people have come from elsewhere to work side-by-side with a CERN specialist using material provided by their own institute, in particular for the design and construction of components for their accelerators. Examples include: magnet design and measurement for the ELETTRA Sincrotrone Trieste research centre in Italy; klystron modulators for the linac of the European Synchrotron Radiation Facility (based on the modulators for the LEP pre-injector); magnet measurements and supply of some magnets for the Italian Hadron Therapy Centre (CNAO); design and measurement of magnets, vacuum technology, beam diagnostics, and RF (based on the latest solution for the PS Booster at CERN [Highlight 3.4]), for the MedAustron facility in Austria [Highlight 10.2].

Knowledge transfer through new companies

In recent years, transferring CERN technology through the establishment of new companies is gaining interest. The Organization encourages the formation of spin-off companies — start-ups whose activity is based wholly or partly on CERN technologies and which wish to exploit these technologies commercially. CERN does this by giving measured support to its personnel and to external entrepreneurs who wish to establish a spin-off company in a Member State.

To assist entrepreneurs in taking innovative ideas using CERN technology or expertise from concept to market reality, a network of Business Incubation Centres (BICs) [14] has been set up in several countries. The Laboratory supports companies in BICs through technical visits to CERN and preferential rates for licences of CERN Intellectual Property. The incubators advise on establishing a business plan and provide logistical, financial and administrative support.

CERN's activities in entrepreneurship and start-up creation have also gained momentum, and first successes are appearing on the horizon. For example, as a result of a strategic partnership with NTNU Trondheim, each year students in entrepreneurship spend a week at the Laboratory to evaluate the potential of CERN technologies for starting a company. In 2014, a team of NTNU students started a company that provides professional services and consultancy for Invenio — software originally developed at CERN to manage the increasing volume of digital information. Invenio has grown into a fully featured digital library platform, and a number of large institutions worldwide are now customers.

Knowledge transfer through licensing

As a principle, CERN has not taken patents on its technological inventions. However, in 1999, this patent policy was revised, not in view of a possible financial return, but rather to protect the Organization's intellectual property and facilitate the transfer of innovations to industry. CERN's patent portfolio now contains a few tens of patents, many of which have been licensed to industry. Most concern the application of CERN's competences and inventions to problems relevant to society. Examples of such transfer through licensing are described in the following sections.

The KT Fund

Part of the revenue originating from knowledge-transfer (KT) activities at CERN (charges for contract research and services, licensing and consultancy) is allocated to the KT Fund. This provides financial support in the early stages of KT projects, regardless of their field of application, provided they are of potential use to society. Proposals for projects that meet the eligibility criteria are submitted to the KT Group and evaluated by the KT Fund Steering Committee. Together with a technical description of the project, proposers must outline the market potential. Proposals are evaluated based on their quality, likelihood of dissemination, potential impact, and scientific value. Gas-electron multiplier (GEM) detectors [Box 4.6] for flame detection and early earthquake prediction, radio-frequency absorbers for energy recovery, and exotic radioisotopes for medical applications are among the projects that have been partially funded under this scheme.

Transfer of CERN technologies

Transfer originating in CERN core activities

The technological innovations described in this book have been motivated by the needs of CERN core activities. For this reason most of these innovations have found applications in laboratories with a mission similar to that of CERN. In these cases, the transfer occurred naturally via scientific publications and communications. However, in some cases the innovations have been transferred to a wider community as described above.

Whenever industry was not able to provide the required solutions for a particular goal, development work was undertaken in-house and then transferred to industry, through a production contract awarded via competitive tendering. As a result of this process, the selected industries were able to offer this same solution to other customers. In this, the most frequent case, the transfer occurred through procurement. Examples of this process are the ISR vacuum gauges [Highlight 4.3] and superconducting quadrupoles [Highlight 4.4], and the LEP niobium-coated RF cavities [Highlight 7.4].

In other cases the initial development was done jointly with industry, which then proceeded with the production phase for CERN (and also, possibly, for other clients). Examples include the flexible cryogenic transfer line [Highlight 4.5], and cryogenic system [Highlight 8.3]. In the case of the LHC superconducting magnets [Highlight 8.2] the joint development work also served to whet the appetite of industry for the forthcoming large series of costly devices.

Finally, when tight schedules prevented industrial involvement, the production was carried out at CERN, the innovative solution was protected by patent and the technology transfer done by licensing. In this case the solution was also made generally available. The getter coatings developed for the LHC [Highlight 8.5] provide an example of such transfer. In another approach, the prototypes and an initial 10% of the production were made at CERN, with the remainder supplied partially by purchase from industry via competitive tender, and partially "in-kind", based on build-to-print specifications [Highlight 8.4].

Transfer motivated by social utility

CERN's expertise can also be valuable in fields outside the Organization's core mission. A small fraction of the CERN budget is made available for the development of innovative applications in fields that are outside nuclear and high-energy particle physics, but which have particular relevance for society in general, as described in examples below.

Medical applications[b]

The application of CERN expertise in accelerators to the field of cancer therapy, which goes back to the 1980s, has been particularly fruitful. Not only has the laboratory hosted important studies that have contributed to the design of treatment centres that are now operational, it has also been involved in training personnel for the facilities, for example, at MedAustron. With a view to bringing such initiatives more formally into its range of activities, in 2014 the CERN Medical Applications (CMA) Office was set up to stimulate and coordinate activities in medical physics. One major initiative is an open-access facility for biomedical research, OPENMED, to be based on the Low Energy Ion Ring (LEIR), a low-energy accelerator at CERN that supplies heavy ions to the LHC injector chain, for a few weeks each year. Another accelerator-based project is CERN-MEDICIS. Building on a long-standing development of methods of isotope production via the ISOLDE

[b]The editorial team thanks Manjit Dosanjh for her contributions to this chapter.

facility [Highlight 3.8], CERN-MEDICIS will deliver novel radioisotopes for research and applications in therapy and imaging.

Developments in particle detectors have had a big impact on medical imaging, making procedures easier, faster, more accurate and less invasive. Instruments based on the MWPC [Box 4.4], developed at CERN by 1992 Nobel Laureate Georges Charpak, are being used in biology and nuclear medicine. Indeed, Charpak himself immediately saw the possibility of applications in medical imaging. Such activity continues at CERN with the development of the GEM technology [Highlight 4.8] in devices for fast tracking and readout that can be applied to dose-mapping during radiation therapy using X-rays. Medical imaging projects based on other types of particle detector have been initiated at CERN via multidisciplinary collaborations. The Medipix Collaboration [Highlight 10.3] has its roots in hybrid silicon pixel detectors developed for particle tracking in the late 1990s. In one application, a spin-off company in New Zealand is working on using the latest generation of these detectors. The Crystal Clear Collaboration [Highlight 10.4], initiated to develop calorimetry at LEP and the LHC [Highlights 7.9 and 8.8], has adapted these detectors for a variety of medical applications of positron emission tomography (PET).

Computing applications

Data-handling techniques and simulation tools from high energy particle physics now find applications in areas ranging from medicine to financial analysis. In the biomedical field they are applied to help in understanding medical detectors, conducting clinical trials, devising protocols and establishing personalized treatment plans. Grid computing, which allows multiple users to share computing power and storage capacity over the internet, is an ideal tool for a wide range of biomedical fields, from screening of drug candidates, to image analysis, to sharing and processing health records.

FLUKA and Geant4 [Highlight 9.5] are two major simulation packages in particle physics that are widely used in the medical field, amongst others. FLUKA, which is used for calculations of particle transport and interactions with matter, has many applications in high energy physics and engineering, from shielding and detector design to cosmic-ray studies and radiation protection. A number of commercial companies, typically medical hardware and software developers, are FLUKA licensees. The Geant4 toolkit, which allows detailed simulation of particle interactions with matter for a wide range of particle types and energy, also finds applications in medical and space science, as well as in LHC experiments.

Environmental applications

An early transfer of technology to industry was an improved getter technique developed at CERN to fulfil the demanding vacuum conditions for stored beams in colliders [Box 7.3]. This was later applied to vacuum-insulated solar thermal collectors, which are particularly suited to colder, less sunny climates where they are more efficient than traditional solar panels [Highlight 10.5].

In the 1990s, a strategy was proposed to validate new technological concepts (Accelerator-Driven subcritical Systems or ADS) based on the use of particle accelerators to destroy nuclear waste, to produce clean and safe energy, and deliver radioisotopes for nuclear medicine and industry. This led to a trilogy of programmes on basic studies in neutron physics. The First Energy Amplifier Test (FEAT) experiment [15] elucidated the concept of the Energy Amplifier [16], and was followed by the TARC experiment [Highlight 10.6] to study neutron transport in lead and test new ideas for transmutation. The third strand in the trilogy is the CERN neutron time-of-flight facility, n_TOF [Highlight 3.9].

In another, surprising, example of the transfer of CERN know-how — in this case precision engineering expertise — the multi-disciplinary CLOUD Collaboration studies the formation and behaviour of clouds. It uses a beam at the CERN PS to assess the influence of cosmic rays on aerosols and their impact on cloud formation [Highlight 10.7], and is gathering information and developing models that are important for studies related to climate change.

The changing dimension of CERN's mission

This chapter started by asking how the investment in research and education can be best put to use for society. Some of CERN's answers address this through the sharing of technology, knowledge and expertise. However, this raises a related question: is CERN doing this well, too much, or too little?

During the first 40 years, technology transfer occurred naturally through industrial contracts, providing substantial returns for many of the industrial partners. Since then there has been a plethora of initiatives involving technology transfer and knowledge sharing. Is this a distraction from achieving the primary goal of the Organization as defined in the Convention?

To set the scale: the size of CERN's Knowledge Transfer Group is approximately 2% of the CERN staff, with an annual expenditure of the order of 1% of the total material budget. To this one should add the personal effort of knowledge sharing of many of the staff, at the level of some percent overall.

However, the question is not so much whether an organization like CERN, or for that matter a university or an individual scientist, should devote a fraction of the available resources to effective knowledge sharing: today it is considered an integral part of our social responsibility. This sharing and communicating has to happen at all levels: with peers, with the public, with the decision makers and politicians. More than ever, scientific culture is part of our society and scientists have to be at the forefront of fostering this scientific culture. A scientifically literate public will be the best insurance that politicians follow the lead, appreciate the needs, and allocate appropriate funds to scientific education and research.

Summary

The value of CERN as the host laboratory for worldwide collaborations of scientists and engineers that use its various facilities extends far beyond basic research. CERN and the collaborating universities and institutions are together enhancing their efforts in knowledge transfer and education, triggering innovation and communicating the role of basic research as a motor for new technologies of relevance to society.

10.2 Medical Accelerators: A Tool for Tumour Therapy

Kurt Hübner

The interest in using charged particle beams to treat human cancer stems from the observation [17] that the ionization density, i.e. dose deposition, is very high at the end of the track of a proton or an ion stopped in a tissue (Bragg peak), whereas electrons, X or γ-rays and neutrons deposit the dose decreasingly over the whole range (Fig. 10.2). Thus, protons and light ions stripped of all electrons seem to be very well suited for the well-targeted (conformal) treatment of deeper-seated localized tumours with minimal impact on the surrounding tissue. In particular, carbon ions are of interest due to their enhanced Relative Biological Effectiveness (RBE), a measure of the damage by ionizing radiation per unit of energy deposited in biological tissues. For heavier ions the RBE is less favourable [18].

CERN's contribution to this field started with a study of dedicated medical accelerators in the late 1980s. The study team examined two types of accelerator: a cyclotron with a superconducting magnet and a synchrotron similar to LEAR [Chapter 3.1]. It recommended that the best choice would be a synchrotron providing light ions of 400 MeV/u, ranging from carbon to neon [19]. As for other facilities under construction at that time [20, 21], a synchrotron was preferred for its inherent flexibility in energy and for being based on well-known technology.

In the early 1990s, CERN hosted and contributed to a study of an accelerator-driven spallation source, which included a synchrotron providing fully-stripped carbon or oxygen ions up to 425 MeV/u for hadron therapy [22]. After this project was shelved, CERN and partners from Austria, the Czech Republic and Italy

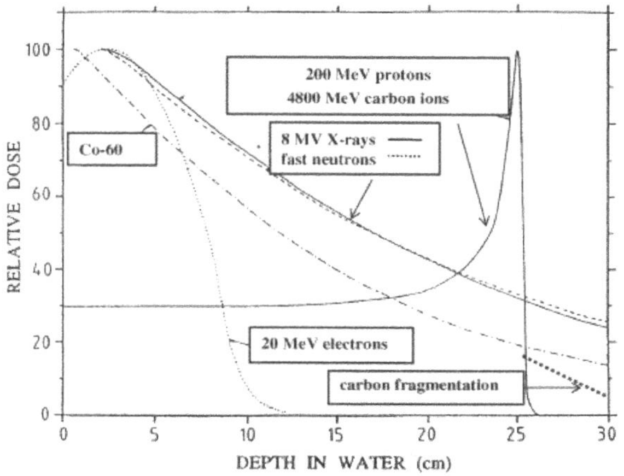

Fig. 10.2. Depth dependence of the deposited dose for photons, electrons and charged particles [18].

decided to concentrate in 1996 on its most promising part and launched the Proton-Ion Medical Machine Study (PIMMS) with the design of a state-of-the-art and cost-effective synchrotron dedicated to hadron therapy as objective. CERN hosted this study group in close contact with GSI, the German research centre conducting similar studies. GSI decided later to pursue its own approach based on experience gained with its SIS synchrotron and associated test facility [23], which was the basis for the successful German facilities HIT [24] and MIT [25], which have been treating patients since 2009 and 2015, respectively.

PIMMS considered an accelerator that would allow the direct clinical comparison of protons and carbon ions for hadron therapy using high-precision scanning [26, 27]. The synchrotron option provided multi-species operation and variable energy as needed for raster scanning. In this process, the beam energy is adjusted to target a slice of the tumour, which is scanned in both horizontal and vertical directions; the beam energy is then readjusted to irradiate the next slice. The maximum energy of the ions determined the size and maximum power of the accelerator while the proton mode determined the design of the injection system and low-energy operation. Applying the lessons learned at LEAR [Highlight 6.4], slow-extraction was chosen for an extended beam spill, allowing on-line dosimetry at the patient. To guarantee a smooth spill, extraction was activated by accelerating the beam into a resonance with a betatron core. This technique has the advantage that it maintains all transverse optical parameters, so that the power converters deliver constant current during extraction, allowing their ripple to be minimized. The betatron core was a 1.5 m long steel toroid around the beam, equipped with a coil changing the magnetic flux that accelerated the beam. Being a high-inductance device, the core smoothed eventual residual ripple.

The PIMMS design was used for the engineering design of the Italian Hadron Therapy Centre (CNAO) [28], which has been treating patients since 2011 (Fig. 10.3). The CNAO technical team is led by scientists who received their training as former junior members in the PIMMS group. CERN also hosted the group performing the magnetic measurements of the dipole magnets, and provided special magnets and expertise with beam diagnostics during design and construction. The design of the accelerator and beam lines of the Austrian Centre for Hadron Therapy (MedAustron) [29], presently being commissioned, is also based on PIMMS, and used parts of the engineering design of CNAO. CERN hosted the design group, composed of senior CERN staff as work-package leaders and staff from MedAustron who were trained on the job and, as for CNAO, went on to assume technical responsibility of the facility. CERN provided substantial assistance for writing specifications and for procurement, as well as space for an injector test facility and for acceptance testing of the magnets and other high-technology components [30]. CERN also contributed to the design, construction and testing of a high-gradient proton linear accelerator operating at 3 GHz. Such a linac in conjunction with a cyclotron could provide a compact facility [31].

CERN's existing antiproton source (AD) [Chapter 6] has served to study the biological effect of antiprotons and so provide a basis for comparison with carbon ions and protons [32]. Now a broader approach is being considered, with an open-access facility for biomedical research based on the existing Low-Energy-Ion Ring (LEIR). This new facility could provide a variety of ions with energies up to 440 MeV/u for testing their effectiveness, dosimetry and cellular responsiveness in humanoid models ("phantoms") and cell cultures [33].

Fig. 10.3. The CNAO synchrotron providing protons up to 250 MeV and carbon ions up to 400 MeV/u with its injector inside the ring (courtesy CNAO).

10.3 Medipix: The Image is the Message

Michael Campbell

The Medipix story began at CERN in the early 1990s with the R&D on hybrid and monolithic silicon micro-pattern detectors, in particular for the LHC. The group developed the first microchips with the typical readout architecture for high-energy physics on each pixel: preamplifier/shaper/discriminator/delay-line and electronic logic that would "freeze" data in time with a strobe signal, so that only strobed hits were read out. The chips were bonded to the detector pixels through "bumps", and to check for missing bonds the chip/sensor assemblies were exposed to radioactive sources, and the data added up to make an X-ray image of the source. The images were not very good as the pixels were rectangular. It also took a long time to make an image, because the chip recorded only the strobed events. It became obvious that one should develop a chip with square pixels and a counter for each pixel, which could be activated with an externally applied shutter (as in a digital camera).

This led to the idea that counting every photon would make more efficient use of the X-ray dose given to patients in medical applications. Several teams began to explore the possibilities, among them an informal collaboration of four research groups (CERN, Freiburg, Glasgow and Pisa). Unlike the others, this team went all the way to making a big chip, which became known as Medipix1 [34]. The Medipix1 chip had 64 × 64 square pixels at a pitch of 170 μm. It proved that counting single photons was indeed possible, and particularly powerful in providing clean X-ray images in a low-rate environment. The pixel size, the limited number of pixels per chip, and the fact that the chip could not be abutted to others — i.e. placed side-by-side with no "dead" edges between — meant that Medipix1 was not widely used, but it did set the stage for further developments.

The big breakthrough came with Medipix2 (Fig. 10.4), developed within a collaboration that was this time grouped formally, in 1999, initially with thirteen members. The 250 μm CMOS process employed for the silicon pixel detector in the LHC ALICE experiment (but without the radiation-tolerant design techniques) was used to make a chip with 256 × 256 pixels on a pixel pitch of 55 μm. Each pixel had a window discriminator and a 14-bit counter on board, and the chip used the same camera-like logic as the first generation. Aimed at medical X-ray imaging, the chip could be butted up on three sides and was therefore better suited than the Medipix1 to covering large areas. The number of uses to which the Medipix2 chip could be put simply exploded over the years and the collaboration peaked at about 17 members [35].

Around 2005, colleagues from the EU-funded EUDet project approached the Medipix2 design team and asked for the pixel readout to be changed so that they could read out a gas detector and obtain track information in three dimensions. This involved sending a clock signal around the chip such that each pixel could be made to count clock ticks from the moment it was hit until the end of the shutter signal. This timing information provided the z-direction for the tracks in the gas volume, the x- and y-coordinates coming from the pixel location. It was obvious to the chip designers that one could also propose adding a third mode to each pixel (after particle counting and arrival time), namely time-over-threshold (ToT), giving the amplitude of the charge deposited in a pixel. This led to the Timepix chip, which has proven to be even more versatile and popular than Medipix.

A significant contribution to the dissemination of this technology has been the miniaturized USB readout and software developed by collaboration members at CTU, Prague. New applications continue to appear: the chip is the main workhorse for the CERN@school project, which began with a visit by a group of UK high school students and their inspirational teacher to the Medipix design team in 2007.

One early idea was to take the first steps towards spectroscopic X-ray imaging, which would permit colour X-ray imaging. However, during the work on Medipix2/Timepix, it became clear that the main obstacle to the extraction of

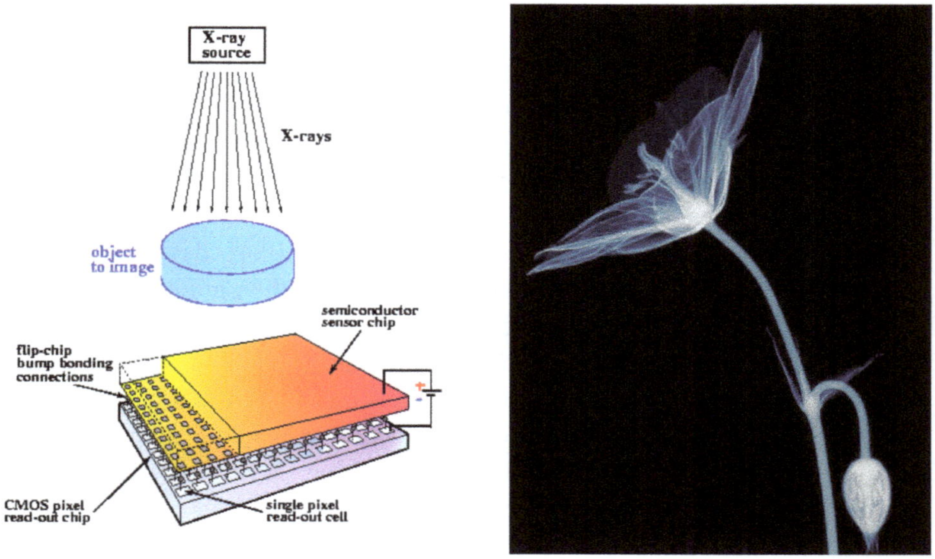

Fig. 10.4. Schematic showing the operation of the Medipix2 chip, left, with an example of the image resolution achievable, right. (Courtesy Medipix2 Collaboration.)

spectroscopic X-ray information comes from charge sharing between neighbouring pixels, caused by charge diffusion during collection (or even the fluorescence present in the heavy materials needed in the diagnostic energy range). One could compensate for the influence of charge sharing by using the Timepix chip in ToT mode and doing cluster analysis offline. However, if there are too many hits per frame, the clusters mingle. To achieve spectroscopic imaging at high count rates, something has to be done at the pixel level in real time. The Medipix3 architecture does just this: it sums charge over neighbouring pixels on an event-by-event basis, but allocates the result to only one pixel — the one with the largest charge from the cluster.

The Medipix3 Collaboration, formed in 2005, has developed two chips: Medipix3, which implements the charge summing and allocation scheme, and Timepix3, which (like Timepix) is intended for use with lower hit rates. Timepix3 uses an entirely different readout approach from the frame-based readouts of Medipix, Medipix2, Timepix and Medipix3. In Timepix3, each time a pixel is hit it generates a data package to send off-chip the hit coordinates, the ToT value (the amplitude of the charge deposited) and the arrival time, with a precision of 1.6 ns. The precise time-stamp recording allows a semiconductor detector to be read out in the same way as a gas-based time projection chamber [Box 4.5].

Recently, through-silicon-via (TSV) technology has become available and can be used in pixel detector readout to take signals away from a face of the chip rather than from an edge. As a consequence, the Medipix4 Collaboration will develop chips that for the first time are buttable on all four sides. This should permit almost seamless tiling over large detection areas. The Medipix4 chip will be aimed at reaching full count rates for human CT scans while Timepix4 will have improved throughput with respect to Timepix3, better time-tag precision and smaller pixels.

10.4 Crystal Clear: From Higgs to PET

Paul Lecoq

The Crystal Clear Collaboration [36] was formed in 1990 with as primary goal to identify and develop a crystalline scintillating material suitable for a high-resolution electromagnetic calorimeter that could detect production of the Higgs boson at the LHC, through its decay into two energetic gamma rays [Highlight 8.8]. The crystal would have to be very dense, with a fast scintillation response and no afterglow, to withstand the 40 MHz beam-crossing rate of the LHC, and resistant enough to the effects of radiation to guarantee the stability of its optical properties within a few percent for at least 10 years of operation. In addition, the collaboration had to address the need to produce about 100 tons of crystals with a consistently high quality within a few years and at an affordable cost. Initial studies concentrated on cerium fluoride (CeF_3), lead tungstate ($PbWO_4$), and heavy metal fluoride glasses.

This formidable list of challenges required a large multidisciplinary effort. The Crystal Clear Collaboration co-opted a number of world experts in different areas of material sciences (crystallography, solid state physics, luminescence, defects in solids) to develop together suitable new inorganic scintillators. Based on Crystal Clear's studies and recommendations, in 1994 both the CMS and ALICE collaborations selected the lead tungstate scintillator for their electromagnetic calorimeters. This choice proved to be judicious and significantly contributed to the discovery of the Higgs boson at the LHC.

At the same time, Crystal Clear succeeded in establishing and motivating a community of scientists to contribute to applied research on the physics of scintillation processes in solids. This attracted scientists and engineers from other areas making use of scintillating materials, resulting in a total community of about 250 scientists worldwide. This community meets every second year at the international SCINT conference, which the Crystal Clear Collaboration has been organizing since 1992.

The ClearPET project [37]

In order to leverage the synergy available within Crystal Clear for further applications, the collaboration decided to focus on medical imaging instrumentation; in particular, to use its high energy physics expertise in integrating multi-thousand-channel detectors to develop a compact high-performance scanner for Positron Emission Tomography (PET), using new kinds of crystals and photon detectors, and low-noise and fast signal processing. The collaboration first proposed to build a PET scanner for small animals, the ClearPET, together with several biologists and medical groups. For this purpose Crystal Clear developed a very dense and fast scintillating crystal (cerium-doped lutetium aluminum perovskite: LuAP:Ce) used in combination with another crystal (cerium-doped lutetium orthosilicate) to reduce the parallax error that limits the spatial resolution in PET scanners with small detector-rings. In early 2000, a resolution better than 1.5 mm over the whole field of view was obtained — a significant breakthrough at the time — and the process was patented and licensed to a German company.

ClearPEM and ClearPEM-Sonic [38]

After the ClearPET success, Crystal Clear decided to develop the ClearPEM, a dedicated breast-imaging PET scanner. It was soon realized that the ClearPEM, which provides metabolic information on breast tumours, could benefit from the combination with another modality for anatomic imaging. The ClearPEM-Sonic is an innovative imaging system that combines a mammography PET and a 3-D ultrasound and elastography system, and thus allows the combination of metabolic, morphologic and structural information for a more refined diagnosis. Phase 1 of the clinical trial, which started in March 2012 on 10 patients at the Hôpital Nord in Marseille (Fig. 10.5), gave excellent results, and was followed by a second, more ambitious clinical trial on 100 patients at the San Gerardo hospital in Milano, which was still running at the time of writing (2016).

ENDOTOFPET-US [39]

The year 2011 saw the launch of ENDOTOFPET-US, a European project for the design of a high-performance medical imaging device to be used in research on pancreatic cancer. The project, funded by the European Union FP7 programme, is a collaborative effort between CERN and 13 partners, including hospitals and companies. The primary objective is the design of a state-of-the-art endoscopic probe associated with an external detector, and the development of biomarkers, with priority for the pancreas and prostate. The ENDOTOFPET-US endoscope is a dual-mode imaging system comprising a miniaturized PET detector-head with excellent 200 ps time-of-flight (TOF) capability, which is attached to an ultrasound probe with a biopsy needle. The ultrasound signals will provide an anatomical image of the tumour, while the PET head will supply highly detailed information

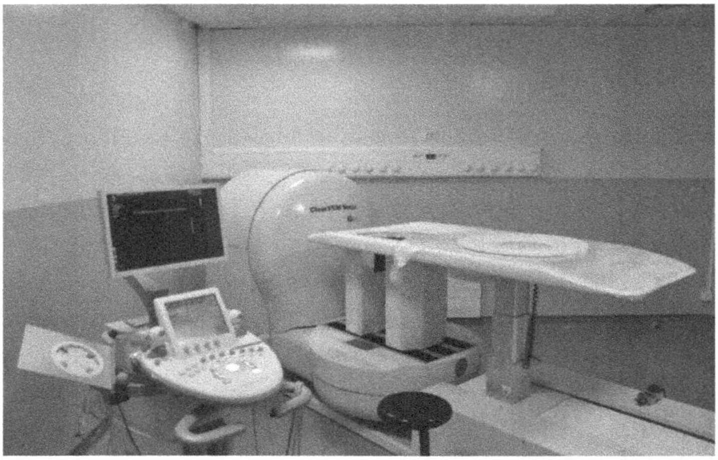

Fig. 10.5. The ClearPEM Sonic, installed at the University Hospital of Marseille for clinical trials in December 2011. (Courtesy Crystal Clear Collaboration.)

about the tumour activity and, indirectly, the associated proteins. The results of the imaging will be correlated with the biological analysis of the tumour. Two versions of the ENDOTOFPET-US have been built, one for the prostate and one for the pancreas. These will be tested at hospitals in Munich and Marseille, respectively.

Photonic crystals [40]

It is not only in physics and astronomy that inorganic scintillators find use in detection systems, but also in applications related to homeland security, medical, imaging, non-destructive industrial testing and more. In all of these detectors, light produced in the scintillator has to be transported towards a photodetector, but the standard optical coupling suffers from inefficient light extraction from the crystal due to the total internal reflections caused by the high refractive index of the scintillator. By using photonic crystals — i.e., by nano-structuring the different surfaces of the scintillator to produce constructive interferences of the evanescent wave near the surface — the transport of the light output can be tailored to optimize the timing performance and light yield of the detector. Some members of the Crystal Clear Collaboration have demonstrated that a factor of two in the light-extraction efficiency can be obtained with crystals commonly used in medical imaging. A European-funded project, TURBOPET, has been launched in collaboration with industry to demonstrate the clinical benefit of such a treatment, using the crystal on the MAMMI breast-imaging PET scanner that has been developed by a Spanish company.

10.5 Solar Collectors: When Nothing is Better

Cristoforo Benvenuti

In a standard solar thermal collector, the air molecules bouncing back and forth between the light absorber (hot) and the front glass window (cold) result in large thermal losses which limit the collector efficiency. To have a vacuum (i.e. nothing) inside the collector would remove this inconvenience.

Despite this well-known fact, evacuated flat-plate solar thermal collectors have never been produced because of the (almost) insurmountable difficulty of making a vacuum-tight joint between the glass window and the metal frame of the collector. A CERN patent [41] shows that this problem may be solved by plasma spray coating the perimeter of the glass with a metallic layer to which a metal joint may be fixed by soft soldering. This CERN patent was licensed in 2005 to a company, SRB Energy, created to produce and commercialize an evacuated thermal solar collector (Fig. 10.6).

The glass-to-metal joint is the main, but not the only, difficulty to be overcome to produce the collector. In addition:

- After the initial evacuation, the vacuum must be maintained despite the continuous outgassing of the collector components;
- The infrared emission from the hot light absorbers must be minimized;
- The pressure of 10 t/m^2 applied to the collector by the external atmosphere, tending to implode the collector glass, must be neutralized.

The vacuum inside the collector is maintained by a non-evaporable getter (NEG) pump [Box 7.3] [43] of which the surface is continuously cleaned by solar heating. The NEG thin-film technology developed for the LHC accelerator [Highlight 8.5] [44] has been used to coat a thin aluminium foil with a getter layer by means of a "roll-to-roll" coating machine developed *ad hoc* by SRB. This machine can coat automatically a few hundred metres of foil per week. To decrease the losses due to emission of radiation, the collectors are coated with a film that absorbs 90% of the solar light but presents a low emissivity in the infrared range (less than 0.07). The external pressure is withstood by spacers placed between the two glass plates.

Two cylindrical mirrors are coupled to the collector, as shown in Fig. 10.7. These mirrors convey the light they receive to beneath the panel, so as to double the incident power on the panel without doubling the cost. It is important to note that although the diffused light cannot be focused, these mirrors transmit it to the collector equally well as the direct light. This feature is particularly important in central Europe, where diffused light may exceed 50% of the total.

The collector with the mirrors may reach a temperature of 400°C in the best sunlight conditions. The available thermal efficiency is shown in Fig. 10.8. The pressure inside the collector varies between 10^{-4} Pa at 400°C and below 10^{-7} Pa

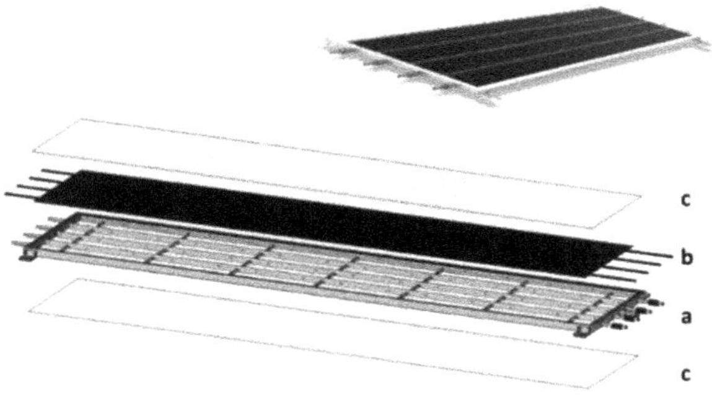

Fig. 10.6. Schematic of the SRB solar panel: (a) frame with spacers, (b) absorbers with cooling pipes, (c) glass windows [42]. (Courtesy SRB Energy.)

Fig. 10.7. The SRB collector with cylindrical mirrors. The collector is 3 m long and 70 cm wide. The mirrors enable the collection area and the power production to be doubled. (Courtesy SRB Energy.)

during the cold winter nights. These pressures, much lower than needed to profit fully from vacuum insulation (10^{-2} Pa), are a by-product of the large amount of NEG required to maintain the vacuum over 30 years.

The collector design was finalized in 2007 and a pilot production plant came into operation at the end of 2008. The SRB collector is multipurpose: it can be used for heating, water desalination, cooling and for the production of electricity. Among the many installations made so far, a good example is that of Geneva airport (about 1200 m^2), where it is used for both heating and air conditioning.

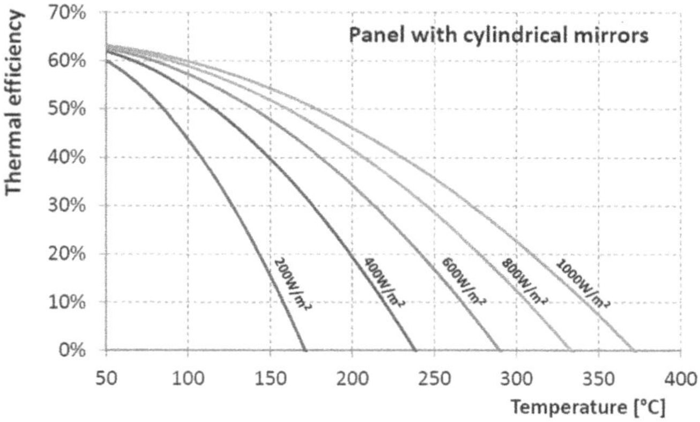

Fig. 10.8. Dependence of the SRB collector efficiency on the incident solar power and on the temperature of operation. The decrease of efficiency at higher temperatures is due to the increase of the thermal losses by IR radiation emission. (Courtesy SRB Energy.)

10.6 The TARC Experiment at CERN: Modern Alchemy

Jean-Pierre Revol

TARC, which stands for Transmutation by Adiabatic Resonance Crossing, was the idea of Carlo Rubbia to make use of the unique properties of lead to enhance, by orders of magnitude, the transmutation rate of long-lived fission fragments, such as ^{99}Tc and ^{129}I (two elements that represent 95% of the radio-toxicity of long-lived fission fragments in nuclear waste). Conversely, the same method may be applied to produce radioisotopes from stable elements, using accelerator-driven systems (ADS) instead of nuclear reactors. The goal of the TARC experiment (PS211), proposed at the CERN PS in 1995 [45], and carried out in 1996 and 1997, was to demonstrate the feasibility of these two attractive possibilities.

The physics of TARC is rather simple and based on a unique combination of properties of neutrons produced inside a volume of lead. Lead is an excellent material for the production of spallation neutrons through the impact of a proton beam; it is an element with one of the lowest cross-sections for neutron capture; it has a short elastic-collision length, $l_0 \approx 3$ cm; and it has a large nuclear mass compared to the neutron mass, which leads to an average fractional neutron energy loss in an elastic collision, quantified by the lethargy coefficient η_0, that is very small, $\sim 9.7 \times 10^{-3}$. In this low-lethargy medium, time and velocity are strongly correlated, such that there is a simple relation between velocity, v, and time ($t \approx 2l_0/\eta_0 v$). In this way, the time at which a neutron is observed in the experiment is correlated with its velocity and, hence its kinetic energy, providing a simple way to determine energy from time measurements [46].

Because of the very low neutron-capture probability, in a sufficiently large volume of lead a 1 MeV spallation neutron will survive for a long time, typically 3 ms, undergoing 1800 scatterings along a 60 m path, before it is eventually captured as thermal energy. Neutrons of MeV energies thus slow down in lead through many very small energy losses. If some ^{99}Tc is incorporated in the lead, such neutrons are certain to go through the energy of a resonance in the neutron-capture cross-section of ^{99}Tc and so enhance by orders of magnitude the transmutation probability of ^{99}Tc, which is radioactive, into ^{100}Ru, a stable element. This process is referred to as adiabatic resonance crossing. The long lifetime of neutrons in lead and the small elastic-collision length have the additional effect of increasing the neutron flux locally, as a neutron is likely to cross the same elementary surface several times during its random walk.

In the TARC experiment, these expected properties of neutrons in lead were beautifully verified with high precision in 334 tonnes of pure lead installed in a CERN PS proton beam line. It was particularly important to check that a sufficiently large fraction of neutrons would survive the capture resonances of lead and reach the capture resonances of interest, for the destruction of long-lived fission fragments (Fig. 10.9). Using $^{6}Li/^{233}U$ target silicon detectors, ^{3}He ionization detectors, and many activation methods, the TARC collaboration measured with unprecedented precision neutron fluxes over neutron energies covering eight orders of magnitude [47].

Tests of transmutation of ^{99}Tc and ^{127}I carried out at various locations inside the lead volume of the TARC experiment, validated both the TARC method and the simulation developed specially for the Energy Amplifier studies [16]. The simulation can therefore be trusted to predict the performance of an industrial system either to destroy long-lived fission fragments or to produce radioisotopes for nuclear medicine and other industrial applications. It was shown that in an Energy Amplifier in which ^{99}Tc or ^{127}I have been diluted in the lead volume, it is possible to destroy these elements at about twice the rate at which they are produced in nuclear reactors [47].

The TARC collaboration also measured the production rate of ^{99m}Tc, an isomer of ^{99}Tc, the most widely used radioisotope in medicine, from the activation of natural molybdenum, and confirmed the feasibility of an industrial activator for

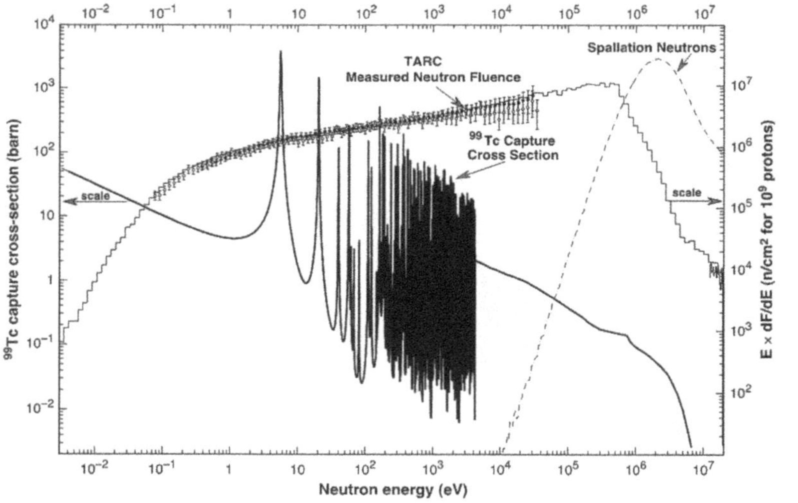

Fig. 10.9. The flux of relatively high energy spallation neutrons (dashed line) is transformed into a neutron flux covering many orders of energy. The resonances in the capture cross-section of ^{99}Tc cannot be missed, as the average energy loss of a neutron through elastic scattering is typically smaller than the width of resonances. Data points are taken from TARC neutron flux measurements for 3.5 GeV/c protons [47]. The histogram (full thin line) is the neutron flux obtained with the TARC simulation [47]. The ^{99}Tc neutron capture cross-section (full thick line) [48] exhibits two most prominent resonances at 5.6 eV and 22.3 eV.

the production of such an element, or of other radioelements used in medical imaging, diagnostics, and therapy. As a result, CERN took a patent [49] on the TARC idea, and various industrial pharmaceutical companies showed interest in its exploitation. The patent, valid until 2017, is very much of interest, as the demand for ^{99m}Tc has exploded and new therapies based on lutetium, rhenium, holmium, etc. are in rapid development.

The TARC experiment was a landmark experiment, which studied the phenomenology of neutrons in lead. It showed that ADS may be used to destroy nuclear waste and provide an alternative to nuclear reactors in the production of radioisotopes. It validated an innovative simulation that is now used for the design of ADS. The TARC concept also led to the design and construction of the CERN neutron Time-Of-Flight facility, n_TOF [Highlight 3.9], with its high rate, high precision neutron flux and low background. It offers unique conditions for the measurement of neutron cross-sections, a necessary input to any reliable simulation and development of new nuclear systems.

10.7 A CLOUD Chamber with a Silvery Lining

Jasper Kirkby

During his first visit to the Ben Nevis Observatory in 1894, the future Nobel laureate C.T.R. Wilson became fascinated by clouds and the beauty of coronas and "glories" (coloured rings surrounding shadows cast on clouds). He returned to the Cavendish Laboratory in Cambridge determined to re-create clouds in the laboratory and study their physical phenomena. This led him to develop the expansion cloud chamber — a detector on which much of the experimental foundation of particle physics was built in the first half of the 20th century.

More than a hundred years later, a new cloud chamber is in operation at CERN. The CLOUD experiment is optimized to study the influence of cosmic rays on aerosols and clouds [50]. CLOUD reproduces atmospheric conditions in a large chamber (Fig. 10.10) to study aerosol particle formation and growth in controlled laboratory conditions. Clouds are generated by adiabatic pressure reductions of humid air parcels, as in Wilson's cloud chamber, but at the much smaller water vapour supersaturations found in natural clouds (a few times 0.1%, compared with around 500% for a Wilson cloud chamber). Depending on the air temperature and the nature of the seed particles, either liquid or ice clouds form.

The primary scientific goal of CLOUD is to answer the question of whether or not cosmic rays exert a climatically significant effect on aerosols and clouds, as suggested by satellite observations first reported in 1997 [51]. This intriguing

Fig. 10.10. Fisheye view vertically upwards inside the 3 m CERN CLOUD chamber. The inner surface of the stainless steel chamber is electro-polished to reduce retention of contaminants. An array of 265 fused silica fibre-optic feedthroughs is installed on the roof of the chamber to provide a uniform column of UV light for simulating atmospheric photolytic chemistry. A pair of magnetically-coupled stainless steel fans (not shown) mounted on the upper and lower manhole covers ensure mixing of the air, trace gases, ions and aerosol particles inside the chamber. Fast response (< 1 s) thermometer strings (not shown) measure the air temperature, which can be adjusted between −70°C and +100°C by a precise thermally-controlled environment. Two transparent high voltage electrodes, supported on partially-conducting ceramic rods, provide an electric field of up to 20 kV/m to clear ions from the chamber for ion-free experiments. Sampling probes (not shown) are inserted through a ring of DN100 ports seen in the mid plane of the chamber. These are connected externally to state-of-art instruments such as mass spectrometers that continuously analyse the contents of the chamber during experimental measurements. The configuration of analysing instruments is modified for each experimental run according to the scientific goals.

possibility could shed new light on the long-sought mechanism for solar-climate variability in the pristine pre-industrial climate [52, 53]. Atmospheric aerosols and their effect on clouds are recognized by the Intergovernmental Panel on Climate Change [54] as the largest source of uncertainty in anthropogenic radiative forcing. A second scientific goal of CLOUD is therefore to quantify the fundamental physico-chemical processes controlling aerosol particle formation and growth in the atmosphere, and thereby help to reduce the current uncertainty in Earth's climate sensitivity and sharpen global temperature projections for the 21st century.

CLOUD is the first — and so far only — experiment to reach the demanding technological performance required to study these processes under atmospheric conditions in the laboratory. Extraordinary care has been paid in the design and construction of the CLOUD chamber and its associated systems — gas, thermal,

UV and electric field — to suppress contaminants at the technological limit. CLOUD experiments have shown that atmospheric nucleation and growth is driven by atmospheric vapours that are present in minute amounts of only a few parts-per-trillion by volume (pptv). The key vapours comprise sulphuric acid, ammonia, amines and highly oxidized biogenic molecules originating from trees [55-59]. Ions from galactic cosmic rays and natural radioactivity play an important role in the formation of aerosol particles under certain conditions (Fig. 10.11), although evaluating their climatic significance requires further measurements.

Why CERN? First, the CERN PS provides an artificial, adjustable source of "cosmic rays" that reproduces, in a chamber, ion concentrations found between ground level and the lower stratosphere. Second, the unrivalled technical performance of the chamber and its associated systems is the result of a great deal of CERN expertise. Finally, there is an important cultural "trading zone" between scientific disciplines. CLOUD is a "general purpose" detector that measures every aspect of the physical process under study. Although familiar to high energy physics, pooling the resources of a large collaboration in this way and building a single, integrated high-performance detector is new to atmospheric science.

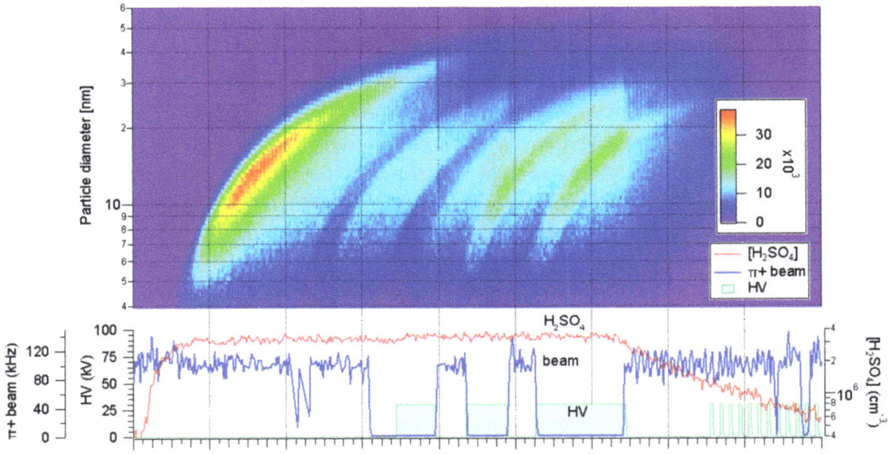

Fig. 10.11. Formation of new aerosol particles from trace atmospheric vapours in CLOUD (upper panel). The aerosol particles grow in size (vertical axis) by condensation over a period of several hours (horizontal axis), reaching sizes sufficient to seed cloud droplets. The characteristic "banana" shapes show fresh bursts of particles that form during periods of high ionization in the chamber due to the π^+ beam from the CERN PS and the absence of an electric clearing field. In this case the trace gases included 0.15 pptv sulphuric acid and oxidized biogenic vapours. The lower panel shows the π^+ beam intensity (blue line), the clearing field (green blocks) and the amount of sulphuric acid (red line).

References

1. H.B.G. Casimir, Technology and World Trade: Proceedings of a Symposium, NBS Misc. Pub. **284**, 22-24 (1966); https://archive.org/details/technologyworldt00symp.
2. C. H. Llewellyn Smith, What's the use of basic science? (1997); https://cds.cern.ch/record/388110?ln=en.
3. O. Barbalat, Technology transfer from particle physics: The CERN experience 1974-1997, *Report CERN-BLIT-97-1* (1997).
4. F. Bourgeois (ed.), Basic science and technology transfer: means and methods in the CERN environment, *Report CERN-BLIT-98-10* (1998).
5. S. Michalowski, The impacts of large research infrastructures on economic innovation and on society: case studies at CERN, OECD (2014), https://www.oecd.org/sti/sci-tech/CERN-case-studies.pdf.
6. S. de Jong, The CERN effect: transferring innovation and knowledge so that cutting-edge research can benefit society, *CERN Courier*, April 2016, 5 (2016).
7. Knowledge Transfer, http://knowledgetransfer.web.cern.ch/knowledge-transfer-report-2015.
8. T. Camporesi. High-energy physics as a career springboard. *Eur. J. Phys*. **22**, 139 (2001).
9. The European Network for Light Ion Therapy, http://enlight.web.cern.ch/.
10. IdeaSquare, http://ideasquare.web.cern.ch/.
11. H. Schmied (ed.), *A study of the economic utility resulting from CERN contracts*, CERN-1975-005 (CERN, Geneva, 1975). http://dx.doi.org/10.5170/CERN-1975-005;
M. Bianchi-Streit *et al.* (eds.), *Economic utility resulting from CERN contracts*, CERN-1984-014 (1984). http://dx.doi.org/10.5170/CERN-1984-014.
12. J.D. Dowell, Particle physics and technological spin-off, *CERN Courier*, April 1985, 91 (1985).
13. https://greybook.cern.ch/greybook/researchProgram/detail?id=OTHER.
14. BIC Network, http://knowledgetransfer.web.cern.ch/bic-network.
15. S. Andriamonje *et al.*, FEAT Experiment: Experimental determination of the energy generated in nuclear cascades by a high energy beam, *Phys. Lett. B* **348**, 697 (1995).
16. C. Rubbia *et al.*, Conceptual design of a fast neutron operated high energy amplifier, *Report CERN/AT/95-44 (ET)* (1995).
17. R.R. Wilson, Radiological use of fast protons, *Radiology* **47**, 487-491 (1946).
18. U. Amaldi and G. Kraft, Photons and hadrons in cancer therapy, *Rep. Progr. Phys*. **68**, 1861-1882 (2005).
19. P. Mandrillon *et al.*, Feasibility studies of the EULIMA light ion medical accelerator, *Proc. 3rd European Part. Acc. Conf.*, Berlin, Germany (1992) pp. 179-181.
20. F.T. Cole, P.V. Livdahl, F.E. Mills, L.C. Teng, Loma Linda medical accelerator project, *Proc. 13th IEEE Part. Accel. Conf.*, Chicago, USA (1989) pp. 737-741.
21. K. Noda *et al.*, New Heavy-ion Cancer Treatment Facility at HIMAC, *Proc. 11th Eur. Part. Accel. Conf.*, Genoa, Italy (2008) pp. 1818-1820.
22. P.J. Bryant, AUSTRON, a Central European Pulsed Spallation Neutron Source, *Proc. 7th Eur. Part. Accel. Conf.*, Vienna, Austria (2000) pp. 108-112.
23. G. Kraft. History of the heavy ion therapy at GSI (2013); https://three.jsc.nasa.gov/articles/Krafts_GSI.pdf.
24. T. Haberer, Cancer therapy using ion accelerators, *CERN Accelerator School, Darmstadt* (2009); http://cas.web.cern.ch/cas/Germany2009/Lectures/PDF-Web/Haberer.pdf.

25. Marburger Ionenstrahl-Therapiezentrum (MIT), http://ukgm.de/ugm_2/deu/umr_ptz/index.html.
26. L. Badano *et al.*, Proton-Ion Medical Machine Study (PIMMS), 1, *Report CERN-PS-99-010-DI, CERN* (2000); http://cds.cern.ch/record/385378/files/ps-99-010.pdf.
27. P.J. Bryant *et al.*, Proton-Ion Medical Machine Study (PIMMS), 2, *Report CERN-PS-2000-007-DR* (2000); http://cds.cern.ch/record/449577/files/ps-2000-007.pdf.
28. M. Pullia, Status Report on the Centro Nazionale di Adrotherapia Oncologica (CNAO), *Proc. 11th European Part. Accel. Conf.*, Genoa, Italy (2008) pp. 982-984.
29. F. Osmič, A. Koschik, P. Urschütz, M. Benedikt, Status of MedAustron - The Austrian ion therapy and research centre, *Proc 5th Int. Part. Accel. Conf.*, Dresden, Germany (2014) pp. 2146-2148.
30. M. Benedikt, U. Dorda, J. Gutleber, The MedAustron project at CERN, *Report CERN.ACC-2013-0227 (2013)*; http://cds.cern.ch/record/1603392/files/CERN-ACC-2013-0227.pdf.
31. U. Amaldi *et al.*, LIBO – a linac-booster for proton therapy: construction and test of a prototype, *Nucl. Instr. & Meth. A* **521**, 512-521 (2004).
32. M.H. Holzscheiter, AD-4/ACE The relative biological efficiency (RBE) of antiprotons. Combined data Analysis for the 2008 – 2012 runs, *Report CERN SPSC-2016-004/SPSC-SR-176 (2016)*; http://cds.cern.ch/record/2120418/files/SPSC-SR-176.pdf.
33. CERN Medical Applications, OPENMED, http://cern-medical.web.cern.ch/content/openmed.
34. http://medipix.web.cern.ch/medipix/
35. M. Campbell, 10 years of the Medipix2 Collaboration, *Nucl. Instr. & Meth. A* **633**, S 1, (2011). See also references therein for much of the Medipix story.
36. http://crystalclear.web.cern.ch/crystalclear/.
37. K. Ziemons *et al.*, (Crystal Clear Collaboration), The ClearPET project: development of a 2nd generation high-performance small animal PET scanner, *Nucl. Instr. & Meth. A* **537**, 307-311 (2005).
38. B. Frisch, ClearPEM clarifies breast cancer diagnosis, *CERN Courier* **53**, July 2013, p. 23.
39. B. Frisch, Combining endoscopic ultrasound with Time-Of-Flight PET: The EndoTOFPET-US Project, *Nucl. Instr. & Meth. A* **732**, 577-580 (2013).
40. A. Knapitsch and P. Lecoq, Review on photonic crystal coatings for scintillators *Int. J. Mod. Phys, A* **29**, 1430070-1430101 (2014).
41. C. Benvenuti, Evacuable flat panel solar collector, CERN patent PCT/EP 2004/000503.
42. C. Benvenuti and V. Ruzinov, The SRB evacuated flat solar panel, *Proc. ECOS 2010* 2 429 (Lausanne, CH, June 2010).
43. C. Benvenuti, A new pumping approach for the Large Electron Positron collider (LEP), *Nucl. Instr. & Meth.* **205**, 391-401 (1983).
44. C. Benvenuti *et al.*, Vacuum properties of TiZrV non-evaporable getter films, *Vacuum* **60**, 57-65 (2001).
45. S. Andriamonje *et al.*, TARC Proposal: experimental study of the phenomenology of spallation neutrons in a large lead block, *Reports CERN/SPSLC 95-17, SPSLC/P291* (1995).
46. The first mention of a slowing-down spectrometer was in a seminar by E.L. Feinberg in 1944, at the USSR Academy of Sciences. See also L.E. Lazareva *et al.*, Lebedev Institute of Physics, JETP USSR 29 (1955); A.A. Bergman *et al.*, Proc. GVA Conf. on Atomic Energy (1955).
47. A. Abànades *et al.*, Results from the TARC experiment: spallation neutron phenomenology in lead and neutron-driven nuclear transmutation by adiabatic resonance crossing, *Nucl. Instr. & Meth. A* **478**, 577-730 (2002);

H. Arnould *et al.* (eds.), *The TARC Experiment (PS211): Neutron-Driven Nuclear Transmutation by Adiabatic Resonance Crossing*, CERN-1999-011 (CERN, Geneva, 1999). http://dx.doi.org/10.5170/CERN-1999-011.

48. JENDL Data Base, OECD/NEA Data Bank, Issy-les-Moulineaux, France (1996).
49. Patent Application No WO98/59347, European Organization for Nuclear Research.
50. CLOUD Collaboration. A study of the link between cosmic rays and clouds with a cloud chamber at the CERN PS. *CERN-SPSC-2000-021; SPSC-P-317 (2000), SPSC-2000-030; SPSC-P-317-Add-1 (2000), SPSC-2000-041; SPSC-P-317-Add-2 (2000), SPSC-2006-004; SPSC-P-317-Add-3 (2006), SPSC-2014-018; SPSC-P-317-Add-4 (2014)*.
51. H. Svensmark and E. Friis-Christensen, Variation in cosmic ray flux and global cloud coverage—a missing link in solar-climate relationships, *J. Atmos. Sol. Terr. Phys.* **59**, 1225 (1997).
52. W. Herschel, Observations tending to investigate the nature of the Sun, in order to find the causes or symptoms of its variable emission of light and heat; with remarks on the use that may possibly be drawn from solar observations. *Phil. Trans. Roy. Soc. Lon.* **91**, 265–318 (1801).
53. J. Kirkby, Cosmic rays and climate. *Surv. Geophys.* **28**, 333–375 (2007).
54. IPCC, Climate Change 2013 - The Physical Science Basis. Working Group I Contribution to the Fifth Assessment Report of the Intergovernmental Panel on Climate Change (Cambridge University Press, 2014).
55. J. Kirkby *et al.*, Role of sulphuric acid, ammonia and galactic cosmic rays in atmospheric aerosol nucleation. *Nature* **476**, 429–433 (2011).
56. J. Almeida *et al.,* Molecular understanding of sulphuric acid-amine particle nucleation in the atmosphere. *Nature* **502**, 359–363 (2013).
57. F. Riccobono *et al.*, Oxidation products of biogenic emissions contribute to nucleation of atmospheric particles, *Science* **344**, 717–721 (2014).
58. J. Kirkby *et al.*, Ion-induced nucleation of pure biogenic particles, *Nature* **533**, 521–526 (2016).
59. J. Tröstl, *et al.*, Low-volatility organic compounds are key to initial particle growth in the atmosphere, *Nature* **533**, 527–531 (2016).

Chapter 11

Managing the Laboratory and Large Projects

Philippe Lebrun and Thomas Taylor

11.1 The CERN Approach: Change and Continuity

The role and governance of CERN

The principal mission of CERN is to provide large-scale facilities for performing and analysing experiments related to high energy particle physics. This European laboratory was founded in 1954 to foster collaboration and rebuild confidence between scientists who until ten years earlier had been confronted in a devastating war. From the beginning CERN was to have the ambition to provide world-class facilities that would allow European scientists to engage in fundamental research on a par with the opportunities existing outside Europe, particularly in the USA. The scale of the accelerators and infrastructure, and the personnel and financial effort required for this kind of research had reached such a level that the nations of Europe had to pool resources to build them and thus remain internationally competitive. The CERN Convention, signed in 1953 between 12 founding member states, entered into force in September 1954. This remarkable and visionary 32-page document, sets out the rules for the governance and the purpose of the Organization [1]: "... to provide for collaboration among European States in nuclear research of a pure scientific and fundamental character, and in research essentially related thereto. The Organization shall have no concern with work for military requirements and the results of its experimental and theoretical work shall be published and otherwise made generally available."

The governing body of the Organization is the CERN Council, consisting of two delegates from each member state. The Council is assisted by the Finance Committee (FC) dealing with all issues of personnel and material budgets, and the Scientific Policy Committee (SPC) advising the Organization on the research agenda. Council allocates the annual budget, with funds provided by the member states in proportion to their Net National Income (capped for any one member state, via a formula, to be less than 25% of the total budget). In order to provide a stable funding profile, to enable planning of the medium and long-term scientific programme as well as the day-to day running of the laboratory, a system of five-year rolling forecasts ("Bannier procedure") is applied. Each year the budget for the following year is established, together with firm estimates for the following two years, and provisional estimates for the subsequent two years. While the delegates are briefed by their ministries to hold a certain line, the CERN Council has maintained the authority to negotiate and take decisions in the interest of the Organization, largely without permanent consultation with the governments.

In order to make the best use of worldwide resources, the CERN programme is harmonized with that of other laboratories. The CERN Council is kept informed by the European Committee for Future Accelerators (ECFA) and the International Committee for Future Accelerators (ICFA) concerning the scientific merit and advisability of undertaking new large projects. Along with the FC and the SPC, these entities are independent of CERN.

The astounding swiftness of the implementation of CERN and the visionary scope set out by its founders still remains, 60 years on, a remarkable achievement.

The CERN Organization

The Laboratory is organized today in four sectors and a number of units, as shown in Fig. 11.1. The Accelerators and Technology, Research and Computing, Finance and Human Resources sectors are structured into departments; the fourth sector covers International Relations. The Beams (BE), Technical (TE) and Engineering (EN) departments provide the particle beams for the experiments; they are centres of excellence that work together to design, build, operate, maintain and develop the accelerator complex, including R&D for new facilities. These departments report to the director of Accelerators and Technology; projects are coordinated via the director's office (DO). The Theory (TH), Experimental Physics (EP) and Information Technology (IT) departments are also mutually beneficial centres of excellence in their respective fields, and through which CERN assists visiting physicists; CERN physicists also collaborate in experiments on an equal footing with the external partners. The departments that handle these activities report to the director of Research and Computing. Finance, human resources and general services are provided by departments reporting to the director of Administration, and provide the regulatory environment for all activities. Certain activities are shared: the main workshops are used by the accelerator/technical and research sector; the information and communication technologies department addresses the needs of the entire laboratory, including users. Regulations on health, safety and environment are applied by an independent unit reporting to the director-general.

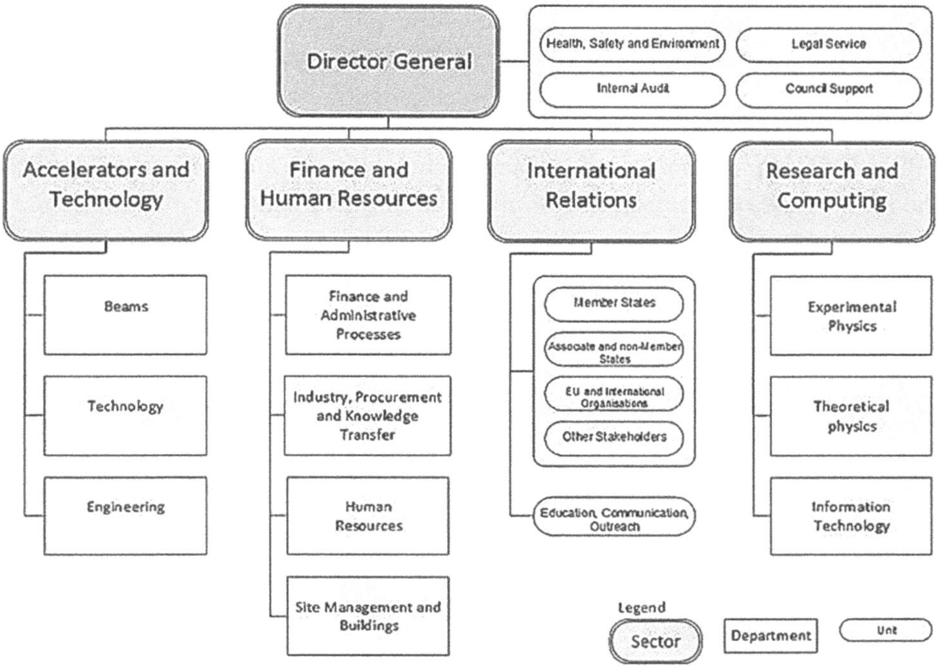

Fig. 11.1. Functional organigramme of CERN in 2016.

Directors, department heads and the director-general, who leads the laboratory, are appointed by the CERN Council. Further information regarding the organization of the laboratory can be found on the CERN web pages [2].

The overall organization of the laboratory has evolved over time; the recent addition of a sector devoted to International Relations reflects CERN's gradual evolution from a solely European entity to a broadening stature in the world. Until the 1980s all projects were administered by the departments (previously called divisions); starting with the LEP project, large accelerator and experimental facility projects are headed by project leaders responsible to the directorate. Until the early 2000s the particle beam facilities (accelerators and colliders) required for the experiments were provided by the respective divisions; subsequently it was decided to group the activities across the different accelerator divisions, to operate all the accelerators from a single control centre, and to assemble the specialists in groups in three divisions (renamed departments in 2004): Beams (BE), Accelerator Technology (TE), and Engineering (EN). Control is accomplished by a system of line management with mainly large groups (~ 100 staff) specialized in the various domains (operations, vacuum, radio-frequency, magnets, cryogenics, etc.). Most sub-projects can be handled within groups, simplifying control and avoiding the perils of matrix management, with essentially self-governing cross-group teams being formed to tackle very large projects. In the research sector the experiments are proposed, and largely staffed, by teams of researchers from external laboratories and universities. CERN groups participate in the experiment collaborations and, coupled with a few technical groups, cover particular needs and do the bulk of interfacing with the CERN infrastructure. The research sector has seen an explosion in the number of users, and the accelerator sector an increase in the complexity of the machines, obviously influencing their evolution. The technical and research sectors benefit from a collaborative administrative sector whose work has also become more complex with time.

CERN is an international organization, with staff drawn mostly far from their countries of origin. This has reinforced the international atmosphere of the laboratory and helped the users to integrate. Importantly, since the beginning, the staff has been motivated by the desire to achieve a common goal, in a constructive and non-bureaucratic collaboration between the sectors, building on their strengths and with a shared commitment to the Organization.

Style of Management

CERN has earned a reputation for developing state-of-the-art technology, the result of the collaboration of creative people in technology and research, covering a large spectrum of competence and coming with different cultural backgrounds. To “lead” and “manage” this staff requires certain talents: done properly it encourages efficiency, and includes the ability to judge when to stop “improving”. Leaders and spokespersons are chosen from those who have earned the respect of colleagues, based on their scientific and personal standing or their technical achievements. In fact, in both the accelerator and research sectors the real motivation is provided by agreeing on a common goal, which can essentially always be achieved by rational discussion on scientific and technical grounds (notwithstanding shows of emotion and passion in certain circumstances!). Thus CERN’s managerial decision model can be qualified as being one of “bounded rationality”, a concept developed by Nobel laureate Herbert Simon [3]. Many of the ideas discussed in this book originated from scientists[a] and technicians actually doing the work, not their hierarchical leaders. Obviously, large accelerator projects and experiments must have a certain level of coordination, but for this to be efficient it must be done by staff respected for their technical competence, and their ability to recognize viable ideas when proposed. In the accelerator sector the

[a]At CERN, professional engineers and research, experimental and applied physicists, enjoy equal status and are referred to as scientific staff.

practice has been to vest group leaders with the necessary authority, and for them to hold the agreed budgets, and bear the responsibility for group activity. It has been found to be important to avoid appointing purely administrative group leaders, unable to provide respected technical leadership. The effective management structure is remarkably flat (especially in the research sector).

A further important aspect is responsible procurement of technical equipment, i.e. aimed at procuring at minimal cost to the Organization while balancing industrial returns to its Member States. How has this been done? The method has consisted of (i) performing a comprehensive cost/performance analysis of all projects, (ii) defining and applying a set of simple, fair and transparent purchasing rules, and (iii) empowering competent individuals or small teams to define goals consistent with the planning of the laboratory, and allowing them to achieve those goals with minimal bureaucracy and cost.[b]

By far the most important element in an organisation such as CERN is the quality of the staff, and this in turn depends on the ability to recruit and retain appropriate personnel, and to provide them with professional perspective. Thanks to its reputation and relatively competitive employment conditions, CERN is able to recruit and retain highly qualified staff.

Evolution of management in the accelerator sector

The management of CERN sectors has evolved over the years to take into account the continuous enlargement of the accelerator complex, and constraints on recruitment following a series of reviews of the Organization by external committees appointed by Council. Similar to other organizations, the staff complement increased rapidly in the period 1955–1970, peaking at about 3600 in 1979. Then, following the recommendations of the external committees, recruitment virtually stopped and numbers were steadily reduced, stabilising around the present complement of 2500. Almost no new staff was recruited for the LEP project, requiring a major redeployment of personnel both within the accelerator sector (closure of the ISR), and from the research sector (Experimental Facilities division) to the accelerator sector (with a consequent reduction in the service for the experiments). Towards the end of the 1980s new recruitment was authorized for about one in three of the posts liberated by an early departure scheme. This had become sorely needed with the appearance of the LHC project, but the approval of the construction of this machine was assorted with a further directive to reduce staff numbers. To face this challenge the accelerator sector

---

[b]In line with this approach CERN has pioneered since the 1990s the electronic issue and handling of administrative documents, aiming at a paperless administration.

underwent a major reorganisation, from being machine-centred to being activity-centred — e.g. having a single vacuum group, instead of separate vacuum groups for the PS, SPS, LEP etc. Similarly, the operation of all accelerators was grouped in a single control centre. This evolution was justified from the standpoint of classical management practice, and necessary for the groups responsible for operation and maintenance, which requires a sufficient pool of staff to provide round-the-clock service. It also purports to ensure perennial expertise within the technical groups in spite of repeated redeployment of personnel to projects. However, the LHC had started, like LEP, with an LHC division that assumed the responsibility for providing the main systems (magnets, vacuum and cryogenics), and despite being later renamed "Accelerator Technology Department" it continued to manage the work via a classical structure, with the department head taking responsibility as *de facto* the technical coordinator/team leader for major LHC work, in addition to providing the services for the other machines. In this way the pitfalls of matrix organization were avoided, and the staff working on the LHC did so as a team of groups, much as the teams on the large experiments, working towards a well-defined common goal. However, whereas for previous accelerator projects those who had participated in the construction continued to work for the machine they had built, taking an interest in its operation (an arrangement that often led to the acquisition of new competencies and the development of improved equipment), operation is now squarely in the hands of the operations team, and contact with the equipment groups is looser and more in the nature of a service. Today, the medium-size project to upgrade LHC luminosity is being handled as if it were a very large future accelerator project, with many collaborations, and in addition has adopted features of matrix-style organization. Time will tell whether this evolution is good for CERN.

Unlike large corporations, CERN is not free to hire and fire. This requires that staff remain flexible in supporting the goals of the organization and adapting to changing requirements. And change there was! The number of user scientists passed from hundreds in the 1970s to thousands in the 1990s and now stands at about 12 000. In parallel the number and complexity of the accelerators also grew: the increase in size, from the 6 m diameter synchrocyclotron to the 8.5 km diameter of LEP/LHC is impressive, but does not do justice to the true magnitude of the evolution. A corporation might have increased staff numbers, but CERN had to respond differently. It developed collaborations with outside laboratories for building accelerators, as was done (on a much larger scale) for the experiments. For the LHC, about 15% of the value of machine hardware was delivered via such collaborations (compared with about 80% in the case of the large experiments). This included the beam transfer magnets (BINP, Russia), the development and production follow-up of main ring superconducting quadrupoles (CEA, France) and cryostats (CNRS, France), the final focus quadrupoles and cryostats (Fermilab, USA, and KEK, Japan) and superconducting corrector magnets (DAE, India). CERN also benefited from the work of contingents of scientists and technicians from DAE, India (to staff the round-the-clock magnetic measurement campaign), and IFJ PAN, Kraków, Poland (to help with the installation and commissioning of the magnet protection system). CERN provided close expert oversight for such work, to ensure timely delivery of quality equipment and conformity to standards. Such arrangements rely heavily on the availability of core competence at the host laboratory and the strong motivation to achieve the goal, be it a working accelerator or working experiment.

In-kind contributions of equipment

The preferred way of acquiring equipment is via competitive tender from industry, using a detailed technical specification, if necessary based on model and prototype work done previously at CERN [Highlight 11.2]. In recent years supply via in-kind contributions from external institutes or laboratories have become more frequent, especially in the research sector, but also in the accelerator sector, as cited above. Although the in-kind supply may be free of charge to CERN it is not "free" for the project: it requires additional coordination, and reduces the degree of control CERN may deem necessary — a risk it has had to learn to take.

Additional monetary contributions from non-member states can be especially efficient, as they allow CERN to enlarge the tendering process. As an example, following Japan's special contribution to the LHC, firms there bid successfully for crucial advanced-technology equipment such as cold hydrodynamic helium compressors, high performance superconductors, and special steel.

While the LHC has so far only produced a few percent of the total number of collisions foreseen, options are starting to be discussed for a next large accelerator project. Such machines would cost much more than the LHC, and would almost certainly require truly worldwide funding. Could the model of the LHC experiments, which were funded at only 20% via the CERN budget, be adopted for financing a new accelerator?

In contrast with the experiments, a major fraction of the cost of a collider is (i) the civil engineering, and (ii) multiple units of a single sophisticated component. The quantities are such that they have to be produced in industry. This reasoning has led the proponents of the International Linear Collider (ILC) to consider in-kind contributions from designated regions that are possibly of a different design but "plug-compatible", bought from regional industry and controlled by regional "hub" laboratories. This ought to be possible, but would rely on there being a strong, competent central group, probably based at the host laboratory. It is generally understood that the civil engineering would be donated by the host region; together with the necessary oversight and central coordination, and procuring some key equipment, the minimum cost to the host region is plausibly close to 50% of the total. This is the starting assumption for discussions on how to fund the ILC; for ITER, hosted in France, the EU is contributing about 45% of the total cost, with the other six regional parties contributing about 9% each [4]. It would arguably be less risky and more economical to manage the funds for building a large new accelerator through the host laboratory, placing contracts worldwide via competitive tender, eventually featuring a degree of fair return on their expenditure. This is discussed in more detail below.

In-kind supply of qualified technical assistance

The testing, installation and quality control of equipment for a large accelerator project involves peaks of activity that call for more personnel than CERN can possibly provide. An efficient in-kind contribution is that of competent staff on secondment for a limited period during these peaks of activity — provided qualified technical supervision is available. This approach was adopted for the LHC magnet testing and the electrical circuit quality assurance referred to earlier.

Collaborations

In the accelerator sector, outside laboratories collaborate increasingly in design and prototyping work. This is clearly important when laboratories have specific expertise in domains not well covered at CERN. In this approach, (i) the collaborative sub-projects have to match the competence and infrastructure of the external laboratory; (ii) there must be effective liaison, recognizing the usual iterative design process; however, (iii) by concentrating on coordination, CERN technical staff is increasingly engaged in dispatching work to others. This has to be balanced with the need to maintain and develop core technical competence [5].

A collaborative response to requests for the transfer of know-how in core activities to external laboratories is part of the mandate of CERN. Occasional secondment of staff to work on projects elsewhere is also part of CERN's mission, and serves to enhance its visibility.

Over the last decade, CERN has become increasingly involved in the EU Framework Programmes (FP) such as CARE (Coordinated Accelerator Research in Europe) and EuCARD (European Cooperation for Accelerator Research and Development) together with a large number of laboratories. The EU FP are an excellent initiative, encouraging small teams to enter into cross-border scientific collaborations, with the possibility of attaining critical mass for specific R&D. There is also a clear sociological dimension. But with it comes a different style of control and reporting, typical of the EU programmes. CERN has shown in the past that it is capable of adapting to changing conditions: one has to be confident that it is able to absorb the additional constraints for the small fraction of activity addressed via EU-funded programmes. For most activities within the accelerator sector CERN can continue to apply the method proven successful over the years, namely to take advantage of in-house technical competence for design and model work, to purchase series equipment through contracts via normal competitive tender, and to transfer technology via close technical follow-up of manufacture.

Coordination

Coordination of large projects is obviously necessary. It is generally recognized that this is best left to those having the technical expertise and leadership ability. In the case of CERN, big projects, such as a new accelerator, are broken down into sub-projects, the leaders of which coordinate the sub-projects, resolve technical issues, and ensure respect of interfaces. Indeed, once the sub-projects have been allocated to competent and responsible technical groups, the remaining problems show up at the interfaces. It is the role of the project leader to organize structured meetings on a regular basis to track progress and manage changes at the interfaces. Between competent staff this goes smoothly with a minimum of meetings and reviews, thanks to a clear definition of the agreed goal.

In contrast with accelerator projects, the role of coordination is somewhat different for large experiments, built up from many collaborating institutes, and where decision-making is essentially via consensus. This requires clearly spelled-out management procedures, enshrined in the "Constitution of the Collaboration". After an initial learning phase this "management by consensus" has proved its worth, witness the swiftness and quality with which the LHC experiment collaboration have produced their scientific results.

Reviews

The use of reviews to examine technical choices and monitor progress of the major accelerator projects started with LEP, i.e. when the control of such projects passed nominally from the divisions to the directorate. However, reviewing the many sub-projects of the main project has only recently been adopted in the accelerator sector. The function was previously within the purview of the machine advisory committees, which reviewed on a regular basis the whole project, including sub-projects. CERN also had to participate in the process through collaborations with US laboratories, where frequent reviews are imposed. While reviews are useful — even essential, oversight does occur. Two instances of failure to detect problems at the LHC come to mind: the cryostats for the high luminosity insertions, and the magnet interconnects. In the first case a design flaw of the support system was not detected in the reviews. It was revealed during commissioning and was corrected (with some difficulty), but did not delay the start-up of the machine. In the second case the inherent weakness of the electrical splice was not pinpointed in the design review, with the well-publicized consequence of the September 2008 incident [6]. Problems may be averted by advice from reviews, but there is a real danger that they dilute responsibility. Reviews do not replace due diligence of project leaders.

The research sector has been accustomed to reviews for several decades. For the LHC experiments, the LHC Committee (LHCC) was established. With members external to CERN and the experiments, it served the important function of monitoring and providing advice, following progress and requesting remedial action if delays were incurred. This committee shares major credit for the remarkable operation of the experiments and the quality of the research results.

Patents

Most of the ideas that were conceived and developed at CERN, some of which even led to the award of Nobel prizes, have been published to make them available as common intellectual property, and not patented. Several studies [7–9] have shown that this policy has led to significant indirect added value, beyond direct commercial interest, for companies involved in producing material for CERN, as well as for society at large. The most dramatic example was the decision of CERN to put the WWW in the public domain. However, there is increasing pressure on publicly financed laboratories to protect technology from being patented commercially, and to provide a measure of their usefulness to society at large. At CERN, while this is still mainly achieved through publication as stipulated in the Convention, the approach with regard to patents is evolving (see Chapter 10).

One should bear in mind that the concept of patenting can itself be questioned, its net utility to society not being so evident [10]. It is well known that the vast majority of ideas develop into usable technology via interaction between members of a team, and for that to happen individuals should not be tempted to keep their ideas to themselves, with the hope of eventual personal profit from a patent taken out by CERN. Added to which it is generally recognized that for institutions like CERN the effort managing a patent portfolio might be such that the cost exceeds the benefit. CERN is vigilant as to the pitfalls of patenting.

Evolution of management in the research sector

The research sector during the "learning years"

Up to about the time of the p–p̄ collider, CERN provided a large fraction of the experimental equipment. At the same time, however, many university institutes acquired the competence to develop, build and operate experimental equipment. Importantly, this helped to establish a base and visibility of particle physics in

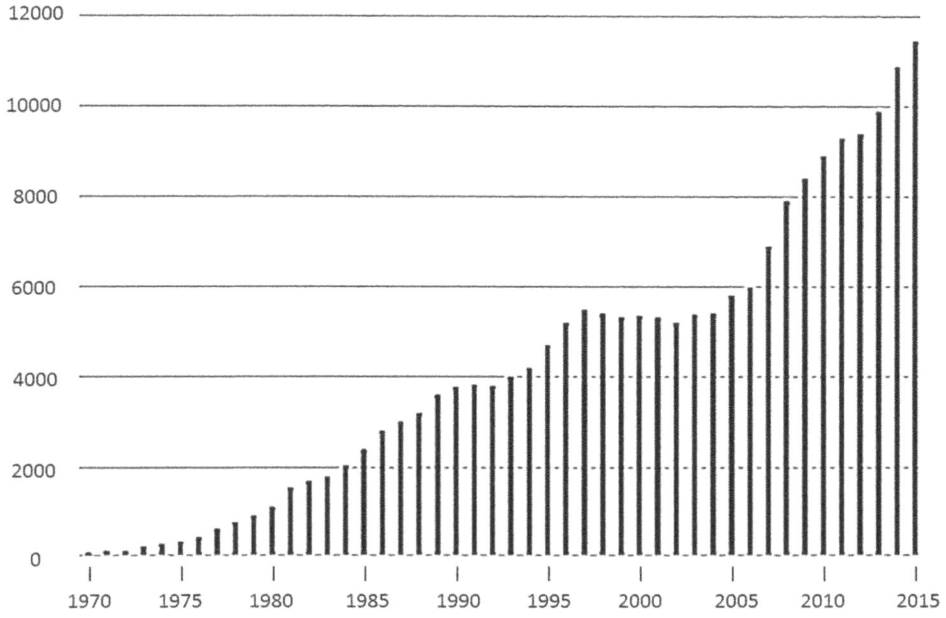

Fig 11.2. Number of CERN users vs. time.

academia. Starting with the p–p̄ collider era, experiments became large collaborative efforts with external institutions taking on a major share in providing, maintaining and operating equipment. This ever-larger involvement of the community is seen dramatically in the rising number of CERN users (Fig. 11.2).

The years to maturity

The four LEP experiments were each a collaboration of about 400 scientists, involving around 50 institutions, with CERN technical coordination and infrastructure. The evolution continued: for the LHC experiments CERN contributed only about 20% to the equipment value of the detectors, with the rest provided by the participating institutes and universities. These new conditions called for fresh ways to design, construct and operate the experiments.

CERN provided for each experiment, in addition to the infrastructure, the technical coordination, interfacing and integration, and financial control. An LHC experiment hosts up to 4000 collaborators coming from over 100 institutions. The funding of the collaborating institutions is provided by the national funding agencies in various forms and sometimes on an annual basis. It was impractical, if not impossible, to draw up legally binding contracts. Instead, the collaborations were (and are) held together via Memoranda of Understanding (MoUs). These are "best effort" agreements between stakeholders to supply selected items of equipment, cash (into a common fund) and associated personnel. Surprising as it may seem, it has worked remarkably well! The strong common interest of the stakeholders to reach the goal, and their ability to motivate and mobilize the experienced scientists, post-docs and students, were certainly important factors, but it should be stressed that the organizational framework and structure provided by CERN, the LHCC, the Resources Review Boards (RRB), and their sub-committees, have been crucial to the success of the LHC experiment projects [11].

Research at the global scale

The LHC experiments are represented by an elected spokesperson and "coordinated" (significantly the terms "managed" and "led" are avoided) by the spokesperson, aided by a technical coordinator (a recognized technical expert who takes responsibility for technical coordination and interfaces), a resource manager (who concentrates attention on funding issues), and elected scientists designated to coordinate the activities that are spread over the many collaborating institutes. The technical coordinators and resource managers are CERN staff. While it is only natural that there can be disagreements, the system is basically self-governing where governance is provided by the consensus derived from rational discussion among stakeholders, and crucially held together by the overriding desire to achieve the common goal of building a working experiment.

Apart from the experimental cavern, which is CERN-supplied infrastructure, the largest single-cost item of a detector is the experimental magnet, representing typically 30% of the value of the experiment. Up to the time of LEP these magnets had been designed and procured by CERN in much the same way as accelerator equipment. For LEP the design and fabrication of the two superconducting solenoids was outsourced (to CEA, France and to RAL, U.K.), with some CERN oversight. The normal-conducting solenoids were built at CERN. For the LHC an attempt was made to completely outsource the design and follow-up of the supply of the magnets. This turned out to be problematic for the magnets of all four major experiments, and closer control and collaboration of CERN was re-established.

In-kind contributions

The detectors of the experiments are complex but can mostly be sliced into packages of reasonable size; most of the equipment was developed and assembled in university laboratories, but sometimes it, was purchased by the institutes from industry. For the LHC, the framework for the tendering process via the common funds was provided by CERN, which often helped the collaborating institutes in this respect. Interfacing and integration was assured by the CERN group, and thanks to the effort (both technical and managerial) of the technical coordinators, the endeavour turned out to be successful. As the sources of both funding and manpower were widely dispersed, the experiments were subjected to regular scrutiny by the various committees to keep them on track with respect to technical performance, budget and schedule.

The volume of data generated by the experiments would have been impossible to handle using the computers available at the time the experiments were proposed. Decisions on data-handling equipment were therefore delayed until the last minute in order to take advantage of improving capacity (and decreasing cost), betting on the continuing validity of Moore's Law. The backbone for the data management was provided by CERN through the development of the Computing GRID, a software driven network of sharing the data and using the computing capacities of the collaborating institutes, distributed around the globe [Highlight 9.7].

Externalities

While the single-minded determination to succeed in the design, assembly and running of the very large experiments was essential, the congenial and fertile environment provided by the long-established infrastructure at CERN, its prescient Convention, the constructive support of the CERN Council and the national funding agencies, CERN's status as a leading research institution, and its location in an internationally-oriented city, have also been important factors. This should not be forgotten when trying to apply the successful formula elsewhere.

11.2 Building Large Accelerators with Industry: Lessons from the LHC

Philippe Lebrun

High energy particle accelerators are among the largest scientific instruments built by man. From their invention as table-top physics instruments a century ago — the cathode-ray tube with which J.J. Thomson discovered the electron in 1896 rested on a laboratory bench and the beam chamber of the first cyclotron built by E.O. Lawrence and S. Livingston in Berkeley in 1930 fitted in the palm of a hand — they have developed over the years in size, performance and complexity to become large technological systems, installed in multi-kilometre underground tunnels, federating the work of thousands of physicists, engineers and technicians for their construction, operation and maintenance, and relying on the series production by industry of advanced components that meet demanding specifications at market prices. Sustaining such a development over many orders

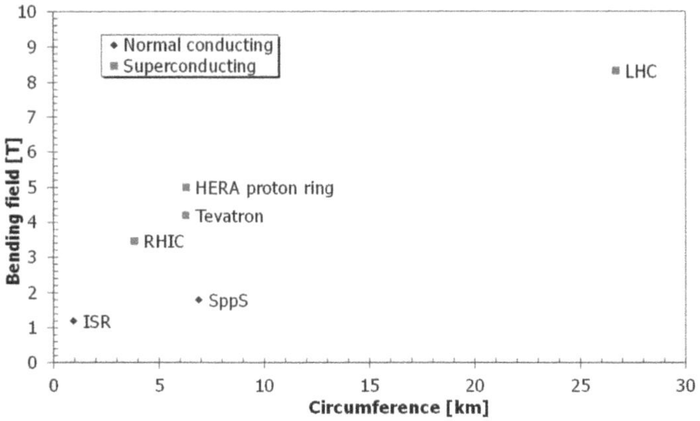

Fig. 11.3. Circumference and bending field of hadron colliders.

of magnitude in performance at affordable cost could only be achieved by a combination of larger size and more advanced technology (Fig. 11.3), leading to a continuous decrease in specific cost (Fig. 11.4) — a trend which could, fortunately, be maintained up to the largest and most advanced machine to-date, the Large Hadron Collider (LHC) at CERN. With more than one hundred major procurement contracts in advanced-technology industry, the LHC constitutes a comprehensive reference in the domain of industrialization and industrial procurement of components and systems for a large accelerator [12, 13]. To draw lessons from this experience can be instructive in the way to involve industry in large scientific projects and to ensure both scientific and industrial success of future endeavours.

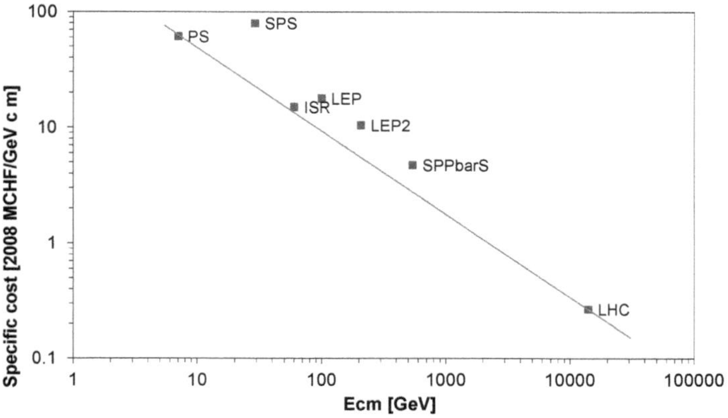

Fig. 11.4. Specific cost vs centre-of-mass energy of CERN accelerators.

Project governance, funding and procurement rules

In view of application to future projects, the feedback from experience referred to above should however be taken *cum grano salis*, due to possible differences between the technical, organizational and institutional context of such projects and those of the LHC. Although it was globalized through special contributions from CERN non-member states — Canada, India, Japan, Russia and the USA — the LHC accelerator remained a CERN laboratory-centred project. It was to a large extent (about 85%) funded by the CERN yearly budget and as such, subject to the regulatory framework and procurement rules and practices of the CERN Organization, focusing on its European member states. Most important, the project team had direct control — within the regulatory framework — over the largest fraction of the budget. In contrast, it is likely that industrial procurement for future large accelerator projects will occur on a worldwide basis, possibly with a major proportion of in-kind contributions arranged by regional funding agencies and handled through regional "hub laboratories", according to something resembling the so-called "ITER model" [14]. A critical issue is then to estimate the value of the in-kind contributions from the different partner agencies in a way that reflects their worth to the project, without departing too much from industrial market prices which will eventually have to be paid by these agencies upon procurement of their contributions. This can only be achieved provided the procurement rules and the conditions of commercial competition are known: lowest price or best-value-for-money, world market or imposed national/regional returns, wide open and perennial market versus monopole/oligopole or monopsone/oligopsone (one-of/limited market). Even when this is settled, differentials between the estimated value and the real price of the in-kind contributions may still appear in the execution of the contracts, due to scope and interface changes as well as technical, commercial or organizational difficulties. In any case, a process has to be established and agreed among the partner agencies and the project governance to manage these differentials. Maintaining a sufficient common fund with the project governance to handle such contingencies must be part of the process.

Aiming at the right level of technology

The history of particle accelerators has shown that adopting more advanced technology is the best way to sustain their development in performance while containing increase in size and cost. For each type of supply, the main technical decision is that which aims at the "right" level of technology, i.e. bold enough to break through and achieve performance at minimal cost, and conservative enough to be compatible with large industrial production and acceptable risk to the project.

A typical example of this is the choice of Nb-Ti superconducting alloy operating in superfluid helium for the LHC magnets, an alternative to magnets made of Nb_3Sn operating in "conventional" liquid helium at normal boiling temperature. This decision was considered conservative enough to warrant industrial feasibility and acceptable cost for series production of superconducting magnets, while requiring novel developments in cryogenic refrigeration and cooling schemes. Although both types of superconductors exist since the early 1960s and model magnets using both technologies were built in the early years of R&D for the LHC, only Nb-Ti was an industrial product, manufactured and commercialized at market prices by several companies in Europe, America and Asia at the time of project approval. Industrialization of superfluid helium cryogenics [Highlight 8.3] was deemed easier to achieve than that of Nb_3Sn superconducting magnets, a decision which proved the correct one *a posteriori*: twenty years after the approval of LHC construction, there is still no example of a Nb_3Sn magnet operating in a particle accelerator, and this technology is now still under development for future projects, including LHC upgrades [15]. Conversely, high-temperature superconductors, discovered only a decade before the LHC project was approved, bore the promise of so large benefits for reducing cryogenic heat loads in electrical current feedthroughs that CERN launched a vigorous — and successful — development and construction programme which culminated in the installation of more than 1100 such components in the machine (Fig. 11.5), where they have been smoothly operating ever since [Highlight 8.4], saving the capital expenditure of an additional large helium cryogenic plant and several MW electrical power in operation.

Fig. 11.5. High-temperature superconductor current feedthroughs in the LHC tunnel.

Fig. 11.6. Modular switch-mode 12 kA, 8 V power converter for the LHC.

The benefits of technological developments may also come from existing industrial products which can be adapted to the specific needs of the project, as shown in the following example. Powering superconducting magnet circuits in a particle accelerator requires high-current, moderate-voltage power converters operating with high precision, controlled at the part-per-million (ppm) level. These current and voltage ranges are characteristic of those used for arc welding, which however requires much less precision. Many of the LHC power converters were procured — for the power part — from companies manufacturing modular arc welding systems, complemented by high-precision measurement and control of the delivered current provided by CERN (Fig. 11.6), resulting in a win-win situation: technical performance was as specified, final cost to the project very competitive, and the companies could learn and assume leading positions in novel markets.

Defining an industrial procurement strategy

The basic decision for procuring technical components is whether to go to industry or start a production line in the laboratory: in this respect, CERN policy has been to go to industry whenever possible, and to foster industry's interest through incentive actions in marginal cases. Only in a limited number of specific cases — absence of industrial competence, lack of interest by companies, impossibility of specifying a supply or forming a price, difficulty of defining or managing interfaces, or failure of a contractor — did CERN undertake industrial-type series activities. One such case was the construction of a cryogenic test station for the series superconducting magnets of the LHC [16], which was gradually designed and constructed by CERN from components procured from industry, integrating feedback from experience with prototype test benches and building up testing capacity from two to 12 benches running in parallel. An essential tool for quality and performance control of the 1250 main dipole and 400 main quadrupole magnets upon delivery from industry, this station operated round-the-clock for several years and remains today a unique facility (Fig. 11.7).

Even when technical constraints favoured or imposed construction activities to take place on its premises, CERN specified and outsourced the work to industrial companies that came and operated in the laboratory. This was the case of the assembly of the LHC superconducting magnets into their cryostats [17]: once assembled, road transport of delicate and expensive 15 m long, 37 t cryo-magnets across Europe would have constituted a technical and financial risk to the project. After designing the cryo-magnets, developing their assembly methods and procuring components and specific tooling, CERN entrusted an industrial contractor to perform the task — with a total volume of 425 person-years — in a dedicated on-site assembly hall (Fig. 11.8).

Fig. 11.7. Cryogenic test station for reception tests of LHC main magnets.

Fig. 11.8. Assembly of LHC superconducting magnets into their cryostats.

Another clear-cut case of on-site industrial contract involving advanced technology is that of the electrical and cryogenic interconnection of the cryo-magnets after installation in the accelerator tunnel [18]. Again the methods and specific tooling, driven by the technical requirements, were developed, validated and specified by CERN and implemented by the contractor. Induction-furnace soldering and ultrasonic welding were used for the 65 000 electrical interconnections which require very low residual resistance (down to below one nΩ) and ground insulation at the kV level, while TIG orbital welding was retained for performing the 40 000 helium-tight cryogenic pipe junctions. In all cases, the processes were rendered as automatic as possible to ensure repeatability in the field and achieve reproducible quality. Still, in spite of several lines of defence implemented against quality drifts — process validation, operator training, process tracking by means of periodic checks of witness samples and equipment reproducibility, quality audits — a few electrical interconnections were outside tolerances, leading in particular to the incident of 19 September 2008 in which powering at high current resulted in destruction of such an interconnection, electrical arcing and damaging of some 50 cryo-magnets which had to be removed from the tunnel for inspection and repair [6].

Once the decision to go to industry for a given supply has been made, comes the choice of single or multiple sourcing. While the desire to procure at the lowest bid price on a competitive market, to benefit from economy of scale in large series and to ease contract follow-up tend to favour single sourcing, multiple sourcing may well be the only solution to obtain adequate consolidated production capacity; it also brings other important benefits such as security of supply and leverage on balancing returns to different regions/countries, an important consideration for an international project. For all these reasons, the LHC project generally practised multiple sourcing, e.g. for the procurement of superconductors from seven companies in Europe, the USA and Japan and for production of the main dipole magnets, contracted to three companies or consortia in France, Germany and Italy [Highlight 8.2]. A few cases of single sourcing, driven by circumstances such as a single lowest bid from a company aiming at obtaining the whole production, have met diverse levels of success in execution, making it difficult to draw general recommendations.

Knowing the number of suppliers and factory sites, the delivery rates per production line can be calculated and the level of automation in the production techniques adapted to minimize overall costs. For large accelerators up to and including the LHC, the series numbers for most components were low enough (a few hundred to a few thousand, sometimes with a large number of variants) not to warrant automation in the workshop: only subcomponents produced in large numbers (e.g. punched laminations) called for automatic production (Fig. 11.9).

Such statements may however need to be revised for the next generation of high energy accelerators beyond the LHC: producing complex, high-precision accelerator devices in much larger series may well be achieved more efficiently, at higher quality and lower cost on automated production lines.

In the light of experience, it appears essential to decide on the industrial strategy prior to launching procurement for each type of component, based on technical criticality, maturity of design, series numbers to be produced, quality assurance requirements, market structure and production follow-up capabilities.

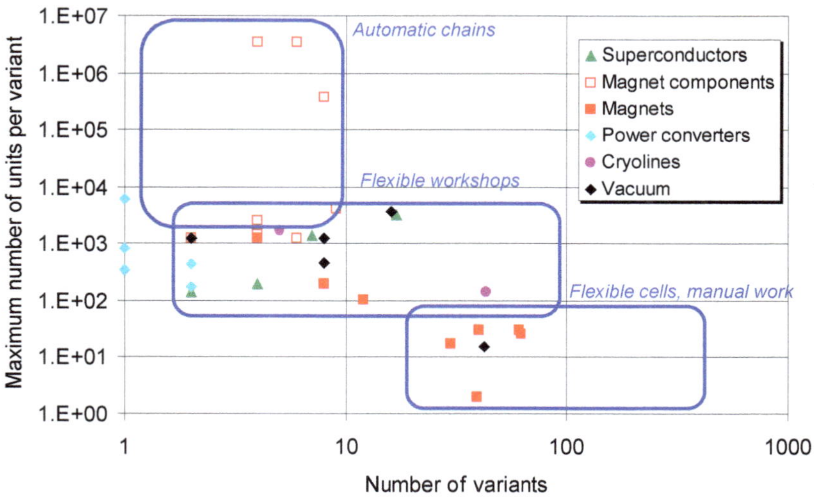

Fig. 11.9. Series numbers and production techniques of LHC components.

Involving industry efficiently and successfully

Once the policy lines are established, one can proceed to industrial procurement proper, starting with the identification and selection of companies. This is done in steps, starting with distribution of information about the requirement in terms of technology, quality, and schedule, as well as the applicable procurement rules and constraints [13]. Such information can be channelled via symposia explaining the main industrial stakes of the project to representatives of industry, summary documents to industrial liaison officers, and presentations to professional societies in the field of interest, all of which was used in the years preceding approval of LHC construction. Companies can then be pre-qualified on the basis of previous experience, capacity and declared interest, as expressed in their responses to market surveys: it is essential at this stage to ensure pre-qualification of a sufficient number of companies so as to maintain commercial competition at the time of the final invitation to tender, especially if one aims for multiple sourcing in the execution phase. A powerful tool in this phase is the best use of industrial models and prototypes, as a means not only to develop and validate technology and build know-how among potential future vendors, but also to foster partnerships, involve companies early and maintain their interest over preparatory periods which may appear very long by their standards [19]. This approach is exemplified in the successive periods of R&D, model building, prototyping and pre-series which led to the series production of the main dipole magnets for the LHC (Fig. 11.10).

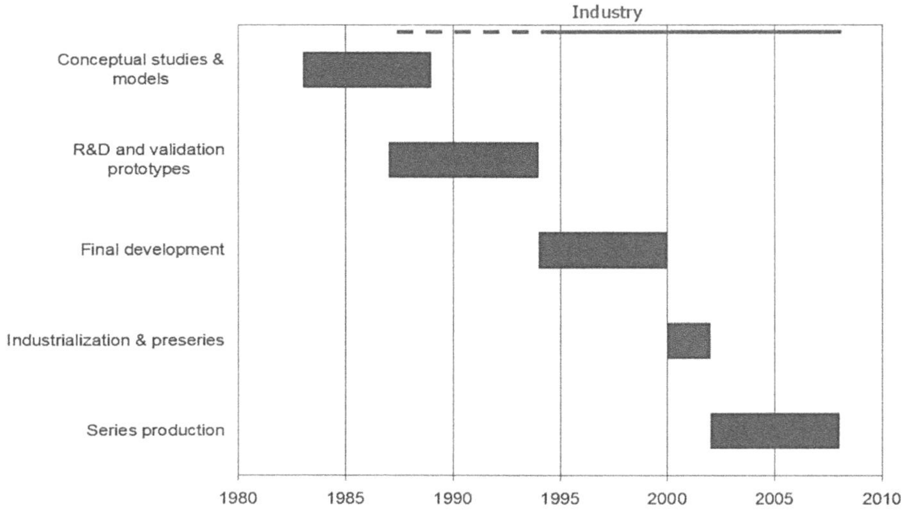

Fig. 11.10. History of LHC main dipole development and production.

Another interesting example of a project fostering industrial collaboration is provided by the cold hydrodynamic compressors providing power refrigeration at superfluid helium temperature for the LHC. The state-of-the-art previous to the project was based on a few small-capacity, lower-efficiency and high-cost machines developed by two vendors, in Europe and Japan. CERN stimulated a development programme, via design studies and the procurement of prototypes of different technologies, resulting in adoption of new technical solutions and enlargement of the industrial basis to six companies worldwide, some of them already working in cooperation. At the time of procurement, contracts for the 28 cold compressors of the LHC were placed with two groups of companies previously involved in prototyping, operating in a consortium or in contractor/ subcontractor mode. For such components, which required specific development in order to meet the demands of the project, it was therefore important to start cooperating with industry from a very early stage, while ensuring that the conditions of commercial competition were maintained up to final procurement.

The technical specification remains the essential document for launching procurement: in the case of the LHC, they were of two different types, depending on the type of technology. Build-to-print specifications were used in cases where CERN owned the technology and the associated technical risk (e.g. superconducting magnets). Functional-&-interface specifications were used for components and systems normally available in industry. In all cases, the guiding principle was to ensure that the risk be taken by the party who is most knowledgeable: as an example, LHC magnets were specified to industry in terms of electro-technical equipment (mechanical tolerances, electrical continuity, ground and inter-turn insulation), while the specific risks of superconducting magnet technology, e.g. quench performance or field quality, were taken by CERN. On a different register, the use of performance incentives, e.g. bonus/malus on measured performance, can be an efficient way to stimulate technical progress. The functional-and-interface technical specification for the LHC cryogenic plants included an adjudication formula based on the sum of investment and operating costs integrated over ten years, thus favouring higher-efficiency designs; in order to make sure that real performance would be in accordance with quoted values and thus "close the loop", a shared bonus/malus was applied on the difference of effectively measured versus quoted performance. As a result, the plants not only show record efficiency, approaching 30% of the theoretical maximum (Carnot cycle), thus saving on operation costs, but they could also be built smaller for the same refrigeration output, resulting in savings on investment.

Another important question concerning series production is that of intermediate component supply: is this left to the main industrial suppliers, or does the project management also act as general contractor procuring these components centrally and providing them to the main suppliers? The latter solution brings the advantages of ensuring better technical control on quality and homogeneity of intermediate supply for critical components (e.g. magnetic and non-magnetic steels, superconducting wire and cable in the case of LHC), economy of scale (e.g. punched laminations), balanced industrial returns, and mutualization of supply logistics, at the cost and risk of having to follow more contracts, of handling, storing and dispatching more material, and of taking direct responsibility in the timely feeding of production chains and — conversely — in their accidental stoppage. During LHC construction, CERN thus ran a "component centre" with several storage sites and associated follow-up, quality control and logistics: with appropriate levels of resources and effort, no rupture of supply occurred and not a single day was lost in the downstream production lines.

Once the procurement contracts are in place, their execution needs to be carefully monitored from the point of view of production rate and quality. Industry will not let production lines roll until all series methods, procedures and tooling are in place and ready, which may lead to initial delays in production ramp-up (Fig. 11.11). The end of series production, including that of sufficient number of spares while the production lines are still running, also requires particular attention and sometimes careful negotiation with the contractors, anxious to redeploy the best operators to new projects.

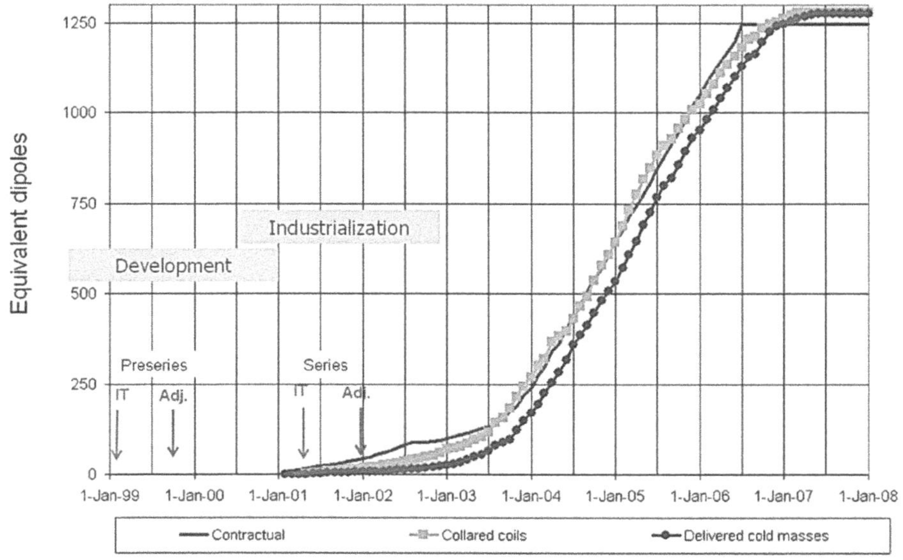

Fig. 11.11. History of the series production of the LHC main dipoles. IT: invitation to tender; Adj: contract adjudication.

Keeping quality under control starts with the project Quality Assurance Plan, which sets the formal framework in terms of procedures and documents. Each component must be accompanied by a Manufacturing and Test Folder containing all relevant data, and enabling to trace back the sources of non-conformities detected upon delivery and reception. In the case of the LHC, all such documents were informatics files in the project Engineering Data Management System, a unique repository of LHC project data kept up-to-date and permanently accessible via the World-Wide Web from any type of platform. Manufacturing and test data were in most cases transferred electronically from the production and testing sites, enabling swift detection of drifts, diagnostics and corrective action. Proper execution of the quality assurance procedures was monitored at production sites by resident and itinerant inspectors from a company, in support of staff from CERN and contributing laboratories. Formal quality audits were also conducted at intervals, or when drifts in production quality had been detected. CERN ensured that such actions were perceived as a help, and not a load by the contractors.

Finally, the best organized production is also subject to the hazards of the industrial world, be it technical (drift of quality, breach of the supply chain), organizational (company mergers), social (work stoppage) or financial (insolvencies and bankruptcies) which are bound to happen over the long time scale of large accelerator projects, and did happen in several instances throughout LHC construction. The project management must then be prepared to react through several types of actions, including — at worst — the taking over of tasks which were originally part of the contractors' duties; this can be done with limited impact on the project schedule provided a minimum amount of core technical resources are kept available, and can be rapidly redeployed in the host laboratory.

Building on capitalized expertise in the laboratory

The facts reported and argumentation developed in the preceding pages may give the impression that sound project management, resting on adequate organizational structure and making proper use of the competency, skill and production capacity of industry, are the keys to the success of large accelerator projects. An absolutely essential component is also the involvement of a numerous, experienced and dedicated personnel in the laboratory, at all professional levels, working in a spirit of collaboration towards a common goal. As an example, construction, installation and pre-operation of the LHC accelerator required some 7000 person-years activity of CERN staff over a period of 14 years, about 40% scientific/engineering and 60% technical. Beyond the bare numbers — and certainly even more important — is the expertise developed over the years by these physicists, engineers, technicians and administrators with the construction, operation and upgrade of previous accelerators. The success of the LHC also rests on the complex RF and beam manipulation techniques developed at the PS synchrotron, on the physics of colliding beams learnt at the ISR and SPS colliders, on the culture of very large projects stemming from the LEP collider, on the technology of superconducting magnets and RF cavities, large-capacity helium cryogenics, "cold" ultra-high vacuum, distributed computer controls, pioneered on previous CERN projects and made perennial through the expertise acquired by the personnel. In order to prepare for the future it is important for CERN to continue to provide the opportunities for staff to accumulate expertise through in-house development of existing and new technology, and thereby acquire the skills needed to interact credibly and efficiently with industrial partners.

11.3 Building LHC Detectors: Collaborations that Span the World

Markus Nordberg, Achille Petrilli and Thomas Taylor

The ATLAS, CMS, LHCb and ALICE experiments at the LHC show that large experimental facilities can be successfully designed, procured and assembled, in a timely manner and close to budget, by large collaborations of scientists. Key to this success was, and is, the quality reference provided by CERN [11].

The first discussions on the possible LHC and detectors took place in the mid-1980s. Possible designs of detectors, and associated R&D started in the early 1990s followed by consolidation of proposals with mergers and withdrawals, with the major experiments taking shape at the time of the demise of the SSC and the increasing likelihood of getting approval of the LHC. The technical proposals for the complete experiments were peer-reviewed and approved in 1994–95, followed by Technical Design Reports (TDRs) for each subsystem from 1997 onwards.

Each experiment formalized its collaboration by drawing up a Constitution stating the rights and obligations of participating institutes, and a Collaboration Board (CB) made up of their representatives — the "Parliament" of the experiment.

In 1997, in order to provide a level playing field the LHC Resources Review Board (RRB) set the budgets for each of the two large, general purpose experiments, ATLAS and CMS, at 475 MCHF (1995 value), called the CORE value.[c] The CORE value does not include the cost of home institute infrastructure, or salaries. The figure for the total CORE value came from careful estimation of expected expenditure, numerous discussions with collaborating institutes, and by paring down the original requests for around 500 MCHF to new agreed values.

---

[c]CORE refers to the LHC COst REview committee.

Where would the money come from? A system was drawn up that functioned as follows: Memoranda of Understanding (MoUs) were established between the experiment and collaborating institutes for the supply of components satisfying performance and interface specifications for an agreed budgetary cost. Though not legally binding, this arrangement by "best-effort agreement" is very much lighter from a bureaucratic point of view, and can work thanks to an intense and shared motivation to build the experiment with the objective of obtaining otherwise unobtainable scientific data — possibly leading to ground-breaking discoveries, and the mutual pressure on collaborations to strive to achieve the agreed goals (not to mention a degree of perceived competition with the other large experiment). No funds were included for institute manpower or contingency, which meant that funding agencies accustomed to including salaries, overheads and contingency in their estimates had to separate these out from their contribution to the CORE cost. By 2001 the estimated cost had increased to 515 MCHF, and the final cost when the accounts for the construction of the experiments were closed in 2009 was about 540 MCHF each for CMS and ATLAS, corresponding to the original estimated cost plus a notional intervening escalation of 2% per annum on uncommitted funds (2% was the figure applied for the accelerator). This result was possible thanks to continuous tight control, the absorption of some cost overrun by collaborating institutes, and the staging of less urgent and/or critical components.

The responsibilities were divided and delegated, with the nominated project leaders having to optimize funding and execution locally. Common funds (about 15% of the total for CMS, 44% for ATLAS) were established to cover projects that had to be controlled centrally: these were funded globally by the collaboration, with contributions being monetary or in-kind, and funds pooled for payment of specific contracts. Expenditure was monitored by the experiment oversight committee and the RRB.

This scheme of things is sometimes referred to as an "adhocracy", which works thanks to the sharing of a common goal and common scientific understanding: problems can be sorted out through rational discussion, and once a consensus is reached the different agents fall naturally into line and get on with the job. If the discussion gets too drawn out the spokesperson is called upon to arbitrate. For such a bottom-up approach to work it is nevertheless essential that those in charge are respected for their scientific/technical expertise and human qualities rather than theoretical management skills. Analysis has revealed that the management of these detector projects would probably not benefit from being approached from a more classical "professional" business viewpoint [20]. Rather the contrary, it could bring increased risk to schedule and cost overruns. Motivation is the essence.

The management of the finances of the large experiments, for which a major input was "in-kind", depended on the arrangements made with the suppliers, which ranged from a truly collaborative effort in which equipment is delivered regardless of the effort, to equipment supplied by institutes acting as commercial partners, which chalk up costs for changes and design oversight with little regard to the impact on the overall budget. This brought plenty of opportunities for creative intervention by spokespersons, technical coordinators and resource managers, needing to keep a cap on the cost. For example, while the possibility for the large experiments to tap into manpower reserves associated with collaborating institutes is a clear advantage, such labour comes at a financial cost to the experiment. However, overcoming the bureaucratic hurdles is made possible by the special status of CERN as an international organization, and the institutes could supply temporary specialized labour at more affordable rates than those applied locally (if indeed such qualified personnel could have been found), but attractive in the home locations. Without such arrangements budgets would have suffered.

To summarize, there are certain perceived advantages and disadvantages of the approach taken for the LHC to construct the large experiments [21]. The advantages of the approach are:

- Technical problems can be solved where the core competence resides;
- It enables collaborating institutes to utilize/maintain/develop skills;
- It involves students and provides a top level educational experience;
- Institutes share technical and financial risks;
- Outreach and economic returns are enhanced due to wide involvement;
- Light financial reporting (enabled by the CORE value arrangement).

But there are disadvantages:

- Management has little power to make collaborators follow decisions;
- Decision making is sometimes slow;
- There risks to be some duplication/waste of resources between institutes;
- The system is tributary to stable conditions of host state services.

Maintenance and Operation

From the outset it was made clear to the collaborations that they would have to continue to support the experiments while they were operating. There would have to be a flow of people (for data-taking) and money (for repairs and maintenance), in addition to funds required for likely upgrades. It was not readily understood by some of the funding agencies that, being dedicated to the investment and running costs of the accelerator complex, the CERN budget does not include a post for also carrying the entire experimental programme. A Scrutiny Group reporting to the RRBs was therefore set up to analyse the costs incurred and suggest how best to share the burden, thanks to which they came to agree as to the size (typically an annual 3% to 5% of the cost of the detector) and the sharing of this charge.

Typical organization of an experiment

The ATLAS organization during the construction phase is shown in Fig. 11.12. To provide stability during the construction phase, until March 2009 the spokesperson was elected, after consultation with the CERN management, for a term of 3 years, renewable with a 2/3 majority. Since then the mandate is for 2 years, also renewable with the same terms. Deputies, technical coordinator and resource coordinator are proposed by the spokesperson, endorsed by the executive board, and approved by the CERN management. For the duration of their mandates these officials are CERN staff.

About 60% of the construction capital was allocated to deliverables: institutes and their funding agencies committed to supply as "in-kind" with a recognized CORE value [20]. The nature of deliverables reflected the core competencies of the institutes providing them. The remaining 40% were defined as common items, shared in proportion to deliverables, and of which around 60% were provided as in-kind contributions. The mechanism for these purchases is shown in Fig. 11.13.

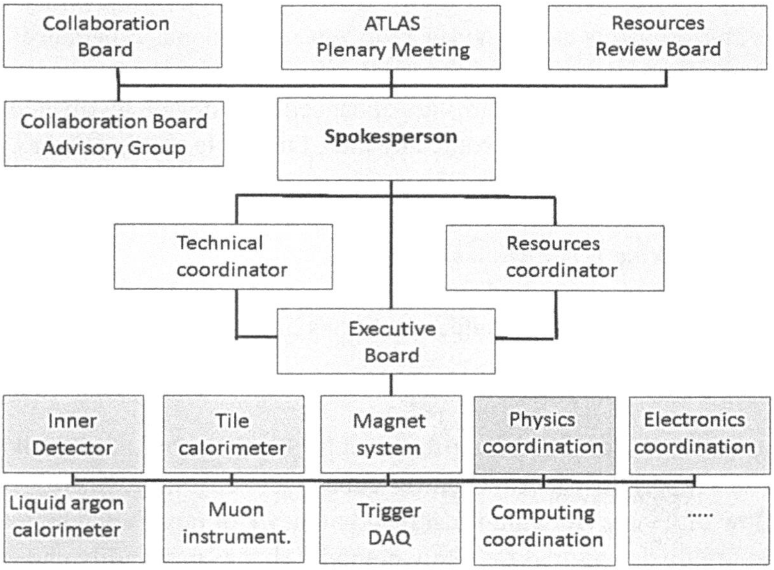

Fig. 11.12. ATLAS Organization during the construction period [21].

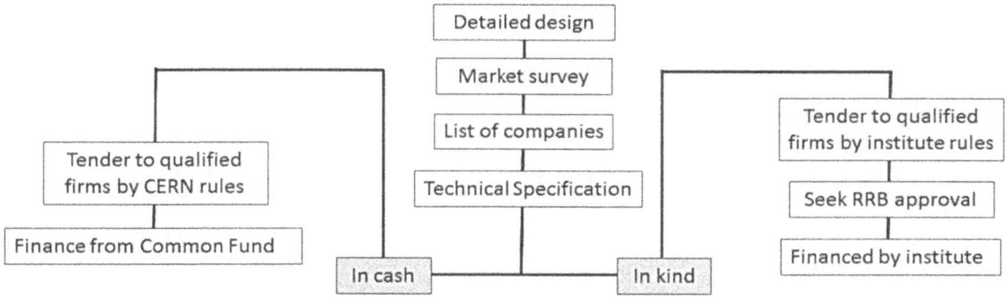

Fig. 11.13. Purchase of equipment via the Common Fund [21].

In the case of CMS, since the installation the spokesperson is elected for a 2-year non-renewable term. As for ATLAS, close assistants (technical coordinator, resource manager, etc.) are chosen by the spokesperson, to be endorsed by the collaboration board. The organizational charts are similar to those of ATLAS.

Concluding remarks

While the RRBs and the LHC Committee did provide a level playing field for the two very different large experiments with full solid-angle coverage, the approach to getting the equipment made and installed was sufficiently similar for observers to isolate plausible macroscopic reasons for the success of the projects:

- Setting up of simple rules and regulations (based on CERN experience);
- MoUs creating peer pressure between suppliers of sub-systems;
- Peer pressure between the experiments;
- The careful selection of competent technical coordinators;
- Problem solving based on technical realities, and a common value scale;
- The common goal of building a viable experiment to probe the unknown.

For accelerator-based high energy physics, a central laboratory staffed with top-level scientists, engineers and technicians is essential [11]. In order to ensure the optimization of complex systems, scientists must be prepared and willing to take a genuine interest in the detail of technical design and manufacturing issues. The smooth interaction between the central laboratory and other laboratories and universities, combined with peer reviews to ensure quality control, are key to the success of the "CERN model". To ensure continuity, it is vital to maintain enthusiasm, and to maintain and renew core expertise. This is best done via a vigorous and ambitious R&D programme.

References

1. https://council.web.cern.ch/council/en/basicdocuments/CONVENTION.pdf.
2. http://home.cern/.
   http://library.web.cern.ch/archives/history_CERN/internal_organisation.
3. H. A. Simon, Rational decision-making in business organizations, *Nobel Memorial Lecture* (1978).
4. http://www.iter.org/FAQ#collapsible_5.
5. The Economist, 20 January 2016, *The collaboration curse,* p. 99 (2016).
6. M. Bajko *et al.*, Report of the task force on the incident of 19th September 2008 at the LHC, CERN-LHC-PROJECT-Report-1168 (2009).
7. H. Schmied (ed.), *A study of the economic utility resulting from CERN contracts*, CERN-1975-005 (CERN, Geneva, 1975). http://dx.doi.org/10.5170/CERN-1975-005.
8. M. Bianchi-Streit *et al.* (eds.), *Economic utility resulting from CERN contracts*, CERN-1984-014 (1984). http://dx.doi.org/10.5170/CERN-1984-014.
9. OECD, The impacts of large research infrastructures on economic innovation and on society: case studies at CERN, OECD 2014.
10. The Economist, 8 August 2015, *Time to fix patents,* p. 9, and *A question of utility*, pp. 43-45 (2015).
11. J. Engelen, The Large Hadron Collider project: organizational and financial matters (of physics at the terascale), *Phil. Trans. R. Soc. A* **370**, 978-985 (2012); doi: 10.1098/rsta.2011.0466.
12. P. Lebrun, Industrial technology for unprecedented energy and luminosity: the Large Hadron Collider, *Proc. EPAC04 Lucerne*, JACoW (2004) pp. 6-10.
13. A. Unnervik, The construction of the LHC: lessons in big science management and contracting, pp. 39-55 in L. Evans (ed.) *The Large Hadron Collider* (EPFL press, Lausanne, 2009).
14. B. Foster (ed.), Revised ILC Project Implementation Planning, Rev. C (July 2015), and http://ilc-edmsdirect.desy.de/ilc-edmsdirect/document.jsp?edmsid=*1115765.
15. G. Apollinari *et al.*, High Luminosity LHC project description, CERN-ACC-2014-0321 (2014).
16. J. Axensalva *et al.*, Cryogenic infrastructure for testing of LHC series superconducting magnets, *Proc. ICEC20 Beijing*, (Elsevier, 2005) pp. 1015-1018.
17. A. Poncet *et al.*, Assembly and quality control of the LHC cryostats at CERN: motivations, means, results and lessons learned, *Proc. PAC07 Albuquerque*, JACoW (2007) pp. 338-340.
18. P. Fessia *et al.*, The LHC continuous cryostat interconnections: the organization of a logistically complex worksite requiring strict quality standards and high output, *Proc. EPAC08 Genoa*, JACoW (2008) pp. 2428-2430.
19. R. Perin, Superconducting magnets for the LHC, a report of CERN's collaboration with industry, *Europhysics News*, **21**, 2-4(1990).
20. M. Boisot *et al.* (eds.), *Collisions and Collaborations* (Oxford University Press, 2011).
21. H. Gordon, Project management in the Atlas international collaboration, http://sites.apam.columbia.edu/fusion/BP_PAC_FNL/Gordon_Atlas_Management.pdf.

Chapter 12

R&D for the Future

Kurt Hübner, Daniel Treille and Daniel Schulte

12.1 The LHC and Beyond

The discovery of the Higgs (BEH) boson [Box 8.2] in 2012 at CERN revealed the cornerstone of the so-called Standard Model (SM) in particle physics, a consistent description of the fundamental constituents of matter and their interactions [Box 6.4]. While the SM represents a spectacular synthesis, it is manifestly an incomplete description. It does not offer an understanding of the cosmic dark matter or the dark energy, supposed to fuel the expansion of the universe. It fails to provide a complete explanation of the apparent matter-antimatter asymmetry in the universe. These are some of the key questions that shape the research agenda in particle physics, astrophysics and cosmology.

New physics phenomena beyond the SM (BSM) can be found "directly", by producing and detecting new particles or by observing processes forbidden in the SM. It can also be revealed "indirectly", e.g. at colliders, through accurate measurements of properties of known particles (couplings, decay branching ratios, etc.), especially of top quark and Higgs boson, if they deviate significantly from the SM predictions. The former case, the only one traditionally considered a "discovery", is the major hope for the future, but Nature will decide. The latter would reveal that new physics is at work virtually [Box 5.1] and may give a hint of its mass scale, potentially much higher than the centre-of-mass energy (E_{CM}) of the colliding particles, and its identity. Both directions should be pursued as far as possible, calling for maximizing the available energy and the luminosity [Box 6.1] of the collider, to produce the heaviest particles and explore the rarest phenomena possible, see also Box 12.1. These goals motivate the worldwide R&D in cutting-edge technologies.

CERN focuses R&D on areas of its expertise, predominantly in accelerator and detector technology. Although R&D is being conducted at CERN in the field of low and intermediate energy and of neutrino physics, this chapter emphasizes

CERN's major R&D activities directed towards the high energy (HE) frontier concerning new very large circular and linear colliders. Other HE options, presently under discussion, may influence this programme, as new results emerge: these are also briefly mentioned.

Exploitation of the LHC is the first priority of the European particle physics community and of CERN. A vigorous effort is made to increase and exploit its integrated luminosity, planned to rise from the present 4 fb^{-1} obtained in the first year (2015) with protons at 6.5 TeV to a total of about 300 fb^{-1} at 7 TeV, the nominal LHC energy, by 2022. In parallel, a significant upgrade, called High Luminosity LHC (HL-LHC) is in preparation, which aims to achieve a total integrated luminosity of 3000 fb^{-1} in pp collisions [1] by the second half of the 2030s. With this substantially higher integrated luminosity the accuracy of measurements, e.g. the Higgs couplings to bosons and fermions, will be improved and the discovery reach, e.g. the search for Supersymmetry, extended [Box 7.2]. The successful ion-ion programme will also be continued.

At the heart of the HL-LHC project is the replacement of the current focusing magnets in the interaction regions, based on NbTi technology, with stronger magnets based on the better performing Nb_3Sn superconductor, to reduce the beam size at the collision points [Highlight 12.2]. This progressive development, initiated for fusion reactors, is also relevant for future hadron colliders. The beam intensity will also be increased. These measures will yield a peak luminosity of 5×10^{34} $cm^{-2}s^{-1}$ [1], a five-fold increase over the present luminosity. The latter already is a far cry from the initially foreseen 10^{33} $cm^{-2}s^{-1}$ which was the aim until 1987. At that time, it was realized that new prospects in physics, in particular Higgs physics, needed a factor 10 more in luminosity, which, thanks to vigorous and sustained R&D programmes both in accelerator and detector sectors, is presently achieved and exploited.

The peak luminosity is limited by the energy deposition of the collision products in the superconducting quadrupoles next to the collision point. A further practical constraint is the number of inelastic interactions μ per bunch crossing due to limited capability of the detectors to unambiguously associate particles with a given interaction. It is mitigated by "levelling" of the luminosity: the interaction rate is reduced through a small separation of the beams at the interaction point; as the beams loose intensity, the separation of the beams is gradually reduced, such as to keep the luminosity constant. This procedure maximizes the integrated luminosity subject to the constraint on μ in the detectors. For example, the number of μ is close to 30 at the present nominal LHC luminosity of 1×10^{34} $cm^{-2}s^{-1}$ increasing to about 140 at $5 \times 10^{34} cm^{-2}s^{-1}$ [1]. (This number μ ignores fluctuations.)

For the HL-LHC the detectors will have to be significantly improved. New levels of radiation hardness will be required. Tracking systems and parts of the calorimetry in the forward region will have to be replaced. Trigger and data acquisition systems will need major upgrades [Highlight 12.6]. Overall, the quality of the measurements will be increased in line with the new physics agenda.

The choice of new projects for the post-LHC era will depend on the findings of LHC and HL-LHC and potentially other experiments. Therefore, a multi-pronged R&D approach has been put in place to prepare for this later choice of the optimum facility. Currently, two main R&D directions are pursued at CERN, coordinated with the European physics community. One is a very high energy e^+e^- collider, the Compact Linear Collider (CLIC); the other is a very high energy proton collider, the so-called Future Circular Collider (FCC-hh), considered to be the next "discovery facility".

The e^+e^- collider would complement LHC by offering high precision measurements, in particular those associated with the Higgs boson, which could hint and/or reveal physics phenomena at a higher energy scale. However, direct searches for new phenomena or particles would be limited by its relatively low E_{CM}, although its precision will allow indirect searches up to much higher energy.

For the study of e^+e^- collisions, linear colliders are being proposed to avoid the strong increase of synchrotron radiation of circular colliders. In linear colliders the beams are accelerated in two linear accelerators (linacs) pointing at each other, avoiding any bending of the trajectory and, thus, synchrotron radiation. The particle beams enter in collision once and are then disposed of.

Linear collider R&D concentrates on several issues. First, very high accelerating fields have to be achieved to obtain the high collision energy in a practical length of the collider. Second, high luminosity requires maximizing the number of colliding particles and squeezing the beam to very small transverse sizes at the collision point. A strong constraint is the maximum power which can be taken from the electric grid: efficient generation of RF power is mandatory to produce these high accelerating fields.

Two different approaches to linear colliders are currently being pursued. A worldwide collaboration is developing the Compact Linear Collider (CLIC) [2]. which can be constructed and operated in energy stages reaching at the first stage 0.38 TeV, the optimum for Higgs and $t\bar{t}$ studies, and ultimately 3 TeV E_{CM}. Figure 1 shows a possible location close to the CERN site. In its final stage, each linac would be 21 km long with an overall length of the facility of 48 km. An alternative approach is the International Linear Collider (ILC), also being developed by a worldwide collaboration, including CERN, aiming at E_{CM} of 0.5 TeV with an option for 1 TeV [3].

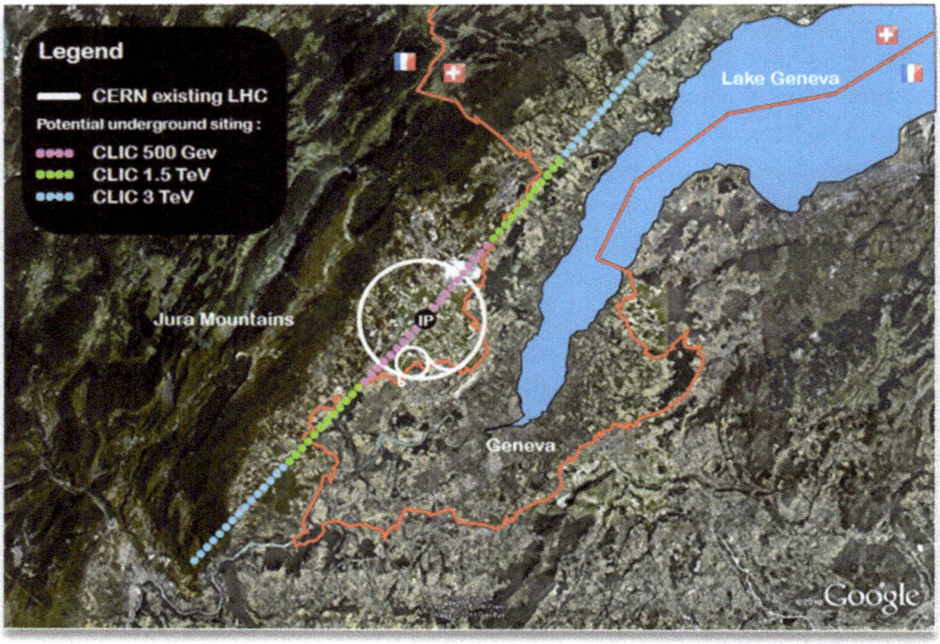

Fig. 12.1. A possible location of CLIC in the Geneva area close to CERN [2].

CLIC is based on high-frequency normal-conducting accelerating structures in the main linac operating at 12 GHz and with high accelerating gradients, the target being 100 MV/m [Highlight 12.4]. This very high accelerating gradient is motivated by the goal of reaching very high collisions energies, while keeping the overall length of the collider within manageable bounds. The salient point of the design concept is the use of an electron drive beam to generate the RF power by passive devices instead of a plethora of klystrons with their power supplies, requiring considerable maintenance.

The ILC plans to use superconducting accelerating cavities operating in pulsed mode at 1.3 GHz. This offers an important advantage compared to CLIC. The beam pulses can be much longer, which makes the generation of the RF by conventional klystrons straightforward. Also the time between bunch crossings in the interaction point is much longer, about 370 ns, which eases the detector design. The main drawback with respect to CLIC is that the achievable accelerating gradient — the aim is 31.5 MV/m — is about a factor three lower. This imposes a practical limit on the length and hence the ultimate energy of the collider.

A different idea is to use plasma waves as the accelerating mechanism in the main linac instead of normal or superconducting accelerating RF structures. These waves can be generated either by an intense electron, proton or laser pulse in a gas-filled volume. The pulse will separate electrons and positively charged ions in the gas generating a strong accelerating electric field. A test of this novel acceleration method using a proton pulse is under preparation at CERN [4].

The detector designs for e^+e^- linear colliders are addressed by a worldwide R&D programme [Highlight 12.5]. The emphasis on precision measurements requires extreme performance of the detectors, concerning e.g. momentum resolution, jet energy resolution and impact parameter accuracy at the vertex to identify short-lived particles. This needs detectors with fine granularity and high resolution in space and time for the reconstruction of the individual secondary particles generated in an event and for background rejection. Vertex and tracking detectors have to be designed with low-mass materials to minimize scattering and loss of particles. The radiation levels are orders of magnitudes lower than for LHC, facilitating the use of new and more performing sensors, electronic and integration technologies. However, the detector designs have to take into account the specific machine-dependent backgrounds due to the extremely high fields produced in the collision of such tiny beams. Furthermore, an excellent coverage of forward emitted $e\pm$ is mandatory, since they provide the best measurement of luminosity. The studies consider mainly two different concepts of detectors distinguished by the tracking instrumentation: a Time Projection Chamber or all-silicon sensors; both concepts put the accent on vertex detectors and very fine-grained calorimetry. The different time structures of the collisions in CLIC and ILC lead to specific requirements and features in the detector design.

Considering future proton-proton colliders, the option to equip the LHC with new dipole magnets capable of producing at least twice the field has received quite some attention; called HE-LHC it would offer collisions at twice the centre-of-mass energy, and possibly up to 33 TeV [5]. However, to make a large step into unchartered territory a very large hadron collider seems to be the tool of choice. Thus, in the CERN tradition of planning far ahead into the future, a study of a Future Circular Collider (FCC-hh) providing hadron collisions with E_{CM} at 100 TeV has been initiated. The design luminosity of 5×10^{34} cm^{-2}s^{-1} is similar to that of HL-LHC, with a possible increase to about 2×10^{35} cm^{-2}s^{-1}. This collider would have a seven-fold higher energy reach than the LHC, a step equivalent to that from the collider preceding the LHC, the US Tevatron, to the LHC. The production cross-sections of particles increase by large factors, e.g. more than 15 for the 125 GeV Higgs production. The search for a possible substructure inside quarks could be extended down to distance scales of 10^{-21} m [6].

FCC-hh [7] would take advantage of the technologies developed for the HL-LHC, but these technologies would be pushed beyond their present limits and new ones must be developed. The proton–proton E_{CM} is determined by the product of the radius of the machine and the magnetic field of the dipoles that bend the beam. To achieve 100 TeV FCC-hh will use a tunnel that is about 4 times longer than that of the LHC (about 100 km vs. 27 km) and dipole magnets of about 16 T, i.e. twice the field of the LHC dipoles. A possible location of FCC is shown in Fig. 2. It is close to the LHC which would become the injector, following the well-established CERN practice of re-using the latest flag-ship accelerator as injector for the next generation.

The magnets are based on the same Nb_3Sn technology used for the HL-LHC but pushed to higher performance. Their development is the core technological challenge for the project. In addition, new superconductors that can operate at higher temperatures are being developed for the use as superconducting links between magnets and their power supplies, as are high-temperature superconductors (HTS) for use in insert coils to boost field [Highlight 12.3]. Having the potential to be used for coils that provide better performance than Nb_3Sn coils, this could pave the way to magnets with fields of 20 T and more.

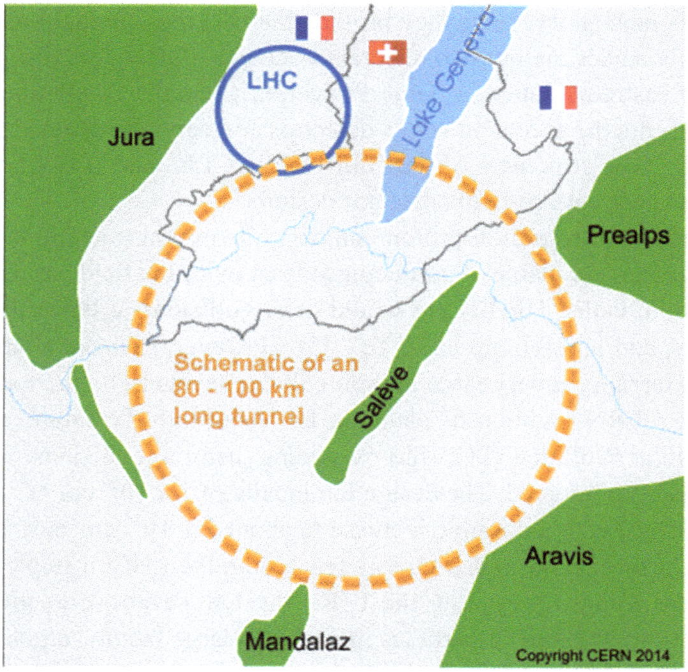

Fig. 12.2. A possible location of FCC near CERN.

At this very high energy even protons start to emit a significant amount of synchrotron radiation. This is on the one hand beneficial since the size of the circulating proton beam will shrink with time increasing the integrated luminosity. On the other hand the magnets need to be protected efficiently from radiation that could heat them to levels exceeding the critical temperature for superconductivity. A novel beam screen design is being developed to cope with this issue.

The kinetic energy of each circulating beam is about 8 GJ, enough to melt 40 t of copper, about an order of magnitude larger than in HL-LHC. This can lead to critical situations, particular during beam injection or extraction from the machine, causing sometimes fractional beam loss. Higher demands are also placed on the collimation system, which removes unwanted tails of the circulating beam and protects the accelerator against accidental loss. A robust active and passive machine protection system must be developed.

The main requirements for the detectors come first from the energies of the particles produced. For tracking, one considers increasing the momentum resolution by about the same factor seven as the energy increase compared to LHC. This will require increased position resolution of the tracker and higher magnetic fields. Coverage of the very forward region, down to 2° polar angle for tracking and to 0.3° for jet detection, is planned. Calorimeters must contain fully the most energetic showers, hence be 12 interaction lengths deep, and must be precise at the 1–2% level. Moreover, all detectors need very high granularity to cope with the high level of interactions "piling up" at each beam crossing and to decipher the internal structure of jets, when these are actually the signature of very energetic particles, such as top quark, W, Z, and Higgs. The existing LHC detectors (Chapter 8) provide a good model for extrapolation, in particular when considering the magnetic structure to be adopted, and identifying potential "showstoppers".

In each experiment the colliding beams will produce high event rates due to the high luminosity concurrently with a high rate of background events, which will produce collision products of up to about 500 kW per experiment. They will be emitted mainly close to the beam direction, so the focusing magnets near the detector need to be carefully protected.

The performance of a high luminosity e^+e^- collider (LEP3) replacing LHC in its tunnel and serving as a "Higgs-factory" has been evaluated [8]. Should a 100 km long tunnel be constructed for FCC-hh, it could be used for a powerful lepton collider either before (similar to the sequence LEP to LHC) or after FCC-hh. This option, FCC-ee [9], could provide high luminosity electron-positron collisions up to E_{CM} of about 350 GeV, albeit far below the energies of CLIC it would allow a similar physics programme as ILC [3]. The circular collider has the important advantage over the linear colliders that several experiments can be installed for simultaneous data taking; in a linear collider two detectors could be considered but only one detector can be operational for data taking at any one time.

The energy of FCC-ee is limited by beam physics rather than by technology, which is well understood. The principal difficulty of the FCC-ee is achieving a luminosity of the order of 10^{34} $cm^{-2}s^{-1}$ at the highest energies. The intensity of the circulating beams is limited by the power due to synchrotron radiation losses. As this power per particle increases with E_{cm}^4, the number of circulating particles has to decrease with energy to avoid excessive RF power consumption. The colliding beams need, therefore, to be very dense to produce sufficient luminosity. This leads to important intra- and inter-beam collision effects, which cause particle loss reducing the beam lifetime. The collider design must limit the inter-beam losses, which increase with luminosity and number of collision points.

The beam injection would differ from that of LEP, where the beam was injected at low energy and accelerated in the collider ring up to the collision energy. In FCC-ee, plagued by short beam lifetime, it is foreseen to have a booster ring in the same tunnel as the collider, by-passing the detectors. The beam is injected at low energy into this ring and accelerated to collision energy. Then it is injected into the collider ring on top of the already circulating beam. This practically continuous top-up injection has been previously used in smaller e^+e^- colliders and storage rings operating as light sources.

A study group, pursuing similar conceptual studies as done for an e–p option with LHC [10], is also considering the merit and feasibility of e–p collisions in FCC. In both cases, a recirculating linac would provide 60 GeV electrons for collisions with the protons in a single interaction point.

CERN is also attentive to other opportunities, such as, for example, the international programmes in neutrino physics, for which it provides R&D support, and has not lost sight of a Higgs factory based on a muon collider, having attractive features but requiring exceptionally challenging R&D.

Whatever the final LHC results, future accelerators will arguably play an instrumental role in the understanding of particle physics. CERN, conscious of the long lead time between concept and operation of large facilities, assumes a leading role in the international R&D collaborations preparing the HL-LHC, CLIC and FCC, while contributing in parallel to ILC. The future LHC results will tell us which of the facilities under study will be the optimum choice. CERN's multi-pronged R&D programme will put it in a position to be among the proponents of the next facility, hopefully a machine that opens a new energy frontier, or, if not, a factory focusing on the most relevant energy domains.

Physics prospects through the crystal ball **Box 12.1**

"Particle physics is at the most exciting crossroads in its history" [1]. The Higgs boson, the first spin 0 (seemingly non-composite) elementary particle, with its associated scalar field, is new to fundamental physics. Its self-coupling (SC) is a key property. The SM does not predict its mass and its low value is "unnatural" [Boxes 7.2 and 5.1]: one can therefore expect new physics at not too high an energy scale, making it "natural". Will such a scenario, such as SUSY, extra dimensions, etc., appear in LHC Run 2, or elsewhere? If so, more luminosity — HL-LHC, and/or energy — HE-LHC and FCC-hh, will allow its detailed study. At HL-LHC, besides possibly discovering what the first phase of LHC did not reveal, a major effort will be put into improving the measurement of the properties of the top quark and Higgs boson. But open issues will remain, calling for much higher energy.

FCC-hh 100 TeV is not just "a quantitative upgrade of the LHC". It would explore fundamental issues, probe distance scales down to 10^{-21} m, produce ~10 times heavier masses and much higher rates of basic processes. Accumulating 20 ab^{-1} (1 ab = 10^{-42} cm^2) would lead to ~10^{10} Higgs bosons, 400 times more $t\bar{t}H$ events than HL-LHC, and ~10^{12} $t\bar{t}$ pairs. FCC-hh will study the Higgs sector and its dynamics, with key inputs to its top coupling and to its self-coupling, the latter measured with ~5% accuracy, probing the shape of the Higgs potential around the vacuum of the Universe. Beyond the SM Higgs sector, Higgs exotic decays, such as μ-e, could be sought, as well as new states produced through the "Higgs portal", connecting the SM world to one possibly hidden. For *dark matter*, the TeV sector of weakly interacting particles (WIMPs), from SUSY [Box 7.2] or any theory offering a lightest stable candidate, would be probed very well. FCC-hh would also allow precise measurements of di-lepton final states, for example, in search of heavy W', Z', i.e. new forces, and explore uncoloured physics (less evident because only subject to the electroweak force) at the TeV-scale.

For *CLIC* [2] *and FCC-ee* [3], the former benefits from higher E_{CM}, the latter from running simultaneously ≥ 2 experiments, and much higher L when used as a Z and W factory. With ILC, there could be a similar program at E_{CM} = 350 GeV, measuring very precisely top and Higgs properties, reaching, from Higgs radiated off a Z boson, ~ 0.2% uncertainty on its coupling to Z and on its invisible decay branching ratio (see the FCC studies). At higher E_{CM}, $t\bar{t}H$ events would offer an accurate Higgs to top coupling measurement and W^+W^- fusion events would give access to rarer Higgs decay modes. CLIC at 3 TeV would deliver Higgs SC with 16% to 12% precision. Overall its decay width will be known at the sub-% level. With their clean environment and well-known initial state, the e^+e^- machines could search for BSM physics in two ways: (a) direct pair production of new particles up to $m = E_{CM}/2$, relevant example for SUSY partners of leptons and gauge bosons — a difficult topic at hadron colliders, and (b) indirectly, from precise SM measurements, sensitive to much higher masses and scales.

An e–p collider would focus on a more restricted domain of physics. A Higgs factory muon collider would allow the direct production of the boson in $\mu^+ \mu^-$ collisions.

[1] arXiv:1511.06495, arXiv:1606.09408 [2] arXiv:1202.5940 [3] arXiv:1601.06640

12.2 Accelerator Magnets with Ever-Higher Fields

Lucio Rossi

In 2008, with the LHC installation and initial commissioning finished, the CERN Magnet group started an R&D programme on high field magnets for the next generation of accelerators [11]. Its main scope is to develop advanced technology based on the use of Nb_3Sn superconductor. Nb_3Sn is not an alloy like Nb-Ti, which is relatively easy to handle and has to date been the superconductor of choice for use in accelerators (Tevatron, HERA, RHIC and LHC), it is a compound, capable of much higher performance with the potential to reach fields of almost twice the level of Nb-Ti. However, Nb_3Sn poses formidable challenges, the most severe being that it must be formed via a solid-state reaction, which requires heat treatment at 650°C in a controlled atmosphere and lasting many days, making it extremely brittle. This means it cannot be wound into a coil after reaction and every coil component must withstand such temperatures. In particular, the insulation cannot be organic (it would be converted into carbon) and the entire coil structure, composed of various materials, must cope with thermal excursion of almost 1000°C, from 650°C down to −270°C. The aim is to reach a critical current density [Box 4.3] of 1500 A/mm^2 at a field of 14–15 T (for Nb-Ti such current density is only achieved in fields of up to 7–8 T), while meeting the quality of accelerator magnets — implying uniformity of performance and small effective diameter of the superconducting filaments. Small filaments are necessary to avoid remanent fields at low field when the beam is injected [12]. The cross-section of a strand (an assembly of superconducting filaments embedded in a copper matrix) that is being developed in collaboration with a supplier of material for cables for the LHC upgrade is shown in Fig. 12.3. Also shown is a newly developed cable assembled from these strands; the white insulation sleeve is a braid of glass-fibre having a very low organic content to limit decomposition at high temperature.

Fig. 12.3. Left: cross-section of a superconducting strand (photo Bruker). Right: cable, showing the braided glass-fibre insulation.

In the present R&D phase, CERN is using simple flat racetrack coils, 400 mm long, which require a relatively small quantity of Nb_3Sn superconductor and have a short construction time, in order to assess conductor properties and coil technology representative of conditions occurring in a full-size magnet.

To develop a suitable conductor is obviously essential, but the mechanical structure of the magnet also requires attention. In the quadrupoles that will squeeze the beam at the interaction points for the LHC upgrade project for higher luminosity (HL-LHC), the field on the conductor will be 50% greater than in the present system, and forces and stresses on the windings will increase by more than a factor of two [13]. To cope with these stresses, CERN and collaborating laboratories (LBNL, FNAL and BNL in the US, CEA-Saclay in Europe, and others) are exploring new concepts for the coil layout and support structure. To date magnets were designed and manufactured according to the classical cosϑ and Roman arch concept described in Chapter 4, using locking collars to provide pre-stress, but applying this approach to very high field magnets can lead to an unacceptable accumulation of stress. An innovative concept devised a few years ago at LBNL [14], in which the pre-stress is applied by pressurizing thin-walled components called bladders, is being perfected for the Nb_3Sn magnets of HL-LHC (Fig. 12.4). In this approach controlled pre-stress is applied directly to the coil support structure, rather than applying calculated pre-stress indirectly via the geometry of the assembly (as done for magnets from the Tevatron to the LHC). When the pre-stress is correct, calibrated keys are inserted to block any movement and the bladder pressure is released. With its 150 mm free bore and 12 T peak field the 7 m long quadrupole magnet will be the most powerful accelerator magnet of its type ever constructed.

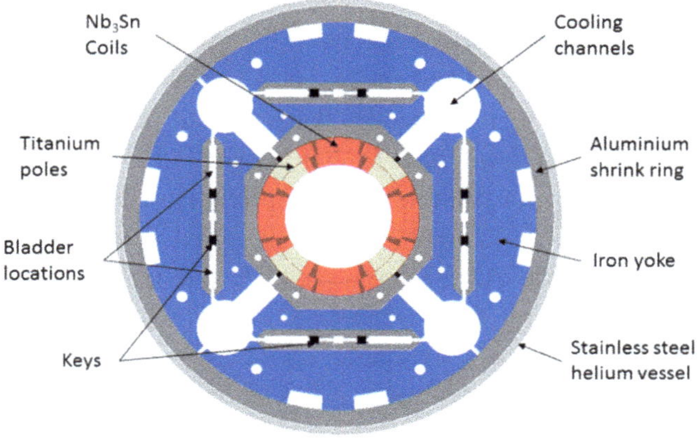

Fig. 12.4. Cross-section of the MQXF Nb_3Sn quadrupole for the LHC upgrade (courtesy CERN-US-LARP collaboration).

The difficulties are even greater for the main dipoles being considered for a Future Circular Collider (FCC) [5], which calls for doubling the LHC field level and entails an increase of a factor four on the forces exerted on the coils. The field required for the FCC is in the region of 15–20 T. It should still be possible to reach the lower end of this range using Nb_3Sn, but the electromagnetic forces become so high that it may be necessary to review the layout of the coil. Detailed analysis and experiments are under way to find and test solutions, and a large test magnet, designed to provide a 13 T dipole transverse field in a 100 mm free bore, and incorporating both bladder-and-key support and racetrack coils with bent up ends, is now nearing completion.

Even more ambitious is the development of a new type of magnet to produce a pure dipole field, wound with the HTS material REBCO [Box 8.4] exploiting the large critical current of these HTS at low temperature in the presence of a strong magnetic field. The objectives are to develop (i) a practical high current flat cable [15], and (ii) a winding geometry where the cable is placed with its broad face parallel to the direction of the magnetic field, in order to minimize the adverse effects of critical current anisotropy in REBCO tape [16]. The initial goal, in 2017, is to generate a 5 T dipole field of a quality suitable for an accelerator; the final aim is to have both a demonstrator cable and winding structure that would open the way to building such a magnet for the 20 T range.

12.3 Teasing Performance from Superconductors Old and New

Amalia Ballarino

Low temperature superconductors (LTS) for use in high magnetic fields

For the FCC, or a possible energy upgrade of the present LHC, it is foreseen to use dipoles with an operating field of up to 16 T for bending the beam around the quasi-circular orbit. The LHC squeezes the most out of standard Nb-Ti material by cooling it to 1.9 K [Highlight 8.2], but because of an intrinsic upper critical field limitation Nb-Ti is not suitable for use in fields greater than 10 T, even when cooled down to 1.9 K (Fig. 12.5) [17]. Magnets for HL-LHC [13] and the FCC are therefore based on the use of the alternative LTS material, Nb_3Sn.

Current density at high field is crucial. For high field solenoids, and large devices such as ITER, space is available, and a critical current density (J_c) in the superconductor of 500–800 A/mm^2 at 12 T is comfortably sufficient, but this would lead to unwieldy accelerator magnets. The target J_c for the Nb_3Sn conductor

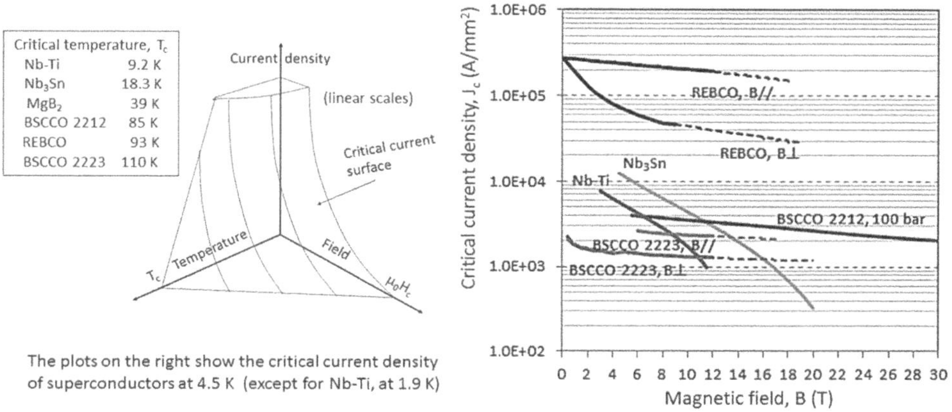

Fig. 12.5. The relative merits of various superconductors at 4.5 K (at 1.9 K for Nb-Ti) for use in high field accelerator magnets. The signs B|| and B⊥ denote magnetic field direction parallel to and perpendicular to the broad face of the tape [17].

of the FCC is 1500 A/mm^2 at 16 T corresponding to an engineering critical current density (J_e), i.e. including stabilizing material and insulation, of about 500 A/mm^2.

Present material performance is sufficient for the HL-LHC magnets [18], but the desired target represents an improvement of about 50% with respect to the state-of-the-art, and sets an extremely challenging goal for the associated R&D.

For an application such as the FCC we would need a large quantity (several thousand tons) of high performance yet affordable material, clearly more than could be produced by one or two companies. Moreover it is expected that the FCC being studied would surely have to be built by a worldwide collaboration of institutes and industry. With this in mind, collaboration agreements have been set up between CERN and industrial partners in Europe and elsewhere to develop the required high performance material, as well as to foster interest in the project.

High temperature superconductors (HTS) used at low temperature

It is apparent in Fig. 12.5 that when used at low temperature HTS materials display far better characteristics at high magnetic field than Nb_3Sn [17]. While one type of BSCCO (Bi Sr Ca Cu oxide) material, BSCCO-2212, can be made in the form of multi-filamentary wire [19], which can be cabled, like Nb-Ti and Nb_3Sn, into convenient, fully-transposed[a] flat (Rutherford) cables, it is also brittle when heat treated to become superconducting. As with Nb_3Sn, coils must be wound prior to

[a]Persistent currents flowing in loops of superconductor give rise to unwanted multipole components of field and reduce transport-current capacity. Persistent currents are minimized by dividing the superconductor in the strands into thin filaments, twisting, and transposing strands within the cable.

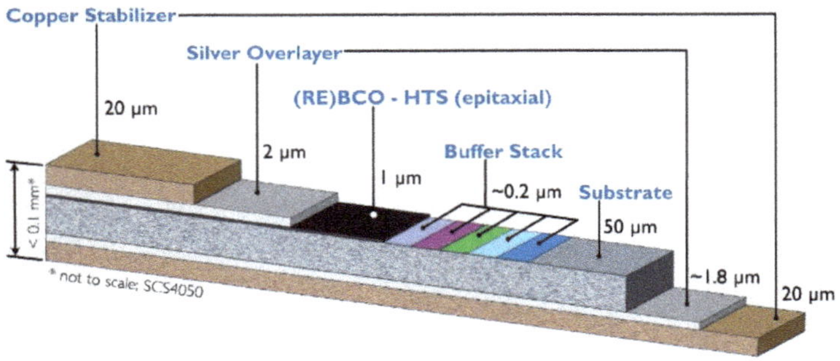

Fig. 12.6. Sketch showing the build-up of a typical REBCO tape [20]. (Courtesy SuperPower.)

reaction (at about 900°C in a high-pressure, oxygen-rich atmosphere) — even more stringent conditions for magnet building than with Nb_3Sn. Besides being weak mechanically, the silver matrix of the material is expensive, so industry is pinning its hopes on so-called "2nd generation" REBCO (rare earth barium copper oxide) as being the most likely HTS material to find widespread practical applications.

REBCO superconductor (the rare earth is usually yttrium) consists of a 1–3 μm thick layer of material deposited on an inert substrate with several thin texturing buffer layers (Fig. 12.6) [20]. Several promising techniques have been adopted by industry for depositing the thin films of HTS onto the metal template, typically a 30–50 μm thick stainless steel alloy. A significant enhancement of the in-field performance is obtained via the creation of artificial pinning centres, by nano-engineering the defect structure in the REBCO film. Electrical stabilization is provided by copper plating one or both sides. Experimental inserts have been made to enhance the field of LTS solenoids from 20 T to 30 T, but for the dipoles and quadrupoles of an accelerator there are two major challenges: how to produce high current cables of the material, and how to reduce its cost. The small dipole insert winding referred to above uses cable derived from the Roebel bar concept (RACC) [15] to reduce magnetization effects (Fig. 12.7). An alternative topology, having the complete transposition of flat Rutherford cable, and which could also lead to better current sharing between strands (RERC), is being studied at CERN [21].

The cost issue will probably not be settled until there appears a regular large-scale commercial application. Although the cost of the raw material is low, the present ball-park price of 30 \$/m (for a standard 4 mm wide tape) barely covers manufacturing cost. For superconductors a more relevant cost-to-performance indicator is expressed in \$/kA m, so improving performance helps to reduce cost. However, for a substantial improvement a large-scale application is still needed: it is recalled that he accelerator community benefited from the low cost of Nb-Ti

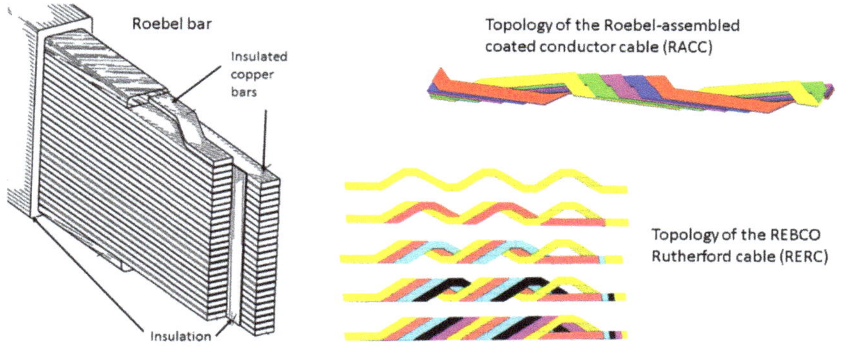

Fig. 12.7. Left: sketch of a Roebel bar, originally conceived (in 1912) to reduce losses in electrical generators by dividing a solid bar into transposed parallel bars. Top right: the concept applied to REBCO tape [15]. Lower right: a tape cable with the topology of flat Rutherford cable [21].

thanks to demand for MRI magnets. Use of the materials in *high magnetic field* is presently a niche application, requiring kilos of material: for a future collider many tons would be needed. At present it is the *high temperature* performance of HTS material that is most appreciated, as liquid nitrogen can be used as the coolant (thanks to increasingly efficient and powerful cryo-coolers, cooling with liquid hydrogen, liquid neon or helium gas is also gaining interest). With this in mind, CERN has been investigating applications that would be useful for accelerators, but could possibly lead to wider use and external demand for the material, and enable a decrease in its cost due to economies of scale.

HTS links for powering superconducting circuits

In view of the success of the development for the LHC of low-loss current leads using HTS [Highlight 8.4], it was decided to look for other possible applications of the new materials to accelerator technology. The proposal of using superconductors for power transmission dates back to seminal work done in 1967 [22], but the advent of HTS saw a regain of interest as it offered the possibility of using liquid nitrogen for cooling. In superconducting accelerators the circuits for powering the magnets require long lengths of interconnecting cable, for which use of the new materials is attractive. The requirement is more exacting than conventional power transmission — several circuits with quasi-DC currents operating at up to 20 kA, instead of single- or 3-phase cables carrying 2 to 4 kA — but the presence of a cryogenic system for the accelerator facilitates the cooling. A study was therefore launched at CERN to evaluate the potential [23], which was later seized upon for HL-LHC [24], where superconducting links would allow relocation of power converters and sensitive electronics to radiation-free areas.

Several types of cable have been studied, and demonstration units built and evaluated [25]. One type (designed to feed twenty-four 600 A circuits) consists of a twisted pair based on the use of standard 4 mm wide tape of BSCCO-2223 (as used for the current leads), REBCO, and magnesium diboride (MgB_2) tape. MgB_2 has a critical temperature of 39 K, is potentially low cost, and is attractive for use in applications where magnetic fields do not exceed 4 T to 5 T. Besides minimizing electrical coupling, the twisting provides sufficient flexibility to allow safe bending within the minimum bending radius (1 m) of the cryostat. This type of cable was new, and cabling machines were developed at CERN to assemble and splice pre-reacted materials. Such cables have been successfully made and tested.

A second type of cable is being developed for the high current circuits. Tens of cables operating at different currents (ranging from 0.12 kA to 18 kA) are contained in the same compact cryogenic envelope with a total current-carrying capacity of over 150 kA. For cooling, the cryogenic system can provide helium gas maintained at below 25 K. Besides the cost aspect, this is interesting because unlike BSCCO 2223 and REBCO, MgB_2 can be made in the form of a round multi-filamentary wire, convenient for making large cables. In 2008 a programme was launched by CERN to focus on the development of round MgB_2 superconductor. A collaborative R&D activity was set up with a manufacturer [25] to develop wire having a critical current of 320 A at 25 K and 0.5 T, and a minimum bending radius of 100 mm, suitable for the development of high-current cables. The first attempts to make MgB_2 wire were based on the architecture of a tape product [26]. After several iterations and the measurement of many samples, the study finally converged on a 1 mm diameter strand that satisfied the electrical and mechanical requirements. The baseline topology of the multi-circuit cable is illustrated in Fig. 12.8. The space taken by the group of cables including its cryostat is less than a quarter of that required for water-cooled bus-bars.

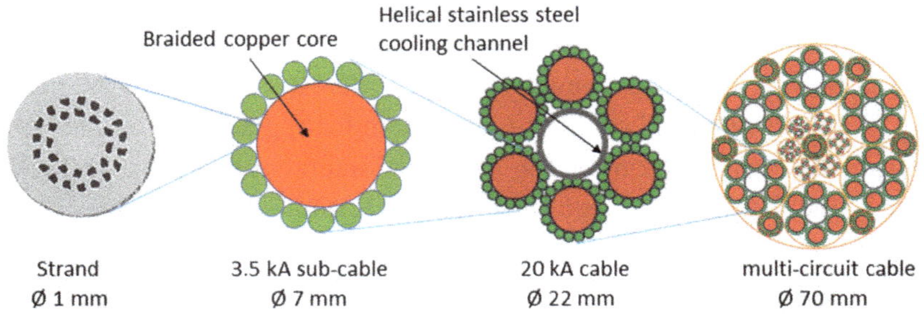

Fig. 12.8. The build-up of the multi-circuit cable proposed for efficient cold powering of the magnet system of the high luminosity upgrade of the LHC. The twisted strands consist of 30 superconducting filaments of MgB_2 embedded in a Monel matrix, copper plated for stabilization and ease of splicing.

A 40 m long MgB2 cable, assembled at CERN, has been successfully tested at up to 20 kA @ 25 K [27], with a peak field on the conductor of 0.9 T.

The activity on superconducting links was performed in the interest of accelerator technology, but such cables can be used in any application requiring the transport of large currents. In fact, the 20 kA prototype tested at CERN also served as a demonstrator for a project proposed for long-distance power transmission [28]. This aspect of the activity was covered in the framework of a collaboration between CERN and the Institute for Advanced Sustainability Studies in Potsdam [29]. Such a cable could also be interesting for application to a pipetron-type accelerator [30], being considered for the injector of FCC-hh, and for FCC-ee [31].

12.4 RF Power for CLIC: Acceleration by Deceleration

Igor Syratchev

The Compact Linear Collider (CLIC) [2] is based on a two-beam scheme. RF power at 12 GHz for acceleration is produced by decelerating a high-intensity but relatively low-energy 2.5 GeV electron beam, the **drive beam**. The **main beam** to be collided with the opposite beam is accelerated with a field of 100 MV/m.

The linac of the main beam is operated in a pulsed mode; every 20 ms each accelerating structure will be fed with a ~250 ns pulse of RF power that generates the accelerating field, with a total peak RF power per burst and linac of about 4.5 TW. During the pulse, a train of 312 bunches, spaced by only 0.5 ns, is accelerated. This burst mode operation is essential to limit the losses of RF power in the accelerating structures, the main issue being the cost- and power-efficient generation of the RF pulses.

The novel design concept uses an electron drive beam to generate the RF power. For each main linac pulse, a long (140 μs), high-current drive beam pulse of 2.4 GeV is produced in a central injector complex for each linac. It is accelerated using long 1 GHz RF pulses that can be produced efficiently with klystrons. During the pulse, the total klystron power is about 10 GW. The design of the drive beam accelerating structures allows to transfer close to 100% of the power to the beam. The long beam pulse, which has a spacing between bunches of 60 cm, is then sent into a system of combiner rings. Here groups of 24 sequential 250 ns long parts of the drive beam pulse are merged to form a single 250 ns long drive beam pulse with only 2.5 cm bunch spacing which are then sent into one decelerator that runs in parallel to the main beam. This decelerator consists of a sequence of Power Extraction and Transfer Structures (PETS) in which most of the beam kinetic energy is turned into RF power. In this concept the initial pulse is compressed in time by almost three orders of magnitude and its peak power increased by the same factor. The concept has been demonstrated in a test facility at CERN [2].

To reach the luminosity needed for physics, the colliding CLIC beams will have a vertical size of only 1 nm, i.e. about ten times the size of a hydrogen atom! At this scale, the natural motion of the ground would displace the beams enough to jeopardize the collision. To avoid this, all of the beam guiding magnets are decoupled from the ground with a novel local feedback system. It consists of inertial sensors, which measure the magnet motion, and of piezo-movers, which can compensate the ground motion. Prototypes of this system manage to stabilize heavy objects to within 0.1 nm. The machine components must be placed with an accuracy of order 10 μm, about an order of magnitude better than the 100 μm typically achieved with the alignment systems used in existing accelerators. A novel alignment system has therefore been developed and successfully tested.

Two key elements are described hereafter: the PETS-structure producing the RF power with the drive beam and the accelerating structure providing the accelerating field for the main beam.

Power-Extraction and Transfer Structure (PETS)

This passive device has to produce 136 MW peak RF power at 12 GHz in 250 ns feeding two accelerating structures of the main linac by decelerating the drive beam (Fig. 12.9).

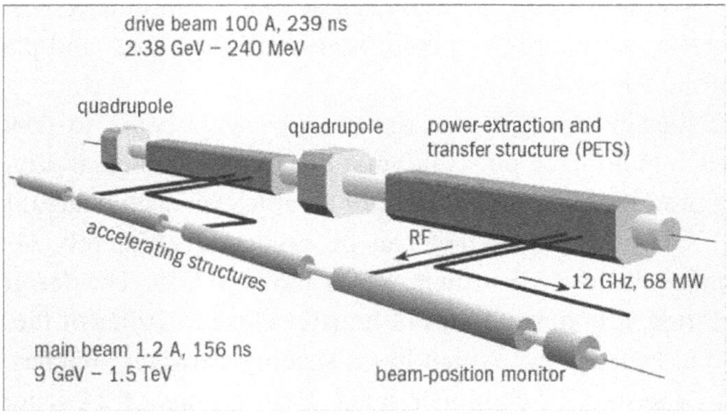

Fig. 12.9. RF powering scheme of CLIC showing the PETS in the drive beam and the accelerating structures of the main beam [2].

The pulse length is longer than the bunch train length in the main beam as the first part of the pulse serves to build up the field in the accelerating structure.

Since the drive beam energy becomes relatively low during deceleration, it is rather sensitive to perturbations, in particular to harmful parasitic higher-mode electro-magnetic fields (HOM) induced by the head of the beam pulse and acting adversely on the tail of the beam pulse by transversely deflecting it. Thus, the design must assure strong damping of these modes.

The structure chosen is a periodically loaded waveguide with 34 cells and 0.21 m active length. It is formed by eight vanes carrying a cog-rail being the outer cell boundary as shown in Fig. 12.10 [2]. Each of the cells is 6.25 mm long with a corrugation depth of 4.28 mm. The electron beam travels in the centre of this structure, generating the RF power in the interaction with vanes of the device. The length of a cell corresponds to one quarter of the 12 GHz RF wavelength.

The vanes are joined by electron-beam welding and separated by 2.2 mm wide longitudinal slots which distort the field pattern of the HOM such that their energy is mainly stored in these slots. A lossy dielectric material close to the slot opening guides the HOM energy to the damping material in the enlargements filled with absorbing silicon carbide. The design keeps the surface field to less than 56 MV/m.

In addition, the operation of the individual PETS has to be controllable either to significantly limit field build-up in case the PETS suffers from breakdowns or to cut power to an accelerating structure broken down. To condition either structure a gradual increase of the power level is also required. These features are guaranteed by a remotely controlled variable reflector in the waveguide after the PETS output coupler terminating the structure [32]. A total of 35 832 PETSs form the RF power generator for each of the 1.5 TeV beams.

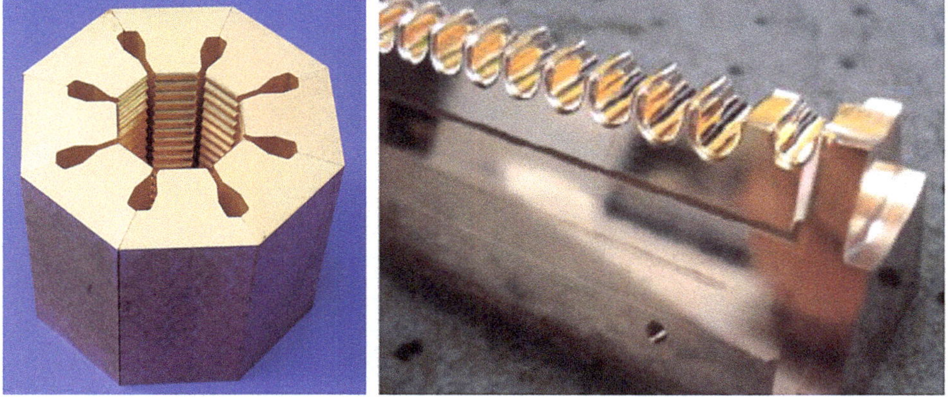

Fig. 12.10. Left: model of PETS showing the octants and the slots leading to the enlargements for the absorber (the absorbent material is not shown). Right: view of an octant.

Accelerating structure

Each accelerating structure consists of an array of 26 copper cells (Fig. 12.11), basically pill-box cavities [Box 3.1], with an eigenfrequency of 12 GHz [33]. The array is preceded by the input coupler which links it to the waveguide coming from the PETS. The RF power is transmitted from cell to cell through the iris. The RF pulse starts before the main beam pulse arrives in order to build up the electric field in all cells and lasts during the beam pulse to replace the energy picked up by the beam (travelling-wave structure).

The final structure parameters including the frequency are the result of an optimization procedure striking a compromise between performance, power consumption and cost. To this end millions of variants were simulated based on a model of investment and exploitation cost as well as taking into account the results from a long series of model testing under RF power [33]. Simply stated, high-frequencies are favoured as this limits the stored energy in the structures leading in turn to their small dimensions.

The resulting cell design maximizes the accelerating field and minimizes high-power RF effects such as RF breakdown (rate $<10^{-7}$ per pulse and per m) and RF induced surface heating (<45 K). Further, it combats the development of HOM, equally harmful in the main linac.

The cell of 80 mm external diameter has a geometry minimizing surface fields (<230 MV/m). The four radial damping waveguides departing from the inner effective accelerating cell are terminated with broad-band loads (not shown) made of silicon carbide to supress the perturbing HOM without affecting the fundamental mode (12 GHz). The iris radius and thickness are tapered along the structure to provide a constant accelerating gradient, which avoids hot spots along the structure. Their average values are 2.75 mm and 1.34 mm, respectively.

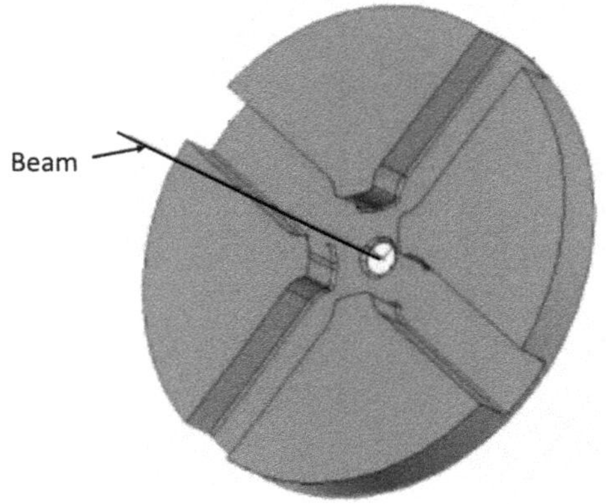

Fig. 12.11. One of 26 cells of the accelerating structure showing the central opening, the iris and the radial waveguides connecting to absorbers of higher modes.

The cell geometry is adapted to the manufacturing process. One side of the cell carries all the features required for the RF, whereas the other side is flat for bonding (Fig. 12.11). The cells are diamond machined, joined by diffusion bonding under hydrogen at about 1000°C and vacuum baked at 650°C. With exacting machining tolerances, typically in the 1 μm range, the tuning of the final structure is avoided.

A number of test facilities is in operation worldwide including those at CERN which houses the only facility modelling the two-beam scheme in addition to a number of test stands powered by klystrons for high turn-around of testing structure variants. Nominal accelerating gradients of 100 MV/m have been achieved in the accelerating structures under development in a worldwide collaboration with the required low break-down rate while the PETSs have been tested at the nominal output power of 150 MW. However, more variants of these two structures have to be built and examined to address HOM damping, lifetime, breakdown statistics, industrial-scale fabrication, and better understanding of the break-down mechanisms.

12.5 The Next Energy Frontier e^+e^- Collider: Innovation in Detectors

Lucie Linssen

Throughout the history of particle physics, both protons and electrons, with their respective antiparticles, have been used extensively for high energy colliders [Box 4.1]. The merits and difficulties of physics research with high luminosity proton–proton colliders are briefly described in the Box 8.3.

Electrons, in contrast to protons, are genuine elementary particles. Apart from initial-state radiation and beamstrahlung effects (see below), they provide very good knowledge of the initial state of the collision. The e^+e^- annihilation cross section (Fig. 12.12) decreases as E^{-2} and, at TeV energies, cross sections for most processes are small, typically of order 10^{-37} cm^2 or lower, and at a similar level as the interesting events in proton–proton collisions. However, the absence of the large physics background that plagues hadron colliders allows for much easier signal selection. If they can deliver sufficiently high luminosity ($>10^{34}$ cm^{-2}s^{-1}), future e^+e^- colliders are therefore the tool of choice for precision measurements.

Two design studies of high energy, high luminosity *e^+e^- colliders* are currently hosted by CERN. The Compact Linear Collider (CLIC) is designed to cover a centre-of-mass energy range from ~350 GeV to 3 TeV, while the e^+e^- option of the large circular collider, FCC-ee, is foreseen to cover energies from the Z-pole

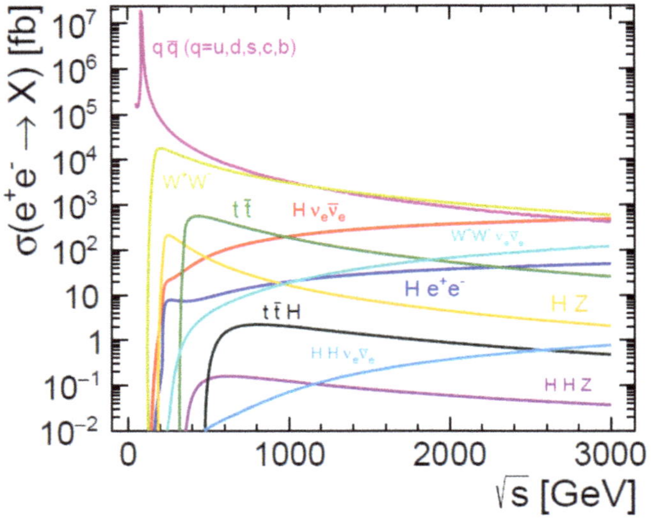

Fig. 12.12. Calculated e^+e^- cross-sections.

at ~91 GeV up to the top pair production threshold near 350 GeV. CERN also participates in the International Linear Collider (ILC) study, a 250–500 GeV collider, upgradable to 1 TeV, which could be realized in Japan.

Beyond their respective *physics discovery potential*, all three machines will allow high-precision measurement of processes involving known particles. Besides a deeper understanding of the underlying processes these studies could provide insight into the nature of new physics at very high energy scales. In this context the most prominent examples are the accurate measurement of the Higgs boson and its couplings to mass, as well as the mass and production characteristics of the top quark. In the case of FCC-ee this is complemented by measurement of W and Z bosons with very high statistics. The centre-of-mass energy at CLIC will give access to the discovery and accurate measurement of pair-produced particles with a mass of up to 1.5 TeV, or single particles up to 3 TeV. In this context new electroweak particles or dark matter candidates are of special interest, as they may escape detection at the LHC. In addition, CLIC energies will give optimal access to double Higgs production, governed by the Higgs potential.

Spin plays a key role in research at an e^+e^- collider. In particular, beam polarization gives access to the spin dependence of the production processes. A polarized e^- beam at a linear collider would be useful for testing the Standard Model (Chapter 7) and "feeling" new physics. While waiting for a broader view on the physics case after LHC, the CLIC baseline design foresees e^- polarization, and reserves space in the design for the more challenging e^+ polarization option.

The superb accuracy potential offered by e^+e^- collisions drives the *detector requirements* [34–36]. Several "must-do" physics measurements, serving as benchmarks, determine specifications for the momentum resolution, energy measurement and quark flavour tagging of the detectors. The vertex and tracking systems have to be significantly more accurate (a factor 3 to 5) than the current state-of-the-art LHC systems, while highly granular calorimeters are needed to obtain the anticipated improvement of a factor 1.5 to 2 in jet energy resolution. In order to achieve this performance, the tracking and calorimeter systems will be located inside a powerful superconducting solenoid (Fig. 12.13).

The detectors also have to meet requirements driven by the *beam conditions*. In a linear collider the transverse size of the bunch is very small — indispensable for achieving high luminosity (12.1 and 12.4). As a result, the electron and positron beams produce strong electromagnetic radiation (beamstrahlung) in the high field of the opposite beam. Such radiative effects change the centre-of-mass energy distribution of the e^+e^- collisions: the electron or positron may radiate a high-energy photon before the collision, resulting in a centre-of-mass energy below the nominal value. As a result, the spectrum of centre-of-mass energies shows a sharp peak at the nominal value and a long tails towards lower energies. This so-called luminosity spectrum can be measured in situ from Bhabha scattering, and subsequently used in the interpretation of the physics results. In principle the tail allows for a crude exploration of the physics which lies below the nominal energy, but with much reduced luminosity.

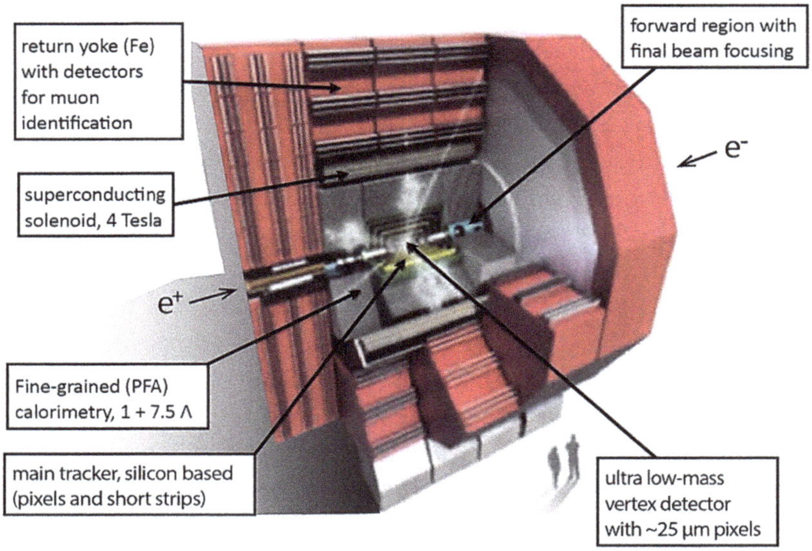

Fig. 12.13. Artist view of a possible CLIC detector.

Beamstrahlung is also a source of additional background in the detectors, which is strongest at CLIC due to its high beam energy and smaller beam size. Background e^+e^- pairs are produced at high rates and very close to the beam direction, while background hadrons with transverse momenta up to ~5 GeV/c reach out to larger polar angles. At 3 TeV the pile-up of beamstrahlung over the entire 156 ns bunch train amounts to ~5000 charged particles and ~19 TeV of energy deposited in the central detector volume. For comparison, at most one hard e^+e^- physics event is produced per bunch train at 3 TeV. Provided the CLIC detectors produce timing information for all hits, with an accuracy of 10 ns for the tracking detectors and 1 ns for calorimeter hits, these beam-induced backgrounds can be suppressed efficiently in the data. At the ILC, beam-induced backgrounds will be less pronounced, due to the lower centre-of-mass energy and larger bunch sizes. Moreover, individual bunch crossings will be separated by several hundred ns. Therefore, the ILC detectors do not need to provide such accurate time information. Reduced occupancies in the inner ILC detector regions will allow the most inner vertex layer to be placed at a radius of 15 mm, as compared to 31 mm for the 3 TeV CLIC. At FCC-ee the dominant backgrounds originate from synchrotron radiation in the bending magnets closest to the experiment. The extent of this background and the resulting demands on the detector are under study.

Given the small e^+e^- interaction cross sections the radiation levels are low. For example, at CLIC they will be typically a factor 10^4 lower than at LHC. Only the small very-forward calorimeter systems will experience LHC-like radiation levels.

In the domain of *vertex and tracking detectors*, semiconductor devices provide the best accuracy. Silicon detectors are extensively used at the LHC and many interesting developments are on-going for the upgrades of the LHC experiments, with special emphasis on reducing the mass of the detectors and increasing their granularity, position accuracy and radiation hardness. Requirements for CLIC push the demands on position accuracy, cell size and low mass even further. For example, at CLIC each layer of the vertex detector should feature a position resolution of 3 μm and a time resolution of 10 ns for a material thickness corresponding to 0.2% of a radiation length X_0. Cell sizes need to be smaller than $25 \times 25\ \mu m^2$ to limit occupancies due to beam-induced background particles. Assuming a minimal amount of material required for supports and services, the design allows for 50 μm sensor thickness and 50 μm readout ASIC thickness. To reduce heat dissipation, the electrical power to electronics will be turned off during most of the 20 ms interval between 156 ns bunch trains. This allows for air cooling, thereby avoiding the material otherwise needed for cooling tubes and liquids.

The semiconductor industry progresses fast, thanks to societal applications such as personal computers, mobile phones, sensors and cameras. Even though particle physics has often quite specific detector requirements, innovations in the wider realm of semiconductors provides many opportunities for our field to progress. An example is the decreasing CMOS feature size with the associated increase in density of readout electronics. Industrial sensor and camera R&D involves integrated (high-resistivity) sensor materials, nanostructures and compact interconnect features, which are also interesting for particle detectors. The current readout ASIC for CLIC is based on 65 nm CMOS technology. Compared to the 0.25 μm CMOS circuits commonly used at LHC, this technology can accommodate about 25 times more logic cells per mm^2, which is very similar to the ratio of LHC to CLIC pixel sizes in the vertex detector. The 65 nm technology has the added advantage of reduced power consumption, with reduced detector mass as a further bonus.

CLIC R&D is also testing so-called integrated HV-CMOS sensors, which comprise a high resistivity sensor layer (for fast signal development) and the electronics circuit in the same thin device [37]. At the current stage of the R&D, the integrated sensor includes two amplification stages. As a result, the detected signal is large enough to be transmitted through a simple layer of non-conductive glue (AC coupling) to the readout ASIC (Fig. 12.14).

At a readout pitch of 25 μm this represents a significant production gain over the delicate traditional bump-bonding (DC coupling) between sensors and readout ASICs. The ALICE experiment is currently developing a related fully-integrated monolithic CMOS technology for their 10 m^2 inner tracker upgrade. It has supreme position resolution, albeit with limited signal speed [38]. In the coming years one can expect such trends in technology to evolve further, and tenacious particle physicists to cleverly exploit such features for their future detector systems. The CLIC R&D effort towards a fast and thin 100 m^2 e^+e^- silicon tracker system with 7 micron single-point resolution and small readout cells will surely explore such opportunities.

For the *calorimeter system* CLIC will make use of the so-called *Particle-Flow Analysis* (PFA) approach, a development that was initiated at CERN by the ALEPH experiment at LEP and was successfully developed further for the CMS experiment at the LHC [39]. The approach will be exploited in future experiments thanks to newly available detector technologies.

Traditionally, the reconstruction of collision data proceeds mostly independently for the tracker system and the calorimeter. The production angle and momentum of charged particles are measured in the tracker, while the energy of nearly all outgoing particles (except neutrinos and muons) is measured independently in the calorimeter system. Particle showers in the calorimeter, in particular in the hadron calorimeter, develop over large volumes. This hampers the

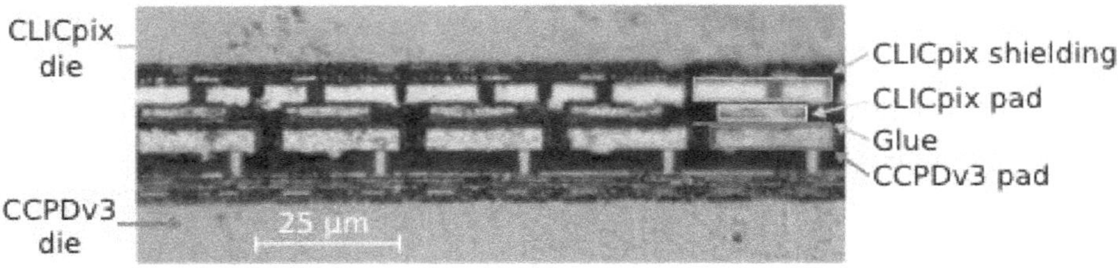

Fig. 12.14. AC-coupled assembly of an integrated HV-CMOS sensor (CCPDv3 PAD) and a CLICpix readout ASIC in 65 nm technology with 25 × 25 µm² pixels. The HV-CMOS sensor includes two amplification stages. The AC coupling is achieved through a thin layer of non-conductive glue.

matching between tracking and calorimeter information when particles are produced close in space. Quark and gluon jets are typically composed of many nearby particles, 60% of which are charged hadrons, 30% are photons (mostly from π^0 decays), and 10% are neutral hadrons. The charged hadrons are normally measured to high accuracy in the tracker (relative accuracy better than 10^{-4} p_T, with p_T in GeV/c). The energy of photons is measured typically better than $15\%/\sqrt{E}$ (with E in GeV) in the first layers of the calorimeter system. Neutral hadrons are not equally well measured, due to the much larger fluctuations in the hadronic shower development. Using the best information available for each of the individual particles in the jet, the jet energy resolution can be improved by a factor 1.5 to 2 compared to traditional calorimeter-only reconstruction techniques. This will allow to reach ~3.5% accuracy in jet energy measurements at high energies and to separate W, Z and Higgs decays reliably in multi-jet topologies.

This detailed PFA can be realized by combining fine-grained calorimetry with powerful event reconstruction software. In case calorimeter cells are small enough, one can assign energy deposits to individual particles in the event reconstruction. At the same time, PFA allows to identify individual particles originating from beamstrahlung and to reject them in the data analysis by using time information [40]. Figure 12.15 shows the PFA reconstruction of a simulated event in the CLIC detector model.

Cell sizes in the active layers are foreseen to be as small as 5 × 5 mm² in the electromagnetic section of the calorimeter and some 3 × 3 cm² in the hadronic section. The electromagnetic section is foreseen to have 30 layers in depth of one radiation length each, while the hadronic section will have ~60 layers for a total of 7.5 interaction lengths in order to fulfil the jet energy resolution goals up to 3 TeV centre-of-mass energy. Compared to the current ATLAS or CMS calorimeters this represents a factor 400 increase in the number of readout cells.

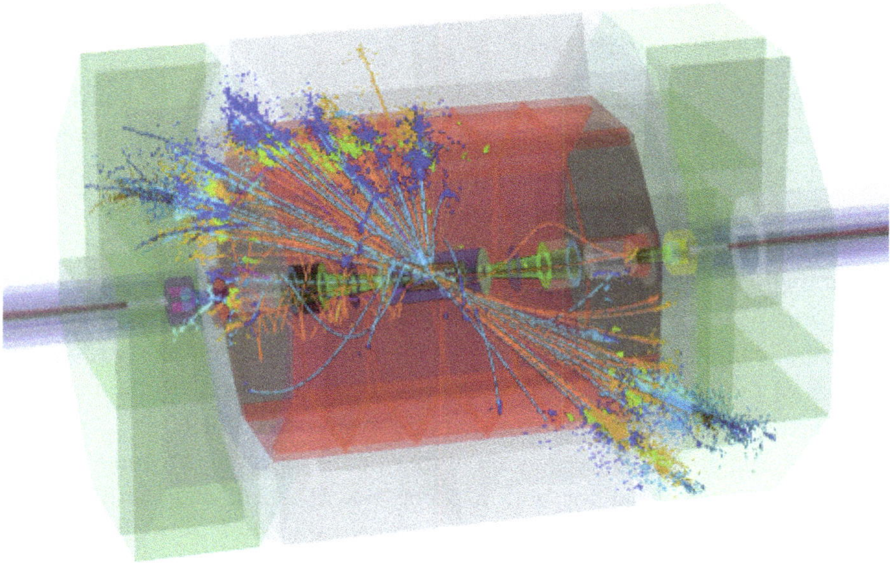

Fig. 12.15. PFA reconstruction of a simulated event in the CLIC detector model. The various colours are used to distinguish different reconstructed objects in the crowded regions.

SiPMs are single photon counting devices composed of an array of avalanche photodiodes (APD) on a common silicon substrate. The miniature devices can host over 10 000 APDs per mm^2, and are insensitive to magnetic fields. Compared to traditional photomultipliers they offer a wealth of new possibilities for particle detection, including highly granular calorimetry.

Prototypes of highly granular calorimeters have been pioneered by the CALICE collaboration, initially targeting calorimetry for the ILC and subsequently also for CLIC and other facilities. Full-size fine-grained electromagnetic and hadronic calorimeters were built and successfully tested. Several options for the active layers and passive absorbers were explored [41]. Such fine-grained prototypes do not only demonstrate technical feasibility and performance capabilities, they also provide invaluable information on calorimetric shower development: a very welcome input for detector simulation tools, such as Geant4.

The demanding performance goals for future e^+e^- collider experiments have motivated advanced detector R&D across the planet. The community is preparing to realize these experiments in a foreseeable future and to explore the physics discovery potential at these new machines.

12.6 Hadron Collider Detectors: A Bright and Energetic Future

Werner Riegler

The LHC programme is at present planned to extend to about 2035. Until 2025 the collider will operate at the nominal parameters, followed by the implementation of the so-called High Luminosity LHC (HL-LHC), which should increase the delivered luminosity by another factor of 10 at the same energy [1].

The LHC represents a 7-fold energy increase over previous colliders. The first phase of the LHC is presently clearly focused on its *discovery potential*. Whatever the outcome of searches for new physics may be, the second phase guarantees a programme characterized by *precision measurements*. As one example, the couplings of the Higgs boson, discovered in the first phase, will be measured to high precision with the high statistics sample from the HL-LHC.

The major upgrades of the general purpose detectors ATLAS and CMS aim at optimizing their potential for both aspects. For HL-LHC they will implement technologies to perform these precision measurements at a luminosity of $5 \times 10^{34}\,\mathrm{cm^{-2}s^{-1}}$ corresponding to close to 4 billion collisions/s, and which must remain operational at unprecedented levels of radiation. The aim of the ongoing R&D is to achieve a detector performance at these increased rates that is at least comparable to, if not better than, the present one [42, 43].

The improvement programmes of the ALICE and LHCb experiments also focus on increased rate capability and measurement precision. They are however not linked directly to the HL-LHC and follow therefore a different time line [44, 45].

The requirements for the LHC detector upgrades are clearly defined: the R&D is targeted at radiation resistance, rate capability, very low-mass sensors for precision tracking and new technologies to implement sophisticated trigger algorithms in computing farms.

Beyond the LHC, detector concepts for a large circular hadron collider with collision energies of about 100 TeV are being developed. More than ever before will the long-term improvements in detector performance be linked to progress in technology of semiconductors, computing and networking. Given the continued close to exponential growth of performance of these technologies, they arguably will be a major contributor to the required detector improvements and strategic changes in experimentation.

Some examples from this long term programme are presented here.

ATLAS and CMS at HL-LHC

For the HL-LHC, the luminosity will be increased from the nominal *peak* value of 10^{34} to a *constant value* of $5 \times 10^{34}\,\mathrm{cm^{-2}s^{-1}}$. At this luminosity an average of 140 inelastic proton–proton collisions are produced every 25 ns at the centre of the detector. The vertices of these 140 proton collisions are spread out along the beam direction with an average separation at the millimetre level and time differences on the scale of 10 ps. Given the time response of present tracking sensors these collisions are therefore *simultaneous*, dubbed *pile-up* in the jargon of the experimenters. The individual collision vertices have to be disentangled with high granularity pixel detectors. However, R&D on sensors reaching timing resolutions in the 20 ps range targets temporal separation of the collisions. Cherenkov detectors, calorimeters with fast readout or silicon sensors with internal charge amplification have the potential to reach such levels of time resolution, the challenge being to cover a large area, where necessary.

The pixel sensors close to the beam pipe will experience radiation loads in excess of 10^{16} hadrons/cm^2 during their lifetime, a factor ten above present day levels. Novel sensor materials and geometries, operated at low temperature will be required to withstand these radiation levels. High quality positively doped silicon material is becoming available, in which the so called n-in-p technology can be implemented and should provide the required radiation resistance.

ATLAS and CMS will replace their entire trackers [Highlight 8.7] for HL-LHC operation in order to cope with the increased rate of particle production as well as to satisfy radiation requirements. They also plan to extend the tracking acceptance to very small forward angles to measure forward jets, signatures of the fundamental "vector boson fusion" processes, i.e. the collision of two (virtual) bosons emitted by the incoming partons on both sides.

The 40 MHz event rate has to be reduced to a manageable 1 MHz detector readout rate. Muon and calorimeter information alone is not sufficient for efficient event filtering, so tracking information must be included to achieve the required selectivity. CMS will use closely spaced silicon sensors which will provide angular and hence momentum information on track segments. High momentum tracks can be selected in the trigger decision to achieve the required selectivity.

The liquid argon calorimetry of ATLAS is well adapted for the high radiation levels of the HL-LHC. It will mainly require electronics upgrades to cope with higher rates. The forward crystal calorimeter of CMS will however not withstand the HL-LHC radiation load. It will be replaced by a high granularity sampling calorimeter with Si-detector readout. Such high granularity calorimeters are required to exploit the *Particle Flow Techniques* (PFA), a novel method for event reconstruction [Highlight 12.5]. Jet energies are measured with the calorimeter combined with the momentum information of the charged particles provided by the high precision tracker. The calorimeters are also used for particle identification and measurement of the neutral component of the jet. In the high pile-up environment of HL-LHC sufficiently high granularity is essential for the proper matching of tracks with the calorimeter energy clusters.

The most extensive part of the detector upgrade concerns the readout electronics. The already quoted close to exponential performance increase of microelectronics, expected to still continue in the near future ("Moore's law"), has a dramatic impact on the strategy for treatment of detector signals, event selection (trigger), data readout and data storage. While e.g. the signals of the ATLAS liquid argon calorimeter are presently buffered in an analogue pipeline and only digitized upon a trigger at a maximum rate of 100 kHz, the expected decrease in ADC power consumption will allow 10 bit digitization at the full bunch crossing rate of 40 MHz by the time of the HL-LHC.

The present readout of the ATLAS and CMS detectors is based on a first level hardware trigger from calorimeters and the muon system, reducing the 40 MHz bunch-crossing rate to a 100 kHz detector event readout rate into a so called High Level Trigger (HLT) computing farm, which performs further event selection, resulting in an event rate of about 100 Hz to permanent storage. For the HL-LHC, an HLT input rate of 1 MHz, corresponding to 5 Terabyte/s (TB/s), and an event storage rate of 10 kHz are considered, shifting a more refined and flexible event selection and larger data samples to offline analysis computer farms.

Reading the full event rate of 200 TB/s into a computing farm without any hardware trigger selection would be not economic, but technically feasible in the near future. The major principle obstacle is the transmission of the electric signals from the sensors over some meters of copper or aluminium cable to the optical conversion electronics for transmission over fibre. The large quantity of metal cables and the cooling infrastructure to remove the heat from the power-hungry high-rate transmission links in the active volume of the tracker impact severely its performance. Therefore efficient data reduction on the detector electronics, radiation hard optical links and low power signal transmission are key to a fully continuous readout of future detectors.

ALICE, LHCb

Towards the end of 2020, ALICE and LHCb will implement important upgrades to prepare the experiments for the HL-LHC era. These experiments will realize dreams that have been keeping the detector community awake for quite a time.

Given the increase in processing power and network performance, ALICE and LHCb plan to no longer rely on *hardware triggers* to reduce the events rate. All events will be transferred at data rates of 1 to 4 TB/s into a computing farm for detailed online event reconstruction. In the case of LHCb the online farm will perform event selection, while in ALICE, the online farm will perform data reconstruction and compression with all events written to permanent storage. Evidently this approach exploits at best the physics potential provided by the LHC.

The ALICE detector is optimized for heavy ion collisions. A requirement of its research programme is to measure very low momentum particles. This implies an ultra-low mass silicon tracker to limit the degradation of measurement precision due to the deflection, i.e. multiple scattering, of particles [Box 3.2]. This will be achieved by developing novel CMOS monolithic silicon sensors. In current silicon detectors, the sensors and the readout electronics are separate components that must be connected by sophisticated soldering techniques. The monolithic sensors used for ALICE integrate the sensor diode and signal processing circuitry inside the pixel on the same silicon substrate. The total thickness can therefore be reduced to 50 μm, compared to the several hundred μm typical of present Si trackers. Together with low mass carbon fibre support structures and aluminium signal buses, a significant reduction in mass can be achieved, improving measurement precision by a factor of 3 over the presently used silicon tracker.

Detection efficiency of >99%, and radiation resistance up to 10^{13} hadron/cm^2 have been achieved on prototypes, well adapted for the heavy ion collision rates of ALICE. The ALICE tracker will use 10 m^2 and 12.5 Gpixels of this revolutionary integrated detector + signal processing technology [44].

The Time Projection Chamber (TPC) [Box 4.5] of ALICE will implement a novel readout technology that transforms the detector from a so-called "gated" into a continuously sensitive mode of operation. The standard operation of a TPC using wire chambers requires a gating grid to be closed for a time of typically a few hundred microseconds after each event. This prevents the ions, created in the gas amplification on the sense wires, from moving back into the drift volume and building up significant amounts of charge. However, this limits the readout-rate capability of the ALICE TPC to about 3 kHz. This will be overcome by replacing the wire chambers with a stack of 4 specifically developed GEM detectors [Box 4.6]. Their design will prevent 99% of the ions from moving back into the TPC volume, so that the TPC can be continuously sensitive without the need for gating. These GEMs will measure the particle tracks with an accuracy of 200 μm. The 100 m^2 of GEM detectors will have 500 000 channels, with their signals continuously digitized at a rate of 5 MHz [44].

The LHCb experiment is focused on the physics of B-mesons and selects events based on the characteristic decay pattern of these particles, the so-called displaced vertices [Highlight 8.6]. In order to achieve this complex selection, the present detector was conceived for an average rate of only one proton collision per bunch crossing in the detector. The collaboration plans to increase this rate by a factor of 10. The new detector readout electronics will operate at the full bunch-crossing rate of 40 MHz and transfer all data to a computer farm where it is possible to select the superimposed events. In addition to the electronics upgrade, some other detector systems will also have to be upgraded.

LHCb will employ a large scale scintillation fibre tracker. Exploiting the attractive features of a fibre tracker as a fast low mass detector has until recently been hampered by the difficulty of deploying a large number of photo sensors in a cost effective way. The recent development of Silicon Photo Multipliers (SiPM) having single photon detection efficiency has however changed this picture dramatically. The LHCb fibre tracker will use 12 detection layers with a total of 12 000 km of fibres and will employ 590 K SiPM channels [45].

The signals from the SiPMs are digitized and transferred to the online farm at the bunch crossing rate of 40MHz. The radiation damage of these SiPMs due to the total hadron fluence in excess of 10^{14} hadrons per cm^2 results in a significant increase of the SiPM dark count rate, and the detector is therefore operated at 233 K (minus 40°C) to mitigate this effect to acceptable levels.

Beyond LHC

One post-LHC collider concept is a 100 TeV circular hadron collider being considered in the Future Circular Collider (FCC) study [7]. This machine is again conceived as a *discovery machine* [Box 12.1]. The programme is to provide proton collisions at 100 TeV, initially with a luminosity of 5×10^{34} $cm^{-2}s^{-1}$ at 25 ns bunch crossing and leading to an ultimate luminosity of 30×10^{34} $cm^{-2}s^{-1}$ at 25 ns or 5 ns bunch crossing.

Radiation levels will increase in proportion to the luminosity. For example, a pixel detector at a radius of 2.2 cm from the beam will be exposed to a fluence of 5×10^{17} hadrons/cm^2, thirty times the already challenging HL-LHC numbers. Clearly, R&D for radiation hard sensors is again key for such an enterprise.

Most ideas about physics beyond the Standard Model are related to the Higgs boson, so Higgs studies can be seen as a portal to new physics and benchmark for a future collider. Due to its low mass, the Higgs particle will be boosted to very small forward angles at high collision energies, requiring an acceptance for precision tracking and calorimetry down to polar angles of 2°.

The present detector concepts consider solenoids of 4 to 6 T and 5 to 6 m free bore radius with forward dipoles or solenoids for momentum spectroscopy. In comparison, the CMS experiment uses a 4 T solenoid with a free bore radius of 3 m [Highlight 8.12]. An iron flux return would be exceedingly heavy ($\sim 10^5$ t), which has motivated seeking alternative solutions. Active shielding with a shielding coil, partial shielding with iron or completely unshielded geometries are presently being studied. A conceptual layout is shown in Fig. 12.16 [46].

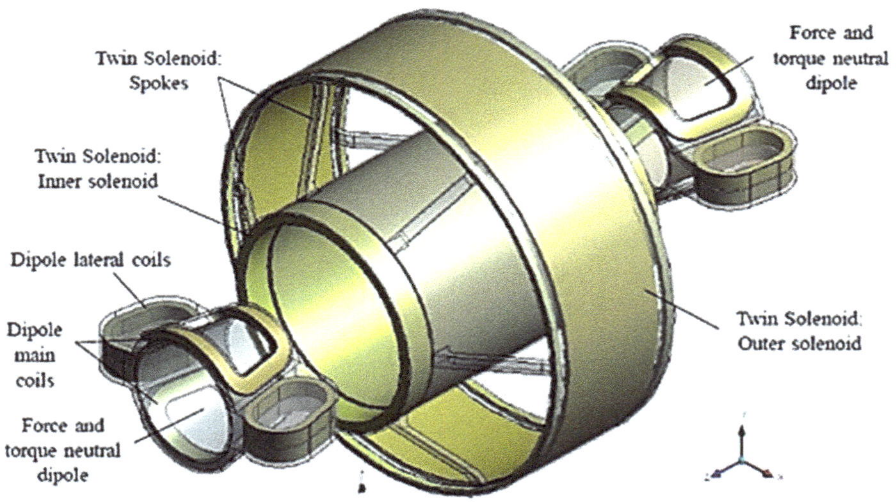

Fig. 12.16. Cold mass of the twin solenoid and dipoles. In the figure the support structure is made partially transparent to reveal the superconductive coils. The outer diameter and total assembly length is 27 and 43 m, respectively [46].

To achieve a resolution of 10–15% for particles with a p_T of 10 TeV (10 times better than LHC) requires a tracker combining long distance of tracking, a large area and very high precision tracking. A prime goal is therefore to continue the development of monolithic silicon sensors into fast and radiation hard sensors that will allow high precision tracking over large areas at affordable cost.

The detector and accelerator studies for this energy frontier are part of the "Future Circular Collider (FCC)" study [7], a comprehensive conceptual design study for post-LHC particle accelerator options in a global context.

References

1. G. Apollinari *et al.* (eds.), *High Luminosity Large Hadron Collider (HL-LHC): Preliminary Design Report*, CERN-2015-005 (CERN, Geneva, 2015). http://dx.doi.org/10.5170/CERN-2015-005.
2. M. Aicheler *et al.* (eds.), *A Multi-TeV Linear Collider Based on CLIC Technology: CLIC Conceptual Design Report*, CERN-2012-007 (CERN, Geneva, 2012). http://dx.doi.org/10.5170/CERN-2012-007.
3. Multiple editors, *ILC Technical Design Report*, Vol. 1 to 4, 2013, ISBN 978-3-935702-74-4; http://www.linearcollider.org/ILC/Publications/Technical-Design-Report.
4. A. Caldwell *et al.*, AWAKE design report: a proton-driven plasma wakefield acceleration experiment at CERN, *Report CERN-SPSC-2013-013* (2013); http://cds.cern.ch/record/1537318?ln=en.

5. E. Todesco and F. Zimmermann (eds.), *Proc.EuCARD-AccNet-EuroLumi Workshop: The High Energy Large Hadron Collider, Villa Bighi, Malta, October 2010*, CERN-2011-003 (CERN, Geneva, 2011). http://dx.doi.org/10.5170/CERN-2011-003.
6. N. Arkani-Hamed, T. Han, M. Mangano and L. Wang, Physics opportunities of a 100 TeV proton-proton collider, *Report CERN-PH-TH-2015-259*, submitted to *Phys. Rep.* (2015); http://cds.cern.ch/record/2104388?ln=en.
7. M. Benedikt *et al.*, FCC-hh Hadron collider – parameter scenarios and staging options, *Proc. Int. Particle Accel. Conf.,* Richmond, VA, USA (2015) p. 2173.
8. A. Blondel *et al.*, LEP3: A High Luminosity e^+e^- Collider to Study the Higgs Boson, *Internal Report CERN-ATS-Note-2012-062 TECH* (2012); http://cds.cern.ch/record/1470982?ln=en.
9. F. Zimmermann *et al.*, Status and challenges for FCC-ee, *Report CERN-ACC-2015-111* (2015); http://cds.cern.ch/record/2057706?ln=en.
10. J. Abeilleria *et al.*, A large hadron electron collider at CERN, *J. Phys. G* **39** (7), 075001 (2012).
11. L. Rossi and L. Bottura, Superconducting magnets for particle accelerators, *Review of Accelerator Science and Technology*, **5**, 51-89 (2012).
12. A. Ballarino and L. Bottura, Targets for R&D on Nb3Sn conductor for high energy physics, *IEEE Trans. Appl. Supercond.* **25**, 60000906, (2015).
13. L .Bottura, G. de Rijk, L. Rossi and E. Todesco, Advanced accelerator magnets for upgrading the LHC, *IEEE Trans. Appl. Supercond.* **22**, 4002008, (2012).
14. S. Caspi *et al.*, The use of pressurized bladders for stress control of superconducting magnets, *IEEE Trans. Appl. Supercond.* **11**, pp. 2272-2275, (2001).
15. W. Goldacker *et al.*, Improvement of Superconducting Properties in ROEBEL Assembled Coated Conductors (RACC), *IEEE Trans. Appl. Supercond.* **19**, pp. 3098-3101, (2009).
16. G. Kirby *et al.*, Accelerator quality HTS dipole magnet demonstrator designs for the EuCARD-2, 5 tesla 40 mm clear aperture magnet, *IEEE Trans. Appl. Supercond.* **25**, 4000805 (2005).
17. A. Ballarino, Prospects for the use of HTS in high field magnets for future accelerator facilities, *Proc 5th Int. Particle Accel. Conf. (IPAC2014)*, Dresden, Germany (2014) pp. 974-979.
18. B. Bordini *et al.*, The barrier-type PIT wire developed for the HiLumi LHC project, presented at the *Applied Superconductivity Conference (ASC)*, Denver, USA, September 2016.
19. D. Larbalestier *et al.*, Isotropic round wire multifilamentary cuprate superconductor for generation of magnetic fields above 30 T, *Nature Materials* **13**, 375-381 (2014).
20. www.superpower-inc.com.
21. A. Ballarino, HTS cables for accelerator magnets, presented at the *Applied Superconductivity Conference (ASC)*, Denver, USA, September 2016.
22. R. L. Garwin and J. Matisoo, Superconducting lines for the transmission of large amounts of electrical power over great distances, *Proceedings of the IEEE*, **55**(4), 538–548, (1967).
23. A. Ballarino, Alternative design concepts for multi-circuit HTS link systems, *IEEE Trans. Appl. Supercond.* **23**, 980-983 (2011).
24. L. Rossi, LHC upgrade plans: options and strategy, *Proc. Int. Part. Accel. Conf. (IPAC11),* San Sebastian, Spain (2011); https://accelconf.web.cern.ch/accelconf/IPAC11/.
25. A. Ballarino, Development of superconducting links for the LHC machine, *Supercond. Sci. & Technol.* **27**, 044024 (2014).
26. G. Grasso *et al.*, Large transport critical currents in unsintered MgB_2 superconducting tapes, *Appl. Phys. Lett.* **79**, 230 (2001); http://dx.doi.org/10.1063/1.1384905.
27. A. Ballarino, Final design report, *CERN-ACC-2015-0134* (2015); https://cds.cern.ch/record/2063726/files/CERN-ACC-2015-0134.pdf.

28. http://cds.cern.ch/journal/CERNBulletin/2014/16/News%20Articles/1693853.
29. http://www.iass.potsdam.de/en/sites/default/files/files/pr_english_2014_11_03_sc.pdf.
30. G.W. Foster and E. Malamud, Low-cost hadron colliders at Fermilab: a discussion paper (1996) http://www.iaea.org/inis/collection/NCLCollectionStore/_Public/28/001/28001544.pdf.
31. A. Milanese, H. Piekarz and L. Rossi, Concept of a hybrid (normal and superconducting) bending magnet based on iron magnetization for 80-100 km lepton/hadron colliders, *Proc. 5th Int. Particle Accel. Conf. (IPAC)*, Dresden, Germany (2014).
32. I. Syratchev *et al.*, High power operation with beam of a CLIC pets equipped with on/off mechanism, *Proc. Int. Particle Accel. Conf.*, New Orleans, USA (2012) p.1852.
33. A. Grudiev and W. Wuensch, Design of an X-band accelerating structure for the CLIC main linac, *Proc. XXIV Linear Accel. Conf.*, Victoria, Canada (2008), p.933.
34. L. Linssen, A. Miyamoto, M. Stanitzki and H. Weerts (eds.) *Physics and Detectors at CLIC: CLIC Conceptual Design Report*, CERN-2012-003 (CERN, Geneva, 2012). http://dx.doi.org/10.5170/CERN-2012-003.
35. M. Bicer *et al.*, First look at the physics case of TLEP, arXiv:1308.6176 (2013).
36. T. Behnke *et al.*, The international linear collider technical design report, *ILC-REPORT-2013-040* (2013); arXiv:1306.6329 [physics.ins-det].
37. N. Alipour Tehrani *et al.*, Capacitively coupled hybrid pixel assemblies for the CLIC vertex detector, *Report CLICdp-Pub-2015-003* (2015); https://cds.cern.ch/record/2048684?ln=en.
38. B. Abelev *et al.* (The ALICE Collaboration), Technical design report for the upgrade of the ALICE inner tracking system, *J. Phys. G: Nuclear and Particle Physics* **41**, (8) 087002 (2014).
39. F. Beaudette, The CMS Particle Flow Algorithm, arXiv:1401.8155 (2014).
40. J. Marshall *et al.*, *Nucl. Inst. & Meth. A* **700**, 153 (2013).
41. F. Sefkow, A. White, K. Kawagoe, R. Pöschl, J. Repond, Experimental tests of particle flow calorimetry, *arXiv:1507.05893* (2015).
42. ATLAS Collaboration, Letter of Intent for the phase-II upgrade of the ATLAS experiment, *Report CERN-LHCC-2012-022* (2012); https://cds.cern.ch/record/1502664.
43. D. Contardo *et al.*, Technical proposal for the phase-II upgrade of the CMS detector, *Report CMS-TDR-15-02* (2015); https://cds.cern.ch/record/2020886.
44. Y. Aystö *et al.*, Upgrade of the ALICE Experiment, *J. Phys. G* **41**, 087001 (2014);
ALICE Collaboration, Upgrade of the ALICE time projection chamber, *Report CERN-LHCC-2014-001* (2014); https://cds.cern.ch/record/1647400.
45. LHCb Collaboration, Letter of Intent for the LHCb upgrade, *Report CERN-LHCC-2011-001* (2011); https://cds.cern.ch/record/1333091.
LHCb Collaboration, LHCb tracker upgrade technical design report, *Report CERN-LHCC-2014-001* (2014); https://cds.cern.ch/record/1647400.
46. M. Mentik *et al.*, Design of a 56 GJ Twin Solenoid & Dipoles Detector Magnet System for the Future Circular Collider, *IEEE Trans. Appl. Supercond.* **26**, 4003506 (2016).

Concluding Remarks

The interplay between research and technology over six decades is documented in some 70 technological highlights reported in this book. Sometimes technology is developed in response to the needs of research. Inversely, technological breakthroughs open new vistas for science.

The underlying scientific theme of the past 100 years is the emerging understanding and description of our visible world, encoded in the theory of relativity, quantum mechanics, and the "Standard Model of particle physics".

The Standard Model, an outstanding intellectual achievement of humanity, has been developed in a worldwide collaboration. A score of intellectual giants — some 60 Nobel Prizes in Physics have been awarded for work in connection with the endeavour — have struggled with it in collaboration with tens of thousands of scientists and engineers for more than 60 years. The splendid success of the Standard Model has allowed the connection with cosmology, opening a deeper understanding of our origins. It has also put the spotlight on a vast scientific landscape that is yet to be explored and will determine the research agenda (not only for CERN) for decades to come.

Since its foundation in 1954, CERN has developed into one of the major actors — a worldwide centre for research in this domain. Today it attracts more than 12 000 scientists from around the globe, who make CERN their research home. About one third of them are students, who write some 1000 doctoral theses per year on CERN-based research. In parallel, technology developed at CERN has spawned advances for society, such as the World Wide Web or instrumentation for medical applications.

The institution that is CERN owes its existence to the vision of its founders. They, with awe-inspiring prescience, imagined both an audacious initial programme of world-leading quality and proposed a unique form of governing structure, the CERN Council. Council has been entrusted by the political leadership of the CERN Member States to guide and shape the Organization in a rather autonomous way. Credit must be given to the Council, which in its vision has provided sufficient stability to allow medium to long term planning, while providing enough control and critical feedback to maintain the quality and standing of the Organization. Council is also vigilant — justifiably — that the considerable financial investments are "returned" in a transparent and fair way to

European industry, which collaborates closely in the development and construction of the infrastructure.

Stability is essential for the professional quality and development of the personnel, enabling them to transmit their expertise — often in unique disciplines — to the next generation. Stability has been essential for coherency in the development and exploitation of the technical infrastructure, which is the service that CERN provides to the research community. Stability also makes possible the wise and economical reuse of the infrastructure: the PS, which started operation in 1959, remains to this day at the core of the accelerator complex. Continuously improved and adapted to the changing needs of research, the PS was and still is the injector for all of the subsequent accelerators. The 27 km long tunnel, which now houses the LHC, was planned in the 1970s for LEP, but with the potential for a machine like the LHC — an investment with a lifetime that is expected to approach some 50 years or more.

It is this unique combination of world-class scientific infrastructure, and the large user community, that is at the basis of CERN's achievements.

Seen from a global perspective, it was crucial that the worldwide scientific community and the leadership of the major particle physics laboratories coordinated their respective research programmes to optimize the efficient and responsible use of resources and scientific output. As new major actors emerge, it will require a new vision and new diplomacy to benefit fully from this enhanced level of scientific competition, while the current institutions aim at providing a similar level of scientific service and support to their own scientific constituencies.

CERN's approach to scientific governance and policy, which fosters the collaboration between scientists and engineers of diverse personal and educational background, is sometimes referred to as the "CERN Model". It has been successfully applied by other European organizations, such as the European Southern Observatory (ESO), the European Space Agency (ESA), or the European Molecular Biology Laboratory (EMBL). The United Nations has started to consider this model with a view to applying it, possibly, to some of the world's most pressing issues, such as climate control or eradication of diseases. The CERN Model has been exported to one of the world's most challenged regions, the Middle East. SESAME, near Amman, Jordan, is a research centre for synchrotron-based fundamental and applied research. It is supported by nine countries of the region, including Israel and Palestine, together with 17 observer states worldwide.

Demonstrating to the world that science and technology is united in the peaceful pursuit of advancing knowledge is perhaps the most important contribution of CERN.

The Editorial Team

www.ingramcontent.com/pod-product-compliance
Ingram Content Group UK Ltd.
Pitfield, Milton Keynes, MK11 3LW, UK
UKHW052114170926
8624IPUK00026B/376

9 789814 749138